U0275582

寰宇文献 Universal Library | SINOLOGY 系列

SELECTED WORKS OF BERTHOLD LAUFER

劳费尔著作集

第九卷

[美] 劳费尔 著

黄曙辉 编

图书在版编目(CIP)数据

劳费尔著作集 / (美) 劳费尔著；黄曙辉编. —上海：中西书局，2022
(寰宇文献)
ISBN 978-7-5475-2015-4

Ⅰ. ①劳… Ⅱ. ①劳… ②黄… Ⅲ. ①劳费尔 – 人类学 – 文集 Ⅳ. ①Q98-53

中国版本图书馆CIP数据核字（2022）第207067号

第 9 卷

129

中国瓷器的起源

FIELD MUSEUM OF NATURAL HISTORY

PUBLICATION 192

ANTHROPOLOGICAL SERIES VOL. XV, NO. 2

THE BEGINNINGS OF PORCELAIN IN CHINA

BY

BERTHOLD LAUFER
Curator of Anthropology

With a Technical Report by H. W. Nichols
Assistant Curator of Geology

Twelve Plates and Two Text-Figures

The Mrs. T. B. Blackstone Expedition

CHICAGO
1917

CONTENTS

The Beginnings of Porcelain in China

INTRODUCTORY

In February of 1910, while in Si-ngan fu, the capital of Shen-si Province, the writer received from Mr. Yen, a Chinese scholar and antiquarian of note with whom he was on very friendly terms, a curious bit of ancient pottery, which at first sight bore all the characteristic marks associated with what is known as Han pottery, but which, on the other hand, exhibited a body and a glaze radically different from that ware (Plate I). Mr. Yen accompanied the object with a written message, explaining the circumstances under which it had been found, and commenting to some extent on its historical value. Following is a literal rendering of his letter: "I once heard dealers say that they had seen 'Han porcelain' (*Han ts'e* 漢磁), but I had no faith in this statement. In the winter of the year *ting wei* 丁未 (1907) I secured a large vase, and suspected that it might be an object of the Han period, but did not dare to be positive about this point. In the spring of last year some one brought to light, from a Han grave which he had excavated, ancient jade pieces and such-like things, together with an enormous iron cooking-stove. On the latter are found, cast in high relief, six characters reading, 'Great felicity! May it be serviceable to the lords!' (*ta ki ch'ang i hou wang* 大吉昌宜侯王). On the top of this stove was placed a small 'porcelain jar.' I lost no time in sending out an agent to effect a purchase, but the stove had already passed into the hands of a merchant. So I obtained only the 'porcelain jar' in question, the material and style of which proved identical with those of the large vase purchased by me years ago. For this reason I now felt positive that the question is here of 'Han porcelain.' Subsequently I acquired also a jar of the type styled *lei* 罍, and big and small vases; in all, four. From that time the designation 'Han porcelain' began to be established in the world.

"Written in Ch'ang-ngan by Yen Kan-yüan 閻甘園 on the day when the flowers sprout forth (百花生日), of the second month of the second year of the period Süan-t'ung (February 27, 1910)."

While I had a deep respect for Mr. Yen's learning and extensive knowledge of archæological subjects, I remained sceptic as to the identification of his jar with what he styled *Han ts'e*, and, though recognizing its intrinsic merit as a piece of evidence filling a lacune in our knowledge of ancient pottery, I did not allow myself to be carried away by the usual wave of enthusiasm over a first discovery (since then six years and a half have elapsed), but decided to hold the matter in abeyance till a thorough analysis, to be made at home, would permit us to base an opinion on facts. Meanwhile opportunities were seized at Si-ngan fu to collect as much as possible of this novel pottery. My first concern, naturally, was to secure the large iron stove mentioned in Mr. Yen's missive. A desire thus expressed spreads in that quaint old town like a prairie-fire; and when the sun had risen and set again, I was the lucky owner of that precious relic. Indeed, Yen's description was by no means an exaggeration. In type and style, this cast-iron stove (Plate II), partly in decay and the iron core having entirely rotted away, exactly corresponds to the well-known Han burial cooking-stoves, and it is the finest specimen of ancient cast-iron that I was able to find. Being posed on four feet in the form of elephant-heads, it is built in the shape of a horse-shoe, and provided with a chimney at the rounded end, five cooking-holes, and a projecting platform in front of the fire-chamber. On the latter is cast an inscription in six raised characters, which read exactly as indicated by Mr. Yen, — a formula typical of the Han and earlier ages, and encountered on many bronze vessels. The style of these characters is in thorough agreement with that of Han writing. The object was discovered in a grave near the village Ma-kia-chai 馬家寨, 5 *li* north of the town Hien-yang, in Shen-si Province. As previously remarked,[1] without laying down any hard and fast rules, there is a great deal of probability in assigning such cast-iron objects to the period of the Later Han (A.D. 25–220), while it is equally justifiable to extend the time of their manufacture over the entire third century of our era. The iron stove thus furnishes a clew to the date of the jug which was found in the same grave with it. Needless to say, I left no stone unturned, and kept on inquiring and hunting for this so-called *Han ts'e* ware in and around Si-ngan. I succeeded in bringing together only eight more pieces (Plates III–X), among these the vessel *lei* referred to in Yen's memorable epistle,[2] and a number of larger fragments and small shards, which are always precious and encouraging acquisitions to the archæologist, as they are not under suspicion, and offer welcome study material.

---

[1] Chinese Clay Figures, p. 216.

[2] The pottery vase of this designation is mentioned in the *Chou li* as holding the sacrificial spirits called *ch'ang*, which were offered to the deity Earth (BIOT, Tcheou-li, Vol. I, p. 468). It is the reproduction in clay of an original bronze-type, frequent among the bronze vessels of the Chou.

It will be noticed that these nine bits, in their forms and decorations, decidedly agree with the mortuary Han pottery,[1] and that, taken merely as ceramic types, they represent archaic types of Han art. On the other hand, however, apart from their technical composition, they have in common some characteristic features which are not found in Han pottery. To these belong the curious loop handles, obviously imitative of a knotted rope or a basketry handle, and the geometric wave patterns. The latter, it will be remembered, occur also in the relief bands on many vases of Han pottery, but are of a different style, in the manner of realistic waves. There is in our collection only one unglazed, gray Han pottery vase with a geometric wave design approaching that in the above group; but it is a much bolder and freer composition, and not so neat and refined as in the porcelanous vases. Even in some shapes, the traditional rules of the Han may not be quite strictly observed; they may be less stern and rigorous, and, while dignified and partially imposing, treated with somewhat greater individual freedom. This, however, is rather a point of sentiment or impression than a ponderable argument. The deviations from the standard Han pottery are insignificant when contrasted with what the two groups have in common. The best tradition and spirit of Han art are preserved in these nine productions.

The comparative scarcity of this ware is notable, and gives food for serious reflection. As the writer was able to secure on his last expedition for the Field Museum many hundreds of pieces of Han pottery of all types and descriptions, while several thousand specimens have passed through his hands during the last fifteen years, and as he could hunt up only nine representatives of this novel (porcelanous) ware, these numbers may be regarded as the relative (certainly not absolute) proportions in which the two classes of pottery are to be found, and, we may add, were made in the past. Two inferences may be drawn from this phenomenon,— this peculiar ware was the product of only a single kiln or of very few kilns; and these kilns did not flourish during the Han period, but either at its very close, or even, and more probably, toward the middle or end of the third century. This point will be more fully discussed hereafter.

---

[1] In speaking of Han pottery, it should be understood that in this case the term "Han" does not refer to the chronologically exact boundaries of a dynastic period, but to an archæological epoch, a certain phase of ancient Chinese art, which is not necessarily gauged by the dates 206 B.C. and A.D. 220. There is naturally an overlapping at both ends, and we have, at least for the present, no means of determining exactly either the beginning or the end of Han art. This much seems certain, that the middle and the latter part of the third century A.D. have thoroughly remained under the influence of Han tradition.

On my return to America, two objects remained to be pursued in connection with this new material,— first, to secure the co-operation of a competent investigator for a chemical analysis of the body and glaze of this pottery; and, second, to search in other museums for corresponding specimens. My colleague Mr. Nichols, assistant curator of geology in the Field Museum, volunteered to undertake the technical task, and he has carried it out with rare devotion and perseverance. His experiments were conducted, and his results were obtained, in 1912. From the date of our publication it will be seen that we were not in a hurry to bring it to the notice of the world. We allowed it to rest and to mature, and discussed the new problems with each other and with ceramic experts at frequent intervals. Their friendly interest and advice at last encouraged us to make known the results of our research, which we trust will be of some utility to students interested in the history of Chinese pottery.

In regard to kindred objects in other collections, I have been able to obtain the following information. Mr. Francis Stewart Kershaw of the Museum of Fine Arts, Boston, Mass., who saw the pieces of pottery in question in the Field Museum, mentioned to me that similar specimens were in the Boston Museum. On sending him some fragments from our material for comparison with that under his care, he wrote as follows:[1]

"The bits of potsherd are quite large enough to tell me their story, and I am very much obliged for them. Except in hardness, they are similar to the clay of three of our pieces, being of the same color, texture, and apparent constituents. Two of our pieces were bought in China by Mr. Okakura, and both were labelled 'Sung' by some Chinese (probably a dealer). Okakura called one (12875) which is covered with a blackish shaded gray-green glaze, opaque and dull, 'Sung.' The second (12865), which is precisely similar in potting, clay, and glaze, to your Han porcelanous jars, Okakura called 'T'ang.' Mr. Freer, by the way, has a vase like 12865, which he calls 'T'ang.'[2] The third of our pieces (12118) was bought from Mr. C. F. Gammon (formerly a lieutenant in the United States Army), who obtained it in Nanking from a cooly, who had unearthed it while digging in a railway cutting in Nanking. The jar was partly full of coins, all alike, of the denomination '*pan liang*' 半兩, issued in 175 B.C. in the reign of the

[1] The letter is published here with Mr. Kershaw's consent.

[2] This object was exhibited in the National Museum of Washington in 1912, when a selection from the Freer Collection was temporarily shown. I then had occasion to see it. It is not a T'ang production, but of exactly the same type as our early porcelanous ware.

Emperor Wen. Mr. Gammon told me that he had bought the jar on the spot where it was found. The jar itself, like the others belonging to us, was welded or coiled up by hand before a summary smoothing-off on the wheel. It had four loop handles, finger-modelled, at the shoulder (two only of these remain), and was glazed in a thin running blackish-green, of which the little that still adheres is for the most part oxidized to dull brownish-ochre. The clay is softer than your shards, and softer, too, than that of 12865 or 12875; but it seems to be quite the same in all other respects. It has the same admixture of black and occasional white particles in the mass of gray, the same unevenly ferruginous surface, and the same occasional thickening of that surface. The jar is much less well potted than your pieces and ours. Perhaps it is more primitive; that is, it may be an early example of the method used so expertly in making your jars and ours. Perhaps, on the other hand, it is simply cruder; that is, the potter may have used a well-known and well-developed method carelessly in making an unimportant vessel. Who knows? I incline toward the latter possibility.

"I dated the jar 'Han' because of the evidence of the coins found in it. Now, emboldened by your ascription of the date to the porcelanous jars, I shall classify No. 12865 in the Han period or shortly after. As regards 12875, because of its different glaze and an obscure device impressed on its shoulder, I am not yet sure."

At my request Mr. Kershaw was good enough to send me for examination the *pan-liang* copper coins, twenty-one all together, found in Mr. Gammon's jar. They all proved to be authentic, as particularly determined by close comparison with numerous corresponding issues in the Chalfant coin collection, and to have been issued under the Han.[1] The presence of this batch of coins in that vessel is, of course, no absolute proof warranting us in assigning the vessel to the early Han period, as these coins may still have been in circulation long after Han times. In 1901 I found in actual circulation at Si-ngan fu Han copper coins with the legend *wu chu*. A collection of twenty-one Han *pan-liang* coins in a single jar would rather hint at a high appreciation of this money, and such is rather more probable in post-Han than in Han times. At any rate, the exclusive presence of a single Han issue, together with the absence of any later coin, would seem to favor a period approaching very closely the age of the Han.

---

[1] Money with this legend, weighing exactly half an ounce (*pan-liang*), was first issued under the Ts'in (see CHAVANNES, Mémoires historiques de Se-ma Ts'ien, Vol. III, pp. 539, 542).

Several similar pieces have been collected by Mr. Orvar Karlbeck, an official of the Tientsin-Pukow Railway, residing at Chu-chou, Ngan-hui Province. This gentleman, in the course of several years' residence in China, has formed a very interesting collection of ancient pottery, that consists of 144 pieces. I did not have occasion to see it, but, judging from photographs and descriptions which he has been good enough to send me, he seems to own several bits such as are here under consideration.

Mr. R. L. Hobson, the prominent expert in pottery of the British Museum, while visiting Chicago in January, 1913, and doing me the honor of studying the collections under my care, called my attention to two early jars of similar glazes which were found at Black Rock Hill in Fu-chou, and are now preserved in the British Museum. They are sketched and described by H. F. HOLT.[1] They are oval-shaped jars, with short necks and straight rims, a pair of loop handles (in one piece double handles) being stuck on to the shoulders. They are described as being made "of a grayish clay resembling almost stoneware, over which a coat of greenish-brown glaze has been coarsely laid; a curved line at the bottom sharply defines where the glazing ended." The further remark, however, that the glaze is quite decomposed and can easily be detached, would rather hint at this glaze being of a character different from that on our specimens, which, owing to its chemical composition, is not capable of decomposition. The great antiquity of these two jars is not doubtful: in shape and style they are true descendants of Han pottery. Holt adduces an interesting piece of evidence as to their age,— the fact that the grave in which they were found was situated within the city-walls; and, as no burial within the latter is permitted, they would seem to have been deposited there at a time prior to the erection of the wall. He refers to the "Geography of the Manchu Dynasty" (*Ta Ts'ing i t'ung chi*) as containing the information that in A.D. 625 Fu-chou was a city of the first class.

Mr. Hobson was also good enough to read in manuscript Mr. Nichols's report, that follows, and to anticipate some of these results in his admirable work "Chinese Pottery and Porcelain,"[2] which denotes decided progress in our knowledge of the entire subject, and is now the best general handbook on porcelain. Referring to Mr. Nichols's analyses of the body and glaze of this pottery, Mr. Hobson states, "The results show that the body is composed of a kaolin-like material

---

[1] On Chinese Cinerary Urns (*Journal British Archæological Association*, Vol. XXVII, 1871, pp. 343–349, Plate XVII).

[2] Vol. I, p. 15 (New York and London, 1915).

(probably a kind of decomposed pegmatite), and is, in fact, an incipient porcelain, lacking a sufficient grinding of the material. The glaze is composed of the same material softened with powdered limestone and colored with iron oxide. . . . The nature of the pottery, in spite of its coarse grain and dark color, which is probably due in part to the presence of iron in the clay, seems to show that the manufacture of porcelain was not far distant."

The report of Mr. Nichols is of sufficient importance and interest to warrant its publication in full. It is divided into two parts. Part I is devoted to a detailed investigation of the ancient porcelanous ware; and, in order to render possible a comparison with the earlier Han pottery, analysis of a green glaze from a bowl of Han pottery follows in Part II.

REPORT ON A TECHNICAL INVESTIGATION OF ANCIENT CHINESE POTTERY

By H. W. NICHOLS

I. PORCELANOUS HAN POTTERY

For the purpose of analysis, one fragment about two inches long and two inches wide, and a number of smaller pieces, were examined. The body of the ware, which is from three-sixteenths to one-quarter of an inch thick, consists of a gray vitrified porous substance which contains a few scattered black specks of minute size and glassy lustre. The body is coated on the outside with a very thin opaque red slip, and on the inside with a white engobe and a thick transparent greenish-yellow glaze.

CHEMICAL CHARACTERS OF THE BODY.—An analysis of the body from which both the inner and outer glaze and engobe coats had been removed, but with the black specks included, was made in the Museum laboratory.

ANALYSIS OF BODY

Silica, SiO_2	71.61
Alumina, Al_2O_3	18.67
Iron oxide, FeO	3.57
Lime, CaO	0.59
Magnesia, MgO	0.33
Soda, Na_2O	4.43
Potash, K_2O	1.37
	100.57

When this is compared with other analyses, it must be remembered that there are small ferruginous specks scattered through this body, so that the iron content shown by the analysis is higher than that of the true body substance.

TABLE SHOWING ANALYSIS OF ANCIENT CHINESE POTTERY
In comparison with that of modern Chinese and Japanese porcelains

	A	B		C				
Silica, SiO_2	71.61	74.53	71.31	69	70	73.30	69	70.50
Alumina, Al_2O_3	18.67	16.09	19.74	23.60	22.20	19.30	21.30	20.70
Iron oxide, FeO	3.57	1.03	0.73	1.20	2	3.40	0.80	0.80
Lime, CaO	0.59	0.06	0.17	0.30	0.80	0.60	1.10	0.50
Magnesia, MgO	0.33	0.25	2.04	0.20	trace	trace	trace	trace
Soda, Na_2O	4.43	1.19	0.10	3.30	3.60	2.50	3.40	6.00 (Soda and Potash combined)
Potash, K_2O	1.37	4.37	4.04	2.90	2.70	2.30	1.80	

EXPLANATION OF TABLE

A.—Ancient porcelanous Chinese pottery in question, analysis by H. W. NICHOLS.

B.—Modern Japanese porcelains, analyses by H. A. SEGER (see his Collected Writings, Vol. II, p. 686).

C.—Modern Chinese porcelains, analyses by A. SALVÉTAT, contained in the work of S. JULIEN, Histoire et fabrication de la porcelaine chinoise, p. LXXXVI (Paris, 1856).

The analysis proves that this body has all the chemical characters of a true porcelain. Its resemblance to the analyses of Japanese porcelains made by SEGER[1] is remarkable.

The silica and alumina both fall within the rather narrow limits set by Seger for this ware. The important deviations from the composition of Japanese porcelain are precisely those which characterize modern Chinese porcelains. These are: the high content of iron, in this instance of little significance; the high alkali content; and the excess of potash over soda. An important feature in the composition of porcelain and pottery bodies is the silica-alumina ratio. The ware presents, in this feature, a decidedly Japanese aspect. The Chinese porcelains analyzed by Salvétat generally are higher in alumina, and lower in silica, than this specimen and the Japanese bodies. The analyses of Chinese porcelain indicate a decidedly variable composition, as might be expected from Julien's description of the rather haphazard way in which the mixtures are made. In respect to this silica-alumina ratio, which sharply distinguishes Oriental from Occidental porcelains, the ancient bit of pottery under consideration comes distinctly into the Oriental class.

The quantity of alkali is essentially the same as in Salvétat's analyses of modern Chinese porcelains. Salvétat's average is 5.59%, while this ware contains 5.80%. The quantity of iron in some of Salvétat's specimens is essentially as great as that of this specimen. The variation among themselves of the analyses of modern Chinese porcelain is fully as great as the difference between these and the pottery under discussion. As the chemical composition of the ware is that of a good porcelain, the reason it failed to make a fine ware must be sought in those physical features which are consequent on the handling of the materials during manufacture, and not in any qualities inherent in the nature of the materials themselves.

PHYSICAL CHARACTERS OF THE BODY.— The body is composed of a gray vitrified material, with the slightly greasy lustre characteristic of some varieties of vitrified ware. Under an ordinary hand magnifying-glass, it appears as a kind of solidified froth composed of pores enclosed by thin walls of a translucent porcelain-like substance. These pores are elongated, so that there is a well-defined laminated structure. There are numerous inclusions of a black and glassy iron slag. Each of these glassy inclusions surrounds a minute spherical bubble. Throughout the body there are angular patches of lighter and darker gray which are vestiges of coarse particles in the mixture from which the body

[1] Collected Writings, Vol. II, pp. 687 and 716.

was burned. In thin fragments the material is somewhat translucent. A somewhat thick micro-section transmits light as freely as do many rock-sections, although confusion from the overlapping of much fine detail does not permit a very profitable study of the section.

It is not possible to tell from the examination of any well-burned vitrified ware whether the mixture from which it is burned is of natural or artificial origin. It would not be at all impossible, although perhaps a task of some difficulty, to find along the outcrop of some pegmatite dike kaolin-like material from which a body identical with this might be burned. The Japanese, formerly at any rate, burned their wares from a single clay, while the Chinese use a mixture. This ware might have been prepared either way.

The raw material contained iron-bearing minerals in coarse grains only. Each grain has left its individual splash of glassy black slag. The absence of any marked tone of buff, green, or yellow in the color of the mass indicates that there was no important quantity of finely-divided ferruginous mineral present. A simple and crude washing would have eliminated the iron-bearing minerals. Although the pottery does not look at all like porcelain, the only real point of difference, as far as the body is concerned, is the porosity of the ware. This porosity seems to be due to the use of too coarsely ground material, with not enough fine to fill the interspaces. It is a porcelain froth.

THE OUTSIDE RED GLAZE.— The red glaze on the outside is very thin. Its surface is rough and interrupted by numerous minute black blotches, where ferruginous minerals from the body have penetrated. The glaze is very uniformly distributed. It has not run during firing, nor has it crazed since. It is in as good condition to-day, as on the day it was made. It has, as well as may be determined under a powerful magnifying-glass, the structure, or rather lack of structure, of a uniform, translucent, vitrified mass. It seems to be a simple slip of some good red-burning clay. It is so thin that a sample for analysis could not be obtained. Between the red coating and the body is a white engobe coat. This nowhere exceeds one-tenth of a millimetre in thickness. It differs from the similar coating under the transparent glaze of the inside of the vessel only in its greater thinness and in the possession of a slight pinkish color, apparently absorbed from the overlying glaze. In places this coat becomes very thin and even occasionally disappears.

THE INSIDE GLAZE.— That surface of the fragment examined, which corresponds to the inside of the vessel of which it formed a part, is covered with a transparent glaze upon a porcelain-like engobe. This engobe coat is thicker than that upon the outside of the vessel.

Its average thickness is one-quarter millimetre, but this thickness is very variable. Although it is not pure white in color, it is of a distinctly lighter gray than the body; also it differs from the body, in that it is compact and free from pores. When examined under a hand magnifying-glass, it seems to be very sharply and distinctly separated from the body. When examined as a thin section under the microscope, the sharp line of demarcation disappears, as well as the difference in color. It then seems to be of the same material as the body freed from ferruginous particles and from coarse grains, so that it has vitrified into a dense non-porous body. The object of such a coating as this is twofold: it provides a light-colored background for the transparent glaze, whereby its brilliancy is enhanced; and it provides an impervious support for the glaze, which otherwise might be absorbed into the pores of the body during the firing. The appearance of the material, when viewed in the form of a micro-section, suggests that this coat is merely the result of floating the finer particles of the mix to the surface during the process of forming the vessel. This would ordinarily be accomplished by the friction of the hand or of some tool. But the coating under the more fusible glaze, where its presence is imperative, is much thicker than that under the less fusible glaze, where the necessity for it is much less. The way the coarse particles of the body project through the red glaze is difficult to understand on the theory of a floated surface; and there are no signs of dragging along the surface of those coarse particles which lie immediately under the surface; also it would be difficult to float so much fine material when the deficiency of this matter is such as to leave so many voids in the interior. The preponderance of evidence indicates that this material is an engobe coat put on possibly by dipping, but more probably by spraying. In both its physical and chemical aspect, this coat is a true porcelain.

The glaze is a greenish-yellow glass, brown in the thicker places. It is of variable thickness, as it ran badly during firing. Aside from this serious deficiency, it is a remarkably good glaze. It still adheres firmly to the body, and there has been no chipping or scaling. The crazing takes the form of a fine and uniform network of cracks. The brilliancy is very great, and there is no sign of devitrification. The attainment of these qualities, especially the continued perfect adhesion, which necessitates a very nice adjustment of the coefficients of expansion of body and glaze, indicates that the potters had already attained a high degree of skill. Running of a glaze of this type during firing is a condition unusually difficult to contend with. The color almost certainly identifies this glaze as a lime-alumina-iron silicate, and this is verified by an analysis made in the Museum laboratories.

ANALYSIS OF THE GLAZE

Silica, SiO_2	54.17
Alumina, Al_2O_3	14.16
Iron oxide, FeO	4.36
Lime, CaO	19.05
Magnesia, MgO	2.04
Soda, Na_2O	5.49
Potash, K_2O	0.00
	99.27

This is obviously an alkali-lime-iron-alumina silicate glaze. This is so purely a Chinese type, that it is useless to compare it with any but Chinese glazes. Even the Japanese glazes differ materially from those of the Chinese, being intermediate in character between these and the European. Those Chinese porcelain glazes the analyses of which have been examined are all white, and hence free or nearly so from iron. The influence of iron on a glaze is very great, and extends to nearly all its properties. Hence, in modifying a yellow glaze to a white one, there is much to do in the way of readjusting the proportions of all the elements, besides removing the iron. Therefore the close correspondence which appeared among the several body analyses will not be found to hold between the yellow and the colorless glazes, even if one has been derived from the other.

COMPARATIVE TABLE OF CHINESE GLAZES

	A	B	C
Silica, SiO_2	54.17	68	64.1
Alumina, Al_2O_3	14.16	12	10.2
Iron oxide, FeO	4.38	traces	traces
Lime, CaO	19.05	14	21
Magnesia, MgO	2.04	not determined	
Alkali, Na_2O, K_2O	5.49	6	5

EXPLANATION OF TABLE

A.—Ancient Chinese pottery glaze, analysis by H. W. NICHOLS.
B and C.—Modern Chinese porcelain glazes, analyses by A. SALVÉTAT (*l. c.*, p. 132).

The glaze on porcelain is thin, and Salvétat evidently had difficulty in securing enough material for a thorough analysis. The examples given in the table are sufficient to show that all these glazes are of the same character.

A comparison of the compositions of glaze and body suggests that the glaze has been prepared by mixing the material of the body with pulverized limestone. A brief calculation of the quantitative relations between the several elements of body and glaze confirms this impression in such a manner that there can remain no doubt as to the mode of preparation of the glaze. It must have been made by the addition of approximately one part of limestone, or the lime burned from it, to two parts of the clay from which the body was prepared. It is also possible, but not certain, that small quantities of soda and oxide of iron were added to rectify minor defects.

The calculation follows: It is assumed that the limestone is a pure, more or less magnesian, limestone, such as would naturally be employed. The limestone is taken to be somewhat magnesian, partly from inspection of the analyses, and partly because a non-magnesian limestone is rather an unusual rock. As such a limestone is practically free from silica, the silica of the glaze must come from the clay, and the ratio of the silicas in body and glaze will give a measure of the quantity of clay used in the mixture. As the body contains 71.61% silica, and the glaze 54.17%, it is evident that, ignoring for the present losses in burning, 75.66 parts of clay were used per 100 parts of glaze. The following table may then be readily calculated:

TABLE SHOWING RELATIONS BETWEEN THE COMPOSITION OF THE GLAZE AND OF A MIXTURE OF 75.66% OF THE POTTERY BODY WITH 24.34% OF LIME

	BODY	75.66% OF BODY	GLAZE	DIFFERENCE	LIMESTONE	EXCESS
Silica, SiO_2	71.61	54.17	54.17	0.00	....	0.00
Alumina, Al_2O_3	18.67	14.12	14.16	—0.04	....	—0.04
Iron oxide, FeO	3.57	2.70	4.36	1.66	....	1.66
Lime, CaO	0.59	0.45	19.05	18.60	18.60	0.00
Magnesia, MgO	0.33	0.25	2.04	1.79	1.79	0.00
Soda, Na_2O	4.43	3.35	5.49	2.14	....	2.14
Potash, K_2O	1.37	1.04	0.00	—1.04	....	—1.04
Carbonic Acid, CO_2	....	....	....	....	16.54	....
	100.57	76.08	99.27	....	36.93	....

In the column marked "excess" are recorded the differences between the actual and computed compositions of the glaze. These differences are trifling. The absence of potash from the glaze is in line with the known volatilization of potash from the surface of wares subject to the kiln fires.

The slight excess of iron oxide and soda in the mixture is not surprising, as crude, untreated earths of the kind used are by no means uniform in composition, and greater discrepancies than this are to be expected in analyses of consecutive batches of such material. Especially common is such an interchange of potash and soda as appears in this instance. The correspondences between figures and theory are, in fact, so close, that it is probable that the material employed was carefully selected by such physical characters as color, texture, etc.

It is of course possible that the potters had learned to adjust the qualities of the glaze by small additions of alkali and iron oxide. Slight variations in the quantity of either of these substances greatly influence the physical properties of the glaze.

This table cannot give more than a rough approximation of the quantities of the two ingredients of the mixture, as the losses of volatile matter in both limestone and clay during burning cannot be computed with accuracy. The table suggests that not far from one part of limestone to two parts of clay were employed. We may safely conclude that this glaze was made by adding pulverized limestone, lime, or milk of lime to the material from which the body of the pottery was made. The modern Chinese glaze for porcelain is made by mixing lime with one of the two ingredients of which they make the body. This process seems to be peculiar to China.

CONCLUSIONS.— At the time this ware was made, the potters had already acquired a high degree of dexterity. Many of the things that they accomplished in the fabrication of this pottery required technical skill of no mean order. The engobe coat, without which no satisfactory glaze could be made upon so porous a ware, was used. The expansion of the glaze has been very accurately adjusted to that of the body. The glaze is remarkably brilliant for one free from lead. The glaze has no large bubbles, nor are small bubbles numerous enough to cloud the ware. On the other hand, they made the glaze too thick, and they could not prevent it from running during the firing.

With potters as skilful as these, the discovery of methods of overcoming the porosity of the ware, and thus making it a true porcelain, should be only a matter of time. As the engobe coat is porcelain, it is quite possible that the knowledge was not lacking even at that time. They may not have realized that a dense ware would be worth the great expense involved in grinding the materials to the necessary fineness by the crude methods then available, and in the control of the drying and firing methods to prevent distortion of the ware.

II. ANALYSIS OF A GREEN GLAZE FROM A BOWL OF HAN POTTERY

This is a brilliant glassy glaze of a bottle-green color from a Han pottery bowl (Cat. No. 118578). It is thickly applied over a red porous body.

It is believed that the material selected for analysis correctly represents the original unaltered glaze. The glaze with its red backing was crushed to fragments of about a millimetre average size, and clear unaltered fragments were selected after scrutiny under a powerful glass. These fragments were freed from the adhering films of red earthy matter by use of forceps and a fine file. As finally prepared, the glass showed no altered material, nor any but a few unweighable traces of earthy matter.

The analysis gives:

Silica, SiO_2	29.91
Lead oxide, PbO	65.45
Iron oxide, FeO	0.81
Copper oxide, CuO	2.60
Lime, CaO	0.94
Alkalies, Na_2O, K_2O	0.00
	99.51

This gives the molecular formula:

$$1 \text{ RO} : 1.4 \text{ SiO}_2 \quad \text{or nearly } 5 \text{ RO}. \ 7 \text{ SiO}_2.$$

The traces of iron and lime are obviously impurities.

This is a simple lead silicate colored by copper, and is utterly unlike any glaze of which I have any analysis, the nearest approach to it being the alkali-lead silicate which seems to have been an ordinary glaze in all countries. The omission of alkali places this glaze in a very different class. It could be easily and simply compounded, as there are but three ingredients,—some lead salt (perhaps red lead or white lead), a pure white sand, and a small quantity of some copper compound for coloring.

Professor R. T. Stull, Acting Director of the Ceramic Department of the University of Illinois, has been good enough to supply the following additional information on the preceding analysis:

"I am very much interested in the data you present on the early Chinese glaze. I have calculated an approximate empirical formula from the analysis, which gives:

$$\begin{matrix} .827 \text{ PbO} & \\ .093 \text{ CuO} & 1.408 \text{ SiO}_2 \\ .049 \text{ CaO} & \\ .031 \text{ FeO} & \end{matrix}$$

"This approximates closely the theoretical formula:

$$\left.\begin{matrix} .9 \text{ PbO} \\ .1 \text{ CuO} \end{matrix}\right\} 1.5 \text{ SiO}_2 = 2\text{RO}. \ 3 \text{ SiO}_2$$

A glaze can be made by mixing the following materials, which would be very similar to the Chinese glaze when first made:

Red lead	205
Copper oxide	8
Potter's flint	90

It is quite probable that the Chinese glaze was originally made by mixing together three ingredients,—a lead compound, a copper compound, and a form of silica. The iron and lime present were probably impurities existing in the raw materials used in making the glaze. A glaze of this type (which is in reality a glass, since glazes generally contain alumina) fuses at a very low temperature, is very brilliant, has a high specific gravity, high index of refraction, and high coefficient of expansion; and is easily dissolved by chemical agents (comparatively so). Owing to the high coefficient of expansion, the glaze is very susceptible to crazing. The glaze could be improved by the addition of alumina in the form of clay, which would lower the coefficient of expansion, thus reducing crazing, and would make the glaze more resistant to the weathering action or to chemical agents. In good glaze practice, it is customary to introduce an alkali in some form, although good glazes can be produced without the use of alkali. One glaze being used for glazing roofing tile has the formula:

$$\begin{matrix} .9\ PbO \\ .1\ CuO \end{matrix}\ .15\ Al_2O_3\ 1.6\ SiO_2,$$

which is very similar to the Chinese glaze plus Al_2O_3. A mixture which will produce this glaze is:

Red lead	205
Copper oxide	8
Ball clay	39
Potter's flint	78

If the Chinese glaze has been disintegrated by long exposure, the alkalis would naturally be leached out partially, if not entirely."[1]

[1] The material for analysis was carefully-picked, unaltered fragments [H.W.N.].

HISTORICAL OBSERVATIONS AND CONCLUSIONS

The preceding report of Mr. Nichols leaves no doubt that the pottery in question, as confirmed by Mr. Hobson, is a porcelanous or porcelain-like ware, as regards the composition of both body and glaze. It is a forerunner of true porcelain; it represents one of the initial or primitive stages of development through which porcelain must have passed before it could reach that state of perfection for which the Chinese product gained fame throughout the world. The history of porcelain has been singularly exposed to misrepresentations and misunderstandings, chiefly for the reason that Chinese accounts of the subject are obscure, enigmatic, and, moreover, disappointingly meagre and unsatisfactory. In his eminently critical and excellent work, Hobson has done a great deal to eradicate many of the old superstitions. It was obvious that the problem of the origin of porcelain could be solved only by archæological, not by philological, methods; and it is due to the investigations of Mr. Nichols that we may now for the first time formulate certain opinions regarding the beginnings of porcelain, which are grounded on matter-of-fact observation, and not on a more or less arbitrary interpretation of texts. Therefore the question may first be discussed from an archæological viewpoint; and then it remains to be seen whether, with the result thus obtained, Chinese traditions may not be better and more profitably understood.

Before attempting to determine the date of the "Han" porcelanous ware, it will be useful to raise the question whether there is now a possibility of dating the first manufacture of true porcelain. I shall not insist on the evidence deduced by Bushell and Hobson from Chinese sources, to the effect that porcelain was made under the T'ang dynasty (618–906) as early as the beginning of the seventh century. Reference will be made to only one source which has not yet been enlisted for the study of the question, and then we may proceed to archæological evidence.

An incontrovertible proof for the existence of porcelain in the seventh century is contained in the memorable account of the Buddhist pilgrim I-tsing (635–713), who visited India from 671 to 695. In discussing the utensils to be utilized by the monks of India, I-tsing speaks also of Indian earthenware vessels, and remarks, "In India, there was originally neither porcelain (*ts'e* 瓷) nor lacquer. Porcelain, if glazed, is no doubt clean. Lacquered articles are sometimes brought to India by traders."[1] It is evident beyond cavil that I-tsing understands the word *ts'e* in this passage in the sense of porcelain with which he was familiar in his native country. He could most assuredly not mean to say that pottery was originally unknown in India, for in more than one case he himself refers to Indian pottery or earthenware (*wa* 瓦), which could not escape the attention of a keen observer like him. He expressly avails himself of the word *ts'e* in this passage, advisedly in contradistinction to the word *wa* used previously, and connects it with another characteristic product through which China then became widely known,— lacquer. He does not state explicitly that porcelain, in the same manner as lacquer-ware, was then imported from China into India; but this fact may be inferred from the statement made in the beginning of Chapter VI, that "earthenware and porcelain (*wa ts'e* 瓦瓷) are used for the clean jar" (that is, the jar containing the water for drinking-purposes).[2] This passage is sufficient evidence for the fact that porcelain was then found in India; and also his statement that porcelain did not originally exist in India seems to imply that it occurred there at the time of the author's visit. He does not speak of porcelain as a new, but as a familiar, production; and he must certainly have seen it in China before the year 671, the date of his departure for India. Judging from I-tsing's memoirs, porcelain, accordingly, must have existed in China during the latter half of the seventh century. At the same time, it was exported into India; and this harmonizes with the observation made in the *T'ao shuo*, that porcelain bowls were widely distributed abroad from the time of the T'ang dynasty (618–906).[3]

The testimony of the Arabic merchant Soleyman, who in 851 wrote his "Chain of Chronicles," must be regarded as one of the most weighty to prove the existence in China of true porcelain in the age of the T'ang, during the ninth century. In the translation of

---

[1] J. TAKAKUSU, A Record of the Buddhist Religion as practised in India by I-tsing, p. 36 (Oxford, 1896); Japanese edition of the text, Vol. I, p. 17 a.

[2] *L.c.*, p. 27; text, Vol. I, p. 12 a.

[3] *T'ao shuo*, Ch. 5, p. 2 b (edition with movable types, published 1913); S. W. BUSHELL, Description of Chinese Pottery and Porcelain, p. 104.— According to W. CROOKE (Natives of Northern India, p. 136, London, 1907), common clay pots, owing to their perishable character, are little valued in India, "and caste prejudices prevent the use of the finer kinds of pottery. Hence no artistic industry like that of china has flourished in India, although kaolin and other suitable kinds of clay are in some places abundant." We have a formal judgment on Indian pottery from the Buddhist monk Yüan Ying, who in his *Yi ts'ie king yin i* (Ch. 18, p. 7; see p. 115), written about A.D. 649, remarks that the state of culture is so low in the Western Regions that finer pottery cannot be made there, and that only unburnt bricks and vessels fired without glaze are turned out.

M. REINAUD,[1] he reports that "there is in China a very fine clay from which are made vases having the transparency of glass bottles; water in these vases is visible through them, and yet they are made of clay."[2]

The presence of china in the India of the seventh century, and the acquaintance of the Arabs with transparent porcelain in the ninth century, based on literary sources, naturally raise the question whether this documentary evidence is corroborated by any archæological facts. Such have heretofore been lacking; but an important discovery due to the excavations of F. Sarre and E. Herzfeld in the ruins of Samarra, the former residence of the Caliphs, is fortunately apt to settle satisfactorily this much-disputed question. The report of these remarkable finds has recently been published.[3] According to F. Sarre, who carefully figures and describes these objects, they belong to a period which is well determined by the years A.D. 838 and 883. The ceramic specimens exhumed in Samarra fall into two classes,— those imported from eastern Asia, and those potted locally for home-consumption. Among the former we are confronted with a material which in general must be designated as stoneware, but which, to use the words of Sarre, partially approaches porcelain to such a high degree that it may straightway be styled "porcelain." In the latter case, the body of the vessels cannot be scratched by steel, is almost white, transparent in thin places, the shards being dense, and hard like shell. The smooth and brilliant glaze is evenly applied, and so closely linked with the body that both can but have been fired simultaneously,— characteristic qualities of genuine East-Asiatic porcelain. Besides fragments of more or less coarse and shallow bowls, whose low rim around the bottom is ground off, those of finer ware have also come to light; thus, for instance, a fragmentary oval cup decorated with a fish in relief, surrounded by wave designs and birds on the wing. Judging from the author's description and the very excellent illustrations, there is no room for

---

[1] Relation des voyages faits par les Arabes et les Persans dans l'Inde et à la Chine, Vol. I, p. 34.

[2] The report of Soleyman is in full accord with the Chinese notices of T'ang pottery. In the beginning of the T'ang dynasty (618), vases of a white clay, with thin body of white and brilliant color, were made by a potter of the name T'ao, in the village Chung-siu, belonging to King-te-chen; they were styled "imitation jade utensils," and sent as tribute to the Court. Similar vessels were turned out simultaneously by Ho Chung-ch'u from the village Tung-shan (*King te chen t'ao lu*, Ch. 5, p. 1 b; JULIEN, Histoire, pp. 81, 82). It is notable that both potters were rural residents, and that their work possessed sufficient quality to earn imperial approbation.

[3] F. SARRE, Die Kleinfunde von Samarra und ihre Ergebnisse für das islamische Kunstgewerbe des 9. Jahrhunderts (*Der Islam*, Vol. V, 1914, pp. 180–195, 4 plates).

doubt that the piece in question is of real, white porcelain, and that it affords an example of the hitherto lost porcelain of the T'ang period. T'ang porcelain is thus raised into the rank of plain fact. Soleyman's testimony proves true.

The date of this specimen is indubitable, and meets a welcome confirmation from two green and white glazed dishes of pottery[1] secured in the same locality. Without having any clew to their provenience, the writer, who through his researches in China is somewhat familiar with this and similar ware, would not hesitate for a moment to diagnose them as Chinese productions of the epoch of the T'ang. Mr. Sarre is perfectly correct in calling attention to the fact that pieces of identical technique are preserved in the Imperial Treasury of Nara in Japan, and that T'ang clay statuettes are formed of the same material. Another discovery of no less importance, for which we are indebted to Mr. Sarre's energy, is a group of celadon-like stoneware, one of which, bearing the design of a fish scratched in under the glaze, is reproduced in his report. The facts brought out by Mr. Sarre's researches are of such far-reaching consequence, that he is entitled to a just claim to our lasting gratitude. Above all, he has succeeded in safely establishing the fundamental fact that porcelain was made in China under the T'ang; and that Chinese porcelain, as well as non-porcelanous pottery, was exported in the ninth century into the Empire of the Caliphs. These conclusions embolden us and justify us in regarding the word *ts'e*, whenever it appears in T'ang documents, as conveying the notion of true porcelain, and in giving full credence to the account of I-tsing, that India possessed Chinese porcelain during the seventh century.[2] Consequently it is at some earlier date that the beginnings of porcelain — those initiatory and preparatory steps finally leading up to the perfection of the ware — must be sought for. Porcelain has been discovered in Turkistan by Sir AUREL STEIN.[3]

Our previous knowledge of references to T'ang porcelain was chiefly based on the two modern works, the *King te chen t'ao lu* (first edition, 1815) and the *T'ao shuo* (1774). It remains to be ascertained, however, from the contemporaneous records of the T'ang, whether these extracts

[1] On Plate II in the article referred to.

[2] As shown by I-tsing, a clear distinction between common pottery and porcelain is made in T'ang literature. This is further evidenced by the frequent occurrence of the compound *ts'e wa* 瓷瓦 ("porcelain and stoneware"), for instance, in the *Yu yang tsa tsu* (Ch. 11, p. 7 b; ed. of *Pai hai*) and in the *Ta T'ang sin yü* 大唐新語 (Ch. 13, p. 9; ed. of *T'ang Sung ts'ung shu*).

[3] Ancient Khotan, Vol. I, pp. 461, 464 (see also HOBSON, Chinese Pottery and Porcelain, Vol. I, p. 149). It would be desirable that analyses be made and published of Sarre's and Stein's porcelains.

are reliable and correctly reproduced. In the geographical chapters of the T'ang Annals we find under each locality an enumeration of the taxes in kind annually sent to the Court, and the *T'ai p'ing huan yü ki* of Yo Shi gives a still more extensive list of the products of the empire during that period. The following localities are known as having produced porcelain under the T'ang:—

1. Hing chou 邢州 (modern Shun-te fu in Chi-li) turned out white porcelain vessels 白瓷器 (*T'ang shu*, Ch. 39, p. 6; and *T'ai p'ing huan yü ki*, Ch. 59, p. 5), which were accepted as taxes.

2. Ting chou 定州 in Chi-li (*T'ai p'ing huan yü ki*, Ch. 62, p. 4 b); the T'ang Annals do not mention porcelain among its products.

3. Yu chou 幽州 (modern Yung-p'ing fu in Chi-li), according to *T'ai p'ing huan yü ki*, Ch. 69, p. 6.

4. Jao chou 饒州 in Kiang-si (*T'ai p'ing huan yü ki*, Ch. 107, p. 3).

5. Yüe chou 越州 (modern Shao-hing fu in Che-kiang), according to *T'ang shu* (Ch. 41, p. 4 b) and *T'ai p'ing huan yü ki* (Ch. 96, p. 5).

6. Ho-nan fu (according to *T'ang leu tien*, Ch. 3, p. 4 b, ed. of *Kuang ya shu kü*, 1895; and *T'ai p'ing huan yü ki*, Ch. 3, p. 8b).

As may readily be seen from JULIEN's translation (pp. 28 and 6), only two of these localities (Nos. 1 and 5) are mentioned in the *King te chen t'ao lu* as having produced porcelain under the T'ang (not, however, Nos. 2–4); while several others are so designated, which cannot be verified from coeval documents.[1]

As established by archæological evidence, porcelain was an accomplished fact under the T'ang (618–906); and there is further good reason to assume that it existed in the latter part of the sixth century.[2] It is futile, of course, to look for an inventor of porcelain, as has been done by E. ZIMMERMANN.[3] This invention of an inventor of porcelain is a romance, not history. Chinese records know absolutely nothing about such an inventor, simply for the reason that he never existed. Porcelain is not an "invention," that can be attributed to the efforts of an

---

[1] In the writer's forthcoming second part of Chinese Clay Figures will be found a chapter on T'ang pottery.

[2] BUSHELL, Description of Chinese Pottery, p. XII; HOBSON, Chinese Pottery and Porcelain, Vol. I, p. 147. In 1844, during the negotiations preceding the Franco-Chinese Treaty, one of the Chinese envoys, Chao Chang-li, well acquainted with the antiquities of his country, assured N. Rondot that the manufacture of porcelain could be traced back only as far as the middle of the sixth century (see *Journal China Branch Roy. As. Soc.*, Vol. XXXII, 1897–98, p. 73).

[3] *Orientalisches Archiv*, Vol. II, 1911, pp. 30–34; and Chinesisches Porzellan, p. 24. I strictly concur with HOBSON (*l. c.*, Vol. I, p. 145) in his criticism of Zimmermann's hypothesis.

individual; but it was a slow and gradual process of finding, groping, and experimenting, the outcome of the united exertions of several centuries and generations. We clearly observe a rising development of porcelain from the T'ang to the Sung, Yüan, and Ming periods, till the high perfection of the ware culminates in the K'ang-hi era. It is therefore logical to assume that preceding the age of the Sui (590–617) there was a primitive stage of development which ultimately resulted in the T'ang porcelain. This primeval porcelanous product was hitherto unknown, but, as demonstrated by the researches of Mr. Nichols, its existence is now proved in the nine vessels figured on Plates I and III–X, with analogous specimens in the Boston Fine Arts Museum, the Freer collection, and the British Museum. The tentative attributions "T'ang" and "Sung" (p. 82) were based only on isolated cases, and ventured as personal impressions; they were not grounded on the fact of analytic study. The Han tradition of ceramic forms had completely died out under the T'ang and Sung, to give way to more graceful and pleasing shapes partially conceived under Iranian and Indian influences. As has been shown, the objects in question decidedly breathe the spirit of Han art in forms and decorative motives. There is good circumstantial evidence in the case of the jug on Plate I, discovered in the same grave with a Han cast-iron stove, and in that of the *pan-liang* coins of the Boston jar. Nevertheless I am not convinced that we are entitled to assign these vessels to the Later Han dynasty within its strict chronological boundaries (A.D. 25–220), as the predominant bulk of the kiln-products turned out under the Han was common glazed and unglazed pottery (*wa* 瓦).[1] Moreover, the new term *ts'e* 瓷, applied to porcelanous ware, does not yet occur in the contemporaneous records of the Han, at least such an occurrence has not yet been proved (see p. 102); and this is the main reason which prompts me to the opinion that the pottery in question was manufactured in post-Han times, say, roughly, under the earlier Wei (220–264), or toward the middle or in the latter part of the third century A.D.[2] From a purely philological point of view,

[1] This is the term employed for the burial pottery of the period in the Han Annals (*Hou Han shu*, Ch. 16, p. 3). It is therefore out of the question that the new term *ts'e*, as stated by HOBSON (*l. c.*, Vol. I, p. 141, note), should refer to the glazed pottery of the Han. Credit must be given also to the Chinese for their correct feeling for their own language and their own antiquities: the present-day Chinese style the glazed Han pottery *liu-li wa* (accordingly, with the same term as employed in the Han Annals), while the term *Han ts'e* is applied to the porcelanous ware here described. In this case, Chinese feeling signifies a hundred times more than all the hair-splitting and pedantic subtleties of European sinologues.

[2] It is curious that this result agrees with the opinion of PALLADIUS (Chinese-Russian Dictionary, Vol. II, p. 343), who held that the output of porcelain took its beginning from the Tsin dynasty (263–420).

the term *Han ts'e*, applied to this pottery by Mr. Yen, is not justified. From the standpoint of the archæologist, however, it is perfectly correct; for this pottery, as recognized by Mr. Yen with just instinct or intuition, combines in itself two characteristic features,— the style of Han art, and the technical character of porcelanous ware. It is justifiable to regard it as a very early production, or even as one of the earliest, of the ware styled *ts'e*. We might therefore say that porcelain ran through its experimental stages for at least three centuries; and it seems to me a reasonable conclusion that a development of such a length of time was required until mature and highly finished products should ultimately result.

It is possible also to make a plausible guess at the kiln, where the nine vessels were produced. As has been pointed out, the jug in Plate I was found in a grave near the village Ma-kia-chai, 5 *li* north of the town of Hien-yang 咸陽, the ancient capital of the Ts'in, belonging to the prefecture of Si-ngan. The "Records of the Potteries of King-te-chen" inform us that "under the earlier Wei dynasty (220–264) vases were turned out at Kuan-chung 關中, corresponding to Hien-yang and other places of the prefecture of Si-ngan, and that the output of this kiln was intended for the use of the Court, and offered to the Emperor."[1] Thus it is not impossible that our ware was actually made in the district of Hien-yang, or, taking the wider area, in the prefecture of Si-ngan. If the passage quoted should really be derived from an ancient text, which I am not in a position to prove, it would have another significance, in that it would represent the earliest allusion to pottery deemed worthy of being sent to the palace. Neither in times of antiquity nor under the Han do we hear of any tribute pottery. In the famous Tribute

---

[1] *King te chen t'ao lu* (edition of 1891), Ch. 7, p. 1 b. JULIEN (Histoire et fabrication de la porcelaine chinoise, p. 4), in his translation of these passages, speaks in both cases of "porcelain;" but this is not warranted by the Chinese text, which avails itself of the general term *t'ao* ("pottery"); but *ts'e* belonged to the class of *t'ao*. HOBSON (Chinese Pottery and Porcelain, Vol. I, p. 143) complains of Julien and Bushell having been indiscriminate in the use of the term "porcelain" in their translations from the Chinese. But how about LEGGE, who speaks of porcelain in the era of the *Shi king?* In his translation of this work, we read in two passages (pp. 346 and 502) of a "porcelain whistle," which is entered even in the index. Fortunately this musical instrument of porcelain has escaped the students and collectors of Chinese ceramics; otherwise we should probably meet it in one or another collection, since the collector usually gets what he wants or solicits. What is meant in the passage of the *Shi king* is the instrument *hüan* 壎, a pipe made of baked clay, of the size of a fowl's egg, and perforated by six apertures. Again, we read of "porcelain drums" in a translation of DE GROOT (Religious System of China, Vol. VI, p. 977) from a text of the *Tu tuan* by Ts'ai Yung (133–192), relative to conditions of the Chou period. The text has *t'u ku* 土鼓, which means "earthen drums."

of Yü (*Yü kung*), forming a section of the *Shu king*, pottery is conspicuously absent. In pre-Han and Han times it had not yet reached such a state of perfection that it would have been brought to the immediate attention of the sovereign, or was eligible to take a place in the imperial chambers. It is conceivable that pottery of the class of our porcelanous ware was entitled to admission to Court, and answers to the tribute ware produced at Kuan-chung.

The origin of this mysterious and much-discussed term *ts'e* has been referred to the Han period by several European authors, but nobody has yet furnished any actual proof that the word really occurs in contemporaneous records of that age. Even BUSHELL[1] merely states, "We know that the word *ts'e*, which means porcelain in the present day, first came into use during the Han dynasty, and Mr. Hippisley takes this coining of a new word to designate the productions of that age to be a strong argument in favor of the early date. Others, more sceptical, before reaching any decision, ask to be shown actual specimens of translucent body that can be certainly referred to the period." Seven years later, Bushell became more confident and positive in his assertion of the origin of porcelain under the Han. In his work "Chinese Art,"[2] an assurance to this effect is given in three passages. The word and character *ts'e*, according to him, is first found in books of the Han dynasty. Again he asserts that the Chinese attribute the invention to the Han dynasty, when a new character *ts'e* was coined to designate, presumably, a new substance;[3] and that "still we may reasonably accept the conclusion of the best native scholarship that porcelain was first made in the Han dynasty, without trying, as Stanislas Julien has tried on very insufficient grounds, to fix the precise date of its invention."

The only piece of evidence that has ever been produced to prove the existence of the term *ts'e* under the Han is the citation of this word in the glossary *Shuo wen*. Sceptics will naturally raise the question

---

[1] Oriental Ceramic Art, p. 20 (New York, 1899).

[2] Vol. II, pp. 4, 17, 20.

[3] The fact cited by Bushell on this occasion — that "the official memoir on 'Porcelain Administration' in the topography of Fou-liang says that, according to local tradition, the ceramic works at Sin-p'ing (an old name of Fou-liang) were founded in the time of the Han dynasty, and had been in constant operation ever since"— is not conclusive for a plea on behalf of porcelain at the time of the Han. That tradition, if correct, merely goes to show that kilns for the manufacture of pottery were established in that locality under the Han, while it implies nothing definite as to the specific character of this pottery. The fact that Fou-liang turned out porcelain at a later period does not allow of the inference that what was produced there in the era of the Han likewise was porcelain.

whether the passage was actually contained in the original edition of the work (A.D. 100), or whether it has been interpolated in the numerous subsequent re-editions.[1] The decision of this question may be left to a competent sinologue. It means little for my purposes, as long as no instances of the word are pointed out in authentic books, which may be regarded as contemporaneous documents of the Han period. This much may be said, that the definition given in the *Shuo wen* has not been adequately explained. It has been asserted the definition should mean that *ts'e* is "pottery and nothing more."[2] It means, however, "*Ts'e* belongs to the category of pottery," or "is a kind of pottery." In the definitions of the *Shuo wen*, the word to be explained is defined by a more general word denoting the wider category. It cannot therefore be deduced from that gloss that *ts'e* in ancient times did not refer to porcelain, for porcelain certainly is a variety of pottery. In regard to the specific character of *ts'e*, the definition of the *Shuo wen* is utterly inconclusive. Holding in abeyance the question as to the time when the term *ts'e* sprang into existence, and leaving aside all subtleties, it remains for plain common sense to say that a new term refers to a new matter, and that *ts'e* as a new ceramic term must have denoted a novel production achieved in the ceramic field. Such was the porcelanous ware as here described; and if, from the Sui and T'ang periods onward, the word *ts'e* was applied to true porcelain, it is self-evident that prior to that time it was attached to porcelanous ware, the forerunner of porcelain. The word *ts'e* did not plainly describe any pottery, but porcelanous pottery specifically.

It is known that the character *ts'e* 磁 is now employed also in place of *ts'e* 瓷. From this change of characters F. HIRTH[3] believed he was justified in concluding that the new form, linked with the classifier 'stone' 石, indicates a substitute of material; while in the older form, combined with the classifier 'clay' 瓦, the nature of earthenware should be accentuated. This argumentation is unwarranted, and, as will be seen, does not answer the facts. Likewise the information given on this point in the "Catalogue of Potteries published by the Japan Society" (p. 56, New York, 1914) is misleading. Here it is asserted that from the fact that the city Ts'e-chou produced porcelain, and that the word *ts'e* in the name of the city is phonetically identical with that of the word meaning "stoneware" or "porcelain," a certain confusion in

---

[1] Neither the *Erh ya* nor the *Kuang ya* contains the word; but also this proves nothing, as none of the ancient dictionaries is complete, and they surely lack numerous words which are found in literature.

[2] F. HIRTH, Ancient Chinese Porcelain, p. 130.

[3] Ancient Chinese Porcelain, p. 130, note 3.

the use of the word has arisen; "but there is no such confusion in the mind of the Chinese scholar; the purist never uses it; and all arguments as to the date of the origin of porcelain which have been based on the use of this word are valueless." All these statements are erroneous. No one has ever based any arguments on the use of this word as to the date of porcelain. In fact, the word has no concern whatever with the origin of porcelain. The chief facts in the case could already be gleaned from JULIEN's "Histoire" (p. 29). There is, first of all, a city by the name Ts'e-chou 磁州, which anciently depended on the prefecture of Chang-te in the province of Ho-nan, but which is now assigned to the prefecture of Kuang-p'ing in the province of Chi-li. The city had formerly various other names. The present name Ts'e 磁 was conferred on it in the year 590, at the time of the Sui dynasty. Near the boundary of the district rose the Loadstone Mountain (Ts'e shan 磁山) producing loadstone (*ts'e shi* 磁石), whence the district and town received their name.[1] At the time of the T'ang dynasty (618–906), the district produced nothing but loadstone and magnets made from it; it did not produce pottery of any kind.[2] Only from under the Sung (960–1278) did the locality in question embark on the manufacture of a kind of white porcelain, the choice specimens of which resembled the Ting ware. This particular kind of porcelain, because it originated from the locality of Ts'e, was styled "vessels of Ts'e" (*Ts'e k'i* 磁器). The word *ts'e* in this case, accordingly, denotes nothing but the place of provenience. "At present," the author of the "Records of the Potteries of King-te-chen" adds, "owing to a very common error, porcelain vases are generally designated by the term *ts'e k'i* 磁器; people employing this term are doubtless ignorant of the fact that it applies in particular only to the porcelain of the city of Ts'e." The fact remains that under the Manchu dynasty, and at present, porcelain is invariably termed 瓷 and 磁, the latter character being more frequently employed.[3] True it is, that K'ang-hi's Dictionary does not

---

[1] *T'ai p'ing huan yü ki*, Ch. 56, p. 10 b. The *Pen ts'ao kang mu* extols the loadstone of this locality as excellent (F. DE MÉLY, Lapidaires chinois, p. 106), and loadstone was supplied from there as tribute to the Court (*Ta Ts'ing i t'ung chi*, Ch. 31, p. 12).

[2] The silence of the *T'ai p'ing huan yü ki* and the T'ang Annals in this respect is conclusive, as the localities producing porcelanous ware at that time are expressly named (see above, p. 99). HOBSON (Chinese Pottery and Porcelain, Vol. I, p. 101) also arrives at the result that there is no information on the subject of Ts'e-chou factories earlier than the Sung dynasty, when they enjoyed a high reputation.

[3] Even in the T'ang Annals the term *ts'e k'i* 磁器 appears, although we are not in a position to state that it was thus written in the original edition: the district Kü-lu 鉅鹿 in Hing-chou (now prefecture of Shun-te in Chi-li Province) sent porcelain vessels as tribute in the year 742 (*T'ang shu*, Ch. 39, p. 6); and the fact that

credit it with the meaning of "porcelain," but attributes to it only the proper significance, "loadstone." This, however, means nothing. Chinese standard works, like the great cyclopædia *T'u shu tsi ch'eng* and others, also the Japanese, employ this character throughout in the sense of "porcelain," so that there is no longer the question of confusion. On the contrary, it is a perfectly legitimate usage, even sanctioned by the English and Chinese Standard Dictionary issued by the Shanghai Commercial Press; and for this reason our own dictionaries, like those of Palladius, Giles, and Couvreur, are justified in assigning the meaning "porcelain" also to the character *ts'e* 磁. This was the outcome of a natural development of the language, which no alleged purism can sweep. The original term "porcelain of Ts'e" was simply amplified into the wider notion of porcelain in general, because the word *ts'e* employed in the name of the city bearing that name, and the word *ts'e* for "porcelain," though physically different words, phonetically are homophonous.[1] This history of the subject clearly shows that Hirth's theory is untenable and should be discarded. The new word *ts'e* 磁, in the sense of "porcelain," has no organic and historical connection whatever with the older word for "porcelain" *ts'e* 瓷, but is an independent side-issue of purely incidental character. The alleged evolution from earthenware to stony material cannot be read from the formation of these characters, as they have nothing in common, and move along separate lines. This conclusion settles also the general speculation[2] to the effect that the word *ts'e* in its origin should have meant nothing but common earthenware, and that gradual improvement of the ware resulted in changes of meaning and writing. We now recognize that the genuine character for *ts'e* 瓷 has not been subject to any alterations, and that it was in the beginning exactly the same as it is at present. It is therefore infinitely more probable that this speculation regarding substitutes of material resulting in altered significations of the word is imaginary in its entire range; that is to say, the newly coined word *ts'e*, from the days of its childhood, denoted not simply "earthenware,"

---

the question is here of porcelain is confirmed by the *King te chen t'ao lu* (JULIEN, Histoire, p. 28). In other passages of the T'ang Annals we meet the regular mode of writing 瓷器; for instance, in Ch. 41, p. 4 b, where the porcelain of Hui-ki in Yüe-chou (the present province of Che-kiang) is mentioned. In the *T'ai p'ing huan yü ki* only the form 瓷 is employed. "Porcelain" is expressed by 磁 in the *Liao shi* (Ch. 104, p. 2) and *Yüan shi* (Ch. 88, p. 10 b).

[1] The mental process underlying this transformation may be compared with the extension of our word "china" to porcelains made in any countries outside of China.

[2] HIRTH, Ancient Chinese Porcelain, p. 130 (repeated in his Chinesische Studien, p. 48).

but a higher grade of pottery which shared characteristic features with true porcelain.

Another problem is whether the kind of porcelain manufactured at Ts'e-chou bore any relation to the mineral *ts'e*. The term *ts'e* 磁, as is well known, is the designation of the magnet or loadstone; but, as admitted by the Chinese, it denotes also another mineral which is suitable for the making of pottery. This fact is brought out by several ancient stone sculptures in the Museum's collection, in the votive inscriptions of which it is stated that the material of the sculpture is *ts'e shi* 磁石 ("*ts'e* stone"), which, however, as shown by a very superficial examination, is not loadstone. The "Records of the Potteries of King-te-chen"[1] inform us that "the *ts'e* stone 磁石 is made into a paste serviceable for pottery vessels, but that this stone is not identical with the magnet attracting iron and used for magnetic needles; further, it is a peculiar and distinct kind of stone of white color and of bright and smooth appearance; the vessels made from it are beautiful, but not delicate, and differ from porcelain earth; aside from Ts'e-chou, they are made in Hü-chou 許州 in Ho-nan Province. It is accordingly not magnetic ore which entered into the manufacture of Ts'e porcelain, but a mineral of a different nature, as yet undetermined, apparently not discovered prior to the age of the Sung, and likewise styled *ts'e*.[2] This point is especially mentioned in this connection, because a supposition that magnetic ore might have been mixed with porcelain glaze would not be entirely without foundation.[3]

In fact, however, we have no account of loadstone ever having been used by the Chinese in the making of pottery; and it is therefore impossible to assume any connection between the two words *ts'e*,—the one denoting "loadstone," the other "porcelain." As the written

[1] *King te chen t'ao lu*, Ch. 10, p. 12 b (new edition, 1891).

[2] PALLADIUS (Chinese-Russian Dictionary, Vol. II, p. 343) states under this word, "Magnet; suitable for the eyes; employed in the making of bowls and pillows; porcelain."

[3] According to PLINY (*Nat. hist.*, XXXVI, 66, § 192), magnet-stone was added to glass during the process of making the latter, because it was credited with the property of attracting liquefied glass as well as iron (Mox, ut est ingeniosa sollertia, non fuit contenta nitrum miscuisse; coeptus addi et magnes lapis, quoniam in se liquorem vitri quoque ut ferrum trahere creditur). The correctness of this report has been called into doubt. The Arabic mineralogy ascribed to Aristotle has replaced the magnet-stone by the stone magnesia as being added to glass (J. RUSKA, Steinbuch des Aristoteles, p. 171). In another passage (*ibid.*, p. 129) it is said that glass cannot be finished without the stone magnesia; the latter denotes manganese, which serves for the refinement of glass fluxes. Whether Pliny is guilty of a confusion in the case, or whether he really reproduces a tradition current in his time, can hardly be decided.

symbols are formed by means of different phonetic elements, the greater likelihood is that also the two words, although now phonetically identical, are traceable to different origins. The history of the word *ts'e* 磁 can be established without great difficulty. The earliest form in which it was written is *ts'e shi* 慈石 (that is, "attractive stone"); in this manner we find it, for instance, in the Annals of the Former Han Dynasty.[1] The character 磁, consequently, is a secondary formation based on a contraction of the words *ts'e* and *shi*, the latter assuming the position of classifier, the former that of phonetic element, the original significance of which was bound gradually to disappear. The word for "porcelain," however, is written with the phonetic element *ts'e* 次, which, as an independent word, has the meaning "second, next in order, inferior," etc. It is clear that in composition with the classifier 'clay' (*wa* 瓦) it has no word-meaning whatever, but has merely the function of a phonetic element. Thus far we are entirely ignorant of how this new word may have arisen in the first centuries of our era. In the Sung period the phonetic part seems to have been altered, for the dictionary *Tsi yün* 集韻, published by Ting Tu 丁度 in the middle of the eleventh century, records the two forms 甆 and 瓷 as popular or common at that time. This manner of writing may have come about under the immediate influence of the porcelain of Ts'e-chou, which then sprang into existence.

The preceding remarks on the term *ts'e* are not intended to encroach on the domain of the sinologue. No one feels more keenly than myself that a critical and detailed study of this term (not based on the modern cyclopædias, but on the actual source-works) is required, and should be taken up some day by a competent sinologue who has a taste for researches of this kind.

The previous discussions on the origin of porcelain were chiefly based on haggling about terms, which at times assumed an almost Talmudic character. Students entered into the arena with a dogmatic definition fixed in their minds, of what porcelain is or should be, and, according to their personal standpoint, rejected or accepted this or that period at which porcelain should have come into existence. Thus we face the amazing spectacle that from 1856, the date of the appearance of Julien's celebrated book on Chinese porcelain, down to the present time, almost any period of Chinese civilization has been claimed as the one responsible for its "invention." From its exalted position in

---

[1] *Ts'ien Han shu*, Ch. 30, p. 32 b. By the way, it may be remarked that in A.D. 906 the name of the city Ts'e-chou was changed in writing into 慈州, while in 916 the old character 磁 was restored (*T'ai p'ing huan yü ki*, Ch. 56, p. 10 b).

the Han dynasties proclaimed by Julien, it was relegated to the beginning of the Sung dynasty (A.D. 960) by E. GRANDIDIER;[1] and all this glory ended in its final degradation into as late a period as that of the Ming. Mr. E. A. BARBER, Director of the Pennsylvania Museum in Philadelphia, one of the most serious students of pottery in this country, gives vent to this growing pessimism in the following observation: "The consensus of opinion among conservative students at the present day, after divesting the subject of all sentimental considerations, is that true porcelain first appeared during the Ming dynasty, which would not carry it back of the fourteenth century. No examples of actual porcelain, that can with certainty be referred to an earlier date, are known to collectors; and it is reasonable to suppose that had such ware been produced before that period, some few pieces at least would have survived. Indeed, it is extremely doubtful whether any actual examples antedating the fifteenth century can be found."[2] Mr. Barber, however, frankly admits that the Chinese themselves have classed all wares which possess great hardness and resonancy (which latter is an indication of vitrification) with porcelain, and that it is true that a porcelanous glaze was used to some extent before the general introduction of semi-transparent bodies. This concession points out that the subject may be viewed from different angles. There is, indeed, a twofold point of view possible and permissible, a European-American and a Chinese one. HOBSON,[3] who possesses a large share of critical ability combined with true common sense and sane judgment, has clearly noticed this diversity. "The quality of translucency which in Europe is regarded as distinctive of porcelain is never emphasized in Chinese descriptions," he observes, and goes on to determine the difference between the Chinese and European definitions of the substance. Now, if this be true, every student capable of objective thinking must admit that it is a logically perverse procedure to read "our" definitions of porcelain into what is called by the Chinese *ts'e*, but that for the correct appreciation of this term the Chinese viewpoint exclusively must be made the basis of our investigation. In other words, the point simply is, that we must endeavor to understand what notion in the minds or in the fancy of the Chinese is conveyed by their term *ts'e*. If a bit of pottery is styled by the Chinese *ts'e*, yet is not true porcelain in our conception of the matter, we are obliged to give the Chinese credit for their appellation, and to get at their mode of reasoning. By rejecting

---

[1] La céramique chinoise, p. 16 (Paris, 1894).

[2] Hard Paste Porcelain, Part first (Oriental), p. 7 (Philadelphia, 1910).

[3] Chinese Pottery and Porcelain, Vol. I, p. 148.

this procedure we deprive ourselves of the opportunity of studying and grasping the development of this peculiar ware. By arguing that in the beginning the term *ts'e* connoted nothing but ordinary pottery, we close our eyes to the real issue, and act like the ostrich; in this manner we utterly fail to comprehend the process of evolution of porcelain. The early *ts'e* has now arisen, and is that ware which is the object of this article. I further make bold to say that in any ancient text down to the T'ang period, where the term *ts'e* may be encountered, it will invariably refer to a porcelain-like pottery which has some relationship to genuine porcelain, and that we shall not err in translating it by "porcelanous ware," or a similar expression.

HISTORICAL NOTES ON KAOLIN

A disquisition on the beginnings of porcelain should take regard also of the question as to when and how those elementary materials that compose porcelain made their first appearance. Porcelain is a variety of pottery the body of which consists essentially of two ingredients of earthen origin, that are fired together. These two substances widely occur in nature, and are designated by us with their Chinese names, "kaolin" and "petuntse." The former is a white clay, infusible, lending plasticity to the paste, and forming the body of the vessel. Geologically it originated through a gradual process of decomposition of granite and analogous crystalline rocks.[1] The latter is a hard feldspathic stone, fusible at a high temperature, constituting the glaze and responsible for its transparency.

The fact that kaolin is used in the composition of Chinese porcelain has been unduly emphasized, or even exaggerated, by European historians of porcelain. Kaolin was heralded as a sort of important discovery, that led to the revolutionizing of the potter's art; and an inquiry into the time when Chinese authors begin to speak of the substance was even taken as a test for the beginnings of porcelain itself. This is not a correct conception of the matter. Kaolin is nothing but a natural clay, not of very unusual occurrence, and, in fact, has been utilized by potters outside of China without resulting in any porcelain-like product.[2] Kaolin itself cannot make porcelain, and the presence of kaolin in the composition of a certain vessel does not constitute proof of its being porcelain. Kaolin should not be confused with the kaolinite of which it is composed. The mineral

---

[1] See PRESTWICH, Geology, Chemical, Physical, and Stratigraphical, Vol. I, p. 48.

[2] Thus in India a white earthenware is made from a decaying white granite, which is carefully washed, and kneaded into a clay that produces a porous white ware. . . . This clay is in composition the same as the kaolin of China, and is very abundant in India (H. H. COLE, Indian Art in the South Kensington Museum, p. 201). The Singalese potter (in the same manner as his Chinese colleague during the T'ang period) uses kaolin as a white paint for decorating pottery (A. K. COOMARASWAMY, Mediæval Sinhalese Art, p. 225; see also WATT, Dictionary of the Economic Products of India, Vol. II, p. 364). It is well known that kaolinic deposits are found in England, France, Germany, and North America, and are well known from many other parts of the world. As to America, compare, for instance, the interesting study of A. S. WATTS, Mining and Treatment of Feldspar and Kaolin in the Southern Appalachian Regions (Bulletin No. 53 of the Department of the Interior, Bureau of Mines, Washington, 1913).

kaolinite is the basis of kaolin, and theoretically pure kaolin would contain nothing but kaolinite; but kaolinite is also the basis of nearly all common clays. In these it is mingled with larger or smaller quantities of various minerals by which its properties are more or less obscured. Hence the chemical examination of almost any burned pottery, even of common bricks and the crudest and cheapest of earthenware, will disclose the presence of derivatives of kaolinite which might be, and as a matter of convenience frequently is, interpreted as due to the presence of small quantities of kaolin, instead of larger quantities of ordinary clay containing kaolinite. It is quite certain that the bodies of many early Han pottery bits contain more or less kaolin or kaolinite, yet they are not porcelains. The utilization of kaolin for potter's work on a large scale is not a "discovery," but rests on experience. It was incidentally found, and its employment was gradually extended through a selective progress in the enrolment of materials.

The distinctive structural character of porcelain is based on the combination of three elements,— a porous, opaque skeleton; a transparent, dense bond permeating the skeleton; and a thin, glassy glaze on the outside, which merges imperceptibly with the body. In typical porcelains the opaque, porous body is kaolin or aluminous derivatives therefrom, which, through their resistance to the effects of heat, supply a rigidity that prevents the ware from deforming in the kiln. Also its opacity clouds the transparency of the other elements to translucency. The kaolin skeleton is permeated and bound together by a more fusible glass or enamel-like substance (petuntse), which makes the ware strong, impervious, and translucent. The glaze serves for the perfection and increased lustre of the surface. Kaolin alone makes a ware which is porous, fragile, and opaque. Petuntse alone softens in the kiln, and runs together into a lump.

For the lover of art the salient and distinctive points in porcelain are the glaze and its organic combination with the body. The body, as a rule, is invisible: it is the glaze that is intended to appeal to the spectator and to convey an esthetic impression.

F. HIRTH[1] was the first to call attention to a statement of the Taoist adept T'ao Hung-king (452–536), to the effect that in his time "white clay" (*pai ngo* 白堊), or kaolin, was much utilized in painting,[2]

---

[1] Ancient Chinese Porcelain, p. 131.

[2] What this means has not been explained by Hirth, who translated, "much used for painting pictures." It cannot be understood, of course, that kaolin was a pigment applied in pictorial art to paper or silk. Technically there are but two possibilities: kaolin may have been utilized in architectural painting for the decoration of walls, being applied to a colored background, or it may have been employed

and was low in price. This passage is found in the *Cheng lei pen ts'ao*, a learned pharmacopœia written by the physician T'ang Shen-wei, and first published in 1108. This text allows of the inference that porcelain clay was known in the latter part of the fifth or beginning of the sixth century; but I should not go so far as to conclude with Hirth that T'ao Hung-king "would have surely mentioned the use of porcelain earth in the manufacture of chinaware if in his time it had been so used on an extensive scale," and that "in the sixth century, when he wrote, the use of porcelain earth for pottery purposes was unknown." This argument, drawn from the mere silence of a writer, is not conclusive: it seems preferable to think, that, judging from the trend of his mind and the direction of his studies, the author was not at all interested in the subject of pottery. What attracted him were not the artifacts of men, but the substances and wonders of nature, that might reveal healing-properties for the benefit of his suffering fellow-men. Even in speaking of the application of kaolin to pictorial subjects or decorative designs, he does not mean to offer a contribution to technology, but he incidentally drops this remark by way of definition, in order to render himself intelligible to his contemporaries as to the matter under discussion; for he says literally, "This [that is, the white clay here in question] is identical with that now largely utilized in painting, and low in price. Customarily it is but seldom administered in prescriptions."[1] The subsequent works dealing with pharmacology, while they give some notice to porcelain clay on account of its

---

for the ornamentation of a surface in pottery vessels. The latter process is now well known to us through numerous specimens of the T'ang period. The *Pen ts'ao kang mu* of Li Shi-chen (section on clays, Ch. 7, p. 1) has the reading *hua kia yung* 畫家用 (instead of *hua yung* of the *Cheng lei pen ts'ao*), which means "used by painters."

[1] Hirth pointed out another text in the *Cheng lei pen ts'ao*, which, he stated, is quoted from the *T'ang pen ts'ao*, the pharmacopœia of the T'ang period, compiled about the year 650. In the edition of the *Cheng lei pen ts'ao* before me, issued in 1523 (Ch. 5, fol. 25), the passage in question, however, is cited from a work styled *T'ang pen yü* (that is, "Remains of the T'ang Herbal"), and introduced by the words, "The commentary says." I venture to doubt that this work *T'ang pen yü* is strictly identical with the *T'ang pen ts'ao* described by BRETSCHNEIDER (Bot. Sin., pt. 1, p. 44), especially for the reason that a quite different extract from the *T'ang pen* is quoted in the *Cheng lei pen ts'ao* shortly before this passage, and that in this work quotations from the former are constantly referred to the *T'ang pen* or *T'ang pen chu* (apparently the annotations of the drawings mentioned by Bretschneider). Be this as it may, there is no doubt that the text brought to light by Hirth comes down from the T'ang period. This is also the opinion of Li Shi-chen, who, in his *Pen ts'ao kang mu* (Ch. 7, p. 6 b), attributes the term "white porcelain vessels" (*pai ts'e k'i*) to the *Pen ts'ao* of the T'ang. In the text translated by Hirth occurs a clause which he rendered, "During recent generations it has been used to make white porcelain." HOBSON (Chinese Pottery and Porcelain, Vol. I, p. 146) has proposed a new translation of this passage, which reads, "During recent generations it has been prepared

alleged medicinal properties, yet maintain strict reticence in regard to porcelain vessels, though these were positively known at the time of their publication, for the simple reason that this topic was beyond their scope. Neither the *Cheng lei pen ts'ao* nor the *Pen ts'ao kang mu* discusses porcelain, but both books are content to recommend prescriptions of kaolin for certain complaints. While Su Kung upholds that of Ting-chou, and Li Shi-chen that of Jao-chou (in Kiang-si), as particularly efficient, this is merely the outcome of a more specialized medical subtlety.

It would likewise be preposterous to assume that T'ao Hung-king is the first author to mention kaolin. On the contrary, he is forestalled by at least one predecessor. The work *Pie lu*,[1] which existed prior to his time, as quoted in the *Pen ts'ao* (*l. c.*), states that "white clay (*pai ngo*) originates in the mountains and valleys of the district of Han-tan 邯鄲,[2] and that it may be gathered at any season." This restriction to a single locality certainly does not betoken the scarcity of the material, which is indeed common in many localities: it reflects solely the limitations of local experience. Under the Sung we hear from the lips of Su Sung that this variety of clay was then ubiquitous, and was throughout used by the people for the washing of their clothes.[3] This view is confirmed by Li Shi-chen, who observes that white clay occurs everywhere, and is employed for the baking of white pottery vessels. However common the occurrence of kaolin in China may be, the fact

---

from white ware." From a grammatical point of view this translation is perfectly correct. It is, however, somewhat difficult to understand why the pharmacists of the T'ang period should have extracted kaolin from finished ceramic products, even though it was only from fragments of such, if kaolin could so easily be obtained in nature; or it is conceivable also that kaolin inherent in pottery was vested with more efficient magical and increased healing-power, as it had undergone a transmutation in the furnace. We have to know more about the development of alchemy in China before we may hope to settle many interesting questions and beliefs connected with pottery.

[1] See Chinese Clay Figures, p. 135, note 4.

[2] It comprised what now forms the two prefectures of Kuang-p'ing and Cheng-te, in the southern part of Chi-li Province, and in particular referred to Ts'e-chou. In ancient times it was the capital of the state of Chao (CHAVANNES, Mémoires historiques de Se-ma Ts'ien, Vol. II, p. 92). It is an attractive suggestion of HOBSON (*l. c.*, p. 147), that the kaolinic deposits of Han-tan should have supplied material for the Ting-chou potters.

[3] K'ou Tsung-shi, in his *Pen ts'ao yen i* of 1116 (Ch. 6, p. 1b; ed. of Lu Sin-yüan), makes the same observation, adding that the substance was made into square blocks sold in the capital under the name "white earth powder" (*pai t'u fen* 白土粉). According to the *Ling piao lu i* (Ch. A, p. 4; ed. of *Wu ying tien*) by Liu Sün of the T'ang period, a white and greasy earth was gathered north of the city of Fu chou 富州 (in the prefecture of Wu-ch'ang, Hu-pei) and traded over southern China, where the women used it as a face-powder. This probably was a kind of pipe-clay.

remains that this observation is only the result of later periods, and that in times of antiquity the knowledge of it was much restricted, and attached to but few places. The wondrous book of geographical fables, the *Shan hai king*, mentions it in two passages. One is embodied in the chapter on the "Mountains of the West" (*Si shan king* 西山經), saying that on the south side of the mountains of Ta-ts'e there is plenty of clay.[1] The other contains the notice, in the chapter on the "Mountains of the Centre" (*Chung shan king* 中山經), that "in the midst of the mountains of Ts'ung-lung there are many great valleys in which there is plenty of white clay; apart from the latter, there are also black, dark blue, and yellow clays."[2] Kuo P'o adds that also variegated clay is said to occur. Whether the two texts are of ancient date, I do not venture to decide: they are quoted as early as the Sung period by Su Sung (a distinguished scholar, and editor of the materia medica *T'u king pen ts'ao*), in his discussion of kaolin, which he winds up by remarking that solely the white clay is medicinally employed. Personally I am under the impression that the *Shan hai king*, in the version which is now before us, is not older than the Han period, and doubtless contains also many post-Han interpolations. I would certainly not base on this work any chronological conclusions as to the term *pai ngo*.

The Chinese explanation of the term *ngo* is interesting, because it has led to the formation of a new word. The character 堊 is composed of the classifier 土 ('earth') and the phonetic element 亞. The latter enters also into the formation of the character 惡, which likewise has the sound *ngo* or *ngu* ('evil'). Li Shi-chen[3] is therefore led to the following speculation: "Since the normal color of earth is yellow, white must be considered as an evil color in earth; hence it was called *ngo* [that is, 'evil earth']. Subsequent generations tabooed this word, and changed it into *pai shan* 白善 [that is, 'the white good one']." The notion of "wicked earth" is elicited by punning, the two words 堊 and 惡 being homophonous. This jocular interpretation must have existed as a popular tradition since ancient times, since the result of it, the opposite term *pai shan*, is said to have occurred in the *Pie lu*. K'ou Tsung-shi, whose *Pen ts'ao yen i* was published in 1116, styles kaolin "white good earth." This was under the Sung, when the porcelain industry received a powerful stimulus. The term *pai shan*

---

[1] 大次之山其陽多堊 (Ch. 1, p. 27 b; of the edition printed in 1855 at Shun-k'ing, Sze-ch'uan). The character 堊, according to the commentary of Kuo P'o (276–324), is to be read *ngu* (or *ngo*), explained as "earth of very white color."

[2] 葱聾之山其中多大谷是多白堊黑青黃堊 (Ch. 2, p. 15 b).

[3] *Pen ts'ao kang mu*, Ch. 7, p. 1.

白 善 is met with as early as the T'ang period (618–906), in the mineralogical glossary *Shi yao erh ya* 石藥爾雅, compiled by Mei Piao 梅彪 in the period Yüan-ho (807–821).[1] Here it is given as a synonyme of *kan t'u* 甘土 ("sweet earth"), on a par with other synonymes for this term, which are *pai tan* 白 單, *tan tao* 丹道, and *t'u tsing* 土 精 ("essence of earth"). At an earlier date we find the term *shan* in the Buddhist dictionary *Yi ts'ie king yin i* 一切經音義,[2] compiled by the monk Yüan Ying 元應 about A.D. 649, who explains it as *shan t'u* 善土 ("good earth"), and identifies it with "white clay" (*pai t'u* 白土) and *ngo*. The most interesting point is, that this author cites the *Wu p'u pen ts'ao* 吳普本草 to the effect that the term *pai ngo* has a synonyme in the form *pai shan* 白 墠. According to BRETSCHNEIDER,[3] the *Wu p'u pen ts'ao* was written by Wu P'u under the Wei dynasty in the first half of the third century A.D. If the definition, as handed down by Yüan Ying, was really contained in this work, we should have a formal testimony for the knowledge of kaolin in the third century. The case was presumably such, that in the T'ang era, when the excellent qualities of kaolin were first recognized, the transformation of the word took effect, and ultimately resulted in a new character formed with the word *shan* 善 as phonetic element, and the classifiers 'earth' 土 or 'stone' 石. The taboo announced by Li Shi-chen cannot have taken serious dimensions, for the ceramic authors of the Manchu dynasty perpetuated the word *ngo*, and abstained from the word *shan*.

In a poem of Se-ma Siang-ju, entitled *Tse sü fu* 子 虛 賦,[4] ochre and white clay (*che ngo* 赭 堊) are spoken of as natural products of Sze-ch'uan.[5] The attribute "white" is not in the text, which merely offers the word *ngo;* but Chang Yi 張揖, the author of the dictionary *Kuang ya* 廣雅, who lived in the first part of the third century A.D.,

---

[1] Reprinted in the collection *Pie hia chai* (Ch. A, p. 4).

[2] Ch. 17, p. 2 (edition of Nanking). Regarding this work see JULIEN, Histoire de la vie de Hiouen Tsang, p. xxiii; WYLIE, Notes on Chinese Literature, p. 211; WATTERS, Essays on the Chinese Language, p. 52; BUNYIU NANJIO, Catalogue of the Tripitaka, No. 1605.

[3] Bot. Sin., pt. 1, p. 40.

[4] *Shi ki*, Ch. 117, p. 2 b. The poet died in 117 B.C.

[5] They are likewise mentioned as products of that region (Shu) in the *Hua yang kuo chi* (Ch. 3, p. 1 b, ed. of *Han Wei ts'ung shu*). Under the year 991 there is mentioned in the Sung Annals the pictorial decoration of a palace by means of the same two substances. The same term appears in Lie-tse (WIEGER, Les pères du système taoiste, p. 104), when King Mu built a palace for a juggler, who had come from the farthest west. This chapter of Lie-tse (and probably many others), in my opinion, comes down from the Han period; and this conclusion is confirmed by the term *che ngo* which does not occur earlier than that time. The work of Lie-tse is first mentioned in the *Ts'ien Han shu* (Ch. 30, p. 12 b).

comments on this passage, that *ngo* has there the meaning of "white clay" (*pai ngo*), which, he adds, is identical with the term *pai shan* used in the Herbals (*pen ts'ao*), so that what he means is doubtless kaolin. Also Yen Shi-ku (579–645), annotating the same word in the Han Annals, states that "it is identical with what is now called 'white earth' (*pai t'u*)." It is interesting that these Confucian scholars of the third and sixth centuries respectively were acquainted with kaolin, thus following suit with their Taoist colleagues; but it appears rather doubtful whether the term, as used in the Annals of Se-ma Ts'ien, can really be credited with the significance "kaolin." There is no other testimony to this effect (leaving aside the dubious *Shan hai king*) in the Han period; and, be this as it may, the passage in question is not conclusive, the substance *ngo* being mentioned solely as a product of nature, without any allusion to human exploitation. In the Glossary of the T'ang Annals the term *ngo* is interpreted as "white earth" (*pai t'u* 白土).[1]

In the T'ang period, kaolin formed also a desirable article for tribute or taxes to the Court, which certainly means that it was employed in the manufacture of pottery. The *Wu ti ki* 吳地記 ("Records of the Land of Wu"), by Lu Kuang-wei 陸廣微, written at the end of the ninth century, mentions the mountains of Hang 杭山 as hoarding white earth that resembles jade and is very resplendent, and that the people of Wu, who gathered it, sent as tribute under the name *pai ngo*.[2]

Passing beyond the Han period, we find the word *ngo* employed in times of antiquity, but in a peculiar sense, quite distinct from the later significance "potter's clay." In the early period it was strictly an architectural term, and implied a function falling within the province of a mason. This ancient significance is acknowledged by the dictionary *Erh ya*, which, in its section concerned with the nomenclature of buildings, states that *ngo* is the designation for a whitewashed wall; and the dictionary *Shi ming* 釋名, by Liu Hi 劉熙 of the Posterior Han, is still more explicit on this point, as evidenced by the annotation that the wall is first raised from mud, and then invested with a coating of lime.[3] The *Shuo wen* explains the term as "white plaster" (*pai t'u* 白涂). The principal office of the word was that of a verb, with the

---

[1] *T'ang shu shi yin*, Ch. 5, p. 20.

[2] According to the Gazetteer of the Prefecture of Su-chou (*Su chou fu chi*, Ch. 20, p. 15 b), kaolin is still dug on the Yang-shan near Su-chou to a depth of a hundred feet.

[3] 先泥之次以白灰飾之也 (*Shi ming*, section 5, p. 8; ed. of *King sün t'ang ts'ung shu* or *Han wei ts'ung shu*).

meaning "to plaster or whitewash the floor or the walls of a house." This is particularly evidenced by the verb *yu* 黝 ("to blacken"), its opposite, to which it is closely linked in order to express the performance of a religious ceremony during the period of mourning. The mourner was obliged to dwell in an unplastered earth hut for two years. After the sacrifice in the commencement of the third year, the ground of his cot was blackened, and the walls were whitened,— a rite simply expressed by the compound *yu ngo* 黝堊.[1] In the same chapter of the "Book of Rites" in which this practice is mentioned, the same word *ngo* occurs in a somewhat different usage. The dwelling specially erected for the mourner is styled *ngo shi* 堊室, a term explained as "a hut made of unburnt bricks or earth pisé and not plastered," and used in the *Li ki* four times. The mourner was compelled to divest himself of all comfort, and to relapse into the most primitive habitation of early times. The term *ngo shi*, accordingly, means literally "earth house;" and during the archaic period, *ngo* designated "loam, mud, or clay fit for building-purposes." Simultaneously, however, it was applied also to chalk or limestone, denoting the process of coating a coarse wall with a layer of white. In this sense it is utilized also by Chuang-tse in regard to the whitening of one's nose.[2] Since the word *ngo*, which is still defined by the *Shuo wen* as "white plaster," originally referred to clay and chalk at the same time, the early Chinese do not seem to have clearly discriminated between the two substances. The term *pai ngo*, which adopted the meaning "kaolin" in the post-Christian era, is still used to convey the notion of "chalk," while a stricter terminology formulates for the latter such compounds as *shi ngo* 石堊 ("stone clay"), *ngo hui* 堊灰 ("clay lime"), or *pai t'u fen* 白土粉 ("white earth powder").[3]

One point stands out clearly,— that in the archaic period the word *ngo* signified "loam and chalk used in building," and was appropriate to the activity of the mason, but that it neither denoted potter's clay nor had any relation whatever to the work of the potter. The main point to be borne in mind is, that there is no reference to "white clay" (*pai ngo*) in any authentic document of the Han period,— a fact thoroughly corroborated by archæological evidence. The "white clay,"

---

[1] *Li ki*, ed. COUVREUR, Vol. II, p. 240; translation of LEGGE, Vol. II, p. 192.

[2] Ch. 24, § 5; see the edition of L. WIEGER, Taoisme, Vol. II, p. 420. It is notable that the stage-fool still appears in China with his nose whitened; and the figure of an actor represented by a T'ang clay statuette in the Museum collection is thus characterized.

[3] See F. DE MÉLY, Lapidaires chinois, p. 99; F. PORTER SMITH, Contributions towards the Materia Medica of China, p. 58.

or "kaolin," makes its first appearance in the *Pie lu*, an early Taoist work of uncertain date, and preserved only by way of quotations in subsequent pharmaceutical literature. This lacune in our knowledge, however, is no matter of great concern for the history of porcelain, for that work contains no allusion to pottery. Chang Yi and Kuo P'o of the third century appear to have been familiar with kaolin; likewise Wu P'u, the author of a materia medica under the Wei (p. 115). The medical literature of the T'ang period is, and thus far remains, the earliest source to convey an allusion to white porcelain produced from kaolin. Prior to that time, this substance seems to have found application chiefly in medicine, and as engobe on pottery. It probably played a rôle also in alchemical experiments. There is every reason to believe that it was the nature-loving and drug-hunting professors of Taoism who first experimented with this clay, and this accounts for the fact that the subject has found its way into the pages of the *Shan hai king*. What the share of the Taoists was in the initial stages of porcelanous ware, or whether a share in it is due to them at all, we have as yet no means of ascertaining. That they had a share in it, however, is more than probable, since the preparation of clays and glazes is a matter of chemistry; that is, in ancient times, of alchemy (see also p. 142).

It is obvious that no forcible conclusion as to the date of porcelain can be deduced from a consideration of the history of kaolin. It is notable, however, that it was known at least in the third century A.D.; and this chimes in with my dating of the early kaolinic ware in the same period. Once more we see that for the history of porcelain we have to depend on archæological evidence.

It is unfortunately impossible to outline a similar sketch of the history of petuntse, or porcelain stone; but it is not surprising that the Chinese have preserved no historical notes regarding this substance. It is simply a feldspathic rock, for which no other than the general designation "stone" (*shi* 石) exists. It is a general error to believe that the mass itself is styled by the Chinese "petuntse" (properly, *pai tun-tse* 白不子), an error chiefly propounded by A. J. C. GEERTS.[1] JULIEN[2] was somewhat astonished at the expression, saying that the Chinese authors who wrote on porcelain fail to explain the sense of the word *tun* 不. K'ang-hi's Dictionary does not ascribe to the latter any mineralogical significance; in fact, it has none whatever, and is never used by Chinese writers on mineralogy. The character in question is

[1] Les produits de la nature japonaise et chinoise, Vol. II, p. 376 (Yokohama, 1883).

[2] Histoire et fabrication de la porcelaine chinoise, p. 122.

merely substituted as an easy and convenient abbreviation for *tun* 墩, which means, as Giles rightly says, "a square block of stone."[1] The term *pai tun-tse*, therefore, simply signifies "white briquette," and certainly is one of a purely commercial, not mineralogical character: it relates to the color and shape of these blocks, as they are traded from the places of production to the centres of porcelain manufacture. Our mode of applying the term "petuntse" to the material, therefore, is wrong. The fact that this rock, which enters into the manufacture of porcelain, was roughly known to the Chinese long before the time of this specific employment, cannot reasonably be doubted.

---

[1] In the second edition of his Dictionary, GILES has justly placed the term "petuntse" under this character (No. 12205).

THE INTRODUCTION OF CERAMIC GLAZES INTO CHINA, WITH SPECIAL REFERENCE TO THE MURRINE VASES

We know at present as a fact that glazed pottery first appeared in China during the Han period, and that the process of glazing earthenware was unknown in pre-Han times. The Han potter's art was revolutionized, as we have seen, by the adoption of this new technique, which finally resulted, toward the middle or the close of the third century, in the production of a peculiar porcelanous glaze, the forerunner of true porcelain. Porcelain being universally considered as a truly Chinese invention, the broader question may now be raised, Is the invention of glazing, the technical foundation of porcelain, wholly due to the genius of the Chinese, or was the impetus received from an outside quarter? R. L. Hobson[1] has made the following general reply to this query: "Though supported by negative evidence only, the theory that the Chinese first made use of glaze in the Han period is exceedingly plausible. In the scanty references to earlier wares in ancient texts no mention of glaze appears, and, indeed, the severe simplicity of the older pottery is so emphatically urged that such an embellishment as glaze would seem to have been almost undesirable. The idea of glazing earthenware, if not evolved before, would now be naturally suggested to the Chinese by the pottery of the Western peoples with whom they first made contact about the beginning of the Han dynasty. Glazes had been used from high antiquity in Egypt; they are found in the Persian bricks at Susa and on the Parthian coffins, and they must have been commonplace on the pottery of western Asia two hundred years before our era." I am of the same opinion, that Chinese knowledge of glazing is derived from the West, and propose to discuss this problem on the following pages. I hope to enlist all the available facts in the case, so as to place our theory on a solid historical foundation.

The course of my investigation is as follows. The home of glass, glazed pottery, and faience, was Egypt and the anterior Orient; and the reputation of this ware spread to Rome under the name "murrine vessels." The latter subject, being still of a controversial nature, is of especial importance in this connection, as it shows us the high appreciation and expansion of glazed ware over the Mediterranean area at

[1] Chinese Pottery and Porcelain, Vol. I, p. 8.

a period synchronous with the coming into existence of this pottery in China. This synchronism is not accidental, but is due to the wide fame and diffusion of this novel process in the Far East. It will then be set forth from Chinese records how the Chinese became acquainted with it in consequence of their contact with the Roman-Hellenistic Orient; how the materials required for the technique were propagated to India, Cambodja, and China, and in what manner they were turned to practical use by the ancient Chinese.

If I venture to dwell here at some length on the much-disputed murrine vases of the ancients, the main reason for this invasion of foreign territory is that this subject seems to me to embody an essential chapter in the history of the art of glazing, which allows us to grasp clearly the significance of its eastward migration. My further line of defence rests on various attempts made by older and more recent authors to interpret the murrine vases as having been Chinese porcelain; and in further vindication I may point to two sinologues who in the first part of the nineteenth century participated in the discussion of this problem,— JOSEPH HAGER and ABEL-RÉMUSAT. The former[1] endeavored to prove in a hardly convincing manner that the substance of which the murrines were made was identical with the jade of the Chinese; while the latter[2] combated this opinion, and conclusively demonstrated that Chinese nephrite does not at all correspond to the description given by Pliny of the murrine vases. The chief argument which runs counter to this theory, and which has not been stated by Abel-Rémusat, is that ancient Chinese jade objects have as yet not been traced in any country of classical civilization, and that nothing is on record in regard to such a trade, either in Chinese or classical documents. Moreover, the provenience of the murrines, as indicated by Pliny and the Periplus Maris Erythraei, must not be disregarded: they came from Egypt, Persia, and India, and were chiefly productions of Persia. In none of these countries have we any evidence as to the occurrence of Chinese jade pieces in ancient times.[3]

In a study devoted to the beginnings of porcelain in China, in which an attempt has been made to determine more exactly the first appearance of porcelanous ware on Chinese soil, a word may be permitted

[1] Description des médailles chinoises du Cabinet Impérial de France, pp. 150–168 (Paris, 1805).

[2] Histoire de la ville de Khotan, tirée des annales de la Chine et traduite du chinois; suivie de recherches sur la substance minérale appelée par les Chinois pierre de Iu, et sur le jaspe des anciens, pp. 195–208 (Paris, 1820).

[3] More recently the nephrite hypothesis with reference to the murrines has been reiterated by A. VON NORDENSKIÖLD (Umsegelung Asiens und Europas, Vol. II, p. 230).

with reference to the theory that the murrines might have been porcelain of Chinese origin. This view predominated in Europe for three centuries, till it yielded to still more fantastic ideas in modern times. Jerome Cardan (Hieronymus Cardanus), the Italian mathematician (1501–76), is to be regarded as the father of the porcelain theory. In his work "De subtilitate rerum" (Nürnberg, 1550, p. 119), he made the assertion, "Sunt autem myrrhina ea, quae hodie vocantur Porcellanea," and supported it by the explanation that they had come to western Asia from China, the country of the Seres, and that whatever does not fit in with them in the description of Pliny became subsequently altered in the manufacture of these vessels. Julius Caesar Scaliger (1484–1558) concurred with him in this opinion, and only reproached his predecessor for having advanced his statement in too timid a fashion. His son, the great scholar Joseph Justus Scaliger (1540–1609), inherited and accepted his father's verdict. Whatever we may think of the view of the two Scaliger, it remains interesting, as it was at their time that porcelain gradually became known in Europe; and this fact may certainly have reacted on the shaping of their opinion.

In the eighteenth and the beginning of the nineteenth century, the old opinion that by "the murrines" should be understood porcelain, was revived by P. J. MARIETTE[1] and by E. H. ROLOFF,[2] the latter a physician, whose work is accompanied by notes and additions at the hands of Ph. Buttmann. The theory of Cardanus and Scaliger was here defended afresh and with circumstantial detail, and seemingly with such success that it maintained its place for some twenty-five years, until F. THIERSCH[3] brought about the victory of the mineralogical theory, and replaced the murrines of porcelain by murrines of fluor-spar. Roloff and Buttmann based their argumentation pre-eminently on the famous passage of Propertius in which are mentioned "murrine cups baked in the kilns of the Parthians" (murreaque in Parthis pocula cocta focis), that without any doubt refer to ceramic productions. They utterly failed, however, to furnish any exact and logical evidence for their proposed identification of murrines with porcelain, which was merely a preconceived idea, or nothing more than their personal impression in the matter. They argued that this porcelain must have come from the land of the Seres, China, where it is exceedingly

---

[1] Traité des pierres gravées, Vol. I, p. 219 (Paris, 1750).

[2] Wolf's and Buttmann's *Museum der Alterthumswissenschaft*, Vol. II, pp. 519–572, 1810.

[3] Über die Vasa murrina der Alten (*Abhandlungen der bayerischen Akademie*, 1835, pp. 443–509).

ancient, and must accordingly have been exported as early as in times of antiquity, and certainly to Persia, whence the murrines were imported to Rome. For a brief period it would have seemed as though the alleged discovery of Chinese porcelain bottles in Egyptian tombs might lend support to such an opinion; but for a long time we have known that the whole story amounts to a not very clever fraud.[1]

When the murrine vases were identified with porcelain, European knowledge of the history of porcelain in China was still in its infancy and of the vaguest character; and if a subject is obscure or little known, speculation is usually rife, and the almost incredible is readily accepted. In 1857 BOSTOCK and RILEY[2] still commented on the murrines, that modern writers differ as to the material of which these vessels were composed; that some think that they were of variegated glass, and others of onyx, but that the more general opinion is that they were Chinese porcelain. The last view has never entirely lost its ground, and still counts adherents in this country. In the "New Standard Dictionary," published by Funk and Wagnalls of New York in 1913, we read, under the article "murrine vases," "porcelain vases brought from the East to Rome."

The present investigation allows us to settle this problem definitely. It is out of the question that the murrine vessels were Chinese porcelain, since at the time when the former were traded from the Orient to Rome nothing like porcelain existed on this globe. We have seen that ceramic products with porcelanous glaze do not come up in China earlier than the latter part of the third century A.D., and that anything of the character of true porcelain cannot be pointed out before the sixth century. The *vasa murrhina*, however, are mentioned considerably earlier than these two dates. They were first brought to Rome in 61 B.C. by Pompey, who, after his triumph, dedicated cups of this description to Jupiter Capitolinus. Pompey himself had obtained them from Mithridates. Augustus appropriated a single murrine vessel from the treasure of Queen Cleopatra, which is cited as an instance of his moderation.[3] In the time posterior to Pompey, the murrines became more frequent in Rome, and aroused a passion for them among the upper four hundred. Classical Roman literature does not make

---

[1] Compare S. JULIEN, Histoire et fabrication de la porcelaine chinoise, pp. XI–XXII; F. HIRTH, Chinesische Studien, pp. 45–48; N. RONDOT, On the Chinese Coins and Small Porcelain Bottles found in Egypt (*Journal China Branch R. As. Soc.*, Vol. XXXII, 1897–98, pp. 66–78).

[2] The Natural History of Pliny, Vol. VI, p. 392.

[3] SUETONIUS, Augustus, 71.

any mention of them; they are foreign to the works of Cicero and Varro, as well as to the poems of Horace, Ovid, and Vergil. Propertius (born about 49 B.C.) is the first to make a distinct allusion to them. They are further mentioned by other poets, like Statius, Juvenalis, and Martialis. Pliny is the only one to give a somewhat more detailed, though insufficient, description. The first centuries preceding and following our era, accordingly, were the period when the murrines formed the fashion of the day in Rome; and porcelain was not then made in China. The Chinese records relative to the Roman Orient and Persia are reticent as to trade in pottery; and the fact remains that in Persia, India, Egypt, Greece, or Rome, has never been discovered a specimen of Chinese porcelain of such age that could lay claim to being regarded as murrine.[1]

In the light of our present knowledge, the porcelain hypothesis must be characterized as a failure, and as being doomed to oblivion. The efforts of the men, however, who formulated their thoughts along this line, have not been entirely futile; for, as it so frequently happens, error will ultimately lead us to the knowledge of truth. The champions of porcelain murrines were quite correct in the pursuit of one point of view,— that the murrines were of pottery, not, as has been asserted, of a mineral substance. Their fundamental error lay mainly in the rash manner in which they jumped at the conclusion that Chinese pottery was involved; while we plainly have to adhere to the fact, transmitted to us by the ancients, that the murrine vessels were wrought in the Empire of the Parthians, and that, as stated by Propertius, they were baked or fired in Parthian furnaces. They were consequently products of Iranian pottery; and the peculiar coloration described by Pliny obviously hints at a beautiful and elaborate glazing which was brought out on those vessels. My thesis, accordingly, is that the famed murrines of the ancients were highly-glazed pieces of Oriental,

[1] Even under the Han, the potter's craft, which in that period had without any doubt developed into an art, possessed no more than purely local significance, and merely catered to the home consumption of the small community for whose benefit the produce was turned out. It seems certain that no inland trade in pottery was then developed, still less was there an exportation of the article. It is notable that Se-ma Ts'ien, in his famous dissertation on the "Balance of Trade" (*Shi ki*, Ch. 30, translated by CHAVANNES, Mémoires historiques de Se-ma Ts'ien, Vol. III, pp. 538–604), describing the remarkable efforts of the Han in the second century B.C. toward a regulation of the factors of wealth and commerce, does not make any allusion to potters or pottery as an article of trade. Neither do we meet, in the historical documents of the Han bearing on foreign relations, any mention of such export-ware. The incidental mention by Se-ma Ts'ien of "a thousand jars (*kang*) filled with pickles and sauces," adverted to also in the *T'ao shuo* (BUSHELL, Description of Chinese Pottery, p. 93), is without significance.

that is, Iranian or Persian and Egyptian, pottery. This conclusion directly results from the documentary evidence which the ancient authors have left us. It will be demonstrated at the same time that the substance *murra*, of which the murrine vases were made, cannot have been a mineral of any sort.

The Latin word *murra* (less correctly *murrha, myrrha*), from which the adjectives *murreus* (*murrheus, myrrheus*) and *murrinus* are derived, was adopted from the Greek *morrion* (in Pausanias) and the adjectival form *murrinos*, used in the Periplus.[1] The real significance of this word is as yet unexplained. Certain it is that it is neither Latin nor Greek, but was handed down from the Orient with the objects which it served to designate. Roloff was the only one to attempt an explanation of the peculiar term by inviting attention to a Russian word, *murava*, which denotes "glazed pottery." The defenders of the mineralogical hypothesis have naturally rejected this point of view without giving reasons why it should not be acceptable.[2] Yet this opinion is worthy of serious consideration. If it can be proved that the murrines were glazed pottery vessels, there is a great deal of probability in the conviction that the word *murra* applies to their most striking feature, the glaze. The Russian word pointed out by Roloff indeed exists. It is recorded in all good Russian dictionaries. VLADIMIR DAL,[3] the eminent Russian lexicographer, notes it in the forms *muráva, muravá*, and *mur*, with a dialectic variant *múrom* (or *múrom'*),[4] used in the Governments of Pskov and Tver, and interprets it as the glaze applied to the surface of a pottery vessel. Besides this word, the Russian language avails itself of the loan-word *glazur* (derived from German *Glasur*) and the indigenous word-formation *poliva* for the connotation of the same idea. The words *mur* and *murava*, not to be found in any other Slavic or European language, are not derived from any Slavic stem, but, like other Russian culture-words, are borrowings from an Iranian language. The onomasticon of Ancient Iranian is but imperfectly preserved; and the word *mura* or *murra*, which has doubtless existed in that language, has not been handed down to us in an Iranian literary monument; although a survival of it, in all probability, is preserved in Persian *mōrī, mūrī*, or *mūriš*,

---

[1] The readings *morrinos, myrrinus*, also occur (see the edition of B. FABRICIUS, pp. 42 and 90); but *murrinos* merits preference.

[2] F. THIERSCH, *l. c.*, p. 457.

[3] Dictionary of the Living Great-Russian Language, Vol. II, col. 939 (in Russian only).

[4] The accent after *m* is intended to express the palatalization of the labial nasal *m* (soft or *mouillé m.*)

meaning "small shells" or "glass beads."[1] The conjecture is therefore admissible, that Greek *morrion* (aside from its Greek ending) is an Iranian loan-word, and that the Iranian prototype had the significance "glass paste, glaze."[2]

The earliest author to speak of murrine vessels is the poet Propertius (born about 49 B.C.), in one of his elegies (IV, 5, 26), in which a procuress tries to allure an inexperienced lass by promising her all the wealth of the Orient, like purple robes, dresses from Cos, urns from Thebæ in upper Egypt, and murrine goblets baked in Parthian furnaces,—

> Seu quae palmiferae mittunt venalia Thebae
> murreaque in Parthis pocula cocta focis.

The most biased adherents of the mineralogical hypothesis were obliged to concede that mineral vessels could not be understood in this passage: no one would be likely to say regarding a mineral that it is cooked or baked. Nor is it necessary to press the verb *coquere* into a forced

---

[1] The Persian word *mīnā* signifies "enamel" and "glass, glass bead, goblet." It is very probably connected with Young-Avestan *minav*, "necklace, ornament" (BARTHOLOMAE, Altiranisches Wörterbuch, col. 1186). The Persian *mōrī* ("glass bead") is found also in the language of the Abdāl or Tābārji in northern Syria (A. VON LE COQ, *Baessler-Archiv*, Vol. II, 1912, p. 234).

[2] Also the Russian designation for Chinese porcelain, *farfór*, is derived from Iranian. In the allied Slavic languages we have Ruthenian *faifúrka*, Bulgarian *farfor* and *farforiya*, Polish *farfura* (in dialects *faifura; farfurka, farforka*, and *faforka* with the meaning "vessel, plate of stoneware"). The same word is found in Neo-Greek as *farfuri* (φάρφουρι) and in the same form in Osmanli (in other Turkish dialects, *farfuru:* W. RADLOFF, Wörterbuch der Türk-Dialecte, Vol. IV, col. 1914). The Russian lexicographer Dal is unable to account for the Russian word, and doubtfully refers it to a Turkish source of origin. E. BERNEKER (Slavisches etymologisches Wörterbuch, p. 279) proposes to derive the Slavic words from Osmanli *fagfur*, which means "title of the Chinese sovereign; name of a region in China which was celebrated for its porcelain; Chinese porcelain; porcelain in general, vases made from it." It must be understood, however, that this word is not Turkish in origin, but Persian, and was borrowed by the Osmans from the latter language. For a long time we have known that *fagfur* is the Persian term designating the Emperor of China (D'HERBELOT, Bibliothèque orientale, Vol. III, p. 320), and it was d'Herbelot who first pointed out that the Turkish name for porcelain, *fagfuri*, was adopted from the Persian title *fagfur* (see also YULE's Marco Polo, Vol. II, p. 148). The older form is *pakpur* or *pakur* (in the form Pakurios preserved by Procopius, the Byzantine historian of the sixth century, in his De bello persico, I, 5). Masūdi (translation of A. SPRENGER, Vol. I, p. 326) was familiar with the correct significance of the term, explaining it as "Son of Heaven." It is accordingly a literal rendering of the Chinese title *T'ien-tse* ("Son of Heaven"), claimed by the sovereigns of China since times of old, the ruler receiving his mandate from the supreme deity Heaven and governing the world in his name. Persian *fag* is evolved from *bagh* (corresponding to Sanskrit *bhaga*), and signifies "God" ("Bagdād" signifies "gift of God"); Persian *fūr*, *būr* (Sanskrit *putra*) means "son." Also in Persian, *fagfūri chīnī* and *fagfūrī* relate to Chinese porcelain.

meaning, so as to conform it with a process to which a mineral could be subjected; for, as has been shown by H. BLÜMNER,[1] it is the verb utilized in regard to the burning or baking of bricks and all fictile ware in general.

The fundamental passage in Pliny relative to the murrine vessels runs as follows:—

"The Orient sends the murrine vessels. They are found there in several localities which otherwise have no special reputation,[2] for the most part in places of the Parthian Empire; excellent ones, however, in Carmania. The opinion prevails that the humidity[3] contained in these vessels is solidified by subterranean heat. In size they never exceed the small sideboards (*abaci*); in thickness, rarely the drinking-vessels, which are as large as previously mentioned. Their brightness is not very powerful, and it is a lustre rather than brilliancy. Highly esteemed, however, is the variety of colors, with their spots changing into shades of purple and white; these two tinges, again, result in a third hue resplendent, through a sort of color-transition, as it were, in a purple or milky red. Some laud profusely in them the edges and a certain iridescence of the colors, such as are visible in the rainbow. Others are pleased by oily spots: translucency or pallor is a defect, and likewise are salt grains and warts, which are not projecting, but which, as in the human body, are depressed. Also their odor is commendable."[4]

The account of Pliny is vague. One point is conspicuous and quite certain, that he had no opinion of his own to offer on the subject. As illustrated by the application of such phrases as "putant, sunt qui, aliis placent," he simply reiterates second-hand information which he had picked up from unnamed sources, most probably from oral accounts circulated by traders in the article. Most likely, these stories were

---

[1] Technologie und Terminologie, Vol. II, pp. 19, 44.

[2] Or, in little-known localities.

[3] There is no reason to take the word *umor*, as has been done, in the sense of "moist substance."

[4] Oriens myrrhina mittit. Inveniuntur ibi pluribus locis nec insignibus, maxime Parthici regni, praecipua tamen in Carmania. Umorem sub terra putant calore densari. Amplitudine numquam parvos excedunt abacos, crassitudine raro quanta sunt potoria. Splendor est iis sine viribus nitorque verius quam splendor. Sed in pretio varietas colorum subinde circumagentibus se maculis in purpuram candoremque et tertium ex utroque, ignescente veluti per transitum coloris purpura aut rubescente lacteo. Sunt qui maxime in iis laudent extremitates et quosdam colorum repercussus, quales in caelesti arcu spectantur. Iam aliis maculae pingues placent — tralucere quicquam aut pallere vitium est — itemque sales verrucaeque non eminentes, sed, ut in corpore etiam, plerumque sessiles. Aliqua et in odore commendatio est (XXXVII, 8, §§ 21, 22).

directly imported from the Orient, together with the ware. This assumption is a necessary postulate in the case; and it is evident also that Pliny was ignorant of the real nature of the murrines, for he neglects to state what their actual character was. He fails to give a plain and matter-of-fact definition of the material, or to classify it in any known category of objects. True it is, he placed his article in his book on stones; but this only justifies us in concluding that Pliny regarded the murrine vases as possibly of stone, but not that they really were of stone. The opponents of the pottery theory forget that pottery is composed also of mineral substances, that we ourselves speak of stoneware, and that many a piece of stoneware is so hard that it is difficult enough to distinguish it from stone. Pliny must have been in the same quandary, and therefore did not commit himself to a frank utterance. This attitude of restraint is conclusive, and at the outset is conducive to two inferences. The substance *murra* was neither a mineral nor pure glass, for both were perfectly familiar to Pliny and his contemporaries. Why, if the *murra* plainly was of a mineral nature, should the learned and experienced naturalist not have unequivocally avowed this fact? The *murra* can have been but a most striking and novel material, which heretofore had been foreign to the Romans, and which, owing to the very novelty of its character, greatly puzzled them.

Pliny discusses in this chapter the murrine vessels, as they were sent to Rome from the Orient, in the shape of manufactured articles. In the preceding chapter he dilates on their first introduction and their excessive valuation, and tells of renowned individual cups. Naturally he is now bound to say what these sensational and luxurious objects looked like. He certainly does not intend to describe here the substance *murra*, alleged by some interpreters to have been a species of stone. The same interpreters, however, are agreed that in Chapter 7 the word *myrrhina* (eadem victoria primum in urbem myrrhina invexit) refers to murrine vessels, and not to the mineral of which they are alleged to have been made; and it is therefore obvious, also, that in the beginning of Chapter 8 the same word, *myrrhina*, must refer to exactly the same murrine vessels. Pliny means to convey the meaning that the murrine vessels came to Rome from the East. According to THIERSCH, it was not the vessels, but the mineral, which was imported; but unfortunately he fails to inform us where and how the mineral was wrought. Pliny does not say that the vessels were carved in Rome from an imported substance, but he does plainly state that they were first brought to the metropolis by Pompey. THIERSCH[1]

---

[1] *L. c.*, p. 471.

sets forth the opinion that Pliny opens the description of the "mineral" by speaking of its size and thickness, then passes on to the description of the surface, its brightness, its colors and their play, and winds up with remarks on the properties of the mass. It would be impossible to unite more absurdities in a single sentence. The dimensions, according to Thiersch, are exactly stated by the terms *amplitudo* and *crassitudo;* and the *murra* was a mineral, and, as Thiersch insists, fluor-spar. This mineral, consequently, was quarried in regular blocks of constantly equal dimensions,— a really astounding feat! Fluor-spar or fluorite crystallizes in the isometric system, commonly in simple cubes; this fact could not have escaped Pliny, had he ever had an opportunity of examining this mineral, which is not at all mentioned by him nor by any other ancient writer.[1] There is, moreover, no evidence that fluor-spar occurs in Persia, where the murrine vessels were made. There is no evidence that fluor-spar vessels were ever turned out in Persia, and, above all, no such vessels have ever come to light among classical antiquities. They did not survive, because they never existed, save in the imagination of nineteenth century writers.[2] But does our Pliny, indeed, speak of any mineral? There

---

[1] See this volume, p. 62.

[2] Thiersch himself is not the originator of this fancy. He attributes (p. 495) the germ of the idea to an English scholar signing himself "A. M." in the *Classical Journal* of 1810 (p. 472), who, after having seen vases carved from fluor-spar of Derbyshire in his time, persuaded himself that the murrine cups should have been composed of the same material,— an opinion presented without an iota of evidence. According to the Encyclopædia Britannica (Vol. X, p. 578), F. CORSI, the eminent Italian antiquary, held that fluor-spar was the material of the famous murrine vases; Corsi, however, followed Thiersch. H. BLÜMNER (Technologie, Vol. III, p. 276), reviewing the various opinions, observes that this theory has recently been strongly contested; he himself believes in the mineral character of the vessels, for which weak arguments are given. It is astounding with what high degree of tenacity the unfounded opinion of fluor-spar vessels could hold its position in the face of the bare fact that no such vessels ever existed in ancient Persia, Egypt, or in classical antiquity, and have never come to light. GUHL and KONER (Leben der Griechen und Rōmer, p. 699, 6th ed., 1893) adhere to this explanation, and, while admitting that we do not possess vessels which can positively be identified with murrines, point to a semi-transparent bowl found in Tyrol in 1837, which should probably be one. This supposition, however, conflicts with the fact that the murrines were not at all transparent, as shown by a distich of Martial (IV, 86): Nos bibimus vitro; tu murrā, Pontice: quare! prodat perspicuus ne duo vina calix. In the Century Dictionary it is justly remarked under "murra," "The principal objection to this theory is that no fragments of fluor-spar vases have been found in Rome or its vicinity." M. BAUER (Edelsteinkunde, 2d ed., p. 653) sensibly states that there is no positive and sufficient evidence for the allegation that the murrines were of fluor-spar; but neither is there any more evidence for his own opinion, that they may have been of chalcedony quarried in Ujjain in India. E. BABELON (in Daremberg and Saglio, Dictionnaire des antiquités grecques et romaines, Vol. II, p. 1466) says, "Nous ne savons pas sûrement ce qu'était cette matière précieuse qui servait

is no sense in speaking of dimensions with reference to a raw mineral. Certainly nobody would compare the size of a mineral with a piece of furniture, and its thickness with a drinking-cup. The use of the word *potoria* demonstrates that our author, alluding to the costly vessels mentioned in the previous chapter, understands drinking-vessels likewise in this passage.

Any one who has had any experience in reading Chinese texts relative to pottery or porcelain will be deeply struck by a certain kinship or affinity of terminology that prevails in the latter and in the Plinian tradition of murrines. No statement or attribute used in this text contradicts the opinion that ceramic stoneware is here in question. On the contrary, some words, indeed, are as well chosen as though they were directly derived from a ceramist's vocabulary, and are well apt to uphold my theory. The effect of the changing colors produced by the heavy glaze could not be better described than by Pliny's style. Every lover of Chinese pottery who reads this passage intelligently will confess that he has many times had this delightful experience of observing color changes and transitions, as well as the rainbow iridescence which we so greatly admire in the ceramic productions of the Han. Translucency as a defect is intelligible only in pottery: it refers to a thin glaze that allows of the transparency of the clay body. "Oily spots" (*maculae pingues*) is a felicitous ceramic expression; likewise is "salt grains and warts."[1]

à fabriquer les célèbres vases murrhins. La description quelque peu obscure que Pline donne des vases murrhins . . . est entremêlée de fables et elle ne s'adapte parfaitement bien ni à des coupes d'agate ou de sardonyx, ni à des coupes d'ambre ou de pâtes vitreuses, ni enfin à des coupes de jade, comme le pensent quelques critiques." Leaving aside the vitreous pastes, this statement is perfectly fair.—L. DE LAUNAY (Minéralogie des Anciens, Vol. I, p. 85) quotes a writer on onyx as saying, that, despite the similarity of descriptions, the murrines were not of onyx or sardonyx: "Si l'une ou l'autre de ces pierres avait été le *murrhinum*, les Anciens auraient certainement donné aux vases murrhiens, le nom de vases d'onyx ou de sardonyx, au lieu qu'ils ont distingué expressément les vases murrhiens d'avec ceux faits de l'une, ou de l'autre des pierres susdites." "The onyx has been proposed, but our authorities plainly imply that the onyx was a material akin to but yet distinct from that here in question" (W. SMITH, Dictionary of Greek and Roman Antiquities, Vol. II, p. 182). Other speculations in regard to the murrines were advanced, to the effect that they were made of a gum, or formed from shells. Others referred to obsidian. Veltheim proposed Chinese soapstone. "No mineral has been suggested which answers exactly to Pliny's description, and at present the problem is unsolved" (SMITH, *l. c.*),—sufficient reason for assuming that Pliny's description does not answer to any mineral.

[1] The *sales* (this is the only passage in Pliny where *sal* is used in the plural) were presumably identical with what the Chinese ceramists praise in the Ting porcelain of the Sung period, which exhibited vestiges of tears (JULIEN, Histoire, p. 61); those with tear-marks were even considered as genuine (EITEL, *China Review*, Vol. X, p. 311, and Vol. XI, p. 177; HIRTH, Ancient Chinese Porcelain, p. 141).

As regards the pleasant odor which Pliny accredits to the murrines, this is intelligible only if the question is of pottery; scented minerals or glass are not conceivable. We are informed by Athenæus (XI, p. 464 b) that the clay in the ceramic export-ware of Koptos in Egypt was blended with aromatics before the process of baking; and Aristotle follows him in this account. In the Greek papyri of the second century A.D. are mentioned fragrant vessels (εὐώδη κεράμια) which were possibly turned out in this manner.[1]

In the two chapters following the one in question, Pliny deals with crystal: the introductory sentence contains a reference to the murrines. He adopts the popular notion that crystal is a sort of petrified ice, and occurs only in cold regions where the winter snow freezes intensely.[2] A cause opposite to the one producing the murrines, accordingly, makes crystal which assumes form through a process of somewhat vehement congelation.[3] This observation hints at the previous sentence, "Umorem sub terra putant calore densari." The murrines are a product of heat, crystal is that of cold. This remark shows that murrines and crystals are not allied, but adverse substances; and this contrast believed to prevail between the two may be one of the reasons why they formed a favorite compound of speech.

Passing on to a discussion of amber, our author informs us that this natural product takes rank next among articles of luxury, though the demand for it is restricted to women, and is held in the same regard as precious stones; but whereas no evident reason can be conceived for this appreciation of amber, the reason is manifest for the two former substances, the crystal vases lending themselves to cold beverages, the murrine vases to hot and cold ones alike.[4] The former notion

---

[1] T. REIL, Beiträge zur Kenntnis des Gewerbes im hellenistischen Ägypten, p. 41 (Leipzig, 1913). A reddish, odoriferous clay (Portuguese and Spanish *búcaro*, Italian *búcchero*) was much in use for pottery during the eighteenth century.

[2] This does not restrain him from stating immediately that the Orient sends crystal, and that none is preferred to that of India. The Buddhist monk Yüan Ying (*Yi ts'ie king yin i*, Ch. 22, p. 2; see above, p. 115) was more discriminative on this point. Speaking of rock-crystal, and mentioning the theory that it should originate from ice a thousand years old, he points out that there is no ice in the scorching heat of India, and that accordingly Indian rock-crystal is not a transformation of ice, but merely a kind of stone. See also *T'oung Pao*, 1915, p. 190.

[3] Contraria huic causa crystallum facit, gelu vehementiore concreto (XXXVII, 9, § 23).

[4] Proximum locum in deliciis, feminarum tamen adhuc tantum, sucina optinent, eandemque omnia haec quam gemmae auctoritatem; sane priora illa aliquis de causis, crystallina frigido potu, myrrhina utroque; in sucinis causam ne deliciae quidem adhuc excogitare potuerunt (XXXVII, 11, § 30). Compare J. H. KRAUSE, Pyrgoteles, p. 90. The passage is somewhat equivocal, owing to the uncertainty as to what *omnia haec* is intended to refer. It may point to the various kinds of

directly results from the supposed cold nature of crystal; and *murra*, being the outcome of heat, must be well adapted for holding hot drinks, or, as the case may be, for cool liquids. The distinction here made by Pliny seems to me to add another weight of proof adverse to the opinion that the murrines were of stone; it is not probable, at least, that any stone cups served for hot beverages, while pottery, and heavily glazed pottery in particular, is a material well suited to such a purpose.

Aside from the main chapter, Pliny devotes a brief sentence to the subject (XXXIII, 2, § 5), in his notice on gold, by saying that "from the same earth [where gold and silver are mined] we dug up murrine and crystal vessels, the very fragility of which is deemed to enhance their price" (murrina ex eadem tellure et crystallina effodimus, quibus pretium faceret ipsa fragilitas). The passage has materially contributed to the notion that *murra*, in the same manner as crystal, should be a natural substance extracted from under the ground. "Here," F. THIERSCH (p. 460) remarks, "*crystallina* evidently does not mean crystal bowls and cups, since the latter are not dug out of the soil, but crystal masses from which they are made; and for this reason the parallelism of the words *murrina et crystallina*, as well as the application of *effodere* and *invenire*, compel us to assume that *murrina* is likewise used in Pliny with regard to the substance of the vessels, the *murra;* and Pliny means to say that the *murra*, in the same manner as crystal, is found beneath the earth and dug up." This conclusion is artificial, and by no means cogent. We all know that not only minerals, but also objects manufactured by human hand, are dug up from the soil; and there seems no valid objection why Pliny's words could not be construed to mean that murrine and crystal vases have been turned up from the soil as the result of excavations. This was not necessarily Pliny's own opinion, but it may have been the outcome of a story transplanted directly from the Orient; and in part this report may well have had a foundation in fact. The passage may signify also that the mineral substances employed in the manufacture of the *murra* were dug up from the soil. It must be directly connected with the sentence, "Umorem sub terra putant calore densari," discussed above. The pottery vessels were baked in an underground kiln,

---

amber, as has been translated above; or to the previously mentioned murrines and crystals, with the inclusion of amber. The following *priora illa* would seem strongly to favor the latter point of view. In that case, Pliny would say that murrines, crystal, and amber enjoy the same consideration or esteem as precious stones. It cannot be read, of course, into this context, that the three materials were classified among *gemmae*, and that for this reason *murra* was a precious stone; on the contrary, the passage means that this in fact was not the case, and only that the three were regarded as of the same value as precious stones.

where the humidity of the clayish substance was solidified by artificial heat, and thus they were extracted from the soil (e tellure effodimus); or the vessels, after being perfectly finished, were intentionally buried under ground to produce an oxidation of the glaze, which resulted in that well-known iridescence and the rainbow colors accentuated by Pliny. Much ado has been made by the adherents of the mineralogical hypothesis about the juxtaposition of murrine and crystal vases in the relevant passage and in another to be cited presently: this fact has been regarded as one of the strongest bulwarks of the mineralogical defence, which, however, is purely illusory. The union of the two products, previously alluded to, was mainly dictated by commercial considerations, since both were received from the Orient: this is the opinion of Pliny, and no other motive guided him in the choice of this expression. On concluding his chapter devoted to the murrine vases, he passes on to the topic of crystal, and notes that "the Orient likewise sends us crystal, that of India being preferred, and it originates likewise in Asia."[1] The clause "oriens et hanc mittit," owing to the addition of the particle "et," forcibly points to the beginning of the preceding chapter, "Oriens myrrhina mittit." For the reason that the Orient despatched murrine as well as crystal vessels, they were enumerated and discoursed in close succession and combined in speech into a compound of pleasing rhythm. There is no valid reason why we should conclude, that, because the names of the two products are allied, the murrine vases must have been of mineral character.[2] Similar compounds are found in all languages without giving rise to such forced conclusions. We are wont to speak of the tea and porcelain of China as the most characteristic products reaching us from that country; but no one means to imply that tea must be a substance related to porcelain, or that porcelain must be a kind of tea. The Chinese couple jade with porcelain to denote *objets de vertu* worthy of the collector, and the substances with which both are concerned are as congenial as murrines and crystal. And who will guarantee that the crystal vases shipped from the Orient, according to Pliny, were all of real rock-crystal? They may have been partially of glass as well.[3]

The price of the murrines was enhanced by their frailty,— again an attribute that thoroughly fits pottery, and most assuredly is not

[1] Oriens et hanc mittit, quoniam Indicae nulla praefertur; nascitur et in Asia (XXXVII, 9, § 23).

[2] We shall meet the same alliance in the Chinese texts relative to the Hellenistic Orient, where crystal (including also cut glass) and faience were closely joined in architecture.

[3] H. BLÜMNER, Technologie, Vol. III, p. 250, note 6.

applicable to agate, fluor-spar, or any other stone with which these vessels have thoughtlessly been identified. The murrines were fragile and delicate: Pliny adduces several examples testifying to this fact. A man of consular rank used to drink from a murrine cup, and, from sheer love of it, wore out its edge, resulting in an upward tendency of its value. This good man surely did not possess iron teeth to break through an agate or onyx cup. Pliny himself beheld the broken fragments of a single cup, and tells the story of T. Petronius, who, on the verge of death from his hatred of Nero, broke a murrine basin[1] of great value. In another passage Pliny observes, "With all our wealth, we even at present pour out libations at sacrifices, not from murrine or crystalline vessels, but from plain earthenware ladles."[2] This sentence occurs in the introductory part of a chapter dealing with works in pottery; and the contrast intended by the author between the rustic, unglazed, indigenous Italic earthenware and the pretentious, glazed, imported Oriental pottery is self-evident. The same discrimination is insisted on in the further discussion of the subject when Pliny, expanding on the exorbitant prices paid for fictiles, laments that luxury has arrived at such a height of excess as to make earthenware sell at higher rates than murrine vessels.[3] This comparison cannot be construed, as has been done by THIERSCH,[4] as favoring the opinion that the *murrhina* were fundamentally different from *fictilia*, but it is intelligible only when both were productions of a cognate nature.

Finally, Pliny enumerates murrines among the most valuable products derived from the interior of the earth, on a par with *adamas* (the diamond), *smaragdus*, and precious stones.[5] H. BLÜMNER[6] regards this text as furnishing strong evidence in favor of the murrines being stones. In my opinion it is of no consequence. Also the passage relating to white glass in imitation of murrines[7] is unimportant for our purpose; but it proves at least that the real murrines cannot have been purely of glass, as has been supposed by some authors.

---

[1] Trulla myrrhina, explained also as a ladle or scoop.

[2] In sacris quidem etiam inter has opes hodie non murrinis crystallinisve, sed fictilibus prolibatur simpulis (XXXV, 46, § 158).

[3] Eo pervenit luxuria, ut etiam fictilia pluris constent quam murrina (*ibid.*, § 163).

[4] *L. c.*, p. 470.

[5] Rerum autem ipsarum maximum est pretium in mari nascentium margaritis; extra tellurem crystallis, intra adamanti, smaragdis, gemmis, myrrinis (XXXVII, 78, § 204).

[6] Technologie, Vol. III, p. 276.

[7] PLINY, XXXVI, 67, § 198.

Hitherto the attempt has been made to extract the realities from the ancient traditions, and to interpret them without prejudice. It is more difficult to correctly judge the legendary ingredients by which they are incrusted, as we are unaware of the lore of the Orient which prompted such notions as are echoed in Pliny. An analogous field, however, might contribute a little to aid us in understanding some of this folk-lore. Nothing could better enlighten Pliny's account of murrines than a remembrance of the first experience which Europe had in regard to the newly-introduced Chinese porcelain. If the ancients were deeply impressed and perplexed by the thickly glazed faience of the anterior Orient, and may have mistaken it for stone, an interesting parallel is offered by the fact that in the inventory of the Duke of Anjou (1360–68) is found "une escuelle d'une pierre appelée pourcellaine," and, in that of Queen Jeanne d'Evreux (1372), "un pot à eau de pierre de pourcelaine."[1] In these two cases, Chinese porcelain (corresponding to that of the Yüan period, 1260–1367) is styled "a stone called porcelain."

The beliefs of the ancients in an underground substance from which the murrine vessels were made, receive a curious parallel from the fantastic notions entertained by early European writers as to the composition of Chinese porcelain. BARBOSA[2] wrote about 1516, "They make in this country a great quantity of porcelains of different sorts, very fine and good, which form for them a great article of trade for all parts, and they make them in this way. They take the shells of sea-snails, and egg-shells, and pound them, and with other ingredients make a paste, which they put underground to refine for the space of eighty or a hundred years, and this mass of paste they leave as a fortune to their children." In 1615, Bacon said, "If we had in England beds of porcelain such as they have in China, which porcelain is a kind of plaster buried in the earth and by length of time congealed and glazed into that substance; this were an artificial mine, and part of that substance" . . . Sir Thomas Browne, in his "Vulgar Errors" (1650), asserted, "We are not thoroughly resolved concerning Porcellane or China dishes, that according to common belief they are made of earth, which lieth in preparation about an hundred years underground; for the relations thereof are not only divers but contrary; and Authors agree not herein" . . . These fables were refuted at the end of the sixteenth and seventeenth centuries by travellers who had occasion to make observations on the

---

[1] F. BRINKLEY, Japan and China, Vol. IX, Keramic Art, p. 371 (London, 1904).

[2] YULE and BURNELL, Hobson-Jobson, p. 726.

spot. JUAN GONZALEZ DE MENDOZA,[1] who wrote in 1585, reiterated Barbosa's story, and (in the early English translation) called its validity into doubt; for, if it were true, the Chinese, in his opinion, could not turn out so great a number of porcelains as is made in that kingdom and exported to Portugal, Peru, New Spain, and other parts of the world.[2] J. NEUHOF,[3] who accompanied the embassy of the East India Company of the Netherlands to China from 1655 to 1657, scorns the "foolish fabulists of whom there are not a few still nowadays who made people believe that porcelain is baked from egg-shells pounded and kneaded into a paste with the white of an egg, or from shells and snail-shells, after such a paste has been prepared by nature itself in the ground for some hundred years." The Jesuit, L. LE COMPTE,[4] rectified this error by saying that "it is a mistake to think that there is requisite one or two hundred years to the preparing of the matter for the porcelain, and that its composition is so very difficult; if that were so, it would be neither so common, nor so cheap." These two authors were seconded by E. YSBRANTS IDES.[5] The analogy of the beliefs in the origin of murrines and porcelain is striking; and this fancy has doubtless taken its root in the Orient, whence crafty dealers propagated it in the interest of their business.[6]

It would be presumptuous on my part to state positively what class of Oriental pottery should be understood by the murrines. The decision of this question must be reserved for the specialists in this field. Students of ancient ceramics seem to have already had a premonition of the identity of murrines with pottery.[7] It may be permissible to point,

---

[1] History of the Great and Mighty Kingdom of China, Vol. I, p. 34 (Hakluyt Society, 1853).

[2] This refutation of Mendoza, however, is not contained in the Spanish original, where it is said only, "Y esto fe a visto, y es mas verosimil que lo que dize cierto Duardo Barbosa, que anda en Italiano, que se haze de caracoles de mar, los quales se muelen, y los meten debaxo de tierra a afinarse 100 años, y otras cosas que açerca desto dize. La muy fina, nunca sale del Reyno, por que se gasta en seruicio del Rey, y Gouernadores, y es tan linda que parece de finissimo cristal. La mas fina, es la que se haze en la Prouincia de Saxij" (I. GONZALEZ DE MENDOÇA, Historia de las cosas mas notables, ritos y costumbres, del gran Reyno dela China, p. 25, Roma, 1585). Saxij refers to Kuang-tung.

[3] Gesantschaft der Ost-Indischen Gesellschaft, p. 96 (Amsterdam, 1669).

[4] Memoirs and Observations made in a Late Journey through the Empire of China, English translation, p. 158 (London, 1697).

[5] Driejaarige Reize naar China, p. 165 (Amsterdam, 1710).

[6] E. KAEMPFER (History of Japan, Vol. II, p. 369) alludes to another superstition prevalent in his time (end of the seventeenth century), that human bones should form an ingredient of China ware.

[7] E. FOURDRIGNIER, Les étapes de la céramique dans l'antiquité (*Bull. et Mém. de la Soc. d'Anthr.*, 1905, p. 239); he gives his opinion with great reserve, however.

en passant, to a remarkable find of pottery which offers a fair guaranty of being identical with the murrine vases.

F. Petrie's discovery in 1909–10, at the south end of Memphis, of kilns for baking glazed pottery, with a large number of fragments of vessels, felicitously fills a gap in the early history of glazed ware, and speaks in favor of the presence on Egyptian soil of murrine vessels, and particularly even of Parthian murrine vessels. The date of Petrie's finds is calculated at a period between A.D. 1 and 50, a fragment of a lamp of known type permitting this conclusion.[1] The principal tints of the glazed shards, which are remarkable for their coloring and their design, are a deep indigo blue, lighter blues, manganese purple, and apple green. The designs are almost entirely Persian, showing little, if any, direct Greek influence. Winged bulls, rampant beasts, "sacred tree," etc., all occur; and the problem arises whether this Persian character points to some Oriental revival of the art of making glazed pottery. In Diospolis, according to the Periplus,[2] murrines were imitated in glass; and this imitative manufacture presupposes the existence there of true pottery murrines which were taken as models. The Memphis pottery of Persian style due to Petrie perfectly answers this purpose, as to both its technical properties and its chronology.

Among Greek authors, the murrines are mentioned only by Pausanias and the Periplus. Pausanias (second century A.D.) recalls them merely in a passing manner. In the Arcadica (XVIII, § 5) he speaks of "glass, crystal, murrine vessels, and others made by men from stone."[3] The idea that Pausanias speaks of vessels carved from stone is thoroughly excluded; he hints, on the contrary, at vessels turned out from products and devices of human labor. "Crystal" is probably nothing but cut glass; the union of the terms "crystal" and "murra" has already been discussed. "Glass" indeed belongs to the same category as "murra;" and the passage of Pausanias is sanely interpreted by the rendering, "glass, cut glass, and glazed pottery, and other products made by men from stone."

In the Periplus Maris Erythraei, written approximately about A.D. 85,[4] the murrines are mentioned in three passages. In Chapter VI

---

[1] Compare O. M. DALTON, Byzantine Art and Archæology, p. 608.

[2] See below, and p. 138.

[3] "Ὕαλος μέν γε καὶ κρύσταλλος καὶ μοῤῥία καὶ ὅσα ἐστὶν ἀνθρώποις ἄλλα λίθου ποιούμενα.

[4] Compare the writer's Notes on Turquois in the East, p. 2, note. J. KENNEDY (*Journ. Royal As. Soc.*, 1916, p. 835) is now inclined to date the Periplus at about A.D. 70.

we meet "several kinds of glass and other murrine vases, which are made in Diospolis."[1] The latter city is regarded as identical with Thebæ in upper Egypt. Here the substance *murra* is designated as a kind of glass, but it is "another" kind of glass, different from ordinary glass. There is no doubt in my mind that it denotes here the vitreous paste employed for the glazing of pottery, and this conclusion perfectly agrees with all that we know about the thriving industries of ceramics and glass in Egypt of that period.[2]

Chapter XLVIII of the Periplus mentions the trade of Ozene,— that is, Ujjayinī (Ujjain),— the chief city of Mālvā, in India, whence onyx-like and murrine stones[3] are brought to the port Barygaza on the west coast. In the following chapter it is stated that these articles, among others, are exported from Barygaza. Again, in this case, we have not to understand by the murrine material a pure mineral of uniform character, but an artificial composition of partially mineral origin, turned to glazing-purposes, and introduced into commerce in the shape of cakes, which, on the surface, appeared to the uninitiated as a mineral substance resembling onyx. The Periplus thus opens our eyes to the fact that substances for glazing were traded as far as India, and this is confirmed both by Indian traditions and by the Chinese annals.

The Chinese, indeed, were acquainted with the murra of the ancients; and Chinese records point in the same manner to the home of the substance,— the anterior Orient, styled by them Ta Ts'in ("Great Ts'in"). The glassy paste for the production of ceramic glazes was called *liu-li* 琉璃 (in the Han Annals 流離) or *p'i-liu-li*, derived from Prākrit *veluriya*, Mahārāshṭrī *verulia* (Sanskrit *vaiḍūrya*).[4] The *Wei lio*,

---

[1] Λιθίας ὑαλῆς πλείονα γένη καὶ ἄλλης μουῤῥίνης τῆς γινομένης ἐν Διοσπόλει (ed. of B. FABRICIUS, p. 42).

[2] Compare T. REIL, Beiträge zur Kenntnis des Gewerbes im hellenistischen Ägypten, pp. 37–50. The mass is well described by W. M. FLINDERS PETRIE (Arts and Crafts of Ancient Egypt, p. 117): "Quartz rock pebbles were pounded into fine chips after many heatings which cracked them. These were mixed with lime and potash and some carbonate of copper. The mixture was roasted in pans, and the exact shade depended on the degree of roasting. This mass was half fused and became pasty; it was then kneaded and toasted gradually, sampling the color until the exact tint was reached. A porous mass of frit of uniform color results. This was then ground up in water, and made into a blue or green paint, which was either used with a flux to glaze objects in a furnace, or was used with gum or white of egg as a wet paint for frescoes."

[3] Ὀνυχίνη λιθία καὶ μουῤῥίνη.

[4] PALLADIUS (Chinese-Russian Dictionary, Vol. I, p. 367), our foremost authority on Chinese lexicography, has given as the principal meaning of *liu-li* "glaze" (Russian *glazur*). Several writers accept the term *liu-li* in the too narrow sense of "glass" only, and construe a theory that quantities of glass vessels were imported at the Han time from the workshops of Syria and Egypt (for instance, S. W. BUSHELL,

written in the third century A.D., attributes to Ta Ts'in ten varieties of *liu-li*,— carnation, white, black, green, yellow, blue, purple, azure, red, and red-brown.[1] This extensive color-scale shows us that not a precious stone is involved (and with reference to India *p'i-liu-li* or *liu-li* may well denote a variety of quartz or rock-crystal[2]), but an artificial, man-made product. This is clearly evidenced by other texts, in which the peculiar utilization of *liu-li* in Ta Ts'in is specified. Thus we are informed by the Tsin Annals that the people of Ta Ts'in use *liu-li* in the making of walls, and rock-crystal in making the bases of pillars. The *Kiu T'ang shu* reports that eaves, pillars, and window-bars of the palaces there are frequently made of rock-crystal and *liu-li*.[3] Glazed faience for architectural purposes is doubtless alluded to in these two cases; and we face here the same combination of murra and crystal as we noticed in Pliny.[4] It was almost at the same time, or only a little later, that the knowledge of glazed ware spread to the West and the Far East alike from the same focus. It thus was the knowledge of the highly-developed ceramic processes of the anterior Orient, at their climax in the second century B.C. or earlier, which was transmitted to China, and gave there the impetus to the production of glazes.

The conception of *liu-li* as a precious stone is chiefly upheld in Buddhist texts; but in reading these with critical understanding it is obvious that something else is hidden behind this alleged stone. The *Yi ts'ie king yin i*,[5] written by Yüan Ying about A.D. 649, states that

---

Chinese Art, Vol. II, p. 17). Nothing of the kind, however, is to be found in the ancient Chinese texts, which, with reference to the Roman Orient, never mention any vessels of *liu-li*, but merely speak of a substance of that name, without any reference to objects made from it. This clearly indicates that no vessels of any sort were imported, but only pasty masses of various tinges which could be applied to pottery bodies. That *liu-li* has nothing to do with the production of glass, simply results from the fact that only as late as the fifth century A.D. did the Chinese learn from foreigners how to make glass. If glazed ware makes its appearance under the Han, it is obvious that it bears some relation to the *liu-li* originating from the Roman-Hellenistic Orient.

[1] HIRTH, China and the Roman Orient, p. 73.

[2] See *T'oung Pao*, 1915, p. 198. In the dictionary *Kuang ya* of the third century (Ch. 9, p. 5 b; ed. of *Han Wei ts'ung shu*) *liu-li* is classed with quartz (*shui tsing* 水精).

[3] HIRTH, *l. c.*, pp. 44, 51. Hirth translates *liu-li* by "opaque glass;" but such walls and pillars of glass have not yet been discovered.

[4] In Egypt, as early as 5500 B.C., glazing was applied on a large scale for the lining of rooms. Tiles have been found about a foot long, stoutly made, with dovetails on the back, and holes through them edgeways in order to tie them back to the wall with copper wire. They are glazed all over with hard blue-green glaze (W. M. FLINDERS PETRIE, Arts and Crafts of Ancient Egypt, p. 108).

[5] Ch. 23, p. 12 b (see above, p. 115). This text has been adopted by the *Fan yi ming i tsi* (Ch. 8, p. 12 b; edition of Nanking).

"the name *liu-li* or *p'i-liu-li* is derived from that of a mountain, and is said to be the precious stone of a distant mountain, which is the Sumeru of Buddhist cosmology. This jewel is of green (青) color. Altogether, all jewels cannot be injured, nor can they be melted and cast by means of blaze and smoke. Only the demons and spirits have sufficient strength to break them to pieces. There is further a saying that *liu-li* is the shell of the egg of the bird with golden wings.[1] The demons and spirits obtain it and sell it to mankind." This Chinese text is the reproduction of a theme of Indian lore; and the tradition hints at the importation into India of a substance from abroad, which could be wrought only by demons (that is, foreigners).[2] The allusion to melting shows that it really could be melted; and the comparison with the shell of a bird's egg, which hints at a coating, is the best possible poetical metaphor for a ceramic glaze. It thus seems to me that the Sanskrit term *vaiḍūrya* and its congeners originally denoted some semi-precious quartz-like stone, and were then transferred to the enamel glaze of the anterior Orient.[3]

Chinese tradition refers the earliest employment of *liu-li* to the reign of the Emperor Wu (140–86 B.C.) of the Former Han dynasty. It is said in the Annals of the Han that this sovereign despatched special agents over the sea for the purchase of the substance *p'i-liu-li*.[4] It was likewise known at that period that this article figured among the products of the country Ki-pin (Kashmir), which opened intercourse with China under the same emperor.[5]

It is notable that in the Han period objects were found under ground, said to have been made of *liu-li*, and that we have accounts of objects wrought from *liu-li* by Chinese craftsmen. Since glass was manufactured in China only several centuries later, it cannot come here into question; and from the nature of these objects it follows that they cannot either have been of rock-crystal or lapis lazuli. In the biography of Hu Tsung 胡綜[6] it is narrated that Hu, during the life

[1] The saliva of this bird was believed to produce the gem *mu-nan* (see this volume, p. 70, note 3). It is the fabulous bird Garuḍa.

[2] It is a well-known fact that foreign tribes were characterized by the Aryan Indians as demons under such names as Nāgas, Rākshasas, or Piçācas.

[3] It is possible also that the Indian words are derived from a West-Asiatic language.

[4] In the geographical chapter of the *Ts'ien Han shu* (Ch. 28 B, p. 17 b).

[5] *Ts'ien Han shu*, Ch. 96 A, p. 5. S. W. BUSHELL (Chinese Art, Vol. I, p. 61) dates the appearance of glaze in China only from the Later Han dynasty (A.D. 25–220).

[6] *San kuo chi, Wu shu*, Ch. 62. See also *Yu yang tsa tsu*, Ch. 11, p. 4 (ed. of *Pai hai*).

time of Sun K'üan 孫權 (A.D. 181–252), while digging the ground, found a copper or bronze chest two feet and seven inches long, the cover of it being made of *liu-li* (掘地得銅匣長二尺七寸以琉璃爲蓋). This bronze vessel evidently was of Chinese make; and the only reasonable supposition is that the cover was of glazed ware, the whole affair coming down from the Former Han dynasty. Sun Liang 孫亮, who died in A.D. 260, a son of the aforementioned Sun K'üan, made a screen of *liu-li*.[1]

In the *Han wu ku shi* 漢武故事 (that is, "Old Affairs relating to Wu of the Han Dynasty") it is on record that Wu was fond of the gods and genii, and erected in their honor sanctuaries the doors of which were coated with a white glaze (*pai liu-li* 白瑠璃) that reflected its light afar. The Emperor Ch'eng (32–7 B.C.) built the palace Fu-t'ang 服湯殿 for Chao Fei-yen, and had the doors glazed green.[2] In the same manner, *liu-li* is combined with the names for pottery vessels: thus we read about "glazed wine-cups" (*liu-li chung* 琉璃鍾)[3] and glazed bowls (*liu-li wan* 椀).[4] The Chinese hardly ever made use of glass for practical household purposes. Pottery was always the article they preferred. Wine being taken hot, glass was prohibitive for wine-cups. The same holds good for tea. Glass beads were the only article of practical utility to the Chinese. Those who have written on glass in ancient China, merely by consulting Chinese sources, seem to have never seen antique glass or collections of Chinese glass. When the making of glass became known to the Chinese, they began to cut and polish it in its hard state; that is, they treated it in the same manner as hard stone, and applied to it the principles of their glyptic art. Glass became the domain of the carver, of a rather limited art-industrial importance, but it never had any practical bearing upon the

---

[1] *Ku kin chu* 古今注 (Ch. c, p. 5 b; ed. of *Han Wei ts'ung shu*). A fantastic description of this screen is given in the *Shi i ki* 拾遺記 (Ch. 8, p. 6; ed. of *Han Wei ts'ung shu*). There are several other allusions to such screens of *liu-li*, which in my opinion were made of a thin wall of clay coated with a glaze.

[2] *T'ai p'ing yü lan*, Ch. 808, p. 4. Several writers have conceived the windows and doors of this palace as being made of glass (for instance, A. FORKE, *Mitt. Sem. or. Spr.*, Vol. I, p. 113); but we do not know that window-glass existed at the same time in the Western world. Scanty remains of window-glass have been found only in Pompeii and Herculaneum, but no extensive use was ever made of it in the time of the Roman empire. In western Asia no window-glass was made, and accordingly no export to China could take place. Aside from this point, I would be disinclined to believe in the possibility of transporting window-glass from the Orient to China at that time.

[3] *Tsin shu*, Ch. 45, p. 8.

[4] *Yüan kien lei han*, Ch. 364, p. 31 b; glazed dishes for eating in *Tsin shu* (*T'ai p'ing yü lan*, Ch. 808, p. 4 b).

life of the people. Certainly, the term *liu-li* refers also to opaque glass, especially from the fifth century onward. If in 519, under the Emperor Wu of the Liang dynasty (502–520), Khotan sent to China a tribute gift of *liu-li* pitchers (*liu-li ying* 罌),[1] these may be conceived of as glass as well as of glazed pottery. In other passages the exact significance of the term remains doubtful, as in the case of a saddle of brilliant white *liu-li*, which in the dark emitted light at the distance of a hundred feet, and which is mentioned in the *Si king tsa ki* 西京雜記[2] among presents sent to the Emperor Wu from India. Here we have a fabulous echo of traditions that were exaggerated by later generations.

It is a significant fact that the reign of the same Emperor Wu is characterized by the sudden rise of alchemy and chemical notions and experiments;[3] and this novel line of thought is certainly connected with the western expansion and the newly-opened trade-routes across Central Asia inaugurated by the same sovereign. In the Greek alchemical papyri we meet the oldest technical recipes for the fabrication of glass and enamels, and technical treatises on glass.[4] Aeneas of Gaza, a Neo-Platonic philosopher of the fifth century, represents glass directly as an alchemical transmutation from a baser to a nobler material by observing, "There is nothing incredible about the metamorphosis of matter into a superior state. In this manner those versed in the art of matter take silver and tin, change their appearance, and transmute them into excellent gold. Glass is manufactured from divisible sand and dissoluble natron, and thus becomes a novel and brilliant thing."[5] We have a few intimations to the effect that *liu-li* was appreciated also by the Chinese alchemists. Tung-fang So obtained multi-colored dew and placed it in glazed vessels, which he offered as a gift to the Emperor Wu.[6] The famous alchemist Li Shao-kün 李少君, whose life and deeds have been narrated by Se-ma Ts'ien, is said to have repaired the brilliant-white *liu-li* saddle of Wu mentioned afore, when this saddle was once broken during an imperial hunting-expedition; he availed himself of pieces of bone, which were joined by means of a thin, sticky substance, with such good effects, that no damage could be

---

[1] *Liang shu*, Ch. 54, p. 14 b.

[2] Ch. 2, p. 2 b (ed. of *Han Wei ts'ung shu*).

[3] See particularly CHAVANNES, Mémoires historiques de Se-ma Ts'ien, Vol. III p. 465.

[4] M. BERTHELOT, Introduction à l'étude de la chimie des anciens et du moyen âge, pp. 200, 202; Les Origines de l'alchimie, pp. 123, 125.

[5] M. BERTHELOT, Origines, p. 75.

[6] *T'ai p'ing yü lan*, Ch. 808, p. 4 b.

perceived even in broad daylight.[1] When the ancient Chinese literature on alchemy shall have become as accessible as the Greek, Arabic, and European records of this ancient science, the subject in question will doubtless receive further elucidations.

While *liu-li* was imported into China from the Hellenistic Orient over the established trade-routes across Central Asia, and from Kashmir, another source of supply was represented by Cambodja, which, as we know, was in intimate commercial relations with India, and received from there the products and merchandise of western Asia. In the Calendar or Chronological Tables of the Country of Wu (*Wu li* 吳歷), by Hu Ch'ung 胡沖,[2] it is on record that in the fourth year of the period Huang-wu 黃武 (A.D. 225), Fu-nan 扶南 (Cambodja) and other foreign countries sent envoys to China with gifts of *liu-li*.[3] According to another version of the same text, this event would have taken place in the period Huang-lung 黃龍 (229–231).[4] This text contains the mention of the first embassy from Fu-nan (Cambodja) to China, and allows us to infer that *liu-li* was found there in the beginning of the third century and transmitted to China. Another allusion to the presence of *liu-li* in the countries south of China is encountered in the *Kuang chi* 廣志, written by Kuo I-kung 郭義恭 under the Liang dynasty (502–556), where it is said that *liu-li* is a product of Huang-chi 黃支,[5] Se-tiao 斯調,[6] Ta Ts'in, and Ji-nan 日南 (Annam). Finally *liu-li* was sent also to China from Central India under the Liang dynasty (502–556).[7]

Our most important witnesses certainly are the numerous specimens of Han mortuary pottery glazed in the most varied shades of green

---

[1] *T'u shu tsi ch'eng*, under *liu-li*.

[2] PELLIOT, *Bull. de l'Ecole française*, Vol. IV, p. 391.

[3] *Yüan kien lei han*, Ch. 364, p. 31.

[4] *T'ai p'ing yü lan*, Ch. 808, p. 4 b. Compare also PELLIOT, Le Fou-nan (*Bull. de l'Ecole française*, Vol. III, p. 283). The Wu dynasty, one of the Three Kingdoms (*san kuo*), reigned from 222 to 280.

[5] Presumably on the Malay Peninsula (see Chinese Clay Figures, p. 80, note 2). *Liu-li* is also enumerated among the tribute-gifts sent from Huang-chi to the Chinese Court (*T'ai p'ing huan yü ki*, Ch. 176, p. 2 b). *Pi-liu-li* is mentioned as an article of Huang-chi as early as the Han period (*Ts'ien Han shu*, Ch. 28 B, p. 17).

[6] Probably Java (*T'oung Pao*, 1915, pp. 351, 373). In the latter passage I mentioned a plant *mo-ch'u* as growing in Se-tiao. M. G. Ferrand, Consul General of France in New Orleans, has been good enough to write me that this Chinese transcription corresponds to Javanese *mojo*, the designation of the tree *Aegle marmelos*, and that the emendation of Se-tiao into Ye-tiao is thus assured, and the identification of Ye-tiao with Java becomes a definite result. M. Ferrand himself will soon report about this ingenious discovery.

[7] *Liang shu*, Ch. 54, p. 8.

and brown, and still called by the Chinese *liu-li wa* 瑠璃瓦.[1] The fact that the process of glazing itself is not described in the ancient texts, as pointed out by Hobson, is not of great concern. In fact, we have no ancient description of pottery whatsoever; and no technical treatise, if there ever was any, has survived from the Han period. The subject of pottery began to interest Chinese scholars only as late as the age of the Sung and Yüan; and in the same manner as the old writers fail to record the evolution of porcelanous ware, they are reticent as to glazing and other ceramic processes. It cannot be strongly enough emphasized that our knowledge of the subject should be reconstructed on the basis of actual material before our eyes, and not on literary sources which are still very incompletely exploited, or on philological considerations. It is unreasonable to expect also that literary traditions and antiquities of China should blend into a uniform and harmonious picture: neither is such the case in the archæology of Greece or Italy. We have hundreds and hundreds of Chinese antiquities which cannot be traced to any records, but it would be an absurd procedure to disregard them simply for this reason. Monuments speak their own language, and are entitled to a fair and impartial hearing on their own merits. Both monuments and literature have come down to us only in fragments; and while it is not necessary that one department confirms the other, we must regard ourselves fortunate in seeing one supplemented by the other.[2]

Owing to their lack of interest in technical matters, the notions of Chinese scholars regarding *liu-li* are the vaguest possible. Mong

---

[1] A disk labelled *pi-liu-li* is represented on the Han bas-reliefs among the objects of happy augury. No conclusions can be drawn from this design as to objects made from *liu-li*, as the artist took the first element *pi* in the sense of "disk" or "ring," and based his conception on this interpretation. His work represents merely an art-motive, not a reality. This subject has been well expounded by E. CHAVANNES (Mission archéologique, Vol. I, La sculpture à l'époque des Han, p. 170).

[2] There are several allusions to green-glazed Han pottery in Chinese writings. One is extracted by HOBSON (Chinese Pottery and Porcelain, Vol. I, p. 199) from the Gazetteer of Shen-si Province, and refers to the village Lei-siang in the prefecture of T'ung-chou, where the inhabitants sometimes dig up castaway wares, archaic in shape and style, of green, deep and dark, but brilliant color, some with ornaments in raised clay. The Gazetteer of the District of Hua-yang (forming with the district of Ch'eng-tu the prefectural city of Ch'eng-tu, the capital of Sze-ch'uan) reports (Ch. 41, p. 64), "An ancient pottery censer (古瓦香鼎) is in the Kuang-fa temple (廣法寺), outside of the city, twenty *li* in easterly direction. It is rectangular in shape, posed on four feet, two feet five inches in length, and one foot two inches in width. It is provided with lion's ears [relief designs of animal-heads], and is green and glossy. According to a tradition it is an object of the Shu Han period (221–264)."

K'ang of the third century, commenting on the Han Annals,[1] remarks that *p'i-liu-li* is green in color, like jade. Yen Shi-ku (579–645), however, rejects this generalization, observing that Mong K'ang's definition is too narrow; that the substance is a natural object, variegated, glossy, and brilliant; that it exceeds any hard stones (玉); and that its color is unchangeable. "It is the present practice," he continues, "to prepare it by the use of molten stones, with the addition of certain chemicals to the flux. This mass, however, is hollow, brittle, and not evenly compact; it is not the genuine article."[2] This is apparently an allusion to glass. The notion that *p'i-liu-li* was regarded as a product of natural origin was suggested by the meaning "quartz," which originally adhered to the Sanskrit term *vaiḍūrya*, the prototype of the word *p'i-liu-li;* but this does not mean that vitreous bodies were taken by the ancient Chinese for precious stones, as has been intimated by some authors. The confusion is one of terminology rather than of realities. The parallel with the conception of *murra* as a stone is obvious.

In the *Nan chou i wu chi* 南州異物志, by Wan Chen 萬震 of the third century, we read as follows:[3] "The principal material underlying *liu-li* is stone. In order to make vessels from it, it must be worked by means of carbonate of soda.[4] The latter has the appearance of yellow ashes, which are found on the shores of the southern sea, and are suitable also for the washing of clothes. When applied, it does not require straining; but it is thrown into water, and becomes slippery like moss-covered stones. Without these ashes, the material cannot be dissolved." This is probably a recipe for making a glaze. Compare the Chinese notions on using ashes for porcelain glazes and obtaining such ashes.[5]

At the Court of the Mongol lynasty, four kilns were established in 1276 at Ta-tu for the manufacture of plain, white-glazed bricks and tiles (素白琉璃磚瓦), with an army of three hundred workmen. The so-called Southern Kiln (*nan yao* 南窰) was erected in 1263, the Western Kiln (*si yao* 西窰) in 1267, and that of Liu-li kü 琉璢局 (northwest of Peking) in 1263.[6] The latter was still operated under the

[1] *Ts'ien Han shu*, Ch. 96 A, p. 5.

[2] HOBSON (Chinese Pottery and Porcelain, p. 144) gives only an abridged quotation of Yen Shi-ku's text, as quoted in the *T'ao shuo*, which does not bring out the author's true meaning. The main point is that Yen Shi-ku regarded *p'i-liu-li* as a natural substance, and looked upon the artifacts of his time as poor substitutes.

[3] *T'ai p'ing yü lan*, Ch. 808, p. 5.

[4] *Tse jan hui* 自然灰, literally "natural ashes;" used also with reference to a kind of earth and feldspath (GEERTS, Produits, pp. 404, 416).

[5] JULIEN, Histoire et fabrication de la porcelaine chinoise, p. 131.

[6] *Yüan shi*, Ch. 90, p. 5.

Manchu dynasty, furnishing the well-known glazed tiles and bricks for the palace, official buildings, and state temples of the metropolis. Glazed tiles and bricks, however, were known in China long before the time of the Yüan. They certainly existed under the Sung. Chou Shan, who in A.D. 1177 accompanied an embassy sent by the Sung Emperor from Hang-chou to the Court of the Kin dynasty at Peking, reports that the palace of the Kin was covered with tiles, all coated with enamels, their colors resplendent in the sunlight.[1] Ngou-yang Siu (1007–72) speaks of glazed tiles.[2] Sir AUREL STEIN[3] discovered in the ruins of Ch'iao-tse bricks and tiles bearing in beautiful green glaze scroll ornaments in low reliefs, and employed in a Stūpa constructed during Sung times.[4] Glazed tiles were likewise known under the T'ang. A certain Ts'ui Yung 崔融, who lived in the T'ang era, erected on Mount Sung in Ho-nan, in honor of his mother, a memorial temple covered with glazed tiles (*liu-li chi wa*). The famous poet Po Kü-i (A.D. 772–846) speaks of a pair of white-glazed (*pai liu-li*) vases.[5] Remains from buildings of this period show also the application of glazing for architectural purposes. The bricks and tiles of the Han and Wei periods, as far as we know them, are all unglazed, but it would be premature to assert that glazing was then not applied to them.[6]

The continuity of Chinese tradition is vividly illustrated by the fact that the term *liu-li*, in the same manner as in the Han period, denotes glazed pottery also at the present time. From the T'ang period onward, when porcelain came into vogue as a special class of ceramic ware, a division of nomenclature took place,— *liu-li* remaining reserved for common pottery, tiles, bricks, and other building-material, while a new term was adopted for a porcelain glaze. The porcelain enamel was styled *yu* 油 ("oil"), written also 釉 ,釉,[7] and 泑. As far as I know, this term is first applied by Liu Sün of the T'ang

---

[1] CHAVANNES, Pei Yuan Lou (*T'oung Pao*, 1904, p. 189). Green-glazed tiles were employed in the palace of the Sung Emperors, according to the *Yü t'ang kia huo* written by Wang Hui in 1360 (Ch. 4, p. 4 b; ed. of *Shou Shan ko ts'ung shu*).

[2] *P'ei wen yün fu*, Ch. 51, p. 79 b.

[3] Ruins of Desert Cathay, Vol. II, p. 252.

[4] Many remains of fine glazed pottery were found by Stein on his third expedition in the ruins of Karakhoto (A Third Journey of Exploration in Central Asia, p. 39, reprint from *Geographical Journal* for August and September, 1916). See also the same author's Ancient Khotan, Vol. I, pp. 442, 482.

[5] *T'u shu tsi ch'eng*, XXVII, Ch. 334.

[6] For further notes on this subject see HOBSON, Chinese Pottery and Porcelain, Vol. I, pp. 201 *et seq.*

[7] According to K'ang-hi's Dictionary, this character is first listed in the *Tsi yün* (middle of the eleventh century).

period, in his *Ling piao lu i*,[1] where the making of earthen cooking-kettles in the potteries of Kuang-tung is mentioned: "They were fired from clay and then glazed" (燒熱以土油之). A gloss explains *yu* as 釉. What is meant here is the application of porcelain glazes to earthenware. In ceramic literature the term *yu* refers exclusively to porcelain enamels.[2] It is quite certain also that in the present colloquial language glass is exclusively styled *p'o-li*, never *liu-li*, which strictly refers to glazed ware.

While we recognize that the Chinese received the stimulus for the production of ceramic glazes from western Asia, it must be emphasized at once that it was no more than a stimulus, and that the Chinese were not slavish imitators, but soon applied their own genius to the novel idea. The green glaze of the Han pottery, as analyzed by Mr. Nichols (p. 93), may have its analogies in the West, and a thorough search for corresponding materials would in all probability bring to light a Western recipe of the same composition. The first step to independence, however, is taken by the production of the porcelanous glaze of post-Han times (p. 90), which hardly offers any contemporaneous parallel in the West. From this time onward the Chinese have exercised their own acumen in perfecting the process of glazing and multiplying the scale of beautiful colors. FLINDERS PETRIE[3] has offered the ingenious suggestion that glaze in prehistoric Egypt, where it is found on quartz bases, was probably invented from finding quartz pebbles fluxed by wood ashes in a hot fire; hence glazing on quartz was the starting-point, and glazing on artificial wares was a later stage. Such observations of natural glazes may have also impressed and stimulated the Chinese. The Field Museum owns two earthenware crucibles, obtained by the writer in Si-ngan fu (Cat. Nos. 119076 and 119077), which by purely natural causes, owing to the infusion of molten metals, are colored a sky-blue with red flecks; likewise a melting-pot (Cat. No. 119347), artificially glazed in the interior and in the upper portion of the exterior, while the lower unglazed part has assumed natural colors of fiery-red and dark green from the effect of liquid metals. It is not impossible that this natural process of glazing inspired the imagination of the potters and gave the incentive for certain mottled ceramic glazes.

---

[1] Ch. A, p. 6 (ed. of *Wu ying tien*).

[2] JULIEN, Histoire et fabrication de la porcelaine chinoise, pp. 245, 247.

[3] Arts and Crafts of Ancient Egypt, p. 107.

THE POTTER'S WHEEL

When the clay is on the wheel the potter may shape it as he will, though the clay rejoins, 'Now you trample on me, one day I shall trample on you.'

Sir HERBERT RISLEY, *The People of India.*

Most of the phenomena of Chinese culture have hitherto been studied in splendid isolation. Sinologues have usually been content to gather their information from Chinese sources and to arrange it in chronological order, giving a more or less critical digest of the subject from the Chinese viewpoint; but the question as to what the phenomena actually mean is, as a rule, shunned, their interpretation hardly attempted. It is certainly impossible to grasp any phenomenon within a given culture-zone without understanding the parallel phenomena in other areas, and without setting them in correlation with their concomitant factors. The historical position and development of any cultural idea can be determined only by an attempt to unravel its causal connection with the natural group of related or associated ideas; for no phenomenon is isolated or absolute, but conditional upon others, relative, and cohesive. Whether this method be styled that of comparative ethnology or archæology, or that of culture-science, or something else, does not matter. It is there, and must be applied if we are eager to reach results. How it can be applied I wish to demonstrate by discussing on the following pages the nature of a simple instrument, — the potter's wheel. Its concatenation with other technical elements and with social and religious factors will be pointed out, and may help to show the history of pottery in a new light, and in particular to determine the relation of ancient Chinese ceramic art to that of the West. In a case like this one, the foundation of which reaches back into a prehistoric past, a purely historical method is of no avail, and will lead us nowhere. Thus HOBSON[1] observes, "Unfortunately, none of the [Chinese] writers can throw any light on the first use of the potter's wheel in China. It is true, that, like several other nations, the Chinese claim for themselves the invention of that essential implement, but there is no real evidence to illuminate the question, and even if the wheel was independently discovered in China, the priority

[1] Chinese Pottery and Porcelain, Vol. I, p. 2.

of invention undoubtedly rests with the Near Eastern nations." This indeed is all that from an historical point of view could be stated.

The making of pottery may well be called a universal phenomenon, despite the fact that there are many areas inhabited by peoples not acquainted with the art. It is unknown to the natives of Australia, New Zealand, and all other island groups of the South Sea populated by Polynesians[1] (while it thrives among the Melanesians), to the Negrito of the Philippines, to numerous primitive tribes of the Indo-Chinese,[2] to the inhabitants of the Himalaya (with the exception of the Nepalese), and to many nomadic and hunting tribes of Siberia.[3] It is further absent in the extreme southern parts of South Africa and South America, also in the whole north-western portion of North America. Among the polar peoples, pottery has hardly any importance. Of the Eskimo, only the western group in Alaska makes (or

---

[1] With the exception of Easter Island, where pottery is used for the cooking of certain foods (A. LESSON, Les Polynésiens, Vol. I, p. 457; Vol. II, p. 282). It is difficult to accept the oft-repeated statement that the Polynesians do not make pottery for want of proper clays in their habitats. There surely is workable clay in New Zealand and Hawaii; but whether there is or not, I believe with E. B. TYLOR (Primitive Culture, Vol. I, p. 57), that, "as the isolated possession of an art goes to prove its invention where it is found, so the absence of an art goes to prove that it was never present: the *onus probandi* is on the other side."

[2] Thus the Lo-lo have never produced pottery (A. F. LEGENDRE, Far West chinois, *T'oung Pao*, 1909, p. 611).

[3] It is particularly lacking among the present-day tribes of the Amur, also among the Gilyak and Ainu. Hü K'ang-tsung, who as Chinese ambassador in 1125 visited the Kin or Djurchi, observed that the latter made no vessels of clay, but only wooden cups and plates coated with a varnish (CHAVANNES, Voyageurs chinois, *Journal asiatique*, 1898, mai-juin, p. 395). The same observation still holds good for all Amur tribes, which during historical times appear never to have manufactured pottery. The Japanese traveller Mamiya Rinsō, who visited the island of Saghalin in 1808, reports that the forms of the clay vessels and porcelains of the Gilyak (Smerenkur) resemble Chinese and Japanese ware (P. F. v. SIEBOLD, Nippon, 2d ed., Vol. II, p. 233). The question is here of imported Chinese articles, and the observation is of no great consequence. Nevertheless L. v. SCHRENCK (Reisen und Forschungen im Amur-Lande, Vol. III, p. 448) has based an elaborate speculation on this passage, ascribing the manufacture of crockery and porcelain (!) to the Olcha and Gold on the Amur in the first part of the nineteenth century, and making the Manchu-Chinese Government responsible for the forcible destruction of this industry. This is a fantasy of the worst kind, for which no foundation exists in the history of the Amur tribes. What the Chinese colonists manufactured in Manchuria was only crude pottery; contrary to what is asserted by L. v. Schrenck, porcelain was never made there. The term "porcelain" used in Siebold's translation of Mamiya Rinsō's account with reference to a kiln in the village Kitsi, on the right bank of the Amur, as usual in such cases, rests on a mistranslation. It is of greater importance that the Japanese traveller tells us of earthen pots six to seven inches in diameter, with loop handles on both sides, made at his time by the Ainu of Saghalin. There is indeed reason to believe that the Ainu formerly made a rude and primitive kind of pottery. From the lips of an Ainu seventy years old, on the

rather made) lamps of clay, which ordinarily are turned out of soapstone, and cooking-pots.[1]

A. BYHAN[2] is disposed to assume that pottery is of foreign origin among the Eskimo. The Chukchi, according to BOGORAS,[3] have now forgotten this industry, but it never was more than a sporadic phenomenon among them. The Itelmen of Kamtchatka formerly manufactured clay vessels, chiefly lamps, as shown by finds in ancient pit-dwellings.[4] F. BOAS[5] is inclined to attribute the presence or absence of pottery to geographical location rather than to general cultural causes. Economic conditions have a certain bearing on the question. The production of clay vessels is dependent upon a sedentary mode of life. Pastoral tribes, as a rule, evince no inclination toward the industry, and deem utensils of bark, wood, or metal preferable. In Tibet, with its twofold population of agricultural and nomadic elements, we find the use of pottery only among the stationary settlers, never among the roaming shepherds. Even among the former it is an art introduced from China, as is evidenced by the few kilns in eastern Tibet which are operated by Chinese potters.[6]

The utilization of the potter's wheel is restricted to a well-defined geographical area. It occurs only in the Old World, and belongs to ancient Egypt, the Mediterranean and West-Asiatic civilizations, Iran, India, and China with her dependencies. It is germane to the higher stages of culture only, and is conspicuously lacking among all primitive tribes. In aboriginal American pottery the wheel was never employed.

---

northern Kuriles, TORII has recorded the story of how pots were previously made there, chiefly by women (*Mitteil. d. Ges. Ostasiens*, Vol. IX, 1903, p. 327). As is well known, the Ainu of Yezo have preserved no recollection of pottery-manufacture (J. BATCHELOR, The Ainu of Japan, p. 310), and also on Saghalin and the Kuriles the industry is now wiped out of existence. The prehistoric pottery found in the shell-heaps of Japan likewise must be attributed to the Ainu, who are thus to be classed among pottery-making peoples. See also p. 166, note 2.

[1] J. MURDOCH, Ethnological Results of the Point Barrow Expedition (*Ninth Report Bureau of Ethnology*, 1892, pp. 91–93).

[2] Polarvölker, p. 69.

[3] *Mem. Am. Mus. Nat. Hist.*, Vol. XI, p. 186.

[4] K. VON DITMAR, Reisen und Aufenthalt in Kamtschatka, pp. 246–247. As early as 1695, the first visitor to Kamtchatka, the Cossack W. Atlasov, reported that the inhabitants made wooden and earthen vessels (P. J. VON STRAHLENBERG, Nord- und Östliche Theil von Europa und Asia, p. 435).

[5] The Mind of Primitive Man, p. 183.

[6] W. W. ROCKHILL, in a note to his edition of Sarat Chandra Das' Journey to Lhasa (p. 88), states that, though he never saw the making of pottery in Tibet, he knows that no wheel is used; which is perfectly correct, inasmuch as it is never handled by Tibetans. F. GRENARD (Le Tibet, p. 286) observes, "Pottery is of indigenous manufacture, but the Chinese wheel is utilized."

Our foremost authority on this subject, W. H. HOLMES,[1] makes this observation: "It is now well established that the wheel or lathe was unknown in America, and no substitute for it capable of assisting materially in throwing the form or giving symmetry to the outline by purely mechanical means had been devised. The hand is the true prototype of the wheel as well as of other shaping tools, but the earliest artificial revolving device probably consisted of a shallow basket or bit of gourd in which the clay vessel was commenced and by means of which it was turned back and forth with one hand as the building went on with the other." Of course, if further on (p. 69) Holmes styles the basket used as a support in modelling a clay vessel "an incipient form of the wheel," this is only a figure of speech, for this device bears no relation whatever to the wheel. This remark holds good also for "that simple approximation to a potter's wheel, consisting of a stick grasped in the hand by the middle and turned round inside a wall of clay formed by the other hand," evolved for North America by Squier and Davis,[2] and the "natural primitive potter's wheel," consisting of a roundish pebble, ascribed to the New-Caledonians by O. T. MASON[3] after J. J. Atkinson. Wherever wheel-turned pottery has been found in America on aboriginal sites, it has conclusively been proved either that it is of European manufacture, or that the wheel was introduced there by the white man. Thus it has been disclosed that the wheel-made jars, showing also traces of a brownish glaze, which were reported from Florida and other Southern States, and occasionally were even recovered from Indian mounds, are of Spanish manufacture, having been used in early Colonial times for the shipping of olives to America.[4] The Quichua employ for the making of pottery a very simple lathe, which is justly traced to European influence by E. NORDENSKIÖLD.[5] It is worthy of note also that the distribution of the wheel over the area mentioned has remained almost stationary for millenniums, and that primitive tribes are not susceptible to adopting it, even if surrounded by civilized peoples who make use of it. The Vedda of Ceylon, for instance, fashion pots by hand,[6] while the surrounding Singalese avail themselves of the wheel. Nothing of the

---

[1] Aboriginal Pottery of the Eastern United States, p. 50 (*Twentieth Ann. Rep. Bureau Am. Ethnology*, Washington, 1903).

[2] See J. LUBBOCK, Prehistoric Times (5th ed.), p. 260.

[3] Origins of Invention, p. 161.

[4] HOLMES, *l. c.*, pp. 129–130.

[5] Einige Beiträge zur Kenntnis der südamerikanischen Tongefässe und ihrer Herstellung (Stockholm, 1906).

[6] C. G. SELIGMANN, The Veddas, p. 324.

character of a potter's wheel is known among the inhabitants of the Andaman group.[1] Or, to cite another example, the Negroes of Africa have always remained unacquainted with the wheel, though they might have learned its use from the ancient Egyptians, or at a later time from the Arabs. The sporadic occurrence of the wheel in the Malayan Archipelago indicates its introduction from outside. It is found only in Padang Lawas on Sumatra and on Java;[2] while in all other Malayan regions, including the Philippines, pottery has remained in the stage of handwork, and is the lot of woman. The Yakut, the most intelligent and progressive people of Siberia, never avail themselves of the potter's wheel, nor do they know of any process of glazing vessels. Despite the fact that they intermarry with the Russians, and that on the market of Yakutsk wheel-made Russian crockery is offered for sale, they still adhere to their primitive mode of fashioning vessels solely by hand, the only implement that is used being a half-round or round smooth stone, with which the interior of the pot is shaped and smoothed. Instead of securing Russian ware, they prefer to purchase the raw clay material (at from five to ten kopeck a pound), and entrust it to a skilful woman potter, together with fragments of old broken pots, which are pounded and mixed with the fresh clay. According to SÄROSHEVSKI,[3] to whom we owe a detailed description of the process, also the illustration of a Yakut potter at work, these products come very near to those of the stone age. In their crude technique, they form a curious contrast to the excellent iron-forged work and wood-carving for which the same people are reputed.

While ethnologists have clearly recognized that the pottery-making of primitive peoples is essentially a woman's avocation, it has not yet been sufficiently emphasized that the wheel is a man-made invention, and that, aside from the mere technical difference of the hand and wheel processes, there is a fundamental sociological contrast between the two. Among the Indian tribes of America, the fictile art was woman's occupation, and such it is at present. In discussing the methods of primitive pottery, O. T. MASON[4] observes, "It will be noted that the feminine gender is used throughout in speaking of aboriginal potters. This is because every piece of such ware is the work of woman's hands. She quarried the clay, and, like a patient beast of burden, bore it home on her back. She washed it and kneaded it, and rolled

---

[1] E. H. MAN, On the Aboriginal Inhabitants of the Andaman Islands, p. 154.

[2] Encyclopædie van Nederlandsch-Indië, Vol. III, pp. 321, 322.

[3] The Yakut (in Russian), Vol. I, p. 378.

[4] Origins of Invention, p. 166; see also his Woman's Share in Primitive Culture, p. 91.

it into fillets. These she wound carefully and symmetrically until the vessel was built up. She further decorated and burned it, and wore it out in household drudgery. The art at first was woman's." As regards Africa, we owe a very able investigation to H. SCHURTZ,[1] whose studies of African conditions prompted him to the conclusion that pottery everywhere appears to be an invention of woman, who was more urgently in need of boiling water in the preparation of vegetable food than man in dressing his hunting-spoils. A map constructed by Schurtz, and illustrating the distribution of pottery over Africa, shows at a glance that the largest territory is occupied by female potters; that male potters occur only in Abyssinia, among the Galla and Somali in eastern Africa, and this owing to Arabic influence. In a few other areas men are engaged in the making of the bowls for their cherished tobacco-pipes, while the women produce from clay all domestic and kitchen utensils; and in a few localities only, men and women co-operate in the ceramic industry. In regard to the Khasi in Assam, Major GURDON[2] observes, "The women fashion the pots by hand, they do not use the potter's wheel." On the Nicobars the men take no part in the construction of pots.[3] All over Melanesia, pottery is made exclusively by women. The making of clay vessels is no longer practised by the Chukchi, but their old women (not the men) have a vivid recollection of the clay kettles which were used in former times.[4]

The potter's wheel, however, is the creation of man, and therefore is an independent act of invention which was not evolved from any contrivance utilized during the period of hand-made ceramic ware. The two processes have grown out of two radically distinct spheres of human activity. The wheel, so to speak, came from another world. It had no point of contact with any tool that existed in the old industry, but was brought in from an outside quarter as a novel affair, when

---

[1] Das afrikanische Gewerbe, pp. 13–19.

[2] The Khasis, p. 61.

[3] C. B. KLOSS, In the Andamans and Nicobars, p. 107. According to E. H. MAN (On the Aboriginal Inhabitants of the Andaman Islands, p. 154), the manufacture of pots on the Andamans is not confined to any particular class, or to either sex, but the better specimens are generally produced by men. Compare the same author's Nicobar Pottery (*J. Anthr. Institute*, Vol. XXIII, 1894, pp. 21–27). Also among the Vedda pots are turned out by both men and women (C. G. SELIGMANN, The Veddas, p. 324).

[4] W. BOGORAS, in *Mem. Am. Mus. Nat. Hist.*, Vol. XI, p. 186. The industry of primitive pottery is fast dying out everywhere under the influence of "civilization" (compare, for instance, M. R. HARRINGTON, Catawba Potters and Their Work, in *Am. Anthr.*, Vol. X, 1908, pp. 399–407; and The Last of the Iroquois Potters, in *N. Y. State Mus. Bull.*, 1909, pp. 222–227; as to Africa, see O. BAUMANN, *Globus*, Vol. LXXX, 1901, p. 127).

man appropriated to himself the work hitherto cultivated by woman. The development was one from outside, not from within. All efforts, accordingly, which view the subject solely from the technological angle, and try to derive the wheel from previous devices of the female potter, are futile and misleading.[1] It is as erroneous as tracing the plough back to the hoe or digging-stick, whereas in fact the two are in no historical interrelation, and belong to fundamentally different culture strata and periods,— the hoe to the gardening activity of woman, the plough to the agricultural activity of man. Both in India and China, the division of ceramic labor sets apart the thrower or wheel-potter, and distinctly separates him from the moulder. The potters of India, who work on the wheel, do not intermarry with those who use a mould or make images.[2] They form a caste by themselves.[3] In ancient China, a net discrimination was made between wheel-potters (*t'ao jen* 陶人) and moulders (*fang jen* 瓬人).[4] This clear distinction is accentuated also by Chu Yen 朱琰 in his Treatise

---

[1] E. J. BANKS (Terra-Cotta Vases from Bismya, *Am. Journ. Sem. Langs.*, Vol. XXII, 1905–06, p. 140) has this observation on the making of Babylonian pottery: "From the study of Bismya pottery it is evident that a wheel was employed at every period, yet all of the vases were not turned. No. 43, a form reconstructed from several fragments from the lowest strata of the temple hill, and which therefore dates several millenniums before 4500 B.C., has the appearance of having been formed by placing the clay upon a flat surface, and while the potter shaped it with one hand, he turned the board or flat stone, whatever it was upon which it rested, with the other. This was probably the origin of the potter's wheel; it was but a matter of time when an arrangement was attached to the board that it might be turned with the feet." All this is purely speculative and fantastic, and has no value for the real history of the wheel.

[2] A. BAINES, Ethnography (Castes and Tribes) of India, p. 65.

[3] The social position of the Indian potter is differently described by various authors. H. COMPTON (Indian Life in Town and Country, p. 65) observes that the potter in India is an artist; that he is an hereditary village officer, and receives certain very comfortable fees; that his position is respected; that he enjoys the privilege of beating the drum at merry-makings, that he shares with the barber a useful and lucrative place in the community; and that there is probably no member of it who is happier in his lot, and less liable to the vicissitudes of fortune. H. RISLEY (People of India, p. 130) gives us a bit of Indian popular thought regarding the potter: "He lives penuriously, and his own domestic crockery consists of broken pots. He is a stupid fellow — in a deserted village even a potter is a scribe — and his wife is a meddlesome fool, who is depicted as burning herself, like a Hindu wife, on the carcase of the Dhobi's donkey." According to G. C. M. BIRDWOOD (Industrial Arts of India, Vol. II, p. 146), the potter is one of the most useful and respected members of the community, and in the happy religious organization of Hindu village life there is no man happier than the hereditary potter. The truth probably lies in the midway between these two extreme appreciations. As to ancient times, compare the Buddhist story of the sage potter, translated by E. LANG (*Journal asiatique*, 1912, mai-juin, p. 530).

[4] E. BIOT, Tcheou-li, Vol. II, pp. 537–539.

on Pottery.[1] He justly observes also that the articles made by the wheel-potters were all intended for cooking, with the exception of the vessel *yü* 庾, which was designed for measuring; while the output of the moulders, who made the ceremonial vessels *kuei* 簋 and *tou* 豆 by availing themselves of the plumb-line, was intended for sacrificial use. Also here, in like manner as in ancient Rome, India, and Japan, the idea may have prevailed that a wheel-made jar is of a less sacred character than one made by hand.

Wherever the potter's wheel is in use, it is manipulated by man, never by woman.[2] It is man's invention, it is man's sphere of work. As implied by its very name, it is directly derived from a chariot-wheel, which is likewise due to man's efforts. Such a real cart-wheel with four spokes is still operated by the Tamil potters. It is well illustrated by E. THURSTON,[3] and thus described after E. HOLDER (Fig. 1): "The potter's implements are few, and his mode of working is very simple. The wheel, a clumsily constructed and defective apparatus, is composed of several thin pliable pieces of wood or bamboo, bent and tied together in the form of a wheel about three feet and a half in diameter. This is covered over thickly with clay mixed with goat's hair or any fibrous substance. The four spokes and the centre on which the vessel rests are of wood. The pivot is of hard wood or steel. The support for the wheel consists of a rounded mass of clay and goat's hair in which is embedded a piece of hard wood or stone, with one or two slight depressions for the axle or pivot to move in. The wheel is set into motion first by the hand, and then spun rapidly by the aid of a long piece of bamboo, one end of which fits into a slight depression in the wheel. The defects in the apparatus are, firstly, its size, which requires the potter to stoop over it in an uneasy attitude; secondly, the irregularity of its speed, with a tendency to come to a standstill, and to wave or wobble in its motion; and, thirdly, the time and labor expended in spinning the wheel afresh every time its speed begins to

---

[1] *T'ao shuo* 陶說, Ch. 2, p. 2 (new edition, 1912). Compare S. W. BUSHELL, Description of Chinese Pottery, p. 33.

[2] Woman working on the potter's wheel is a strictly modern artificial reform of our "civilization," which tends to check the "man-made world," with the result that it insures woman's industrial enslavement to perfection. MARY WHITE (How to make Pottery, p. 28) observes, "Until lately, few women potters have worked on the wheel, because the ordinary form of potter's wheel, which was turned with one foot, the potter standing on the other, made the work too difficult and laborious for a woman to attempt. Now, however, a wheel copied from an old French model is in use, which enables the potter to sit while at work."

[3] Castes and Tribes of Southern India, Vol. IV, p. 190. Holder's article is in *Journal of Indian Art*, No. 58, being accompanied by excellent illustrations of potter's wheels and of potters working at the wheel.

slacken. Notwithstanding, however, the rudeness of this machine, the potters are expert at throwing, and some of their small wares are thin and delicate." It should be added, that, as may be seen in the illustration (Fig. 1), the wheel is but slightly above the ground, and that the potter stands bent over the vessel. The apparatus, described by E. A. GAIT[1] for the kilns of Assam, has likewise features in common with the cart-wheel. While the centre consists of a solid disk of tamarind or some other hard wood, about thirteen inches in diameter, there is an outer rim joined to it by means of four wooden spokes, each of these being about six inches in length. The outer rim, about six inches wide, is made of split bamboo, bound with cane, and covered with a thick plaster of clay mixed with fibres of the sago palm. The object of this rim is to increase the weight of the wheel, and thereby add to its momentum.[2] In Assamese as well as in Bengālī, the potter's wheel is simply called *cak* ("wheel," from Sanskrit *cakra*).

In the Çatapatha Brāhmaṇa (XI, 8) the potter's wheel (*kaulālacakra; kulāla*, "potter;" *cakra*, "wheel") is thus alluded to in close connection with the cart-wheel: "Verily, even as this cart-wheel, or a potter's wheel, would creak if not steadied, so, indeed, were these worlds unfirm and unsteadied."[3] A similar association of ideas occurs in the Chinese philosopher Huai-nan-tse, who died in 122 B.C. He compares the activity of Heaven as the creative power with the revolutions of a wheel by saying, "The wheel of the potter revolves, the wheel of the chariot turns; when their circle is completed, they repeat their revolution."[4] In the porcelain-factories of King-te-chen, the potter's wheel is styled *t'ao ch'ê* 陶車 (that is, "potter's chariot") or *lun ch'ê* 輪車 (that is, "wheeled chariot"). Ordinarily the potter speaks simply of his "wheel" (*lun-tse* 輪子). An engraving of about 1540 shows an Italian potter's table in the shape of a regular six-spoked wheel.[5] Technically speaking, the potter's wheel is nothing

---

[1] The Manufacture of Pottery in Assam (*Journal of Indian Art*, Vol. VII, 1897, p. 6).

[2] The Assam potters do not finish their pieces on the wheel, but when taken down and sun-dried, they are placed in a hollow mould of wood or earthenware, in which they assume their final shape by being beaten with a flat wooden or earthenware mallet, held in the right hand, against a smooth, oval-shaped stone held by the left hand against the inner surface. When the required shape has been given the vessel, it is again sun-dried, the surface being then polished with an earthenware pestle or a rag.

[3] J. EGGELING's translation in *Sacred Books of the East*, Vol. XLIV, p. 126. The exact date of this work is not known, but it is believed that it goes back to the sixth century B.C.

[4] CHAVANNES, Mémoires historiques de Se-ma Ts'ien, Vol. V, p. 27.

[5] Encyclopædia Britannica, Vol. V, p. 706.

Fig. 1
INDIAN POTTER'S WHEEL IN THE SHAPE OF A CART-WHEEL
(Sketch after Holder, *Journal of Indian Art*)

but a primitive cart-wheel turning on its axle. The invention presupposes the existence of the wheel adapted to transportation, and in all the great civilizations in which, as stated above, the potter's wheel is found, we indeed meet also the wheeled cart. We further observe, that, wherever the potter's wheel occurs and the wheeled cart does not occur, the former was introduced from a higher culture-zone: for instance, in Japan, to which the conception of the cart is foreign, and which received the potter's wheel from Korea; or among the Tibetans, who have no wheeled vehicles, and in the midst of whom the potter's wheel is only handled by Chinese.[1] Again, the wheeled cart is conspicuously absent in all those culture-areas in which, as has been stated, the potter's wheel is unknown. Wherever original conditions have remained intact and undisturbed by outside currents, the two implements either co-exist, or do not exist at all. Of course, it must not be understood that the idea of the potter's wheel was conceived in a haphazard manner, as though a wheel, intentionally or incidentally, had been detached from a cart, its novel utilization being reasoned out on speculative and technical grounds. Primitive man, and man of the prehistoric past, is not a rationalistic or utilitarian being, but one endowed with thoughts of highly emotional character, and prompted to peculiar associations of ideas that are inspired by religious sentiments. Of the theories which have been expounded in regard to the primeval origin of the wheel, none as yet is wholly satisfactory; but this much is assured, that it was connected with a certain form of religious worship, that in its origin the chariot was utilized in the cult before it was turned to practical purposes of transportation.[2] The symbolism and worship of the wheel in western Asia, prehistoric Europe and India, is so well known that this matter does not require recapitulation. A similar spirit pervades the early references to the potter, his work and his wheel. In the Old Testament the potter's control over the clay illustrates the sovereignty of God, who made man of clay, and formed him according to his will. "O house of Israel, cannot I do with you as this potter? saith the Lord. Behold, as the clay is in the potter's hand, so are ye in my hand, saith the Lord" (JEREMIAH XVIII. 1–6). "Shall the thing formed say to him that formed it, Why hast thou made me thus? Hath not the potter power

---

[1] The wheeled cart is designated in Tibetan *shing rta* ("wooden horse"),— a word-formation which testifies to the fact that the cart is foreign to Tibetan culture. In fact, carts are not employed by Tibetans. We only read in ancient records of vehicles for the use of kings, presumably introduced from India.

[2] E. HAHN, Alter der wirtschaftlichen Kultur, p. 123; and Entstehung der Pflugkultur, p. 40.

over the clay, of the same lump to make one vessel unto honour and another unto dishonour?" (ROMANS IX. 20, 21.) In ancient Egypt, the god Phtah fashions the egg of the world on a potter's wheel, setting it in motion with his feet.[1] According to W. CROOKE,[2] the potter of India regards the making of his vessels as a semi-religious art. The wheel he worships as a type of the creator of all things; and when he fires his kiln, he makes an offering and a prayer. He also makes the funeral jar, in which the soul of the dead man for a time takes refuge. Hence he is a sort of funeral priest, and in some parts of the country receives regular fees. It was a current notion in ancient China that the evolution of Heaven creates the beings in the same manner as the potter turns his objects of clay on the wheel. The potter's wheel was a symbol of the creative power of nature. In the ancient writers in whose works this conception looms up it appears as a purely philosophical abstraction; but it is obvious that the latter goes back to a genuine mythological idea, which, like everything mythical in China, is lost,—the naïve conception of the creator as a potter and thrower (as in the Old Testament). The potter's wheel was used also as a simile with reference to the activity of the sovereign. Yen Shi-ku, in his commentary on the Han Annals, quotes a saying that "the holy rulers by virtue of their regulations managed the empire in the same manner as a potter turns the wheel." It is therefore not impossible that religious speculations, centring around the cart-wheel and the fashioning of clay vessels and figures, might have had a prominent share in associating the wheel with the potter's activity, and given the first impetus to "throwing." If it can be maintained that the ancient Egyptians were the first to employ the potter's wheel, it may well be that the invention is due to the circle of the priests. Be this germ idea as it may, the culture-historical position of the potter's wheel is well ascertained. In view of the vast periods of human prehistory, it is a comparatively late invention, following in time the construction of the wheeled cart, being based on the cart-wheel, and made by man (presumably first by priests in illustration of a myth for religious worship) during the stage of fully-developed agriculture.

In the stage of hoe-culture or gardening, the occupation of woman, the potter's wheel is absent. Wherever it appears, it is correlated with man's activity in agriculture, based on the employment of the ox and plough. This feature is illustrated by both ancient China and India. The Emperor, or more correctly culture-hero, Shun (alleged 2258–2206 B.C.), in his youth, before he assumed charge of the administration of

---

[1] E. A. W. BUDGE, Gods of the Egyptians, Vol. I, p. 500, with colored plate.

[2] Things Indian, p. 389.

the empire, is said to have practised husbandry, fishing, and making pottery jars: he fashioned clay vessels on the bank of the River, and all these were without flaw.[1] The philosopher Mong-tse explained this act by saying that Shun continually tried to learn from others and to take example from his fellowmen in the practice of virtue.[2] Another tradition crops out in the *Ki chung Chou shu*:[3] here the incipient work in clay is attributed to the culture-hero Shen-nung, who, as implied by his name ("Divine Husbandman"), was regarded as the father of agriculture and discoverer of the healing-properties of plants. In this ancient lore we meet a close association of agriculture with pottery, and an illustration of the fact that husbandman and potter were one and the same person during the primeval period.

Likewise in ancient India the potter's trade was localized in special villages, either suburban or ancillary to large cities, or themselves forming centres of traffic with surrounding villages.[4] Thus it is the case at the present day. When the writer, in 1908, passed through Calcutta and desired to see a Hindu potter at work, he was obliged to drive several miles out of the city into a neighboring village. In fact, the potter is a peasant, and attends to his field during the rainy season, when he is unable to pursue his craft; he must have dry weather to harden his pots before they are fired.[5] According to Sir A. BAINES,[6] the potter is one of the recognized village staff, and, in return for his customary share in the harvest, is bound to furnish the earthenware vessels required for domestic use. His caste is associated with the donkey, the saddle-animal of the Goddess of Small-Pox; and his donkey, when the kiln is not in operation, is employed in carrying grain and other produce. In most parts of the country the potters sometimes hold land, and in others take service in large households.

Likewise in ancient China the potter lived in close contact with the farmer, and received from him cereals in exchange for his products.[7]

---

[1] CHAVANNES, Mémoires historiques de Se-ma Ts'ien, Vol. I, pp. 72, 74; compare BIOT, Tcheou-li, Vol. II, p. 462. See also *Shi ki*, Ch. 128, p. 5, where the commentary cites the *Shi pen* to the effect that Kun-wu (this volume, p. 39) made pottery.

[2] LEGGE, Chinese Classics, Vol. II, p. 206.

[3] CHAVANNES, *l. c.*, Vol. V, p. 457.

[4] R. FICK, Die sociale Gliederung im nordöstlichen Indien, pp. 179, 181. Mrs. RHYS DAVIDS, Notes on Early Economic Conditions in Northern India (*Journ. Roy. As. Soc.*, 1901, p. 864).

[5] W. CROOKE, Natives of Northern India, p. 135.

[6] Ethnography (Castes and Tribes), p. 65 (Strassburg, 1912; *Encyclopædia of Indo-Aryan Research*).

[7] According to MONG-TSE, III, I, § 4 (LEGGE, Chinese Classics, Vol. II, p. 248).

The farmer was in urgent need of these articles, which were in large demand; for "a single potter would not do in a country of ten thousand families, and could not supply their wants," and "with but few potters a kingdom cannot subsist."[1]

The potter's particular residence is naturally determined by the sites of suitable clay, and his dependence on clay-digging excludes him from towns and cities. Thus A. K. COOMARASWAMY[2] observes, "The Singalese potters are found all over the country in every village affording the necessary clay, but often aggregated in greater numbers in places where an especially good supply of suitable clay is available. Thence the potter carries his pots for sale to more remote districts in huge pingo loads." The same holds good for China: all kilns are located in the country, and the potters supplying the wants of the villages and towns are farmers themselves.

The modifications brought about in the industry by the application of the wheel were fundamental and far-reaching. Technically they led to a greater rapidity and hence intensity of the process, but, above all, to many new features of form, consigning many others to oblivion. Likewise they resulted in a regularity, symmetry, harmony, and grace of shape, in a refinement and perfection unattained heretofore. The potter's art came in close touch and was set in correlation with other man-made industries, particularly with that of the bronze-founder, who furnished the potter with new ideas of forms and designs.[3] The birth of artistic pottery was thus inaugurated. In passing from the hands of woman into those of man, the whole industry was imbued with a more active and vigorous spirit, and elevated to a higher plane by man's creative genius. It overstepped the narrow boundary of purely domestic necessity and developed into an organized system of carefully-planned and skilfully-directed manufacture on a large scale and with a wide scope. The ceramic work turned out by woman depended on local conditions, and catered to the narrow circle of the

---

[1] MONG-TSE, VI, 2, §§ 3 and 6 (*ibid.*, p. 442).

[2] Mediæval Sinhalese Art, p. 218.

[3] W. HOUGH (Man and Metals, *Proceedings of the National Academy of Sciences*, Vol. II, 1916, p. 125) justly insists on the intimate connection of clay and metal working. The activity of the ancient sovereigns of China is likened not only to that of the potter, but also to that of the founder. Potter and founder 陶冶 are frequently mentioned together (for instance, by Mong-tse: LEGGE, Chinese Classics, Vol. II, p. 248). The correlation of the mortuary pottery of the Han with corresponding types in bronze has been shown by me in detail. The same phenomenon occurs in the prehistoric ceramic art of central Europe, where imported Roman bronze vessels were imitated and reproduced in clay (see particularly A. Voss, Nachahmungen von Metallgefässen in der prähistorischen Keramik, *Verh. Berl. Anthr. Ges.*, Vol. XXXIII, 1901, pp. 277–284).

home community. The widened horizon of man led him to search for clays and other materials in distant localities, and to trade his finished product over the established routes of commerce in exchange for other goods. It was due to the introduction of the wheel that ceramic labor was afforded the opportunity of growing out of a mere communal, clannish, or tribal industry into a national and international factor of economic value.[1]

In the suburbs and villages around Peking, where pottery is manufactured, two kinds of wheel are in use. The two specimens illustrated on Plates XI and XII were secured near Peking by the writer in 1903, and are in the American Museum of Natural History, New York. The one is made of a hat-shaped mass of clay, which is hardened by the addition of pig's hair and straw. This wheel is employed for turning out circular vessels of small and medium sizes, and may be regarded as the common, typical wheel used throughout northern China. The other wheel consists of a weighty stone disk made in the great industrial centre, the town Huai-lu in Shan-si Province. It serves for the making of round and heavy vessels of large dimensions.[2] A round wooden board is placed on the stone disk as support or table on which the mass of clay is shaped. The difference between the clay and stone wheels, accordingly, is one of degree only, not of type; indeed, they represent the same type, and are identical in their mechanical construction. Both wheels revolve on a wooden vertical axis, the lower extremity of which is fixed into a pit, so that the upper surface of the disk lies on the same level as the floor of the shed in which the potter works. The latter squats on the ground in front of the wheel, and sets it in motion by means of a wooden stick, which is inserted in a shallow cavity near the periphery of the stone disk. While the disk continues to twirl, a lump of clay is thrown upon it and worked by the potter with both of his hands: he vigorously presses his thumbs downward, shaping the bottom of the jar, then draws them upward, and it seems as though by magic the walls of the vessel come running out of his

---

[1] With reference to the La-Tène period, these changes are well characterized by H. SCHMIDT in his excellent article Keramik, in the Reallexikon der germanischen Altertumskunde, edited by J. Hoops (Vol. III, p. 36).

[2] Aside from China, stone wheels seem to occur in India, but only occasionally (H. H. COLE, Catalogue of the Objects of Indian Art in the South Kensington Museum, p. 201). H. R. C. DOBBS (*Journal of Indian Art*, No. 57, p. 3) remarks that in the north-west provinces of India wheels are made either of clay, or stone, or wood, but most commonly of clay. The difference is merely one of durability: a clay wheel lasts about five years and can be made in four days without cost to the potter; a wooden wheel lasts for about ten years, being made by a local carpenter for Rs. 1–8; a stone wheel will last a lifetime, and is usually brought from Mirzapur or Indore at an average cost of Rs. 4.

fingers. The procedure is exactly identical with the practice of the ancients, as described by H. BLÜMNER.[1] I never saw a Chinese potter spinning the wheel with his left hand and simultaneously forming a pot only with his right. He will always swing his wheel first, and then use both hands for fashioning the vessel. This point is particularly mentioned, because several authors tell us that the potter at the same time works the wheel with his left hand and fashions the clay with his right. Thus A. ERMAN[2] says, with reference to ancient Egypt, that the wheel was turned by the left hand, whilst the right hand shaped the vessel. The same is asserted with regard to the potter on Sumatra.[3] If these observations should be correct, which may justly be doubted, the potters who behave in this manner can hardly be credited with common sense. If the wheel is once set spinning, a constant revolution of sufficient velocity may very well be maintained for from five to seven minutes, which would afford ample time for a skilful workman to turn out one or even several vessels by the use of both hands. There is no necessity whatever for his left hand to operate the wheel, and how the right hand alone could satisfactorily model a pot is difficult to see. In China, Japan, and India, at all events, the potter will always use both hands in this process; or he has a helpmate to attend to the wheel.

In his description of the porcelain-manufacture at King-te-chen, Père D'ENTRECOLLES has alluded to the employment of the wheel, without, however, going deeper into the subject.[4] In the *King te chen t'ao lu*,[5] the wheel is described as a round wooden board, with a mechanism below, that effects a speedy revolution. The potter is seated over the wheel (literally, "he sits on the chariot" 拉者於車上), pushing it with a small bamboo stick, and moulding the clay with both of his hands. The illustrations reproduced by Julien after the first edition of 1815 (Plates V and VI[6]) show the potter squatting at the end of two low benches, steadying his feet on the latter; but the mode of turning the wheel is represented in a different manner from the description in the text. In one illustration the potter avails himself of an assistant, who bends over a bench, and sets the wheel in motion with his left hand. In the other, the helpmate turns the wheel with his

---

[1] Technologie, Vol. II, p. 39.

[2] Life in Ancient Egypt, p. 457.

[3] Encyclopædie van Nederlandsch-Indië, Vol. III, p. 321.

[4] DUHALDE, Description of the Empire of China, Vol. I, p. 342; or S. W. BUSHELL, Description of Chinese Pottery, pp. 190–191.

[5] Ch. 1, p. 18 b (new edition of 1891); compare JULIEN, Histoire, p. 146.

[6] Those of the new edition are different, and much coarser in execution.

right unshod foot, while supporting himself by means of a rope suspended from the branch of a tree. The wheel itself is a cog-wheel, the projecting teeth being of a rectangular shape.[1] The foot of the turner fits exactly into the space left by two teeth. This arrangement is identical with that of the small lead cylinders fixed around a Roman wheel of baked clay found near Arezzo in 1840, and the pegs attached to the circumference of other wheels discovered in the vicinity of Nancy.[2]

The devices depicted in this Chinese book are obviously those of central and southern China. This is confirmed by an observation of E. S. MORSE, who had occasion to see and to sketch a potter at work near Canton, and who points out the same rope contrivance. "The wheel rests on the ground, and the potter squats beside the wheel. A helper stands near by, steadying himself with a rope that hangs down from a frame above; holding on to this and resting on one foot, he kicks the wheel around with the other foot. The potter first puts sand on the wheel, so that the clay adheres slightly. He does not separate the pot from the wheel by means of a string, as is usual with most potters the world over, but lifts it from the wheel, the separation being easy on account of the sand previously applied. The pot is somewhat deformed by this act, but is straightened afterwards with a spatula and the hand, as was the practice of a Hindu potter whom I saw at Singapore."[3]

Besides the plain wheel, as considered heretofore, another type occurs in China,—a wheel with double disks. In this case, there are two horizontal, parallel disks or wheels connected by a vertical spindle. The lower one, being of considerably smaller diameter, is operated by the feet of the workman, and accordingly turns the upper one, which is reserved as the potter's table. A similar device is described by Jesus Sirach in the third century B.C.[4] The same principle is brought out in a potter's wheel found by Fabroni in 1779 at Cincelli or Centum Cellæ, in the neighborhood of Arezzo, in Italy. It is composed of two disks or tables, both placed horizontally, of unequal diameter, having a certain distance between them, and their centre traversed by a vertical pin, which revolves. The wheel discovered was part of one

---

[1] It is doubtless on this illustration that E. ZIMMERMANN'S (Chinesisches Porzellan, Vol. I, p. 179) description of the potter's wheel is based; but I do not believe that this type is common, at least I never saw it in any of the kilns which I had occasion to visit.

[2] H. BLÜMNER, Technologie, Vol. II, p. 39.

[3] E. S. MORSE, Glimpses of China and Chinese Homes, p. 199.

[4] BLÜMNER, *l. c.*, p. 38, note 3.

of the disks, made of terra cotta, about three inches thick and eleven feet in diameter, with a groove all round the border.[1]

A double wooden wheel is occasionally employed by the potters in the north-west provinces of India and Oudh, but, curiously enough, the upper disk is the smaller one. It is about ten inches in diameter, and on it the clay is worked. The lower disk, two feet apart from the upper one, measures two feet across. The whole apparatus is placed in a pit about three feet deep, the smaller disk being on a level with the surface of the ground. The axle turns on a stone slab at the bottom of the pit, and is kept upright by a crossbeam with a perforation in the middle, through which it runs. The potter is seated on the edge of the pit, and turns the wheel by pressing the lower disk with his right foot. The motion of this wheel is more even and continuous than that of the single wheel, and is employed for the finer kinds of pottery at Rāmpur and Mīrut.[2]

The double wheel is used also in Java, where it is called *prebot*. It is composed of two wooden disks, one placed above the other, the upper one, of somewhat larger size, being revolved on the lower one. The upper one is styled "female board" (*uncher wedok*), the lower one "male board" (*uncher lanang*). The upper wheel, on which is placed a flat board for the clay to be moulded, is set in motion by means of the foot.[3]

F. BRINKLEY[4] describes the contrivance of a double wheel in the hands of the potters at Arita in Hizen. It consists of a driving and a working wheel, fixed about twelve to fifteen inches apart on a hollow wooden prism. On the lower side of the driving-wheel is a porcelain cup that rests on a vertical wooden pivot projecting from a round block of wood over which the system is placed. The pivot is planted in a hole of such depth that the rim of the driving-wheel is slightly raised above the surface of the ground. Beside this hole the modeller sits, and, while turning the system with his foot, moulds a mass of material placed on the working-wheel. His only tools are a piece of wet cloth to smooth and moisten the vessel, a small knife to shape sharp edges, a few pieces of stick to take measurements, and a fine cord to sever the finished vase from its base of superfluous matter.

Sir ERNEST SATOW,[5] describing the work of the potters of Tsuboya, observes that these use wheels of three different sizes. The smallest

---

[1] S. BIRCH, History of Ancient Pottery, p. 556.

[2] H. R. C. DOBBS, Pottery and Glass Industries of the North-West Provinces and Oudh (*Journal of Indian Art*, No. 57, p. 4).

[3] Encyclopædie van Nederlandsch-Indië, Vol. III, p. 322.

[4] Japan, Vol. VIII, p. 68.

[5] Korean Potters in Satsuma (*Transactions As. Soc. of Japan*, Vol. VI, 1878, p. 196).

is formed by two wooden disks about three inches thick, the upper one being fifteen inches, the lower eighteen inches, in diameter, connected by four perpendicular bars somewhat over seven inches long. It is poised on the top of a spindle planted in a hole of sufficient depth, which passes through a hole in the lower disk, and enters a socket in the under side of the upper disk; and the potter, sitting on the edge of the hole, turns the wheel round with his left foot. The largest wheel is about twice the size of the smallest in every way. This description fits very well the illustration of a potter's wheel in the *T'u shu tsi ch'eng* (see Fig. 2), except that the two wheels are here connected by two vertical bars, and that the whole apparatus is above ground, so that the potter is obliged to stand.

Although the real study of Korean pottery remains to be made,[1] the general development of the art in its main features can be clearly traced. We may distinguish four principal periods,— first, a prehistoric or neolithic period prior to the cultural contact of Korea with China, during which primitive vessels without the application of the wheel were turned out, that represent a uniform group with the prehistoric pottery found in the Amur region, Manchuria, Saghalin, and Japan;[2] second, the period of the Silla kingdom (57–924) heralded by the introduction of Chinese culture, in the wake of which the forms of the ancient Chinese sacrificial vessels as well as dishes for every-day use and the potter's wheel made their appearance; third, the Korai period (925–1392), centring around Song-do, where glazed pottery, also porcelain, was produced according to models and traditions of Chinese Sung ware; and, fourth, the modern period after 1392. Here we are concerned only with the second or the first historic period, which is characterized by the novel feature of the wheel and by new and elegant shapes based on Chinese prototypes. We have authentic records in

[1] Compare in particular A. BILLEQUIN, Notes sur la porcelaine de Corée (*T'oung Pao*, Vol. VII, 1896, pp. 39–46); E. S. MORSE, Catalogue of the Morse Collection of Japanese Pottery, pp. 25–31, and the study of P. L. JOUY, quoted below; J. PLATT, Ancient Korean Tomb Wares (*Burlington Mag.*, Vol. XX, No. 106, 1912, pp. 222–230, 2 plates); PETRUCCI, Korean Pottery (*ibid.*, 1912, p. 82, 2 plates), and letter of J. PLATT (*ibid.*, 1913, p. 298); A. FISCHER, *Oriental. Archiv*, Vol. I, 1911, pp. 154–157, plate XXXIV).

[2] As to the Amur region, a great quantity of pottery fragments was dug up by G. Fowke in 1898 (compare his report Exploration of the Lower Amur Valley, *Am. Anthr.*, Vol. VIII, 1906, pp. 276–297); this collection is in the American Museum of Natural History, New York. The Japanese archæologist TORII found similar material in eastern Mongolia and Manchuria (*Journ. of the College of Science*, Tōkyō, Vol. XXXVI, No. 4, pp. 49 *et seq.*, and No. 8 of the same volume, pp. 9, 30–41, 62–64, 71, and plates XIV–XVIII, XXIII). Neolithic Korean pottery is described by Shōzaburi Yagi (*Journ. Anthr. Soc. of Tōkyō*, Vol. XXX, 1915, p. 178).

Fig. 2

CHINESE DOUBLE-WHEEL POTTER'S LATHE

(Sketch after *T'u shu tsi ch'eng*)

regard to the adoption of the latter on the part of the Koreans;[1] and as the greater part of the pottery of this period is turned on the wheel,[2] while that of the preceding ages was fashioned only by hand, it is safe to assume that the introduction of the wheel is due to Chinese influence.

P. L. JOUY writes on the Korean potter's wheel as follows: "The Korean potter's wheel consists of a circular table from two to three feet in diameter and four to six inches thick, made of heavy wood so as to aid in giving impetus to it when revolving. In general appearance it is not very unlike a modeller's table. This arrangement is sunken into a depression in the ground, and revolves easily by means of small wheels working on a track underneath, the table being pivoted in the centre. The wheel is operated directly by the foot, without the aid of a treadle of any kind. The potter sits squatting in front of the wheel, his bench or seat on a level with it, and space being left between his seat and the wheel to facilitate his movements. With his left foot underneath him, he extends his right foot, and strikes the side of the wheel with the bare sole of the foot, causing it to revolve."[3]

A Japanese tradition credits the celebrated Korean monk Gyōgi 行基 (A.D. 670–749)[4] with the invention of the potter's wheel. W. G. ASTON,[5] W. GOWLAND,[6] and F. BRINKLEY[7] have rejected this legend as unfounded by pointing out that the wheel was known in Japan

---

[1] *Hou Han shu*, Ch. 115, and the writer's Chinese Pottery, p. 127. The Wo-tsü in Korea interred in the graves pottery vessels filled with rice. In this respect the Chinese account is of interest, that all the Eastern barbarous tribes, Tung I 東夷 availed themselves of dishes and platters (*tsu tou* 俎豆) for eating and drinking, with the sole exception of the Yi-lou or Su-shen (*T'ai p'ing huan yü ki*, Ch. 175, p. 4 b). See also *Kiu T'ang shu*, Ch. 199 A, p. 1.

[2] P. L. JOUY, The Collection of Korean Mortuary Pottery (*Report of the U. S. National Museum*, 1887–88, pp. 589–596, particularly p. 591).

[3] *Science*, Vol. XII, 1888, p. 144. Mrs. BISHOP (Korea and Her Neighbours, Vol. I, p. 93) says, "The potters pursue their trade in open sheds, digging up the clay close by. The stock-in-trade is a pit in which an uncouth potter's wheel revolves, the base of which is turned by the feet of a man who sits on the edge of the hole. A wooden spatula, a mason's wooden trowel, a curved stick, and a piece of rough rag, are the tools, efficient for the purpose." A Korean drawing showing a potter at work is reproduced in *Int. Archiv. f. Ethnogr.*, Vol. IV, 1891, plate III, fig. 6.

[4] His life is briefly summed up by E. PAPINOT, Dictionnaire de géographie et d'histoire du Japon, p. 152. J. J. REIN (Industries of Japan, p. 457) states only that Gyōgi was the first to introduce the wheel into Japan, which may well be the original tradition, and that this event took place in A.D. 724.

[5] Nihongi, Vol. I, p. 121.

[6] The Dolmen and Burial Mounds in Japan, p. 494.

[7] Japan, Vol. VIII: Keramic Art, p. 9.

long before his time.[1] Of course, Gyōgi is not the "inventor" of the wheel, any more than Anacharsis the Scythian, or Hyperbius of Corinth, or Talus, the nephew of Daedalus. Nevertheless it may be that Gyōgi, who, being a craftsman, was doubtless instrumental in the advancement of the ceramic industry in Japan, brought the specimen of a wheel along on his mission; and, if nothing else, this tradition would at least point to an introduction of the wheel from Korea. This is the natural course of events that we should expect, for the prehistoric pottery of Japan was solely made by hand.[2] The early historic pottery found in the dolmens is wheel-shaped; but whether, with Gowland, it is to be dated in the beginning of our era, is a debatable point. E. S. MORSE[3] has offered another kind of convincing testimony for the fact that the early Japanese potter modelled by hand: the ancient practice is still continued in its prehistoric form in various parts of the empire, where many potters use only the hand in making bowls, dishes, or teapots. The vessels employed as offerings at Shintō shrines are usually made without the wheel, and are unglazed,— a phenomenon that we likewise meet in ancient Rome and in ancient India.

According to Morse, the typical form of the potter's wheel in Japan consists of a wooden disk fifteen to eighteen inches in diameter, and three inches thick. This is fastened to a hollow axis fourteen or more inches in length. A spindle with pointed end is planted firmly in the ground; and on this the wheel is placed, the spindle passing up through the hollow axis, and a porcelain saucer or cup being inserted in the wheel to lessen friction as it rests on the spindle. The wheel itself is on a level with the floor; and the potter, sitting in the usual Japanese position, bends over the wheel, which he revolves by inserting a slender stick in a shallow hole or depression near the periphery of the wheel. With a few vigorous motions of his arm the wheel is set in rapid motion; then, with his elbows braced against his knees, the whole body at rest, he has the steadiest command of the clay he is to turn. As the wheel slackens in motion, he again sets it twirling.[4]

---

[1] I am unable, however, to admit Aston's statement that the text of the Nihongi to which he refers contains evidence of this fact. This evidence is negative or inconclusive, as the text in question speaks only of hand-made (*ta-kujiri*) small jars, which, according to Aston, should lead to the conclusion that "this was exceptional," and that fashioning on the wheel was the common practice of the time. In A.D. 588 the first potters came to Japan from the Korean state Pektsi (ASTON, *l. c.*, p. 117).

[2] E. S. MORSE, Shell Mounds of Omori, p. 9; IIJIMA and SASAKI, Okadaira Shell Mound at Hitachi, pp. 2–5; N. G. MUNRO, Prehistoric Japan, p. 167.

[3] Catalogue of the Morse Collection of Japanese Pottery, p. 6.

[4] Illustrations of the implements used by the Japanese brick-layer and potter may be seen in SIEBOLD, Nippon, Vol. VI, plate IV.

The wheel is termed *rokuro* 轆轤 (Chinese *lu-lu*), which properly means a pulley, windlass, capstan, then further a turning-lathe. The Japanese double wheel has been pointed out (above on p. 165).

If it is correct that the potter's art came to Burma from China rather than from India, and that glazing was acquired there from the Chinese either directly or through the medium of the Shan,[1] it is probable also that the wheel reached Burma from the same centre. In the town of Bassein the double wheel is in use.[2] In like manner it is probable that also the Annamese, who learned the entire process of porcelain-manufacture from their conquerors, the Chinese, adopted the wheel from the latter.[3] The invasion of the outskirts of Tibet through Chinese potters working on the wheel has already been mentioned. They use a plain wooden wheel sunk into the ground, and work it with the foot. China, consequently, was the centre from which the art of wheel-made pottery radiated to all other countries of the East, in accordance with the diffusion of Chinese culture among the same peoples.

The great antiquity of the wheel in China cannot reasonably be doubted. As has been stated, it is alluded to in early writers of the pre-Christian era, and appears to have played a part in mythological conceptions. It is designated by a plain root-word, *kün* 均 or 鈞, which means also "even, level, harmonious." It was the instrument by means of which clay vessels were evenly balanced; it was a sort of "harmonizer." A description of the ancient wheel has apparently not come down to us. A commentator of Se-ma Ts'ien's Annals notes that it was seven feet high and provided with a plumb-line for adjusting the vessels.[4] From Biot's translation of the *Chou li*[5] it would seem as if the wheel were mentioned in that work, for we read, "Tout vase d'usage ordinaire doit être conforme au tour. . . Le tour est haut de quatre pieds. En carré, il a quatre dixièmes de pied." A potter's wheel of course is round, and everybody will be struck by the anomaly that the wheel should be four-tenths of a foot square. In fact, the text does not speak of a wheel, but of an instrument manipulated by the moulders. The passage runs thus: 器中膞膞崇四尺方四寸.

---

[1] Gazetteer of Upper Burma and the Shan States, Part I, Vol. II, pp. 399, 403. In support of this deduction, the fact is cited, that, in proportion to the population, there are more potters' villages in the Shan states than in Burma, and that in many places, notably in Papun, the potters are emigrant Shan.

[2] *L. c.*, p. 400.

[3] A. DE POUVOURVILLE, L'Art indo-chinois, p. 238.

[4] *P'ei wen yün fu*, Ch. 51, p. 77.

[5] Vol. II, p. 539.

The word *po*, as far as I know, occurs only in this text as a potter's term. The commentator Ch'en Yung-chi 陳用之 explains it as "sliced meat" (切肉), saying that the potter's products should be like the latter, that is, as thin and smooth; and that the object of rendering a vessel equally thick and smooth is attained by the application of the instrument *po*, which accordingly may have been a lathe. Cheng Ngo 鄭鍔, another commentator of the *Chou li*, remarks that it was of wood and placed on the side of the potter's wheel (*kün* 鈞), but his further description is not very lucid. At all events, the instrument in question was not, as conceived by Biot, a potter's wheel, which in fact is not mentioned in the text of the *Chou li*.

Almost all the round jars and vases of the Han period have been shaped on the wheel; and these ancient potters exercised considerable skill in its use.[1] The profession of the throwers is emphasized in the ritual of the Chou dynasty (*Chou li*), and distinguished from that of the moulders. Moreover, we now have well-authenticated specimens of pottery of that period, which likewise exhibit the marks of the wheel. A truly neolithic, primitive, hand-made pottery, such as we have from Japan and Korea, has now also been traced in Chinese soil, particularly in southern Manchuria, Liao-tung, and Shen-si. I am inclined to date the use of the wheel in China back to a very remote age. The chief reason which prompts me to this conclusion is, that ancient Chinese records contain no traditions to the effect that pottery was ever the office of woman; on the contrary, they associate the industry exclusively with the activity of man, and these potters were agriculturists. The only ancient industry characterized as a female occupation is that of the rearing of silkworms and weaving. The "invention" of pottery, however, is ascribed to the mythical emperors Huang-ti, Shen-nung, and Shun; and throughout Chinese history we hear only of male potters. In fact, as we observe also at the present time, woman has no share whatever in this business. The potter's wheel, therefore, cannot be simply regarded as borrowed by the Chinese from the West in historical times, but it belongs to those primary elements of culture which the Chinese have in common with certain ancient forms of Western civilization. In our present state of knowledge, it is futile to endeavor to explain the how and why of this interrelation. There can be no doubt, however, that the ancient Chinese wheel has sprung from the same

---

[1] This is also the opinion of so prominent an expert in pottery as J. BRINCKMANN, the late director of the Hamburg Museum für Kunst und Gewerbe, who has written an excellent, though brief, article on Han pottery, especially with reference to its technique (*Jahrbuch der Hamburgischen Wissensch. Anstalten*, Vol. XXVII, 1909, pp. 96–102).

source as that found in the West. Both are identical as to mechanical construction, even in minor points, and as to effect.

A comparatively great antiquity of the potter's wheel may be assumed also for India. Allusion has been made to the early mention of it in the Çatapatha Brāhmaṇa (p. 157). The jar employed for the ritual, as described by Kātyāyana,[1] was solely formed by hand after the fashion of coiled pottery. This does not prove that the wheel was not in use at that time, for jars serving religious purposes were made by hand likewise in Rome and Japan, even after the introduction of the wheel. The case merely goes to show that hand-made ware preceded the wheel-made fabric also in ancient India, and that the concept of a fundamental difference between the two was maintained, the hand-made product being reserved for religious worship.

The potter's wheel is twice mentioned in the Jātaka.[2] In one story it is told how a Bodhisatva went to the king's potter and became his apprentice. One day, after he had filled the house with potter's clay, he asked if he should make some vessels; and when the potter answered, "Yes, do so," he placed a lump of clay on the wheel and turned it. When once it was turned, it went on swiftly till mid-day. After moulding all manners of vessels, great and small, he began making one especially for Pabhāvatī with various figures on it. The potter's work is a favorite simile in Buddhist scriptures.[3]

In this respect the following story is of particular interest: "In the town of Revata, in the north-west of India, there lived a master-potter, who prided himself on his dexterity. He was waiting for the objects which he manufactured to dry on the wheel, and only at this moment he withdrew them. Knowing that the time of his conversion had arrived, Bhagavat (Buddha) transformed himself into a master-potter, and, chatting with the other potter, asked him why he did not withdraw from the wheel the plates and utensils. The potter replied that he would do so, when they were perfectly dry. The Buddha transformed into a man said, 'Also I withdraw them, when they are perfectly dry. You and I follow the same procedure. I, however, have a special method. I withdraw the objects only after they are completely baked on the wheel.' The master-potter retorted, 'You

---

[1] A. HILLEBRANDT, Ritual-Lit., Vedische Opfer, p. 8; L. D. BARNETT, Antiquities of India, p. 176.

[2] Nos. 531 and 546 (COWELL and ROUSE, The Jātaka, Vol. V, p. 151; Vol. VI, p. 188).

[3] For instance, Dīghanikāya, II, 86 (R. O. FRANKE's translation, p. 79); T. SUZUKI, Açvaghosha's Discourse on the Awakening of Faith, pp. 74, 75.

are more skilful than I am.' The Buddha transformed into a man said, 'Not only do I produce on the wheel objects completely baked, but also I can produce objects formed with the seven precious substances.' The master-potter's eyes were opened: he immediately received faith, and was converted. Thereupon Bhagavat, who had transformed himself temporarily into a potter, reassumed his proper body. He expounded the supernatural and subtle law, so that the potter's family was initiated into the four cardinal truths."[1]

In southern India, wheel-made pottery came into general use during the iron age.[2]

The cart-wheel in the hands of the Indian potter has been referred to. This, however, is an exceptional local type, while commonly the wheel is a plain wooden disk. G. C. M. BIRDWOOD[3] describes it as a horizontal fly-wheel, two or three feet in diameter, loaded heavily with clay around the rim, and put in motion by the hand; and, once set spinning, it revolves for five or seven minutes with a perfectly steady and true motion. The clay to be moulded is heaped on the centre of the wheel, and the potter squats down on the ground before it. The Tamil potters (Kusavans) are divided into two classes, northern and southern; the former using a wheel of earthenware, the latter one made of wood.[4] Their badge, recorded at Conjīveram, is a potter's wheel.[5] The Singalese wheel (*pōruva*) is a circular board, about two feet and a half in diameter, mounted on a stone pivot, which fits into a larger stone socket embedded in the ground; the horizontal surface of the wheel itself standing not more than six inches above the ground. The wheel is turned by a boy, who squats on the ground opposite the potter, and keeps it going with his hands.[6]

Ceramic art is very ancient in Iran, being alluded to in two passages of the Avesta.[7] In the latter, mention is made of brick-layer's

---

[1] J. PRZYLUSKI, Le Nord-ouest de l'Inde dans le Vinaya des Mūla-Sarvāstivādin et les textes apparentés (*Journal asiatique*, 1914, nov.-déc., pp. 513, 514).

[2] R. B. FOOTE, Gov. Museum, Madras, Cat. of the Prehistoric Antiquities, p. 111. In regard to South-Indian pottery compare also R. B. FOOTE, The Foote Collection of Indian Prehistoric and Protohistoric Antiquities (Madras, 1914; new ed., 1916); and A. REA, Cat. of the Prehistoric Antiquities from Adichanallur and Perumbair (Madras, 1915). F. W. v. BISSING (*Sitzber. Bayer. Akad.*, 1911, p. 16) seems to overvalue the antiquity of the potter's wheel in southern India; it is certainly out of the question that it should be older there than in Egypt.

[3] The Industrial Arts of India, Vol. II, p. 144.

[4] E. THURSTON, Castes and Tribes of Southern India, Vol. IV, p. 113.

[5] *Ibid.*, p. 197.

[6] A. K. COOMARASWAMY, Mediæval Sinhalese Art, p. 219.

[7] Vidēvdāt, II, 32; VIII, 84.

or potter's kilns.[1] As a rule, the kiln is the natural consequence of the wheel; but it would be premature to conclude from this general observation that for this reason the wheel was known to the Avestans. It is not specifically mentioned in their sacred books; but that it was unknown cannot be deduced, either, from this silence.

The question of the antiquity of the potter's wheel in Babylonia seems not to be settled. PERROT and CHIPIEZ[2] remark that the invention of the potter's wheel and firing-oven must have taken place at a very remote period both in Egypt and Chaldæa; that the oldest vases found in the country, those taken from tombs at Warka and Mugheir, have been burnt in the oven; that some, however, do not seem to have been thrown on the wheel. All that HANDCOCK[3] states regarding the wheel is a reference to the article of Banks, whose theory of the origin of the wheel has already been characterized as unfounded (p. 154). In Palestine the wheel became general from the sixteenth century B.C. Likewise the Israelites were familiar with it, and turned almost all their vessels on the wheel.[4] As has been mentioned, it is alluded to in several passages of the Old and New Testaments.[5]

In the graves of the Siberian bronze age has been found pottery of inferior workmanship, made by hand, of a coarse and badly baked clay. That from the graves of the iron age appears to be wheel-shaped, and abounds in artistic shapes.[6] Its historical position is not yet exactly ascertained, but it appears to bear some relation to Scythian and Iranian cultures.

In ancient Egypt the wheel was known at the earliest epoch of history the sculptures of which have been preserved.[7] It is depicted on the monuments, being of simple construction and turned with the hand.

---

[1] See also W. GEIGER, Ostiranische Kultur, p. 390; and A. V. W. JACKSON, From Constantinople to the Home of Omar Khayyam, p. 234. The Avestan word for the kiln, *tanura* (Middle and New Persian *tanūr)* is regarded as a loan from Semitic *tanūr.*

[2] History of Art in Chaldæa and Assyria, Vol. II, p. 298.

[3] Mesopotamian Archæology, p. 334.

[4] F. VIGOUROUX, Dictionnaire de la Bible, Vol. V, pp. 573–574; S. BIRCH, Ancient Pottery, p. 107. A photograph from Damascus of a potter at the wheel is reproduced in the *National Geogr. Mag.*, 1911, p. 67.

[5] Regarding the use of the wheel in Asia Minor, see W. BELCK, *Z. f. Ethnologie,* Vol. XXXIII, 1901, p. 493.

[6] W. RADLOFF, Aus Sibirien, Vol. II, pp. 89, 90, 129.

[7] J. G. WILKINSON, Manners and Customs of the Ancient Egyptians, Vol. II, pp. 190–192 (new ed., by S. Birch), or 2d ed., Vol. III, p. 163.

It is plausible that the invention spread from Egypt or Crete to Greece, and from there to Italy.[1]

The gradual dissemination of the wheel over Europe is vividly illustrated by the fact that in every culture-area there we encounter a primitive epoch of pottery-making, which shows no trace of the wheel, but a rude hand-made process. Such is found in the earliest stages of Hissarlik, the Homeric Troy, in Italy, central and northern Europe, and in the British Isles. During the second settlement of pre-Mycenæan Hissarlik (presumably before 2000 B.C.) we observe the beginning of the use of the wheel and the covered furnace. Throughout the Mycenæan period, pottery was turned on the wheel. The Swiss lake-dwellers, though capable potters, were unacquainted with the wheel. Likewise it was unknown in the British Isles during the bronze period.[2] In the north of Europe, the potter's wheel appears at a late date in the La-Tène period. Thus the assumption gains ground that Egypt was the centre from which the wheel gradually spread to southern, and ultimately to central and northern, Europe.

In two areas of the Old World, accordingly, we can clearly observe a diffusion of the wheel from one point,— from China to her dependencies Korea, Japan, Annam, and Burma; and from Egypt to Europe. India was perhaps another focus, as far as Sumatra and Java are concerned. A direct transmission of the device from Egypt to India is conceivable, though it is of course impossible to furnish the exact proof. It is inconceivable, however, that the wheels of India and China should be independent from those of the West. Not only is there a perfect coincidence between their constructions and manipulations, but also the culture-associations by which the wheel is surrounded here and there are strikingly identical. The social setting of the wheel and the concomitant culture-elements have been characterized above. The wheeled cart, the highly-developed system of agriculture, bronze casting, and the affiliation of pottery with the latter, are features peculiar to the same area, and absent in other culture-zones. Consequently the presence of the wheel in the East and West alike cannot be attributed to an accident, but it appears as an organic constituent and ancient

---

[1] Regarding details, see H. BLÜMNER, Technologie, Vol. II, pp. 36–40; O. SCHRADER, Reallexikon, p. 868; etc. H. B. WALTERS (Cat. of the Greek and Etruscan Vases in the British Museum, Vol. II, p. 228) describes the medallion of a kylix on which a potter, nude and beardless, is seated before a wheel; on it is a kylix of archaic shape, the handle of which he is moulding. The question as to whether the wheel was employed in Crete at an earlier date than in Egypt, or *vice versa*, must be left to the decision of specialists in this field.

[2] J. EVANS, Ancient Bronze Implements of Great Britain, p. 487; British Museum Guide to the Antiquities of the Bronze Age, p. 43.

heritage in the life of the Mediterranean and great Asiatic civilizations. This well-defined geographical distribution, and the absence of the wheel in all other parts of the globe, speak well in favor of a monistic origin of the device.

The chief results of the present investigation may be summarized as follows. The industry of ancient Chinese pottery, in its principal technical and social features, has exactly the same foundation as the corresponding industry of western Asia, Egypt, and India. This phenomenon is only one of a complex of others with which it is in organic cohesion; that is, the entire economic foundation of ancient Chinese civilization has a common basis with that of the West.[1] It is a reasonable conclusion that identity of apparatus and technical processes must have yielded similar results. Comparative study of forms, however, is futile for the present, as long as we do not have the very earliest prehistoric ceramic productions of China, Central Asia, Iran, and India. This much is evident, that only by co-ordination can the real problem to be pursued be solved, and that isolation or detachment of each particular field will yield no result that is worth while. The incentive for the process of glazing pottery was received by the Chinese directly from the West, owing to their contact with the Hellenistic world in comparatively late historical times. The knowledge of glazing rendered the manufacture of a porcelanous ware possible; yet in this achievement the creative genius of the Chinese was not guided by outside influence, but relied on its own powerful resources. Nothing of the character of porcelain was known under the Han (206 B.C.–A.D. 220). The murrine vases of the ancients were not porcelain, and in fact bear no relation to China. They may have been instrumental, however, in bringing to the notice of the Chinese the beauty and effect of ceramic glazes; hence the manufacture of glazed ware springs up in the age of the Han, more particularly under the reign of the Emperor Wu (140–87 B.C.). It is admissible to place the first subconscious gropings with ware of more or less porcelanous character in the closing days of the Later Han dynasty; and under the Wei, in the middle or latter part of the third century, we see these tentative experiments ultimately crowned with success. Continued till the end of the sixth century and the beginning of the seventh through a long line of experiences and improvements, they gradually resulted in the

---

[1] The details are somewhat more developed in the writer's popular article Some Fundamental Ideas of Chinese Culture (*Journal of Race Development*, Vol. V, 1914, pp. 160–174).

production of a true white porcelain. Porcelain is not an invention, and there is no inventor of it. It is not in a category by itself, but is only a variety of pottery; its diversity from common pottery is one of degree, not of principle.

Finally, the question may be raised as to why Chinese records on all these points are so sparse and unsatisfactory. The same observation holds good for bronze, iron, wood-carving, basketry, and other ancient industries and crafts. The occupation with such themes on the part of Chinese scholars begins as late as the age of the Sung. The ancient professional annalists and chroniclers were not interested in the doings and thoughts of the broad masses of the people. If they recorded with some degree of exactness the invention of rag-paper in A.D. 105, it was for the reason that paper had a direct bearing on the life and work of the scholar. The plain farmer-potter of old led a secluded existence, far removed from the seats of scholarship. The average type of Confucian scholar never took an interest in technical questions, or else looked down upon these without a gleam of understanding. Our hopes for further elucidations of the problems connected with the history of pottery in China must be placed in archæology, not in sinology, which certainly reflects not on the sinologue, but on the character of the scanty source-material that has fallen to our lot.

INDEX

PLATE I.

HAN PORCELANOUS POTTERY (see p. 79).

Small jug. The yellowish-green, vitrified porcelanous glaze covers only the medial portion of the body, inclusive of the two ears or loop handles. The exterior of the neck and the base are unglazed. In the base, nail-marks are left. The bottom is flat and without a rim. The clay appears to contain iron ore. Found on top of a cast-iron stove (Plate II), in a grave near the village Ma-kia-chai, 5 *li* north of the town Hien-yang, Shen-si Province.

Middle or end of the third century A.D

Height, 16.7 cm. Cat. No. 118718.

HAN PORCELANOUS JUG.

PLATE II.

CAST-IRON STOVE (see p. 80).
Side and front views.

In type and style it exactly corresponds to the Han pottery burial cooking-stoves. Posed on four feet in the form of elephant-heads, it is built in the shape of a horse-shoe, and provided with a chimney, five cooking-holes, and a projecting platform in front of the fire-chamber. On the latter is cast an inscription consisting of six raised characters in Han style of writing, reading *ta ki ch'ang i hou wang* ("Great felicity! May it be serviceable to the lords!"); see p. 79. The iron core is entirely decomposed, so that for exhibition purposes the object had to be braced on wooden supports. Found in a grave near the village Ma-kia-chai, 5 *li* north of the town Hien-yang, Shen-si Province. Inserted here as collateral evidence in determining the provenience and date of the pottery jug illustrated in Plate I.

End of Han period (A.D. 220), or, generally, third century A.D.

Height, 35 cm; length, 71.5 cm; width, 40.5 cm. Cat. No. 120985.

CAST-IRON STOVE.

PLATE III.

HAN PORCELANOUS POTTERY.

Small jug. The interior of the neck is glazed in its upper part. Only the upper portion of the body is coated with a thick, lustrous, porcelanous glaze of greenish-yellow tinge, interspersed with small white dots, the glaze running down in streaks over the lower unglazed part. This is the best-glazed piece in the lot. Two rounded ears or loop handles are attached to the shoulders.

Middle or latter part of third century A.D.

Height, 20.1 cm. Cat. No. 118723.

HAN PORCELANOUS JUG.

PLATE IV.

HAN PORCELANOUS POTTERY.

Large globular vase of harmonious proportions, decorated with two opposite animal (tiger)-heads in flat relief, holding dead rings, of the same style as in common Han pottery. In the middle between these heads, but somewhat higher, and opposite each other, are two semi-circular loop handles stuck on to the body of the vessel, obviously for the passage of a cord, by means of which the vase was held and carried. Each handle is bordered by two knotted bands moulded separately in high relief. This feature,— that is, the combination of loop handles with tiger-heads,— to my knowledge, does not occur in ordinary Han pottery. The slip appears to have been lost in part of the neck. The glaze exhibits various tinges of light green, mingled with the deep brown of the slip, and interspersed with black spots, the brown approaching that of maple-leaves in the autumn. The red-brown slip covers one side of the neck and almost the entire base; in the middle portion the porcelanous glaze appears to be laid over this slip. Three bands, each consisting of three concentric grooves, in the same manner as in Han pottery, are laid around the body. The bottom is flat, and has along the rim a broad grayish ring of irregular form and depth. The walls of the vessel are unusually thick, and its weight is almost six pounds.

Third century A.D.

Height, 35.4 cm. Cat. No. 118720.

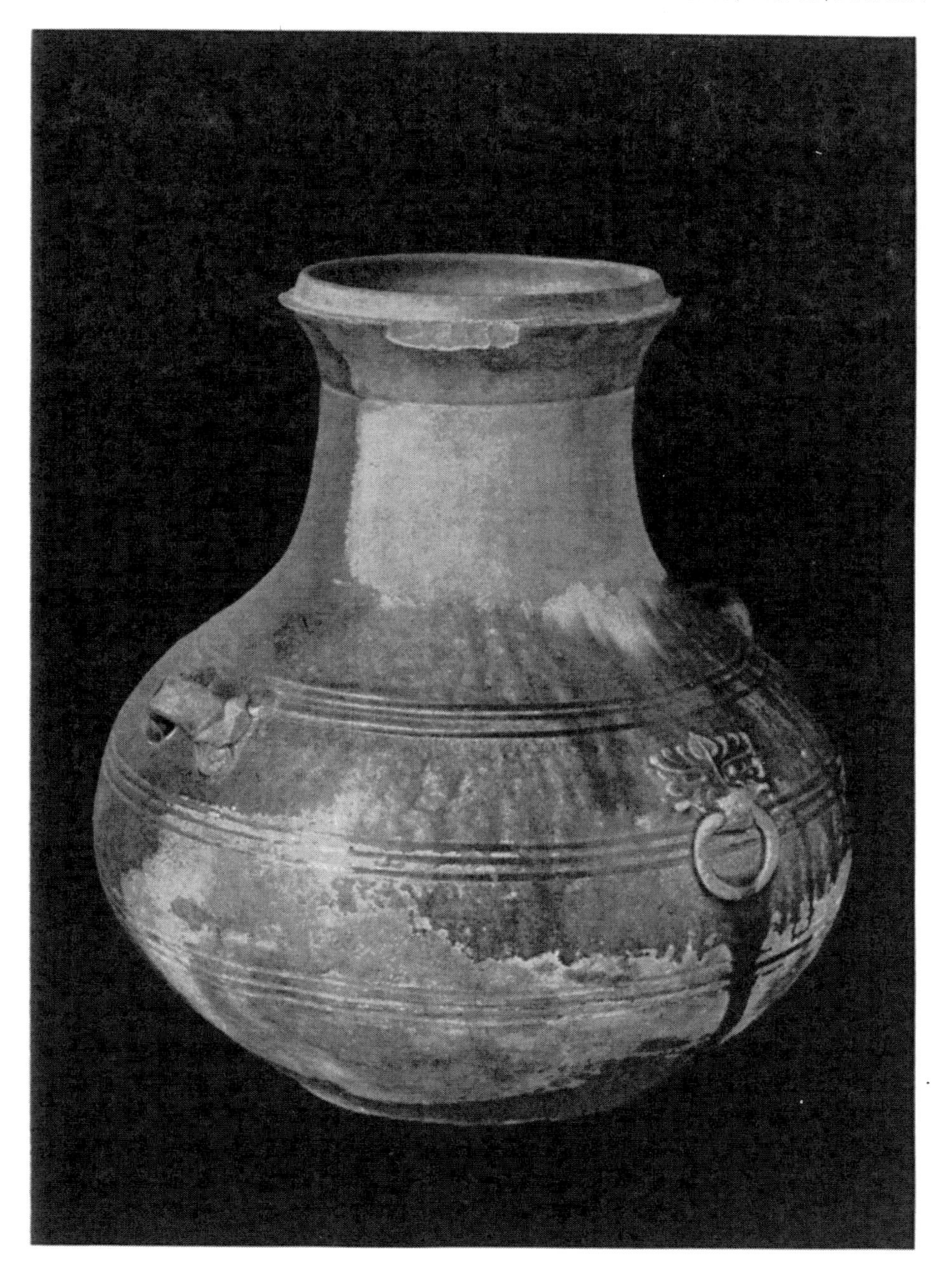

HAN PORCELANOUS VASE.

PLATE V.

HAN PORCELANOUS POTTERY.

Small jar, now unglazed, but originally glazed in its middle portion; when found, covered all over with masses of earth, the glaze having been destroyed by chemical influences under ground, and a white engobe being left in its place. A wave-band, each consisting of five lines, presumably done by means of a roller, runs around the upper rim and the neck. A double knot in low relief is stamped above the loop handles, which terminate in a flat ring filled with incised, radiating lines, apparently the reproduction in clay of a metal ring. The bottom is raised on a rim, about 1 cm high.

Third century A.D.

Height, 21.2 cm. Cat. No. 118717.

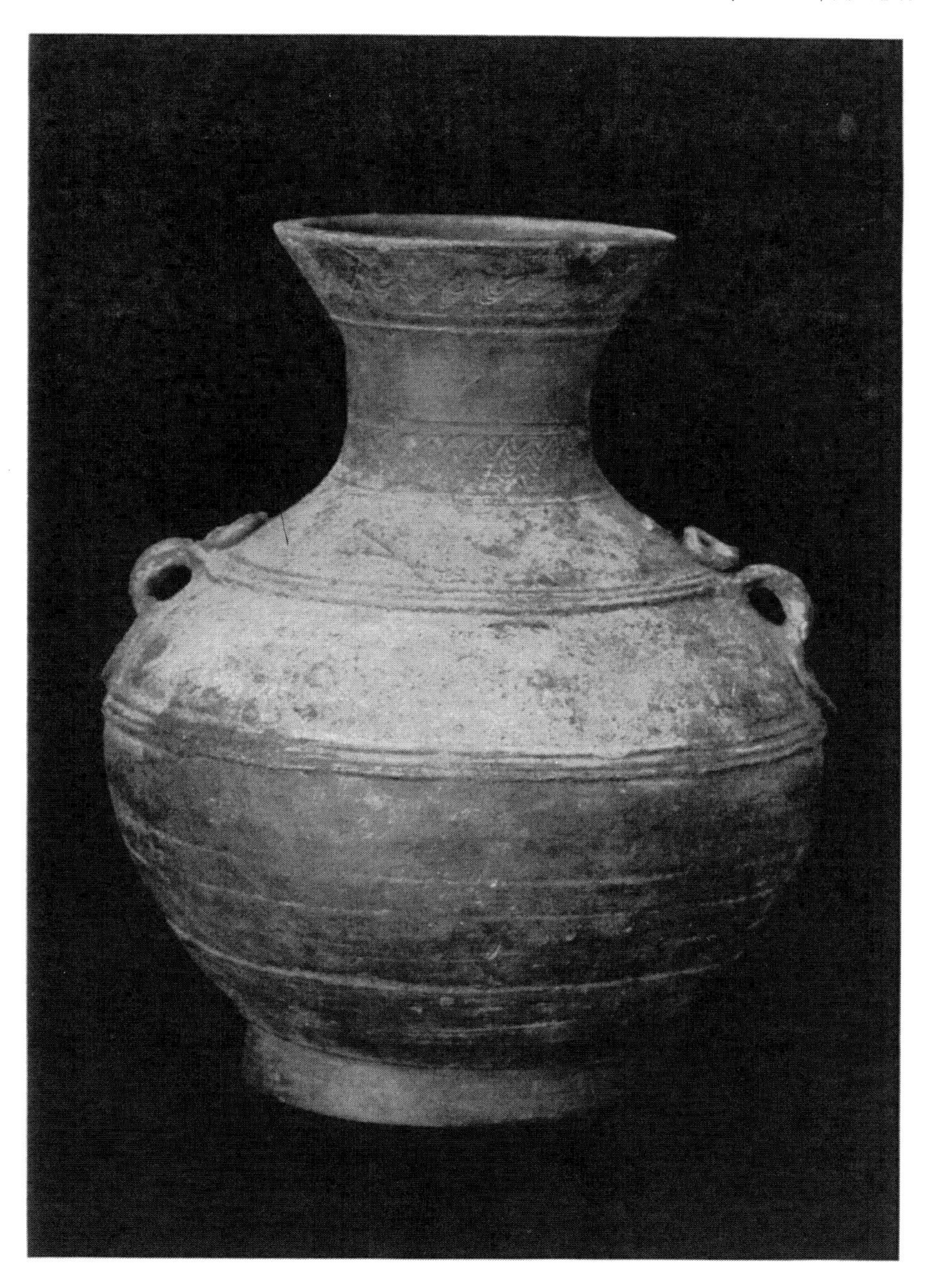

HAN PORCELANOUS JAR.

PLATE VI.

HAN PORCELANOUS POTTERY.

Globular vase, slightly asymmetrical, a narrow medial zone reaching from the neck down to the shoulders being well coated with a uniform, lustrous, yellowish-green porcelanous glaze; the neck and base showing a glossy brown slip. Its interior is glazed over a space of 6 cm. Decorated with three incised wave-bands, bordered by deep grooves, the lower one under the glaze. The almost semi-circular loop handles exhibit a leaf or fish-bone design.

Third century A.D.

Height, 25.2 cm. Cat. No. 118721.

HAN PORCELANOUS VASE.

PLATE VII.

HAN PORCELANOUS POTTERY.

Large globular vase, in its medial portion and inside of the neck coated with a thin, but evenly distributed porcelanous glaze. Wave-band along upper rim, and a broader wave-band of bolder design around the neck. The loop handles show a fish-bone design incised under the glaze. Flat bottom without rim. Of almost perfect workmanship.

Third century A.D.

Height, 34.8 cm. Cat. No. 118722.

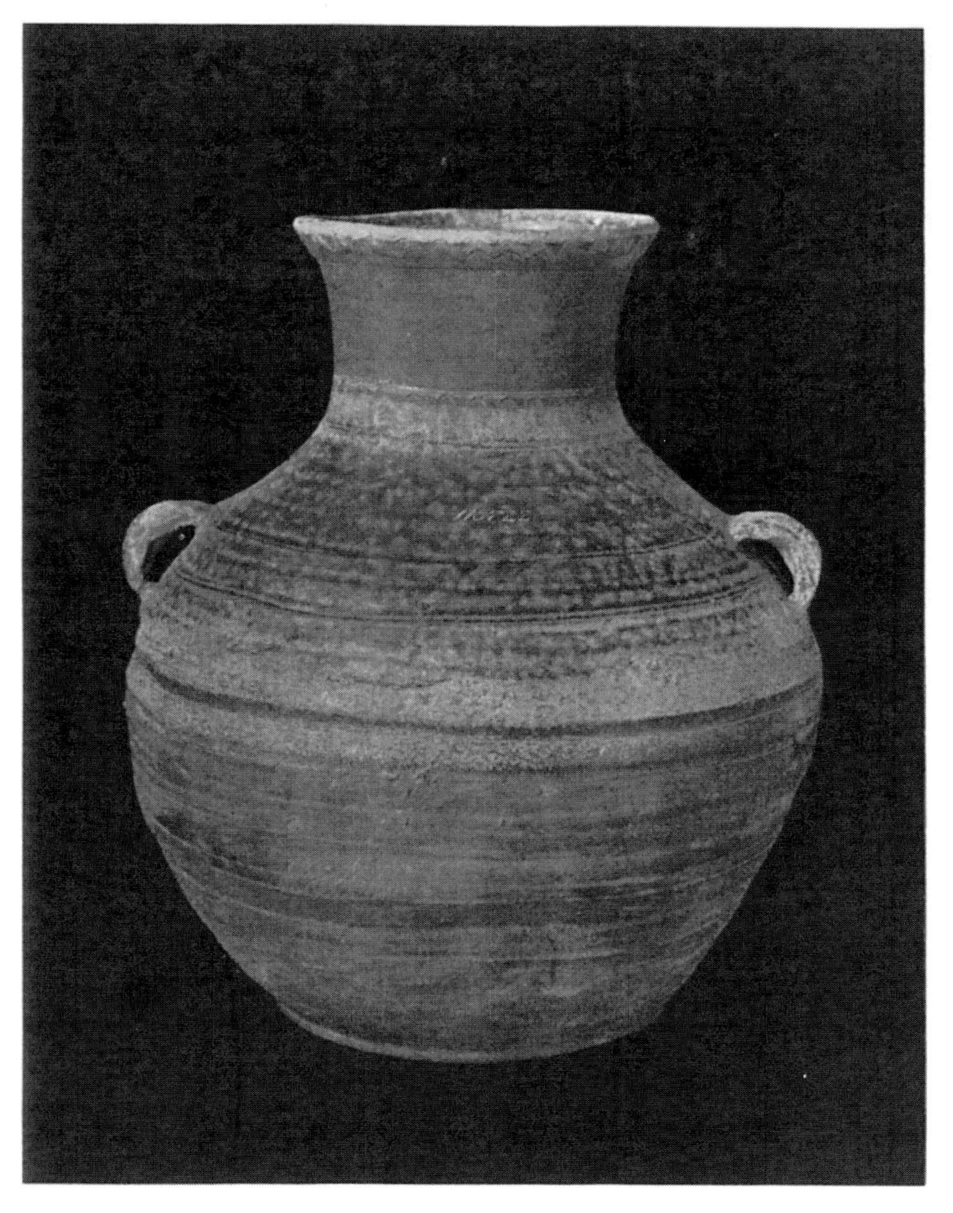

HAN PORCELANOUS VASE.

Plate VIII.

Han Porcelanous Pottery.

Large vase with asymmetrical neck, apparently turned out by an unskilled potter. A large piece is broken out of the neck (found in this condition) on the side of the vase not shown in the illustration. The glaze, covering only the middle portion, is thick and unevenly applied, in some instances forming small warts or globules. Decorated with two wave-bands. Loop handles with fish-bone design. The bottom is raised on a rim 1 cm high.

Third century A.D.

Height, 27.2 cm. Cat. No. 118724.

HAN PORCELANOUS VASE.

PLATE IX.

HAN PORCELANOUS POTTERY.

Large ovoid vase of good proportions, of light-reddish clay, glazed in the medial portion and in the interior of the neck, exterior of neck and base being coated with a brown slip. Two wave-bands. Loop-handles with leaf design of raised lines.

Third century A.D.

Height, 35.6 cm. Cat. No. 118719.

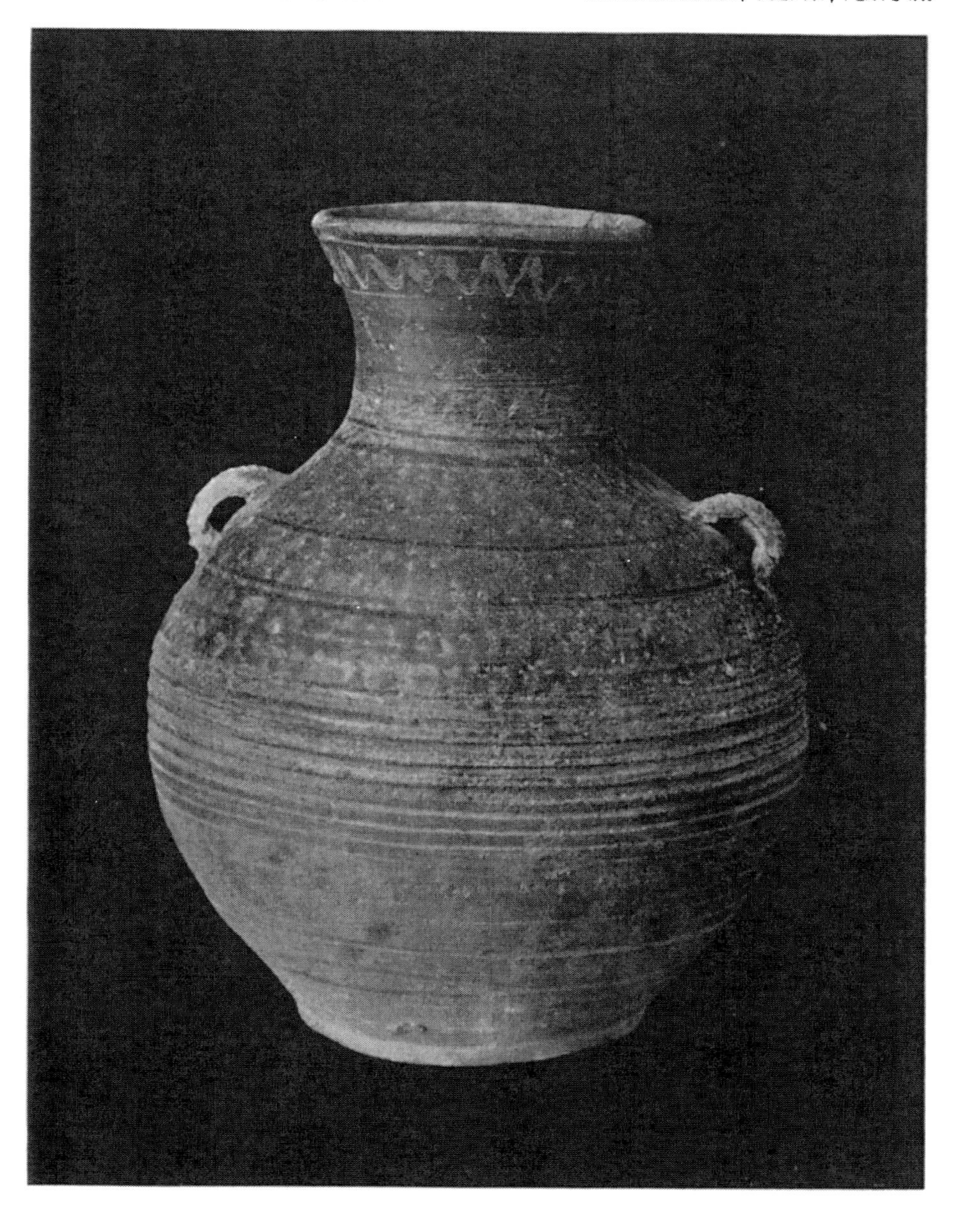

HAN PORCELANOUS VASE.

PLATE X.

HAN PORCELANOUS POTTERY.

Jar of the type *lei* 罍. The bottom inside is glazed. The exterior is glazed as far down as the middle of the body; the base is coated with a brown-red slip. The handles are glazed only in their upper portions. A wave-band is run over the shoulders under the glaze, passing below the loop handles. The latter are wrought into the appearance of an elaborate animal-head of similar style, that is moulded in relief on the body of the vessel.

Third century A.D.

Height, 25.9 cm. Cat. No. 118864.

HAN FORCELANOUS JAR.

PLATE XI.

CHINESE POTTER'S WHEEL (see p. 162).

From kiln near Peking. Table of clay, 52 cm in diameter on the top, 60 cm across the opening below.

In the collections of the American Museum of Natural History, New York. Secured by the writer in 1903.

Height, 1.24 m. Cat. No. $\frac{70}{12797}$.

CHINESE POTTER'S WHEEL OF CLAY.

PLATE XII.

CHINESE POTTER'S WHEEL (see p. 162).

From kiln near Peking. The table is formed by a heavy stone disk 60 cm in diameter and 9 cm thick. On top of it is placed a small wooden table, 35 cm in diameter. The main shaft is of wood and 87 cm high; the two wooden side-supports are 37 cm in length.

In the collections of the American Museum of Natural History, New York. Secured by the writer in 1903.

Cat. No. $\frac{70}{12798}$.

CHINESE POTTER'S WHEEL OF STONE.

130

中国的万向吊架

Holmes Anniversary Volume

Anthropological Essays

Presented to

William Henry Holmes

In Honor of his Seventieth Birthday, December 1, 1916

By

His Friends and Colaborers

Washington
1916

Cardan's Suspension in China

By BERTHOLD LAUFER

HE mechanical contrivance commonly styled "gimbal"[1] is known in the history of science under the term "Cardan's suspension". Although named for the celebrated Italian mathematician Jerome Cardan (Girolamo Cardano, or Latinized Hieronymus Cardanus; 1501–76), both the principle and its application were revealed and practised ages before his day. In fact, Cardan himself lays no claim to originality in this invention, but in his work *De subtilitate rerum*, first published in 1551, he merely describes a chair constructed for an emperor, which permitted him to sit in it during a drive without experiencing the least jolting, and on this occasion observes that the same arrangement had been used previously in connection with oil lamps.[2]

As intimated by Berthelot, Cardan must have encountered the invention bearing his name in the writings on the secret processes of magic, to which he was not foreign. Not only was he a philosopher, a physician, and a devotee of science, but also an astrologer, charlatan, and gambler.[3] We now even possess a clew to this magical document presupposed by Berthelot, as will be seen farther on.

The nearest predecessor of Cardano was Leonardo da Vinci (1452–1519), who left a design showing the suspension of a body, said to be capable of turning itself around three axes perpendicular one to another.[4] This construction, however, is not due to Da Vinci's genius either, for, as first shown by M. Berthelot,[5] the great French historian of chemistry, it was well known during the early middle ages. The document on which Berthelot's opinion is founded is contained in a curious Latin manuscript, the *Mappæ clavicula* (that is, "Key

---

[1] Derived from Old French *gemel*, which is based on Latin *gemellus*, diminutive of *geminus* ("double", "twin"). The New Oxford English Dictionary gives the following definition of the term: "A contrivance by means of which articles for use at sea (esp. the compass and the chronometer) are suspended so as to keep a horizontal position. It usually consists of a pair of rings moving on pivots in such a way as to have a free motion in two directions at right angles, so as to counteract the motion of a vessel."

[2] Compare F. Cajori, A History of Physics, p. 24, New York, 1899.

[3] As is also well known, the solution of the cubic equation known as Cardan's method is not due to his merit, but was the discovery of his countryman Tartaglia, who communicated it to Cardan in 1541, whereupon the latter published the solution under his own name.

[4] E. Gerland, Geschichte der Physik, p. 250.

[5] First in *Comptes-rendus de l'Académie des Sciences*, vol. CXI, 1890, p. 940; then in La Chimie au moyen âge, vol. I, pp. 176, 177, Paris, 1893.

CHINESE BRASS BRAZIER IN CARDAN'S SUSPENSION. DIAMETER 18.7 CM.
(Field Museum of Natural History, Chicago, cat. no. 117679)

to Painting"), treating in the main of recipes giving substitutes for gold and silver, recipes for writing in gold and silver, purple dyeing and technique of pigments. The work is of especial importance for the first indubitable mention of alcohol. It is extant in an English manuscript of the twelfth century edited by Sir Thomas Phillipps in *Archæologia*,[1] and in a German manuscript of the tenth century in Schlettstadt, found by A. Giry in 1877. H. Diels[2] dates the compilation of the Latin work in the time of Charles the Great, on the basis of linguistic arguments, and, like Berthelot, traces its contents to the Alexandrian school of alchemists. The text of the mediæval book in many cases agrees exactly, even in particulars, with the sources of Greek alchemy; and if more of these had been preserved to us, the coincidences would presumably be still more numerous. This point of view should be emphasized, as the ideas emanating from the *Mappæ clavicula* point straight to Alexandria. The passage relative to a suspension à la Cardan is embedded in a series of recipes for magicians or jugglers, and runs as follows: "Given four concentric circles revolving the one about another, so arranged that their diameters coincide; if a vase is suspended in their interior, in whatever way they may be turned, nothing will be spilled." Berthelot thought it probable that this idea would be traceable to Greek physicists, but at that time could not prove his supposition. In this estimate, however, he was not mistaken, for about a decade later the fact was established in the writings of Philo of Byzantium, a Greek mechanician who in all probability lived in Alexandria during the third century B.C.[3] His work on mechanics (μηχανική σύνταξις) is lost, except the fourth book dealing with military engines (the oldest extant work on artillery) and excerpts from the seventh and eighth books. The fifth book on pneumatics is preserved in the Latin translation of an Arabic manuscript and in an Arabic recension. The latter has been edited and translated by Baron Carra de Vaux. Here we read in article 56 as follows: "Construction of the octagonal inkwell, a very elegant apparatus. We make an inkwell octagonal, hexagonal, square, pentagonal, or whatever form may be given to prismatic glasses. This inkwell has on each side a place whence ink may be drawn, and in whatever position this object is placed, its upper face presents you an opening for the introduction of the pen without anything being overturned. The pen is dipped in, it meets the ink, and you write

[1] Vol. XXXII, 1847, p. 187.

[2] Die Entdeckung des Alkohols, *Abhandlungen Preuss. Akad.*, 1913, pp. 6–17.

[3] W. Schmidt, Heronis Alexandrini opera, vol. I, p. lxx; F. Susemihl, Gesch. d. griech. Lit. in der Alexandrinerzeit, vol. I, p. 744. Others assign his lifetime to the second century B.C. The problem is still controversial.

with it. For instance, compare this hexagonal inkwell as you see (fig. 1). In the interior is a collar on a pivot $\alpha\beta$; in this collar is another on a pivot $\gamma\delta$; in this interior of the second chain is a cup on a pivot $\epsilon\zeta$, and this cup forms the inkwell. One might say this is the Jewish style and the construction of the apparatus resembles that of the censer which turns while remaining at equilibrium. Measure this construction well and adjust it carefully, so that every time you place the inkwell on any face, that which is presented for the entrance of the pen is the top of the inkwell." The manuscript of Oxford adds to the figure the following comment which must be a subsequent interpolation: "This movable apparatus is like the throne of Solomon, the son of David. When somebody familiar with Solomon's throne makes use of it and ascends it, he will remain there; a person unfamiliar with

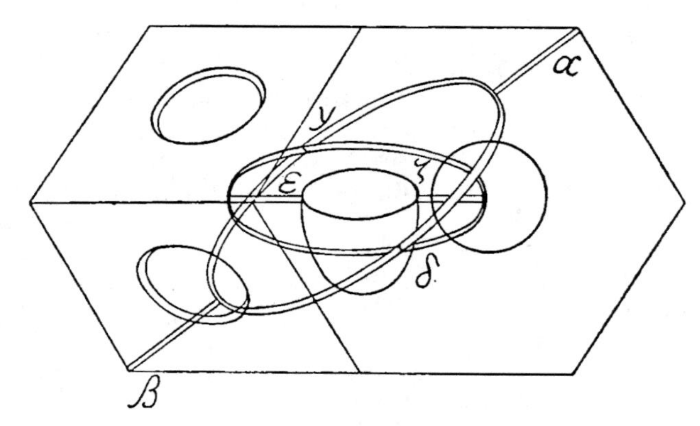

FIG. 1.—Philo's Inkwell in Cardan's Suspension (after Baron Carra de Vaux).

it, who will sit down in it, will fall to the ground. This is very nice."[1] Here we have a magical source like the one from which Cardan may have derived his description of the suspended imperial chair. In his introduction (p. 36) Carra de Vaux discusses the question of possible interpolations in the Arabic version and makes this observation on our passage: "The most curious article is that of the suspended inkwell which is found both in the manuscript of St Sophia in Constantinople and in the anonymous collection of Oxford. This inkwell is based on a suspension à la Cardan. It is somewhat astounding that

[1] Baron Carra de Vaux, Le livre des appareils pneumatiques et des machines hydrauliques par Philon de Byzance (tiré des *Notices et Extraits des manuscrits de la Bibl. Nat.*, vol. XXXVIII, Paris, 1902, p. 195).

this ingenious arrangement has been forgotten by Hero and the subsequent mechanicians; but it is not impossible, after all, that it should have been forgotten. The text says that this mode of suspension was utilized in the Jewish censers; some archeologist versed in Hebrew antiquities might possibly deduce a date therefrom." There can be no doubt that the description of the inkwell was part and parcel of Philo's original Greek text, and that the Arabic translation renders a correct account of it.[1] This is confirmed by the *Mappæ clavicula*, and we are fully entitled to the conclusion that the principle of Cardan's suspension was well familiar to the great Hellenic mechanicians of the Alexandrian epoch.

Whether the allusion to the Jewish censer hints at some Semitic influence can hardly be decided in the present state of our knowledge. What is of particular interest in this connection is the fact that in China also the Cardanic principle has been applied to censers or braziers.[2] Berthelot already had some vague information to the effect that "suspension à la Cardan has been employed in eastern Asia probably from times immemorial, as the Chinese do not change their processes." This point, however, he added, would require further elucidation.

The principle is well exemplified in a Chinese brazier obtained by the writer in 1909 at Si-ngan, the capital of Shen-si Province, for the Field Museum, Chicago (cat. no. 117679). In its outward appearance it presents a hollow sphere of brass, cut out into rosette-like designs in open-work, and composed of two hemispheres (see pl. 1, upper figure), one fitting above the other. If the upper one is removed (lower figure on the same plate), we notice in the interior a round brass bowl, which may be filled with aromatic substances, incense, or burning charcoal. If filled with coal, the brazier conveniently serves as a bed-warmer, as the sphere may be rolled and knocked around *ad libitum*, while the bowl swinging freely and independent of this motion will never turn upside down. The contrivance consists of two brass hoops, so arranged that one is within and perpendicular to the other. The outer hoop is riveted to two lugs, projecting from the inner side of the hemisphere and made in the same cast with it. The inner hoop moves on a pivot connecting it with the outer hoop and encircles the brass bowl. The bowl, accordingly, remains constantly suspended in a vertical position, while the rolling motion of the sphere is not com-

[1] This, so far as I know, is the generally accepted opinion. See, for instance, J. L. Heiberg, Naturwissenschaften und Mathematik im klassischen Altertum, p. 59.

[2] We still apply it to ship's lamps, the mariner's compass, and to barometers. A barometer suspended à la Cardan is figured and described by E. Atkinson, Elementary Treatise on Physics, 13th ed., p. 145, New York, 1890.

municated to it. The whole construction is very similar to the Alexandrian inkwell.

The brazier in question was attributed to the Ming period (1368–1643) by my Chinese informants of Si-ngan, and this date seems plausible. At present, this contrivance is no longer in use, at least not in northern China. According to Chinese tradition it is due to Ting Huan, a famed mechanician of the Han period (206 B.C.–A.D. 220),[1] who excelled in the making of ingenious mechanical devices, and who lived in the capital Ch'ang-ngan (now Si-ngan). He is credited with the invention of censers or braziers styled "reclining on the mattress" or "brazier in the bed-clothes", by the application of four revolving rings, so that the body of the vessel was kept at equilibrium, and could safely be placed on the bedding. This tradition doubtless refers to an instrument like the one described and figured. Its origin in the Han period is perfectly credible, as this was the era when mechanics and engineering awoke in China, and a scientific knowledge of nature gradually began to dawn. This also was the epoch when China was opened to intercourse with the West, and when along the trade-routes leading across Central Asia into the Roman Orient Hellenistic ideas and inventions were conveyed to the Chinese. Hellenism, the first universal civilization in the history of the world, based on a syncretism of Greek, Egyptian, and Semitic concepts, expanded into all directions of the globe, and Alexandria was its spiritual center. A single case certainly lacks convincing force, but the totality of coinciding phenomena with which we are now confronted is so overwhelming that Hellenistic influence on ancient China can no longer be denied. It is most obvious in the natural sciences, particularly alchemy and mineralogy, and in technology. All that is recorded of mechanical innovations in the Han period is traceable to the writings and models of the Alexandrian mechanicians. The application of Cardan's suspension is only one example, but perhaps not the least interesting.

FIELD MUSEUM OF NATURAL HISTORY
CHICAGO, ILLINOIS

[1] The exact date of his lifetime is not known; more probably he lived at the time of the Later or Eastern Han dynasty (A.D. 25–220), rather than under the Former or Western Han (206 B.C.–A.D. 24). The activity of this artisan is discussed in the writer's Chinese Pottery of the Han Dynasty (pp. 196–198), where also the relevant Chinese text is cited.

131

伽波《印度与基督教》书评

AMERICAN ANTHROPOLOGIST

NEW SERIES

ORGAN OF THE AMERICAN ANTHROPOLOGICAL ASSOCIATION,
THE ANTHROPOLOGICAL SOCIETY OF WASHINGTON,
AND THE AMERICAN ETHNOLOGICAL
SOCIETY OF NEW YORK

VOLUME 18

LANCASTER, PA., U. S. A.

PUBLISHED FOR

THE AMERICAN ANTHROPOLOGICAL ASSOCIATION

1916

ASIA

Indien und das Christentum. Eine Untersuchung der religionsgeschichtlichen Zusammenhänge. RICHARD GARBE. Tübingen, 1914. viii, 301 pp. 22×14.5 cm.

This volume presents a collection of essays revised and partially rewritten, which the distinguished author, professor of Sanskrit at the University of Tübingen, had previously published in various journals, and several of which have been rendered accessible to the American public in the *Monist*. The book is divided into two sections,—India's influence on Christianity, and Christian influences on Indian religions. It is accordingly the author's object to study the interrelations of India and the West in the field of religious history. If this work is here brought to the notice of anthropologists, it is chiefly for the reason that it offers an interesting contribution to the methodological study of the vexing question of borrowing or independent origin, which almost daily confronts every one of us in his particular branch of work. It is well known that many a hot battle has been fought over the theory that Buddhism should have exerted its influence on early Christian thought, particularly on the later uncanonical or apocryphal Gospels. R. Seydel (died 1894), who was the first to treat the subject methodically, advanced fifty-one parallels to prove dependence of the Gospels on Buddhist literary sources. Bergh von Eysinga founded his thesis on only six coincidences which he attributed not to written productions, but to orally transmitted themes. Many Indianists and theologians remained skeptical and pointed to the congenial character of the two religions or the similar social conditions in which Christ and Buddha moved. Garbe retains four parallels between Buddhist and Christian stories, as far as the canonical Gospels are concerned: (1) the Buddhist story of the venerable saint Asita and his glorification of the newly-born infant Buddha, compared with the Christian story of Simeon in the temple (Luke, II, 25); (2) Buddha's temptations by Māra ("the Evil one") and Christ's temptation by Satan; (3) Peter's walking over the sea (Matthew, XIV, 25) and Jātaka, No. 190, where a disciple of Buddha, unable to find the ferry, crosses a river on foot, begins to sink in the stream, but reaches the other bank by virtue of his faith in Buddha; (4) the bread miracle of Jesus and Jātaka, No. 78, where Buddha satisfies with a single loaf his five hundred disciples and all inmates of a monastery, so much bread being left that it is thrown into a pit beside the gateway. According to Garbe, the mutual resemblance is less striking for the sake of the miracle itself, but rather owing to the feature that also in the New Testament twelve baskets of crumbs are left and the number five is retained. The smaller number of five hundred in the Buddhist story, as compared with five thousand in the gospels, who are fed with five loaves, according to Garbe, bespeaks the character of originality, and the number five hundred is eminently Buddhistic. The Buddhist text should therefore be regarded as the source of the Christian version. In these four cases, Garbe infers, Buddhistic influence on the New Testament must be acknowledged, and hence the supposition cannot be suppressed that this influence has been evident also in other passages of the Gospels where it cannot be proved with the same degree of probability. It should not be difficult to theology, Garbe thinks, to become reconciled to this idea from which the eternal values of Christianity will not have to suffer.[1]

I am in accord with the author in his assumption of Indian influences upon the Alexandrian Physiologus, and hope to furnish several other contributions to this question (aside from the story of the rhinoceros previously pointed out by me and reproduced by Garbe), but I do not believe that the story of the lion that awakens his stillborn cub with his roaring voice on the third day goes back to a Buddhist notion, as Grünwedel has asserted. At the outset it is not plausible that any symbolical allusion to the lion in the domain of Christian thought is of Indian origin: the lion plays a predominant rôle in the Semitic and classical worlds, in literature as well as in art. The lion is the type of the tribe of Judah (Genesis, XLIX, 9), and since Christ is descended by David from that tribe, he is styled "lion from the tribe of Judah" (Revelation, V, 5) The case of the Physiologus, however, is specific and well-defined, and any corresponding parallel to it does not occur in India; thus the question may be

[1] In discussing the Hebrew and Indian parallels of the story of Solomon's judgment, Garbe (p. 27) goes astray in assuming that, as long as merely the Tibetan version in the Kanjur was known, the opinion would have been justified that the story had penetrated into Tibet through Christian mediation; again, on p. 29 (note), he asserts that it migrated from India by way of Tibet into China. The reverse, however, is the case: the Chinese rendering of the story is the older one, and from the Chinese it found its way into Tibetan literature. The Tibetan story is incorporated in a collection of Buddhist Jātakas, which has become well known through the edition and translation published by I. J. Schmidt in 1843. It is now perfectly assured that this Tibetan work was translated from the Chinese by Chos-grub, a Tibetan scholar, who lived in the first half of the ninth century. The Chinese original itself is rendered after a work edited in the year A.D. 445, or shortly before, at Turfan, Turkistan, by Buddhist monks. The starting-point of the story for the east Asiatic world, accordingly, was India, whence it migrated to Khotan and Turfan, and finally to Kan-chou in Kan-su, the domicile of Chos-grub.

raised as to whether the notion of the lion awakening his stillborn cub has not existed in the western part of the world in pre-Christian times, for we meet curious conceptions regarding the parturition of the lioness in Herodotus (III, 108), the first ancient author who discussed this subject.

While Buddhist influence on the canonical Gospels must still be regarded with restraint, as long as we lack the documentary evidence showing how such literary or oral transmissions from India into Syria could be brought about, there can be no doubt of an extensive Buddhization of the so-called apocryphal Gospels coming down from the third to the sixth century and swarming with fantastic and adventurous miracle stories. Most of these books are of Gnostic origin, and as has been recognized at an early date, the Gnostic sects have largely borrowed from Buddhism. Garbe makes the interesting point that the sparse Buddhist stories adopted into the canonical Gospels (if this be correct) belong to the original and older form of Buddhism, the Hīnayāna, whereas the source for the uncanonical Christian writings is presented by the Mahāyāna, the new development of Buddhism in northwestern India during the first century before and after our era. It is generally known that a Bactrian monk of the sixth century utilized the legends of Buddha for a Christian romance under the title Barlaam and Joasaph, which has found its way into all European literatures. Garbe recapitulates in particular the Christian legends of St. Eustachius and St. Christophorus, which without any doubt are traceable to Buddhistic motives. It seems to me, however, that the story of Sutasoma was not the direct prototype of St. Christophorus, but that we have to look for a missing link in western Asia which is still unknown.

With many others Garbe believes also in Buddhist influences on the forms of worship in the Christian Church: the convents with their monastic life and division of novices and ordained monks and nuns, celibacy and tonsure of the clergy, confession, cult of relics, rosary, church-towers, employment of incense and bells, are to be ascribed to an impetus received from corresponding institutions of Buddhist India, which transmitted to Persia, Bactria, and Turkistan came in contact with Christian sects. The possibility and probability of such derivations cannot be denied, but for each case a profound and detailed investigation will be required. The nimbus has not its origin in classical antiquity, as asserted by the author, but in Babylonia, whence it spread to Iran and India on the one hand, and to the west on the other.

In the second portion of his book, which deals with Christian influences on Indian religions, Garbe is less fortunate than in the first. In his critical discussions he shows at his best, and his analysis of the Thomas legend and the history of the Thomas Christians in southern India forms one of the most attractive chapters. I do not comprehend how Garbe (p. 179) gets at the statement that in Chinese art Buddha is represented as a fisherman with rod and hook, and that this is due to a transmission of Christian symbolism, because fish-catching is an un-Buddhistic action. What is known to me only are representations of the Bodhisatva Avalokiteçvara (Kuan-yin) as either traversing the ocean on a fish or carrying a fish in a basket (see illustrations, for instance, in *Open Court*, 1911, pp. 388, 389). This deity is also the protector of mariners and the lord of the sea, hence a sea-fish has become his attribute, and this surely has nothing to do with Christian ideas.

The author's remarks on the influence of Christianity on the Lamaism of Tibet are weak and ill-founded. Only secondary and partially dubious sources have been consulted, and even those not very critically. Garbe (p. 184) repeats after Grünwedel that Odoric of Pordenone visited Lhasa in 1330 and found there already Christian missionaries and some proselytes. However, the reviewer has shown in an article in the *T'oung Pao* (1914, pp. 405–418) that Friar Odoric had never been in the interior of Tibet, and that it is out of the question that he ever reached Lhasa: he himself in his account lays no claim to this honor (the very name Lhasa is not even mentioned by him), and he has nothing whatever to report concerning Christian missionaries in Tibet. How Grünwedel could ever advance such a statement, is a mystery. This, then, cannot form a link in the chain of alleged evidence for Christian influence on Tibet, still less than the assertion that the possibility of such interaction on the Buddhism of Tibet and China exists *nachweislich* from 635; for, according to Garbe, in this year a Nestorian mission is attested which under the guidance of A-lo-pen betook itself "into those countries." This fact is brought out in the famous Nestorian inscription of Si-ngan fu, erected in A.D. 781; and he who knows what an immense literature has been produced on this unique monument will be not a little surprised to note that Garbe draws this fact from such a compilation as Waddell's *Buddhism of Tibet*. A-lo-pen's journey from Syria to Ch'ang-ngan, the then capital of China, cannot be invoked, as has been done by Garbe, as capable of proving Christian influence on Tibet; for the Nestorian A-lo-pen did not touch Tibet, but took the route common at that time by way of Turkistan. Through failure in verifying his sources, the author has unfortunately become the victim of a much more lamentable error: he thinks that the mission of A-lo-pen was received in the year 639 in northern India by the celebrated King Çīlāditya. The authority cited for this statement is a note in Takakusu's translation of the journey of the Chinese pilgrim I-tsing, where it is remarked, "Dr. Edkins says that Çīlāditya received the Syrian Christians, A-lo-pen and his companions, A.D. 639 (*Athenaeum*, July 3, 1880, p. 8)." The article referred to contains a review of Edkins' well-known book, *Chinese Buddhism*, in which on p. 117 occurs the sentence, "The same emperor, T'ai-tsung, received with equal favor the Syrian Christians, A-lo-pen and his companions, who had arrived in A.D. 639, only seven years before Hüan-tsang's return." Edkins, accordingly, lisps no word about the Indian King, Çīlāditya, but only speaks of the reception which A-lo-pen met with in the capital of the Chinese Emperor T'ai-tsung. The date 639, however, is wrong: A-lo-pen arrived at Ch'ang-ngan, according to the Nestorian inscription, in 635. What happened in 639 was the promulgation of an imperial edict issued in favor of the Nestorians and granting them permission to build a Syrian church in the metropolis of China. Consequently, in that year, A-lo-pen resided at Ch'ang-ngan, and cannot be located anywhere in India. The alleged fact of the presence of Nestorians at the Court of King Çīlāditya in 639 does not exist: Takakusu has wrongly copied the passage of the *Athenaeum*, and Garbe was unfortunate enough to rely upon him, the more so as the error of his informant had already been detected and exposed by M. L. Ettinghausen (*Harsa Vardhana*, p. 92, 1906) and G. A. Grierson (*Journ. Roy. As. Soc.*, 1913, p. 144). Still more unhappily, Garbe (p. 259) is ready to base another far-reaching conclusion on this alleged "first Nestorian mission in mid-North-India." This concerns Christian influences on the Krishnaism of India, which, for that reason, are asserted to have taken effect in the first part of the seventh century. This entire speculation is naturally bound to collapse. There is no evidence whatever that such influences operated, or might have operated, in the India of the seventh century; and, in the reviewer's opinion, they belong to a much more recent, nay, to a very recent period. Be this as it may be, Garbe's attempt to prove a contact of Tibetan culture with Nestorians is no more felicitous than that of any of his predecessors along this line: not a scrap of matter-of-fact evidence has as yet come to the fore to warrant the conclusion that Nestorians have ever set foot on Tibetan soil. Allegations made to this effect are speculative dreams of people who seem to have time to waste. In the same manner as several earlier writers, Garbe is likewise inclined to assume an extensive influence of Catholicism on the cult and rites of the Lamaist Church. The reviewer who has devoted some attention to the study of Lamaism for twenty years is now more skeptical than ever before as to any coincidences pointed out in this direction. Above all, it should be borne in mind that we have at the present time merely a surface knowledge of Lamaism, and that the whole ceremonial has never been seriously and intensively investigated. Without the solid basis of such research, historical and critical, founded on indigenous sources, the ventilation of this question is hopeless. The outward resemblances and affinities on both sides may well exist; but every ethnologist is now sufficiently schooled to appreciate the value of such fortuitous coincidences. We have witnessed too many failures and downfalls of bulwarks of comparative ethnology to be easily ensnared now in such traps. A close study of Lamaist rituals and ecclesiastic history and organization will in all probability bring out the fact that their psychological foundation, their origins and developments, are totally different from any corresponding affairs of the Christian Church. Even now it is perfectly safe to venture the positive assurance that the office of the Dalai Lama, for instance, is as remote from the papal institution as the moon from this globe. The alleged Manicheism of the Tibetans, which has been discussed again by Chavannes and Pelliot (*Un traité manichéen retrouvé en Chine*, pp. 274–278) with negative result, is not touched upon by Garbe, nor has he any knowledge of some real Christian traces found in certain Tibetan legends of Padmasambhava and Milaraspa, to which Grünwedel and the writer have repeatedly called attention.

The alleged Christian elements in the Mahābhārata treated in the following chapter are no more convincing than the invasion of Tibet. The Indian epic tells a legend concerning the Çvetadvīpa ("White Island," or "Island of the Whites"), located over 32,000 miles northeast or north of Mount Meru, on the northern border of the Milky Sea. The white, bright-shining inhabitants of this country are supernatural beings without sensual organs, living without food, fragrant and sinless, blinding with their splendor the eyes of wicked men. Being prompted by supreme love of the one invisible god Nārāyaṇa, they adore him with gently murmured prayers and folded hands. None of them occupies a prominent position, but all enjoy the same authority. These latter data impress Garbe as being due to the contact with a Christian community. Though he regards Christian influence in this case only as probable, he construes an elaborate theory, identifying the Milky Sea with Lake Balkhash and establishing there Nestorian settlements in the sixth century A.D. Unfortunately, this assertion is not proved; at that time, the Nestorians had not yet advanced as far as Lake Balkhash, but were restricted to eastern Iran (see now Pelliot, "Chrétiens d'Asie centrale et d'Extrême-Orient," *T'oung Pao*, 1914, pp. 623–644). The whole theory is a romance which must be decidedly rejected. The Nestorians have to cover a multitude of sins, and like the gypsies in Europe, are employed in Asia on the scholarly chessboard to explain movements of which we are still ignorant.

Garbe's work should be read by our folklorists, not for the results achieved, but for its stimulating qualities. It is a well-written summary of the present state of the problem, and his discussions are always interesting and suggestive.

B. LAUFER

Dravidian Gods in Modern Hinduism; A Study of the Local and Village Deities of Southern India. WILBER THEODORE ELMORE. (University Studies, published by the University of Nebraska. Vol. XV, No. 1, January, 1915.) Pp. 1–149.

This study, in part based on personally gathered data, supplies the student of Indian religions with a considerable mass of new and valuable material. While scattered observations on the local and village deities of southern India are available, few general discussions of them and their place in the Hinduism of this part of India have hitherto been attempted.

In the first chapter, in which a brief outline is given of the Aryan conquest of Dravidian India, the author's statements are in several cases open to criticism. Thus, he is far too sweeping in denying any literature to the Dravidians, and entirely in error in stating that they "left no monuments which throw light on their origin." Southern India, as is well known, contains great numbers of prehistoric sites of archeological importance for the history of the early Dravidian population, although as yet these cromlechs, graves, and other types of remains have not been systematically or scientifically investigated. The author is also unfortunate in giving the impression that there were in India before the Aryan immigration, only Dravidian-speaking people. The whole Munda group—once occupying a large part of central and northern India—is entirely omitted from consideration, and its part in the development of modern Hinduism is wholly neglected.

When the author comes to deal with the specific subject matter of his study, he is on much surer ground, and deserves much credit for the care with which he has gathered and marshaled his material. Pointing out that the characteristics of these deities of the common people (as contrasted with those of the orthodox Hindu pantheon) lie in their local

132

伯格《绿松石》书评

modern adaptation, where the teeth acquire retrograde characters on account of foodstuffs being prepared so as to require minimum effort. Such changes are closely correlated with those of modes of life, the two contrasting poles of which are the arboreal and the domesticated forms, also conditioning very decided changes in skeletal adaptation. The retraction of the face and the loss of the prehensile character of the hallux are the most significant ones.

Dr. Gregory's book is a distinguished example of the approach of phyletic problems by minute and comparative description, supported by explanatory illustrations. He consciously does away with methods of more exact representation by means of diagrammatic and mathematical devices. That such a solution is possible to a fair degree in a field of investigation dealing mostly with transitory fossil forms in view of a, for the time being, fixed biological final form, as represented by man in the phyletic sense, is fully shown in the author's work. It is in fact the most conscientious, resourceful and up-to-date comprehensive work, containing a wealth of acute morphological observations and clever deductive argumentation, that has come to my knowledge for some time. A few disdainful remarks on sciences that have to rely to some degree on exact methods, preeminently anthropology itself (pp. 250, 333, 341) cannot detract from the distinct value of his exposition. They only serve to show, besides forming an unnecessary diversion, a fatal miscomprehension of the object and view of anthropology, whose chief task consists in analyzing a living form, as it were, even if it refers to historically extinct races. That this has to be accomplished with a consideration of all the biological perspectives that pertain to man's physical existence goes without saying. Paleozoological endeavors must naturally cease on the threshold of this science with its complexity of phenomena, but their excellent work in tracing the phyletic connections of man's ancestors should be appreciated without reservation, especially in the work here reviewed.

Bruno Oetteking

MISCELLANEOUS

The Turquois. A Study of its History, Mineralogy, Geology, Ethnology, Archeology, Mythology, Folklore, and Technology. Joseph E. Pogue. (Memoirs of the National Academy of Sciences, Vol. XII, Part II, third memoir, Washington, 1915), 162 pp, 22 plates, one of which is colored, and 1 frontispiece in colors. 30×24 cm.

This splendid publication, the fruit of many years of assiduous and painstaking research, is perhaps the most complete monograph that we possess on any precious or half-precious stone. In eight chapters, Dr. Pogue treats of the history of turquois, its mineralogical properties, its geographical distribution, its origin and use, the chalchihuitl question, finally the importance of the stone in mythology and folklore, and its technology. A carefully drawn-up bibliography and a good index complete the volume.

It is certainly impossible for the reviewer, who is not a mineralogist, to review the mineralogical portion of the work; he must restrict himself to offering a few remarks on the history of our knowledge of the turquois. With a great amount of industry Dr. Pogue (pp. 9 *et seq.*) has lined up sources and pseudo-sources relating to real and alleged turquois in ancient times, but the history of the subject in Europe is not worked out with desirable lucidity. Galenus (pp. 12, 35) must decidedly be eliminated from the list of ancient authors who mention the turquois of Nīshāpūr in Persia, for this statement emanates only from Ibn al-Baitār (1197–1248) and ranks on the same level with all other data attributed by Arabic authors to Aristotle or Galenus. Galenus, of course, never discussed the turquois, still less could he speak of Nīshāpūr, as this city, founded by Shāpūr II (A.D. 309–379), did not yet exist in his lifetime (A.D. 129–199). No Greek author mentions the turquois or any stone that might be interpreted as such; above all, Dioscorides is reticent about it. The only ancient author who has been credited with a knowledge of turquois is Pliny; but in the reviewer's opinion Pliny's *callaina* and *callais*, which have been taken as such, in fact, have nothing to do with the turquois. Dr. Pogue also feels that the alleged correspondence is far from satisfactory. There is an excellent criterion that may guide us in identifying Plinian stones, and this is the perpetuity of tradition in the East and West alike, as has been shown, for instance, by the writer in the case of the diamond. The Plinian tradition in regard to the stone *callaina* is perfectly isolated, however, and was not taken up, reproduced, or continued, by any Oriental or Occidental mineralogists. The Arabic records regarding turquois make no reference whatever to classical authors, as they do in regard to so many other stones, but are plainly traditions which originated in the Orient itself. Pliny's successors, first of all, C. Julius Solinus, who has adopted nearly the complete list of his precious stones, and has contributed much to hand their knowledge down to the middle ages, has passed the *callaina* over in silence; indeed, he does not mention any stone that could be interpreted as the turquois. In accordance with this fact, the French Bishop Marbodus (1035–1123), in his *De lapidibus pretiosis*, is also silent in regard to the subject. Likewise the early French and German stone-books do not mention the turquois. Dr. Pogue (p. 12) informs us, however, that Isidorus of Sevilla (*circa* 570–636), in book 16 of his *De natura rerum*, alludes to the frequent use of turquois in the ears of Orientals. Unfortunately the source for this statement is not cited, while in all other cases Dr Pogue very conscientiously quotes his authorities. Isidorus' work *De natura rerum* deals with cosmography and geography and consists of a single book, divided into forty-eight brief chapters. Chapter XVI treats *de quantitate solis et lunae* and contains nothing about the turquois or any other precious stone, nor does the remainder of the work. It is equally doubtful to me whether the definition of the turquois taken from an anonymous and undated Latin *lapidarium* (p. 13) can be really attributed to Albertus Magnus (1193–1280), as Dr. Pogue inclines to think Albertus Magnus has written a treatise *De virtutibus lapidum*, inserted in his work *De secretis mulierum item de virtutibus herbarum lapidum et animalium* (the edition before me was published at Amsterdam, 1669), the turquois does not occur there, and here was the occasion to deal with it if Albertus had known it. Knowledge of the turquois in Europe did not spread earlier than during the thirteenth and fourteenth centuries. Du Cange (*Glossarium mediae et infimae latinitatis*) quotes a document dated 1347 as the oldest source for the word *turchesius* or *turcica gemma*. According to the *New Oxford English Dictionary* the earliest reference to turquois in English literature occurs in a work of 1398. We meet it also in the famous list of stones enumerated in Wolfram von Eschenbach's *Parcival* (791, 18). The Dictionary of the Spanish Academy (*Diccionario de la lengua castellan*, vol. VI, p 379, Madrid, 1739) quotes as the earliest author to mention turquois (*turquesa*) Gomez de Tejada in his *Leon Prodigioso*. The earliest European writer revealed by Pogue is Arnoldus Saxo in his *De virtutibus lapidum*, where the stone is described as of a yellow color, verging on white The question as to why the stone is called yellow is not discussed by Pogue, but this error surely testifies to the fact that these medieval writers knew the stone merely from hearsay The statement of the early English author of 1398 is identical with that of Saxo and apparently derived from him. His text runs thus:

> Turtogis that hatte Turkeis also is a yelow white stone and hath that name of the contrey of Turkeis. This stone kepeth and saueth the siyt and bredeth gladnes and comforte.

The derivation of the name of the stone from Turkey leaves no doubt that the first knowledge of it was transmitted from that quarter; and it is further evident that European medieval knowledge of the stone hailed directly from the Orient, and was not connected with any tradition of classical antiquity. Consequently, the question as to whether Pliny knew the turquois or not, is irrelevant, as the subsequent generations owed him nothing along this line, but derived their knowledge exclusively from Oriental peoples.

The callean stone (καλλεανὸς λίθος) mentioned in the *Periplus* has no relation to Pliny's *callaina*. The identification of the two names rests on bad philology. It is quite certain that the term of the *Periplus* goes back to a Sanskrit prototype of the form *kalyāna* and means "excellent stone," or may refer to the city Kalyāṇa near Bombay, mentioned by Cosmas Indicopleustes under the name Kalliana (see Burnell, *Indian Antiquary*, vol. III, p. 310). It is equally certain that at the end of the first century A.D., when the *Periplus* was written, the turquois was wholly unknown in India and in all probability even in Iran. It was only the Mohammedans who introduced the stone into India, not earlier than the latter part of the tenth century.

Garcia da Orta in his *Coloquios do simples e drogas* (Goa, 1563) was the first to introduce the Arabic-Persian term in the forms *ferruzegi* and *puruza*, as he writes, and to interpret it correctly as the turquois, simultaneously refuting the previous error that this word should refer to the emerald. It was known to him that there was a great quantity of turquois in Persia, and with regard to its medicinal employment he comments that he was told by some people that it figures in the pharmacopeia among the Gentios (that is, Hindu), by others, however, that it does not. Among the Moors (that is Mohammedans) all say that it is used in medicine (see C. Markham, *Colloquies on the Simples and Drugs of India by Garcia da Orta*, pp. 358–359). In fact, the medical utilization of the stone originated among the Mohammedans of western Asia, who introduced the practice into India. Garcia's statement shows that in the latter part of the sixteenth century the turquois was not yet officially admitted into the pharmacopeia of India, and that its medical employment was reduced to a minimum.

In regard to the antiquity of the turquois in Iran and the history of the mines of Nīshāpūr I feel obliged to maintain strictly my former position in this question, and am not convinced by Dr. Pogue's purely speculative considerations to the contrary (p. 35). The point in historical problems is not what might have been, but what has been, and only facts and data carry convincing force. It is somewhat surprising to note how Dr. Pogue can advance the statement

That the deposits were worked about 2100 years B.C., is suggested by the name of one of its openings, called Isaac's Mine on account of a tradition that it was discovered by Isaac, the father of Israel,

after I characterized this as a legend without historical value (*Notes on Turquois*, p. 42, note 2). Such modern legends connecting famous sites with names of the Old Testament exist by the thousands among the Mohammedans. In writing on the cultivation of the apple one might as well invoke Eve's apple as good evidence for the great antiquity of its cultivation.

In my *Notes on the Turquois* I dated the first acquaintance of the Chinese with the stone in the period of the Mongols, but there is now reason to believe that the latter were preceded by the Khitan (usually classified among Tungusians), who ruled China as the Liao dynasty from 907 to 1123. Officials of that dynasty are said to have worn girdles adorned with gold, jade, rock-crystal, and turquois. Thus far I have found this statement only in the *Su wen hien t'ung k'ao*, written in 1586, but it remains to be traced in the contemporaneous records of the Liao dynasty, before it may be retained as a well-assured fact.

It is regrettable that Dr. Pogue (p. 84) has not had the opportunity of examining the alleged turquois beads found at certain neolithic stations of France and Spain. It would be interesting to see this vexed problem solved, as on the one hand we have the theory of Aveneau de la Grancière (*Les parures préhistoriques et antiques en grains d'enfilage et les colliers talismans celto-armoricains*, p. 147) that these stones were imported from the Orient in a crude state, and on the other hand the opinion of Comte de Limur that this material was brought to light from the tin mines of Montbras. A single analysis made by Damour in 1864 demonstrates that the stones in question more nearly approximate variscite: he also bestowed on it the name *callais*. As no analysis on a large scale has as yet been conducted, the evidence remains inconclusive. O. Montelius (*Chronologie der altesten Bronzezeit*, p. 204) seems to incline toward the belief that all these beads, also those found in Spain and Portugal, are real turquois.

The author asserts that the Nile was named in reference to its blue waters from the Sanskrit word *nīla*, meaning blue (p. 110). Latin *Nilus* is the reproduction of Greek *Neilos* which either goes back to an Egyptian word, or whose origin must be regarded as obscure, but which cannot be sought for in Sanskrit. By the way it may be remarked that the development of color-sense cannot be traced from linguistic arguments, and that Geiger's study cited by Pogue (p. 68) is thoroughly antiquated, nor is it true that Chinese lacks words for blue. Defects of color nomenclature are merely defects or limitations of language, not of color-sense.

To the bibliography may be added Robert de Berquen, *Les merveilles des Indes orientales et occidentales ou Nouveau traité des pierres precieuses et perles*, pp. 51–53 (Paris, 1669), and H. H. Hayden in *Memoirs Geological Survey of India*, vol. 36, pt. 2, 1907, p. 65 (brief reference to turquois at Lhasa).

The preceding observations bear only on some details of Dr. Pogue's monograph, and most assuredly do not detract from the intrinsic value of his magnificent work. My own limitations prevent me from rendering it full justice. It will remain a classic in the hands of all students interested in mineralogy, ethnology, and archeology, and occupy a place of honor in the publications of the National Academy. It is a cyclopedia giving an intelligent summary of all we know at present about the turquois. The attention of Americanists may be specially called to the interesting chapter on the chalchihuitl question. The illustrations are well selected, and the reproductions are excellent.

B. LAUFER

SOME NEW PUBLICATIONS

De Booy, Theodoor. Notes on the Archeology of Margarita Island, Venezuela. (Contributions from the Museum of the American Indian, Heye Foundation, vol. II, no. 5.) New York, 1916. Pp. 28, 8 pls., 15 figs.

Eaton, George F. The Collection of Osteological Material from Machu Picchu. (Memoirs of the Connecticut Academy of Arts and Sciences, vol. V.) New Haven, Connecticut, 1916. Pp. 96, 39 pls., 50 text figs., 2 tables, 1 survey map.

Emmons, George T. The Whale House of the Chilkat. (Anthropological Papers of the American Museum of Natural History, vol. XIX, part I, pp. 1–33, 4 pls., 6 text figs.) New York, 1916.

Gregory, William K. Studies on the Evolution of the Primates. (Bulletin of the American Museum of Natural History, vol. XXXV, art. XIX, pp. 239–355, 1 plate, 37 figs.)

Harrington, John Peabody. The Ethnogeography of the Tewa Indians. (Bureau of American Ethnology, Twenty-ninth Annual Report, pp. 29–636, pls. 1–21, maps 1–29, 29A, 30, diagram 1.) Washington, 1916.

Haeberlin, H. K. The Idea of Fertilization in the Culture of the Pueblo Indians. (Memoirs of the American Anthropological Association, vol. III, no. 1, Jan.-Mar. 1916.) Pp. 55.

133

藏语中的借词

T'OUNG PAO

通報

OU

ARCHIVES

CONCERNANT L'HISTOIRE, LES LANGUES,
LA GÉOGRAPHIE ET L'ETHNOGRAPHIE
DE
L'ASIE ORIENTALE

Revue dirigée par

Henri CORDIER
Membre de l'Institut
Professeur à l'Ecole spéciale des Langues orientales vivantes
ET
Edouard CHAVANNES
Membre de l'Institut, Professeur au Collège de France.

VOL. XVII.

LIBRAIRIE ET IMPRIMERIE
CI-DEVANT
E. J. BRILL
LEIDE — 1916.

LOAN-WORDS IN TIBETAN

BY

BERTHOLD LAUFER.

CONTENTS.

The following abbreviations are employed in this paper:

Chandra Das = Das, *Tibetan-English Dictionary.*

A. von Le Coq, Turfan = *Sprichworter und Lieder aus der Gegend von Turfan.*

French Dict. = *Dict. tibétain-latin-français par les Missionnaires catholiques du Tibet.*

Huth, B. M. = *Geschichte des Buddhismus in der Mongolei.*

Jäschke = J., *Tibetan-English Dictionary.*

Kovalevski = K., *Dict. mongol-russe-français.*

Mong. Pharm. = *Mongol Pharmacopœia* (see p. 443).

27

Pol. D. = Polyglot Dictionary, i. e., *Yü č'i se ts'i ts'iṅ wen kien* 御製四體清文鑑, 36 vols.

Ramsay = *Western Tibet: a Practical Dictionary of the Language and Customs of the Districts included in the Ladák Wazarat*, by Cap. H. Ramsay (Lahore, 1890).

Roero = *Ricordi dei viaggi al Cashemir, Piccolo e Medio Tibet e Turkestan in varie escursioni fatte da* O. Roero dei Marchesi di Cortanze, 3 vols. (Torino, 1881).

Tib. Pharm. = Tibetan Pharmacopœia *Fan Han yao miṅ* 番漢藥名 (see p. 438).

Vigne = *Travels in Kashmir, Ladak, Iskardo*, by G. T. Vigne, 2 vols. (London, 1842).

INTRODUCTORY.

Originally it was my intention to deal only with Chinese loan-words in Tibetan, but closer investigation brought out the fact that such a limitation would not render full justice to the subject. The mere existence of Chinese elements in Tibetan does not yet signify that these penetrated into Tibet straight from China. The chaos of languages which we encounter in central Asia, and which have been more or less subjected to Chinese influence, is such that Chinese words have also reached Tibet by way of Turkistan through the medium of Turkish idioms: thus *pag-ši* (No. 157), Chinese in its foundation, was adopted by the Tibetans from the Uigur; while *yam-bu* (No. 164), likewise of Chinese origin, came to them from Eastern Turkī. Other Chinese terms transmitted to Iran reached the Tibetans even from Persia or India (No. 120). On the other hand, Tibetan words have been claimed as Chinese, like *stod-ja* (No. 107), which in fact is derived from an Indian vernacular. A word like *a-p'im* ("opium;" No. 61) may easily be mistaken for Chinese, while it actually hails from India. This complex state of affairs led to the conclusion that a thorough investigation of all loan-words to be found in Tibetan would be required if satisfactory results were to be insured. Hence Indian, Persian, Arabic, Uigur, Turkī, Mongol, Manchu, Chinese, and finally Portuguese, Anglo-Indian, and English loan-words in Tibetan, are discussed on the following pages. Only those, however, which are generally used in the Tibetan language, have been treated here. Each dialect naturally has peculiar loan-words, which remain to be studied as our knowledge of dialectic lexicography advances. The outpost languages along the western and southern border-lands are influenced by the respective Indo-Aryan forms of speech prevailing in those regions. According to A. Cunningham, Tibetan *t^cul* ("egg") is a Kāçmīrī word, and doubtless there are more Kāçmīrī words to be found in Ladākhī. On the southern frontier of Bhūtān, many words and idiomatic phrases have been adopted from the Assamese, Bengālī, and Hindustānī, while the language spoken on the northern frontier represents a purer form of Tibetan.[1] Also the non-Aryan languages spoken in the Himālaya may have influenced Tibetan dialects. The origin of a word like *be-da* (written *bhe-da* and *ḥbe-dha*), designating a class of professional musicians, usually Mohammedans of Balti extraction, is still unknown.

In the *Li-šii gur k^caṅ*, a lexicographical work written by bZod-pa and Kun-dga don-grub and printed in 1741, three words are recorded as being derived from the language of Žaṅ-žuṅ (the ancient name of the province of Guge in western Tibet),—

1. *sle-tres* (*Mong. Pharm.*, No. 49: *sli-tri*; Mongol *liduri*), according to Jäschke, "name of a creeper or climbing plant;"

---

[1] RENNIE, *Bhotan and the Story of the Dooar War*, p. 25. Thus we have a Tibetan word *la-ta* or *la-da*, designating a kind of yarn imported into Tibet from Assam and Bhūtān; and *abu-ras* or *beu-ras*, the name of a coarse raw silk, likewise imported into Tibet from Assam by Bhūtān traders.

it is identified with Chinese *k'u-šön* 苦參, Sophora flavescens, the bitter root of which is employed in medicine; Rehmann's description of the root called by him *lidri* agrees with this. [1]

2. *tres-sam* (or *tre-sam*), a medicine in the shape of a powder (equivalent of *p'ye-ma*). The element *tres* occurs also in *kru-kru-tres*, cited in the same work and identified with Sanskrit *citraka*.

3. *ćur-nis* (JÄSCHKE: *ćur-ni*), flour (only in medical writings). Chandra Das thinks that this word is derived from Sanskrit.

Among the loan-words received from India, those of purely Sanskrit origin are of secondary importance, while those adopted from the Indian vernaculars are of prime interest. The Apabhraṁça loan-words are here clearly distinguished for the first time.

The Iranian loan-words in Tibetan are all derived from Persian, not from Pahlavī or Sogdian. [2] The only traceable Pahlavī prototype may be the tribal name Tādžik (Persian Tāzī) that underlies the Tibetan transcription Ta-zig or in more Tibetanized writing sTag-gzig (the Ta-ši [*džik] 大食 of the Chinese). The Tibetan name usually relates to Persia and the Persians, designated, though more rarely, also as Par-sig. Tāranātha employs Ta-zig also with reference to the Moguls of India. It is difficult to determine the route over which the Persian loan-words have migrated. Many may have come from Hindustānī; others may belong to an earlier period, and be due to a direct contact of Persian civilization with western Tibet. Arabic words were partially borrowed from Hindustānī, partially from Persian, or were spread also by Mohammedan traders

---

[1] See H. LAUFER, *Tib. Med.*, p. 58.

[2] The only Tibetan word thus far known which is derived from the Pamir languages is *ću-li* (see this volume, p. 82, note 4).

in Tibet. Several loans in Tibetan were formerly taken for Mongol. Now that, owing to the discovery of fresh material in Turkistan, our knowledge of Uigur has considerably advanced, we know that Mongol has borrowed a great deal from the Uigur, and it is certain that several alleged Mongol loan-words in Tibetan are in fact of Uigur origin.

The relation of Tibetan to Mongol words is not always perspicuous. Both languages have a stock of words in common; but while Mongol has borrowed much from Tibetan, and Tibetan a small proportion from Mongol, it is not easy to decide in every case which is the borrower or which the debtor: thus, for instance, in the case of Tibetan *ža-la*, *žal*, *žal-ba*, *šal-ba* ("floor of chunam, earth-pisé"), and Mongol *šala* (Manchu *jelehen*; *Pol. D.*, 20, p. 3).

In the case of Chinese loan-words a clear distinction has been made between Old and Modern Chinese. The former group exhibits the ancient initials and finals, as they were still characteristic of the Chinese language of the T'ang period, and which have apparently survived in Tibetan ever since that time. These words are living witnesses of the former conditions of Chinese phonetics, and are of primary significance for Chinese philology. The attention of sinologues may specially be called to the presence of a final liquida in Tibetan representing Chinese final *t* or *k*: *pir* for **bit* 筆 (No. 229), *par* and *spar* for **pat* 八 (No. 230), *yol* for **yüet* 月 (No. 227), *gur* for **gwok* 國 (No. 228), and the transcription *sar* for **sat* 薩 (see below, p. 419). The same phenomenon is found also in allied words of the two languages, as in Tibetan *dar*, compared with Chinese **dat* 達 (see above, p. 119). While the early loans in Tibetan are fixed according to rigid rules, there is much arbitrariness in the transcription of modern Chinese; many of these are derived from colloquial Chinese, chiefly from the dialect of Se-č'uan. Most Chinese traders in Tibet hail from this province.[1] It is of particular interest that we possess several loan-words in two forms, — in an ancient and a modern style of transcription, — thus illustrating the very development of Chinese phonology. Neither in this nor in any other department is an effort made toward completeness: many loan-words had to be set aside for the present, as their parentage has not yet been traced.

In *T'oung Pao* (1914, p. 89) Tibetan *mog* in *mog-ša* ("mushroom") has been compared by me with Chinese *mo-ku* 摩菇. This equation, however, is improbable, as the element *mo* 摩 was always devoid of a final explosive. The combination *mog-ša*, moreover, is restricted to western Tibet, where Chinese influence could hardly be expected, while the common Tibetan term is *ša-mo*, colloquially also *ša-mań*, which appears to be a genuine Tibetan word. Further, Chinese *mo-ku* is probably not of Chinese origin, but, as proposed by P. P. Schmidt,[2] may be borrowed from Mongol *mugu*, Kalmuk *mögö* (Solon *mogu*, Tungusian *mogo*, Manchu *megu*). The best mushrooms are still supplied to China from Mongolia.

It is not likely that Tibetan *se* (or *seu*)-*ạbru* ("pomegranate," Punica granatum)[3] is a transcription of Chinese *ši-leu*, as I endeavored to explain in *T'oung Pao* (1914, p. 90). Chinese *ši* 石, formerly with final guttural explosive, yields Tibetan *šik* (see No. 226; and above, p. 118, No. 45). In the Mahāvyutpatti (section of trees), moreover, we find as translation of Sanskrit *dāḍimva* (*dāḍima*)-*vṛiksha* ("pomegranate") Tibetan *bal-poi seu-šiṅ*; that is, *seu* tree of Nepal.

---

[1] Th. Manning, who visited Lhasa in 1811, refers to "the common vulgar Chinese at Lhasa, speaking the Szechuen dialect" (C. R. Markham, *Narratives*, p. 260).

[2] *J. Peking Or. Soc.*, Vol. IV, p 36; and *Mandarin Grammar* (in Russian), p. 94.

[3] Written *ze-ạbru* in the *Hua i yi yü*.

The plain *seu* is credited by Jäschke with the significance "pomegranate," but the element *se, seu*, has a wider application in the nomenclature of Tibetan botany. Its principal meaning is "thorn, thorny shrub:" *se-šiṅ* ("shrub good for hedges"), corresponding to *šan kou nai* 山枸柰 (Manchu *sira mō*, Mongol *ukhana sibtur*; *Pol. D.*, 29, p. 29; not identified, but Chinese 山 indicates clearly that the question is of a wild plant, not of a cultivated species); *se-luṅ*, "thorns" (*Si yü tʿuṅ wen či*: E. von Zach, *Lexic. Beiträge*, Vol. III, p. 111; *ibid.*, p. 127, we have *se* in the sense of "fruit-tree"); hence the derivations *se-ba* ("rose") and *g-ze-ma* ("caltrop," the spiny fruit of Tribulus terrestris, Chinese *tsi-li* 蒺藜: *Pol. D.*, 29, p. 15). The term *seu šiṅ*, further, is identified with Chinese *tu li* 杜李 (a kind of wild pear or berry, Pyrus baccata; Manchu *uli*, Buryat *ulir*, Mongol *üril*: *Pol. D.*, 28, p. 54). Moreover, there are *seu dmar čʿuṅ* ("small red *seu*," that is, cherry), answering to Chinese *yiṅ tʿao* 櫻桃, Manchu *ingduri*, Mongol *ingdōr* (*Pol. D.*, *ibid.*); *rgya seu* ("Chinese *seu*"), answering to Chinese *hua huṅ* 花紅 ("apple," Pyrus malus), Manchu *nikan uli*, Mongol *kitat üril* (*ibid.*); *dreṅ siu šiṅ* ("hazel-nut"), Chinese *čen-tse* 榛子;[1] and *sba siu* ("a red currant"), Chinese *mao tʿiṅ tse* 茅篠子, Mongol *ükär-ün nidu* ("ox-eye"), Manchu *jali*; *siu nag* ("mulberry-tree"), Chinese *saṅ* 桑 (*Pol. D.*, 29, p. 22). In view of these various names, we may well assume that *se* or *seu* is an indigenous word, and that *se-ḥbru* means "*se* seeds." The term given in the Mahāvyutpatti may indicate that the pomegranate was introduced to the Tibetans from Nepal.[2] Rockhill (*Diary*, p. 340) gives for the fruit a Tibetan

[1] Tibetan *dreṅ*, *ḍeṅ*, is a transcription of Chinese *čen*. In the modern transliteration, Chinese palatals, if followed by *e* or *ö* and *a*, are reproduced in Tibetan by means of the cerebrals: thus *ṭeṅ* for *čeṅ* 正, *ṭʿaṅ* for *čʿaṅ* 昌. More examples are given farther on.

[2] The following names of products are labelled "Nepalese:" *bal-po gur-gum* ("Nepalese

word *supoṅ* that I cannot explain. The Mongols use the Tibetan word in the form *simbru;* besides, they have adopted *anar* from Persian *anār* انار.

Tibetan *sraṅ* ("ounce") is not necessarily to be connected with Chinese *liaṅ* 兩. As *sraṅ* means also "balance, steelyard, weight," it may be derived from the verb *sroṅ-ba* ("to straighten, to render straight"), and accordingly be a genuine Tibetan word. The curious term *sa-tᶜe*, hitherto traced only in Tāranātha, cannot be derived from Chinese *siao tᶜaṅ* 小堂, as GRÜNWEDEL [1] is inclined to think; a Chinese *tᶜaṅ* would be reproduced only by a corresponding *tᶜaṅ* in Tibetan. The term decidedly is not Chinese.

I. SINO-TIBETAN NOTES.

Owing to the geographical position of the country and historical agencies resulting from it, the intellectual culture of Tibet is of a dualistic character in its absorption of foreign ideas which have flooded over its southern border from India, and which have penetrated eastward from China. Of these two currents of outward influence, the infusion of Indian ideas has always been strongly emphasized and placed in the foreground of scientific interest, so much so that Tibetan studies are usually regarded as a mere side-issue or auxiliary department of Sanskrit philology. The investigation of the Chinese share in the framing of Tibetan culture has almost wholly been neglected, and has not yet received the attention which it merits. Even before the days of Csoma, the importance of this subject was

---

saffron"), *bal sraṅ* ("Nepalese pease"), *bal glaṅ* ("ox of Nepal, elephant"), *bal mda* ("match-lock made in Nepal"), *bal ṭam* ("Nepalese silver coin;" in opposition to *rgya ṭam*, "Indian rupee"), *bal ñiṅ* or *sñiṅs* ("Nepalese rupee"), *bal dril* ("bell," also "cotton stuff made in Nepal"), *bal čᶜol* or *sbug* or *bal-poi sbug čᶜal* ("cymbal manufactured in Nepal").

[1] *Tāranātha's Edelsteinmine*, p. 163 (Petrograd, 1914).

pointed out by the far-sighted KLAPROTH,[1] who held that Tibetan literature was formed by a blending of the literatures of Tibet and China, and that many Chinese books on history, astronomy, medicine, and other sciences, have been translated into Tibetan, and that the arts came to Tibet from China together with printing. Dr. F. W. Thomas, librarian to the India Office, wrote me in April, 1915: "Information concerning Tibet which is obtainable from Chinese sources is a matter which presents difficulty to those of us who approach Tibetan questions mainly as Indianists or Sanskritists, and in several matters I have myself felt the need of enlightenment from the Chinese side: one cannot but feel that the Chinese influence must in reality be very great." In the same manner as there is a Tibetan-Sanskrit philology, there is also a Sino-Tibetan philology, some problems of which are here briefly discussed by way of introduction to the subject of Chinese loan-words. The sources utilized for this purpose will be indicated at the same time.

Above all, Chinese influence is manifest in Tibetan historiography. India transmitted to Tibet religious thoughts, but, lacking herself the historical spirit, had no lesson to inculcate in the methods of recording history. The Chinese, with their keen sense for the chronicling of events and dates, became the teachers of the Tibetans in historical matters. W. W. ROCKHILL[2] expressed the opinion that the Tibetans commenced writing their national history in the reign of Ral-pa-čan (816—838), that he was the first sovereign who appears to have paid any attention to the annals of his country, and that he had all the events of his reign recorded according to the Chinese system

---

[1] *Nouvelles annales des voyages*, 2me série, Vol. XIV, 1829, p. 275.

[2] *The Life of the Buddha*, p. 203.

of chronology.[1] This may now be more exactly determined according to the Tibetan Annals (*rGyal rabs*, fol. 96).[2] There we find the following account embodied in the history of King Ral-pa-čan:

"The book dealing with the history of China and Tibet (*rgya bod lo-rgyus deb-tʿer*) was compiled by the chronicler Su Kʿyi-han,[3] who lived at the time of King Tʿai-dzuṅ[4] of China, the events being recorded in the order of their succession. At a later date this work was rendered into Tibetan in the great monastery Šiṅ-kun by the Chinese translator (*lo-tstsʿa-ba*) U Gyaṅ-dzu.[5] As, however, there were some dates not in harmony in the Tibetan and Chinese histories, and also some discrepancies in the names of persons of that period, the Lama Rin-čʿen Grags with the title *kuo ši* (國 師),[6] at the time of his sojourn in China, collated the Tibetan and Chinese annals as far as the history of Sino-Tibetan

---

[1] *The Life of the Buddha*, p. 224. The source for this statement appears to be the *Bodhi-mör* translated by I. J. SCHMIDT (*Geschichte der Ost-Mongolen*, p. 361), but there are several misconceptions in his translation.

[2] This text closely follows the one translated by the writer in *Tʿoung Pao*, 1914, pp. 70—72.

[3] A Chinese name in Tibetan transcription.

[4] Thus transcribed in Tibetan. From a phonetic point of view we should regard this name as that of the Emperor Tʿai-tsuṅ 太宗 (627—649). It is not likely, however, that under his reign a book on Tibetan and Chinese relations was written, when these were in the initial stage: the year 634 is the first chronicled in the Tʿang Annals with regard to Tibetan events. It seems more probable that, with the assumption of a slight clerical error, the Emperor Tai-tsuṅ 代宗 (763—779) is here understood.

[5] This transcription would seem to correspond to a Chinese *Wu Kiaṅ-ču*; but only a Chinese source could yield the characters with which the name is written. SCHMIDT (*l. c.*, p. 361) writes this name *Uk'jangtschi*. A Stūpa named Šiṅ-kun is situated in Nepal northwest of Yam-bu (VASILYEV, *Geography of Tibet*, in Russian, pp. 61, 65).

[6] Tibetan *Gu šrī Rin-čʿen Grags* (SCHMIDT: *Kuks'ri Erdeni*). Chinese *ši* 師 is often confounded or adjusted with Sanskrit *çrī*, and is hence thus written; regarding this title see Glossary (No. 289). In the text the question is not of "Tibetan and Chinese yearbooks," as translated by SCHMIDT, but a single, individual book is understood; nor did U Gyaṅ-dzu continue the work and write it in Chinese and Tibetan, but he merely acted as translator into Tibetan of a Chinese book.

interrelations was concerned: in the *wood female fowl* year [1] he published his results in a book in the great monastery Šiṅ-kun, and rendered this work accessible to all. What is offered here [2] is merely an extract; he who desires to read in detail the history of Sino-Tibetan interrelations and the annals of the Son-in-law and Father-in-law, [3] may look up that very book."

The adjustment of Tibetan and Chinese history seems to have taken place under Kᶜri gtsug lde btsan Ral-pa-čan, for in the chapter devoted to his reign the *rGyal rabs* inserts a somewhat lengthy dissertation on Tibetan events in the same manner as represented in the Chinese annals. In fact, this is a very succinct abstract, ranging from the time of King gNam-ri sroṅ-btsan down to Ral-pa-čan, which in all probability is based on the book of Lama Rin-čᶜen Grags, and partially is in striking agreement with the Tᶜang Annals. While Su Kᶜyi-han must have utilized for his compilation original documents and state papers of the Tᶜang dynasty, the Lama Rin-čᶜen Grags, in his work of collation, seems to have had recourse to the text of the Tᶜang Annals as we know them at present. The best evidence in support of this opinion is furnished by the date given in this text for the death of King Sroṅ-btsan sgam-po, which is fixed in the *iron male dog* year (*lčags pᶜo kᶜyii lo*), answering to the year 650; and this is exactly the year indicated in the Tᶜang Annals. [4] The Tibetan Annals allow him to live much longer. Sanang Setsen has him die in

[1] The cycle, as usual, is not indicated. The first year that could come here into question is 865 (as 805 is prior to the reign of Ral-pa-čan), and the latest date to be considered would be 1285, the *rGyal rabs* having been written in 1328.

[2] That is, the preceding account in the *rGyal rabs* regarding the relations of Tibet with China

[3] That is, the sovereigns of Tibet and China, who were allied by marriage.

[4] Bushell, *Early History of Tibet*, p. 12 of the reprint.

698.[1] It is noteworthy, however, that the date derived from the Chinese has not exerted any influence upon Tibetan historians, and that also the chapters of the *rGyal rabs* treating of that king's reign do not contain it; this account, indeed, reflects the purely Tibetan tradition. The case, therefore, is not such that the Tibetan Annals (I speak only from my experience with the particular work *rGyal rabs*, and do not mean to generalize on so large a subject) should present a medley of Tibetan and Chinese traditions, but that each group of traditions is dealt with separately. In the beginning of the episode mentioned, a Chinese work[2] is quoted under the title *žu tᶜu han čᶜan*, as furnishing the information that "1566 years after the Nirvāṇa of Buddha the Chinese sovereigns called Tᶜang arose, contemporaneous with the Tibetan king gNam-ri sroṅ-btsan." The latter clause was presumably not contained in that Chinese book, as this king, the father of Sroṅ-btsan sgam-po, is not mentioned in the Tᶜang Annals. Then follows in the Tibetan text the passage regarding King Sroṅ-btsan sgam-po and his war against the Tᶜu-lu-hun formerly translated by me.[3] The concordance of this text with the corresponding account of the Tᶜang Annals is so striking, that on renewed examination of the matter the conclusion seems to me inevitable that the Tibetan notice is copied from the Chinese. This is confirmed by the fact that it is followed by the Chinese date of Sroṅ-btsan sgam-po's death, as mentioned above. It is noticeable, however, that there are two statements in the Tibetan account which differ from the two *Tᶜaṅ šu*,—the one that the king marched to a place called Tᶜuṅ čiu, that is, Tᶜuṅ čou;

[1] SCHMIDT, *l. c*, p. 37. This is the date given in Bu-ston's *Čᶜos abyuṅ* (see *dPag bsam ljon bzaṅ*, p. 153, line 20).

[2] Styled a *rgya deb-tᶜer čᶜen-po*, "great Chinese book."

[3] *Tᶜoung Pao*, 1901, p. 450.

the other that he put his army in charge of an official named γYa-tʻuṅ, who defeated the Tʻu-lu-hun (Tʻu-yü-hun).[1] The Chinese text speaks in this place of the tribe Yaṅ-tʻuṅ 羊同, who united their army with that of the king, and together with him attacked the common enemy. The case is difficult to decide; the similarity of the two names is obvious, but, after all, may be also accidental. Whereas everything else in the Chinese text is correctly understood by the Tibetan, it seems preposterous to turn out against him an indictment of misunderstanding in this particular case. He may have interpolated here a national tradition, or may have fallen back on another Chinese source than the Tʻang Annals.

The first work of Tibetan literature which was edited in Europe, *mDzaṅs-blun*, a collection of Jātaka, presents itself as a translation from the Chinese, as expressly stated in the Index of the Kanjur,[2]

---

[1] De-nas blon-po γYa-tʻuṅ bya-ba-la dmag bskur-nas | tʻu-lu-hun-gyi yul ạjoms-su btaṅ-ba-la.

[2] Ed. of I. J. SCHMIDT, No. 339: rgya-nag-las ạgyur-ba snaṅ-ṅo. In the Berlin Kanjur (H. BECKH, *Verzeichnis der tibetischen Handschriften*, Vol I, p. 67 b) the colophon has been omitted. The colophon in the Index Schmidt does not mean, as erroneously translated by Beckh in imitation of Csoma, "scheint eine Übersetzung aus dem Chinesischen zu sein," but means only, "it has been translated from Chinese." The verb *snaṅ-ba* never has the meaning "to seem," but signifies "to be evident, to appear, to be in a certain condition," it always implies a positive fact, and takes the place of the copula *yod-pa*. In this sense it is still employed in the dialect of Amdo and in eastern Tibet; *zer-ba snaṅ* in Tāranātha, as quoted by Jäschke, means, "it is said." — The Kanjur Index of Derge (fol. 135 b) adds the following colophon to the work: "Translated from Indian and Chinese books by the Locāva Čʻos-ạgrub of ạGos." This personage is well known: he appears in the Index of Derge as translator from the Chinese of Ratnakūṭa, Vol. 2, No. 2; and of Sūtra, Vol. 5, No. 3 (likewise in the Berlin version: BECKH, p. 30 b). The writing Čʻos-ạgrub is but a variant of Čʻos-grub. Čʻos-grub (the name is thus written in this passage of the Derge Index, fol. 136), further, translated from the Chinese the Sūtra in Vol. 32, No. 8; from the Chinese only, according to Derge, but from Indian and Chinese books, according to the Berlin version (BECKH, p. 70). Also the rendering of the Suvarṇaprabhāsa-sūtra (Tantra, Vol. 13, No. 12) from the Chinese into Tibetan is due to him (the colophons of Berlin and Derge closely agree); moreover, three Dhāraṇī relating to Avalokiteçvara (BECKH, p. 124). P. PELLIOT (*Notes à propos d'un catalogue du Kanjur, Journal asia-*

and, according to the latter, repeated by CSOMA, [1] and SCHIEFNER. [2] Schiefner recognized also the Chinese influence exerted on the Tibetan renderings of Sanskrit names, and noticed a few Tibetan words as transcriptions (styled by him "corruptions") of Chinese; but, in view of the fact that the phonology of ancient Chinese was wholly unknown in his time, he did not arrive at any result, and in some cases was even led into error. J. TAKAKUSU [3] attempted to prove at length that, contrary to his former opinion, according to which the Chinese may be a translation of the Tibetan, the Tibetan text is rendered after the Chinese. He compared a number of Tibetan and Chinese names in the two texts as proof of their interdependence. These comparisons have a decided value from a point of view not entertained by the author. The Tibetan renderings give us interesting

---

tique, 1914, p. 143, juillet-août), very felicitously, has identified this Tibetan personage with Fa-č'eṅ 法成, of whom he has discovered numerous translations in the caves of Tun-huaṅ. Fa-č'eṅ, "a subject of the great Tibetans," lived in the Temple of the Sūtra (Siu to se 修多寺) of Kan-čou in Kan-su in the first half of the ninth century. It was accordingly at that time that the translation of the *mDzaṅs-blun* from the Chinese into Tibetan was accomplished. As shown by PELLIOT (*T'oung Pao*, 1912, p. 355; and *Journal asiatique*, 1914, p. 139, juillet-août), the Chinese text was translated in 445 from a work edited at Kao-č'aṅ (east of Turfan) by eight Buddhist monks, who had memorized the legends at Khotan. We are fortunate in now having the bibliographical history of this work clearly outlined; for the Tibetan work, frequently utilized in Schmidt's translation, has given rise to several unwarranted conclusions. For instance, in discussing the Hebrew and Indian parallels of the story of Solomon's Judgment, R. GARBE (*Indien und das Christentum*, p. 27) expressed the opinion that, as long as merely the Tibetan version in the Kanjur (that is, the *mDzaṅs-blun*) was known, it would have been justifiable to argue that the story had penetrated into Tibet through Christian mediation. Again, he asserted (on p. 29, note) that the story migrated from India by way of Tibet into China. As we now recognize, the route of migration led from India to Khotan and Turfan, from there to Kan-čou, the domicile of C'os-grub; and from the Chinese the story found its way into Tibetan literature. M. WINTERNITZ (*Geschichte der indischen Literatur*, Vol. II, p. 221) wrongly states that our work has been translated from Sanskrit, and that the Sanskrit original has not been preserved.

[1] *Asiatic Researches*, Vol. XX, p. 480.

[2] *Tib. Lebensbeschreibung*, p. 85.

[3] *Tales of the Wise Man and the Fool* (*J.R.A.S.*, 1901, pp. 447—460).

examples of the Tibetan method of transcribing Chinese sounds; and if we did not know the fact that the Tibetan translation was made in the first part of the ninth century, the very form of these transcriptions, which are recorded in full harmony with the phonology of Chinese as it prevailed in the T^{c}ang period, would entitle us to the conclusion that the Tibetan translation goes back to the same age. If, for instance, Chinese Po-lo-mo-ta 婆羅摩達 (Sanskrit Brahmadatta) is transcribed in Tibetan Ba-la-ma-dar, [1] we have the actual phonetic state of these syllables as it obtained during the T^{c}ang — as exactly, at least, as it could be rendered by Tibetan writing. Unfortunately we have as yet no critical edition of the Tibetan text; Schmidt's edition is very deficient, and the supplementary notes and emendations of Schiefner are not yet the last word in the matter. It would be a real feast to have for comparison a Tibetan manuscript of this work of the ninth century and a Chinese contemporaneous edition. But, with all the mistakes in the present Tibetan editions, it is not a hopeless task to get at, to a certain degree, the primary condition of affairs. The separating dot (*ts^{c}eg*) is often placed wrongly in Tibetan writing: the name Tsuu-ba-na-ta [2] must evidently be written Tsuu-ban-ta (probably even Tsuu-pan-ta), [3] and renders Čou-li-pan-to 周利槃多 (Sanskrit Cūḍapanthaka). [4] The same holds good for A-na-ta, which is to be corrected into An-ta, rendering *An-da (An-t^{c}o) 安陀, Sanskrit Andhra; and Kan (written ka-na)-ja-ni-pa-li,

---

[1] Erroneously taken by SCHIEFNER (*Ergänzungen*, p. 53) for a "corruption" of Sanskrit Balamitra.

[2] SCHMIDT's edition, p. 320, line 15.

[3] Takakusu's transcription Cūwa-na-ta (*l. c*, p. 456) is not correct: the double *u* is not intended for *ū*, but is employed to express the diphthong *ou* of the word *čou* 周.

[4] Thus SCHIEFNER, *Lebensbeschreibung*, p. 85. Takakusu proposes Çuddhipanthaka. Probably we have to read 般 *pan* instead of 槃 **bwan*.

rendering *Kan-ja-ni-pa-li 虔闍尼波梨, Sanskrit Kañjanī-pālī.[1] In the latter example the Tibetan transcription *ja* for 闍 is of especial interest, inasmuch as it affords additional evidence for the pronunciation *ja*, peculiar to this character during the Tʻang period. Other Tibetan names can be correctly restored on the basis of the Chinese equivalents. SCHIEFNER[2] was greatly puzzled over the name Ba-mi-su-tra, though he himself, as well as Schmidt, conjectured it as Vasumitra. The corresponding Chinese rendering 婆修密多羅 *Bwa-su-mit-ta-la shows us that the Tibetan must have been originally Ba-su-mi-tra, and that only a later copyist has reversed the syllables *su* and *mi*. In his very valuable supplementary notes on Schmidt's edition, Schiefner has attempted to restore the readings of several names on the basis of their Sanskrit equivalents. It now turns out that these emendations are not always acceptable, and that the names, as offered in Schmidt's text, are correct when viewed in the light of the Chinese models of which they are a transcript. SCHIEFNER[3] proposed to read Ši-bi instead of Ši-byi on the ground of the Sanskrit form Çibi. This would be very well, but the Tibetan translator had before his eyes Chinese 尸毘 **ši-b'i*, and with perfect correctness transcribed in Tibetan *ši-byi*. This is by no means a single case, but there are at least three more where Chinese 毗 **b'i* (*pʻi*) is reproduced as *byi* in the Tibetan text.[4] These transcriptions tend to confirm the opinion that the subscribed letter *y* serves the purpose of expressing the palatalization of the consonant to which it is attached.[5] The reading Ši-byi, though rendering Sanskrit Çibi,

---

[1] *Hien yü yin yüan king* 賢愚因緣經, Ch. 1, p. 3 (edition of Nanking). In Schmidt's edition the name is erroneously printed Ka-na-ši-ni-pa-li.

[2] *Ergänzungen*, p. 9.

[3] *Ibid.*, p. 5.

[4] TAKAKUSU, *l. c.*, p. 456.

[5] *Tʻoung Pao*, 1914, p. 97; 1915, p. 287, note 5.

is therefore legitimate, and was no doubt contained in the *editio princeps* of the work. For the same reason, I should think, the name Ka-byin or Ka-pyin need not be restored into Ka-pʿi-na [1] (Sanskrit Kapphiṇa), but most probably into Ka-pʿyi-na. Chinese 賓 *pin* in Piṇḍoladvāja is transcribed *pʿyin* in Tibetan.

The most interesting of the Tibetan transcriptions is *sar pʿag* for Chinese 薩薄 **sat-pak* or *sat-bak* (*sa po*). Schmidt treated this word as a proper name. SCHIEFNER, [2] consulting the Mongol version, encountered there *yäkä sartawaki*, and hence inferred a Sanskrit mahāsārthavāha ("wholesale-dealer"). Takakusu, who did not avail himself of Schiefner's work, arrived from Chinese *sat-pak* at an hypothetical Sanskrit *satpati* ("good lord"), and took the Tibetan in the sense of "householder" or "lord." At any rate, the Tibetan is a transcription of Chinese, and perhaps was originally *sar pag* (the letters *p* and *pʿ* being easily confounded). Schiefner's conclusion is corroborated by the geographical catalogue contained in the Mahā-māyūrī vidyā-rājñī, edited and commented by S. LÉVI. [3] Here we meet the name Sārthavāho, [4] to which corresponds in the Chinese version Sa-tʿo-pʿo-ho 薩陀婆訶, defined as *šaṅ ču* 商主 ("chief of the merchants"), and rendered into Tibetan as *ded dpon* ("head of a commercial concern"). [5] Hence it follows that the above Sa-po is abbreviated for Sa-tʿo-pʿo-ho, and is indeed modelled after Sanskrit sārthavāha. [6] This Sanskrit word has left

---

[1] *Ergänzungen*, p. 38.

[2] *Ibid.*, p. 24.

[3] *Journal asiatique*, 1915, janv.-févr., pp. 19—138.

[4] *L. c*, p. 37.

[5] This term means also "sea-captain" (for instance, *mDzaṅs blun*, p. 29, line 4; and Avadānakalpalatā, prose ed., p 57, line 13) At present *tsʿoṅ dpon* is usually employed in the same sense as *ded dpon* above

[6] Compare also PELLIOT, *Bull. de l'Ecole française*, Vol. IV, p 356, note 1.

28

a trace in the modern Tibetan word *ša-po* or *ša-bo*, designating a trade agent, especially applied to Tibetan women living with Chinese merchants and transacting their sales. This word bears no relation to the East-Tibetan word *ša-bo* ("friend"), as intimated by ROCKHILL, [1] which would not be applicable to women (*ša-mo*, "female friend"). Compare also *tsʿoň ša-po* ("associé de commerce") in the *French Dictionary*, p. 982.

In chapter XXXVII of Schmidt's edition (text, p. 264; translation, p. 331) a king of Jambudvīpa is mentioned by the name Ba-si-li. A special interest is attached to this name, inasmuch as SCHIEFNER [2] was inclined to equalize it with Greek βασιλεύς, but recognized that this conclusion was somewhat rash. In the Chinese version [3] the king is styled 波塞奇 Po-sai-ki (*Pwa-sik-ki; Sanskrit Vāsuki?), so that it is justifiable to regard the Tibetan transcription Ba-si-li as a modern error for Ba-si-ki. At all events, there is little chance for Schiefner's conjecture to survive. There are several other names which in the modern Tibetan versions appear more or less mutilated,—thus Leu-du-či or Leu-du-čʿa, name of a Brāhmaṇa, identified by SCHIEFNER [4] with Rūdrāksha, the Chinese equivalent being Lao-tu-čʿa (*Lo [ro, ru]-dak-dža) 勞度差. [5] Here we have a good example for the fact that, contrary to the opinion of M. Pelliot, [6] 度 is indeed used in transcriptions with the phonetic value *dak*. The name Kyun-te [7] has been identified by SCHIEFNER [8] with Sanskrit

---

[1] *J. R. A. S.*, 1891, p. 123.

[2] *Ergänzungen*, p. 56.

[3] Ch. 11, p. 24 b.

[4] *Erganzungen*, p. 4

[5] Ch. 1, p. 3 (edition of Nanking).

[6] *T'oung Pao*, 1915, p. 24 (compare *ibid.*, p. 422).

[7] SCHMIDT's translation, p 168

[8] *Lebensbeschreibung*, p 85; *Ergänzungen*, p. 33

Cuṇḍa. This is not plausible, for phonetic reasons. Indeed, the Chinese version [1] offers Kiün-tʿi (*K'ün-de) 均提, which would rather seem to go back to a Sanskrit prototype Kuṇḍī, Kuṇḍina. The name of the animal *kun-ta*, [2] not yet traced to its Sanskrit original, is written in the Chinese text [3] *kü-tʿo* (*ku-ⁿda) 鋸陀; hence we may presume that the primeval Tibetan transcription was likewise **kun-da*.

It is thus demonstrated that for a future edition of the *mDzaṅs blun* the Chinese text must be carefully utilized, in particular for the spelling of the proper names. Unfortunately the Chinese text, as we have it at present, has undergone some alterations, and, as Takakusu thinks, also corruptions. I do not concur, however, with this scholar in the opinion that "the Chinese original used by the Tibetan translator seems to have been pretty corrupt, and contained some miswritten characters peculiar to Chinese." [4] Still less do I believe that his heading "transcriptions by the Tibetan translator done without understanding the original Chinese" is justified. What Takakusu here offers are for the greater part very satisfactory Tibetan transcriptions in harmony with ancient Chinese phonology. If the Tibetan has Šiṅ-rta čʿen-po ("Great Carriage") as the name of a king, where Chinese now has 摩訶羅檀那 (Sanskrit Mahāratna), the chances are that the Chinese version was also formerly based on the reading Mahāratha. The examples quoted have convinced me that the Tibetan transcriptions of the Buddhist translators were made according to a regular system fixing actual phonetic conditions

---

[1] Ch. 5, p. 13b

[2] SCHMIDT's translation, p. 101.

[3] Ch. 3, p. 1.

[4] The examples cited by Takakusu in support of this verdict, in my opinion, are to be explained as misunderstandings or carelessness on the part of the later Tibetan copyists, but were not necessarily contained in the original edition.

of the Chinese language of the T'ang period; and I am therefore disposed to believe that an early Tibetan edition of the *mDzaṅs blun*, collated with a correspondingly early Chinese version, would reveal a perfect coincidence between the two texts, which in the course of time have naturally become impaired. I hope these observations will prove also that a knowledge of Chinese is indispensable for a successful study of the Tibetan Buddhist works translated from that language. [1]

Besides *mDzaṅs blun*, several other works of the Kanjur have been translated from Chinese. [2] H. Beckh, [3] in his useful catalogue of the Berlin manuscript Kanjur, has paid some attention to the Chinese titles given in Tibetan script, and has endeavored to restore them with the assistance of F. W. K. Müller and Hülle. As a matter of principle, Beckh is right in retaining these titles exactly in the form in which they are recorded in the Kanjur edition studied by him. However, we must not halt at that point; when these titles have been correctly restored to their Chinese model, it is always possible to emendate the Tibetan transcriptions, which have certainly been disfigured by a host of copyists. The most frequent and flagrant error committed by them, as already mentioned, is the wrong insertion of the separating dot (*ts'eg*). The interesting fact brought out by the Chinese titles in the Berlin Kanjur is that they belong to a more recent form of transcription than the corresponding titles as given in the Index edited by I. J. Schmidt and in the Analysis of Csoma. [4] The Tibetan title in Sūtra, Vol. 32,

---

[1] As to Chinese words occurring in the *mDzaṅs blun*, see Glossary, Nos. 218—220.

[2] Laufer, *Dokumente*, I, pp. 51—52 Add Nos. 690, 872, 873.

[3] *Handschriften-Verzeichnisse der Königl Bibliothek zu Berlin*, Vol. XXIV: *Verzeichnis der tibetischen Handschriften*, Vol. I, p. VII.

[4] This portion of my article was written in the summer of 1914, eight months before

No. 7, of the Berlin edition, *Tai p'aṅ pe na hwa pao din giṅ*, corresponds to 大方便佛報恩經 (BUNYIU NANJIO, No. 431).[1] Hence it follows that the Tibetan title is to be restored to *Tai p'aṅ pen hu* (= *hwa*) *pao ṅin* (*ṅen*) *giṅ*. Tibetan lacks the fricative *f*, and renders it by means of *p'* or *p'h* when the vowel following is *a*, and by *h* when the vowel is *u* (or *a*). Tibetan *hwa* (pronounced *hu*) transcribes Chinese *fu* 佛, and, accordingly, was recorded at a time when 佛 was no longer sounded *but*, but *fu*. In the Index of Schmidt (No. 351, p. 53) we find, in the same title, this word transcribed *p'ūr*[2] and by Csoma *p'ur*. F. W. K. MÜLLER[3] has invoked this transcription as proving the articulation of a final *r* in 佛. At the end of the Tibetan version of the Aparimita-āyurjñāna-mahāyānasūtra from Tun-huaṅ, a manuscript from the latter part of the ninth century, an edition of which is being prepared by me, we meet the formula *na-mo 'a-myi̱-da*[4] *p'ur*, apparently transcribed from the Chinese 南無阿彌陀佛 ("salutation to Amitābha Buddha"); *p'ur*, accordingly, was in the T'ang period the Tibetan transcription of Chinese **but*, which may have been articulated also *fut*, *fur*.

Pelliot's study *Notes à propos d'un catalogue du Kanjur* appeared, or at least reached me in Chicago. Pelliot has dealt with the same titles and their transcription in Tibetan. I leave the text of my manuscript as it stood, omitting only such matters as have been sufficiently cleared up by Pelliot. Our studies supplement each other.

[1] The translation of the title of this Sūtra, as given by BECKH, is hardly correct. The phrase *t'abs-la mk'as-pa č'en-po* cannot be construed as depending upon *drin lan bsab-pa* ("great dexterity in the retribution of the friendliness of the Buddha"), but is an independent adverbialis, as shown also by the reading *č'en-pos* in the Berlin text and by the use of the punctuation-mark after *č'en-po* in the Narthaṅ edition, and as is confirmed by the Chinese version. Dr. Beckh should not have failed to consult Nanjio's translations.

[2] Not *phū-ra*, as erroneously written by Beckh, and after him by Pelliot.

[3] *Uigurica*, II, p. 94 (*A. P. A. W.*, 1911).

[4] By *i̱* the inverted *i* is understood. Aparimita is written throughout in this manuscript *apari̱mita*, where the vowel *i* certainly is short. M. Pelliot's generalization based on another manuscript, that the inverted *i* serves for the reproduction of Sanskrit *ī* (*T'oung Pao*, 1915, p. 2), does not hold good. The case will be fully discussed in my forthcoming edition.

It is more likely that the initial Tibetan aspirate served for the reproduction of Chinese *f*, for Chinese *but* or *bur* could have been easily written and pronounced *bur* by the Tibetans. Index Schmidt and Csoma, which mirror the older stage of transcription, have *byan* (instead of *pen*) for 便, *pou* for 報, *'in* for 恩, and *kyeṅ* for 經.

In the Berlin copy the title of the Sūtra is accompanied by the words *ju* (followed by the letter *ṅ*) *p'i ma ti yi*, which have defied the acumen of F. W. K. Müller and Hülle.[1] It is possible, however, to identify these words. In the Index of the Kanjur printed in the monastery of Čo-ne in Kan-su, the Tibetan title closes with the statement, *bam-po bdun-pa leu dgu-pa*, "containing seven sections and nine chapters." This exactly agrees with the statement of Nanjio, "seven fasciculi, nine chapters."[2] It is therefore obvious that *p'i-ma* is to be read *p'im*, as correctly written by Csoma and in the Index Schmidt,[3] and corresponds to Chinese *p'im (*p'in*) 品, and that *ju p'im* is intended for 九品. The writing *sū* in Csoma and Schmidt, however, seems to be intended for *sü* 序. The words *ti yi* in the Berlin text answer to *dei 'yir* in Csoma and Schmidt.[4] Hence the conclusion is manifest that the Tibetan transcriptions represent Chinese 第一 *di yit (*ti yi*), the Berlin copy reproducing the modern, the editions of Csoma and Schmidt the ancient, pronunciation of the T'ang period. Again, *dei* is an old transcription of 第, while *ti* of the Berlin copy represents the modern pronunciation.[5]

---

[1] On p. 192 this addition is indicated by an interrogation-mark.

[2] But Nanjio is not correct in stating, "deest in Tibetan"

[3] To which BECKH (p. 69 b, note 2) wrongly ascribes the reading *phyim* (also adopted by Pelliot).

[4] Beckh transcribes *ayir* in the one case, and *iyir* in the other, but the way of writing the word is identical in both texts (the letter *'a* with subscribed *y* [*ya-btags*] and superscribed *i*, this graphic combination is never found in the writing of Tibetan words: indeed, this is the only example of its occurrence known to me).

[5] This distinction between two types of transcription is justly upheld also by Pelliot.

On p. 102 of Beckh's Catalogue we find the transcription of the Suvarṇaprabhāsa given thus: *tai č^ciṅ gi ma gvaṅ miṅ tsui šiṅ waṅ gyi*, to be corrected into *tai č^ciṅ gim* [1] *gwaṅ miṅ tsui šiṅ waṅ gyiṅ*, transcribing 大乘金光明最勝王經. In the transcriptions, as given in the Index Schmidt and by Csoma, older forms are partially preserved: thus, *śin* and *šiṅ* for 乘, *myau* for 明 (compare Japanese *myō*), where Csoma gives *med.* [2] *Tsai* in Schmidt for 最 is merely a slip for *tsui*, the vowel-sign *u* being dropped; Csoma's *jwai* for this character is very curious; the *va zur* is the semi-vowel *y* or *q*, so that the Tibetan is to be read *jvi*. The character 最 appears to have had an initial palatal sonant in the T^cang period. Again, 大 is rendered by *dei* in Csoma, *tā* in Schmidt; 經 by *kyaṅ* in the former, *kyin* in the latter.

The fact that the Chinese transcriptions of the Berlin copy of the Kanjur reflect recent phonetic conditions and have consequently been made anew, is clearly attested by the title *Pī du ts^ci ziṅ giṅ* for 北斗七星經, compared with the older *Bī du ts^cid ziṅ giṅ*, [3] where *ts^cid* corresponds to Chinese *ts^cit* 七, while *ts^ci* of the Berlin text presents the modern phonetic condition. As the Berlin Kanjur hails from the Lama temple Yuṅ-ho-kuṅ in Peking, it is very probable that it was also written there, and that the Chinese transcriptions were somehow adapted to the Peking pronunciation.

The only point on which I dissent from Pelliot is that I am inclined to identify the *ju* of the Berlin Kanjur with 九 rather than, as proposed by him, with 序.

[1] *Gim* (= 金 *kim*), not *gin*, as Beckh prefers to restore.

[2] F. W. K. MÜLLER (*Uigurica*, I, p. 11, *A. P. A. W.*, 1908), who was the first to restore the Tibetan title on the basis of Csoma's reading, conjectures *miṅ* for *med*; in view of the fact that *ṅ* and *d* are constantly confounded in Tibetan writing, this is quite plausible, but it may not be necessary to change the vowel into *i*; an older transcription *meṅ* for 明 seems conceivable

[3] *T^coung Pao*, 1907, p. 392.

A chapter from the Laṅkāvatāra-sūtra (Kanjur, Sūtra, Vol. 5, No. 3), according to the colophon, as given in the Index of Derge (fol. 124), has been translated from the Chinese [1] by Č'os-ạgrub of ạGos, [2] in correspondence with a commentary written by the Chinese teacher rBen-hvi. The latter name is written in the Berlin version Wen-hvi, in the Index Schmidt Wan-hvi. This points to a Chinese Wên-hui. In the same manner as Pelliot, [3] I have searched in vain for the Chinese personage with whom this name could be identified.

There are probably even more translations from the Chinese in the Kanjur than appear from several editions of it. The Berlin version and Index Schmidt have no colophon to the Atajñānasūtra (Sūtra, Vol. 10, No. 3); while the Derge Index annotates, that, according to the Index of lDan-dkar of the ninth century, [4] it was translated from the Chinese. This goes to show again that as early as the first part of the ninth century Chinese Buddhist works were rendered into Tibetan; and the Derge colophon of the next treatise, which is without a Sanskrit title and a translation from the Chinese, attests the fact that it was originally draughted in "old language."

The Saddharmarāja-sūtra (Sūtra, Vol. 22, No. 1) was likewise translated from the Chinese into "old language," as stated in the Index of Derge (fol. 131 b); but, as it was not transformed into new language, there were those who had their doubts about it. The Dharmasamudra-sūtra (Sūtra, Vol. 22, No. 12), according to the Derge Index, was translated from the Chinese and edited in

---

[1] This fact is not contained in the colophon of the Berlin version (Beckh, p. 30 b).

[2] See above, p. 415.

[3] *Journal asiatique*, 1914, juillet-août, p. 129.

[4] *Dokumente*, I, p. 51. This index is contained in the Tanjur, section Sūtra, Vol. 126 (P. Cordier, *Catalogue du fonds tibétain de la Bibl. Nat.*, troisième partie, p. 493).

new language.[1] Again, the Jinaputra-arthasiddhi-sūtṛa (Sūtra, Vol. 32, No. 5) was first translated from Chinese into "old language."[2]

In the Tanjur we find a Tibetan translation of the *Yin miṅ žu čeṅ li lun* 因明入正理論 (BUNYIU NANJIO, No. 1216). In the Palace edition the Chinese title is transcribed in Tibetan *g-yen miṅ g-žai* (with following *-g*) *čiṅ lii lun*.[3] This attempt is modern; if ancient, the final *p* of 入 would appear in Tibetan. The writing *yen* with the prefix *g* indicates the high tone of 因, while the plain *yen* is deep-toned. The presence of the same prefix, however, in transcribing 入, is an anomaly, as the latter has the deep tone corresponding to initial Tibetan *ž*; the vocalization also is doubtful (double *e* [*ai*?] with following *g*). The double *i* of *lii* seems to mark the rising tone of 理. The *Ta č'öṅ pai fa miṅ mön lun* 大乘百法明門論 (BUNYIU NANJIO, No. 1213) is likewise translated in the Tanjur under the title *Tai č'iṅ pai hā miṅ mun lun*,[4] again a transcription of recent times, as neither the initial palatal sonant *dž* of 乘, nor the final *k* of 百, is indicated. The letter *h* represents Chinese *f*, a sound which is lacking in Tibetan. Of other works in the Tanjur translated from Chinese, the titles are given only in Tibetan and Sanskrit.[5]

Besides the authorized works of the Canon, there are also uncanonical Buddhist writings translated from the Chinese into Tibetan. J. WEBER and G. HUTH[6] have edited and translated a Sūtra with the Tibetan title Saṅs-rgyas-kyi č'os gsal žiṅ yaṅs-pa snaṅ brgyad

---

[1] The Tibetan title is given there in the form *ap'ags-pa č'os rgya mts'oi mdo*, which is preferable to that of the Berlin version (BECKH, p. 55 b).

[2] Tibetan: sṅon rgya las ḥgyur-bai brda rñiṅ-pa-čan.

[3] P. CORDIER, *l. c.*, p 435.

[4] *Ibid.*, p. 386.

[5] *Ibid.*, pp 322 (No 27), 352 (No. 71), 478 (No. 16).

[6] *Das buddhistische Sūtra der "Acht Erscheinungen"* (*Z.D M.G.*, Vol XLV, pp. 577—591)

čes bya-bai mdo, "Sūtra, called the eight detailed phenomena, explaining the dharma of Buddha." [1] The Tibetan title is preceded by the words *rgya gar skad-du | pañ rkyañ rkyeñ*. As Huth recognized the latter as Chinese, he corrected *rgya gar* ("India") into *rgya nag* ("China"), and restored the Tibetan transcription to *p'āñ kiāñ hīñ*, [2] which he says corresponds to Tibetan *yañs-pa, gsal, snañ*. At the close of the treatise an additional note appears, from which it follows that Huth had meanwhile become doubtful of this supposition. He had encountered No. 463 in the "Verzeichnis" of Schmidt and Böhtlingk, where a work with the Sanskrit title āryapadayañyadarta (with the addition of an interrogation-mark) and the Tibetan title ạp'ags-pa gnam-sa snañ brgyad čes bya-ba t'eg-pa č'en-poi mdo ("Venerable Mahāyāna-sūtra, called the eight phenomena of heaven and earth") is listed. If this work should be identical with the above, Huth argued, the above title might be, after all, corrupted Sanskrit; and the work might have been translated from this language, not from Chinese. But he hastened to add that his former supposition is more probable, that the title was originally Chinese, that *rgya nag skad* ("Chinese language") was afterwards confounded with *rgya gar skad* ("language of India"), and that in this manner a scribe was induced to Sanskritize the Chinese title. This puzzle, however, is capable of solution. The Sūtra translated by Weber and Huth plainly reveals Chinese influence, which has been indicated by HUTH himself: [3] Heaven and Earth are opposed to each other, [4] and their harmonious union is alluded to, etc.

---

[1] A work of this title is contained in the Kanjur (Index ed. SCHMIDT, No. 1041), but whether it is identical with the Sūtra of Weber-Huth remains to be determined.

[2] The Chinese characters are not given, and I do not know what Chinese words Huth had in mind

[3] *L. c.*, p. 588.

[4] The translation "des Himmels Höhe, der Erde Rechte" cannot be defended; neither

The work, therefore, is a translation from Chinese, and Huth had no reason to yield to another suggestion. The work pointed out by him in the "Verzeichnis" is well known to me from two Peking editions. It differs in contents from the Sūtra of Weber and Huth, but it is likewise a translation from the Chinese. This is well attested by the fact that the Eight Trigrams (*pa kua*) are enumerated in the course of the work, — and these could not have occurred in a Sanskrit production, — and again by the mention in the Tibetan title of heaven and earth (*gnam sa*), which savors of Chinese philosophy. In the edition before me, the book opens with the words, "in Chinese" (*rgya-nag skad-du*), followed by the title *ₒārya par yaṅ gyad rta* or *ₒārya pa-ra yaṅ gyad rta*. Now, this "Chinese" title is apparently identical with the alleged Sanskrit title of Schmidt and Böhtlingk *āryapada-yaṅyadarta*, as given above: we have only to substitute the letter *d* for *r*, both being so frequently confounded, and to insert a dot after *gya*, and the identity is established. What to make of this dog-Sanskrit may be left to the decision of the Sanskritists; for myself, I do not believe that it is Sanskrit at all, except the first word, *ārya*. The rest is Tibetanized Chinese, maltreated by ignorant copyists. The resemblance of the two titles to the Chinese title of Huth is apparent. Huth's *paṅ* corresponds to our *pa-ra*, or *par*; that is, Chinese **pat* 八 ("eight"), which we meet with in the Tibetan translation of the title as *brgyad*. Consequently the next word *yaṅ* (Huth's *rkyaṅ*), perhaps 陽, must be the Chinese equivalent of Tibetan *snaṅ*. Instead of *gyad* we have to read *gyaṅ* (corresponding to Huth's *rkyeṅ*), which is doubtless intended for Chinese *kiṅ* 經 (see Glossary, No. 288). What Chinese word is

γyas nor *γyos* is the correct reading, but *yas* "below." Read, "Heaven above, and Earth below."

intended by *rta,* I do not know; in all probability, it is only an addition of the scribe, as the word 經 concludes the title.

In the Tibetan wooden tablets discovered by A. Stein in Turkistan, and, as it seems, chiefly referring to administrative matters, Chinese influence is likewise conspicuous, as far as can be judged at present from the stray notes published by A. H. FRANCKE. [1] Chinese, surely, are the "many words the meaning of which is still quite uncertain." "To mention only one instance," Francke remarks, "we do not yet know how to explain the local names Bod, Tibet, and Li, Khotan, when they are connected with numerals, — *bod-gnyis, li-bzhi,* etc., — as is often the case." Here we certainly face Chinese names of measurements, *li* being a reproduction of Chinese *li* 里 ("league, mile"), while *bod* may represent *pu* (**bu*, *bo*) 步 ("pace, a land measure of five feet"), although the latter never had a final explosive dental. [2] The word *žaṅ* does not mean "uncle," [3] but is a transcription of Chinese *šaṅ* 尙. [4] Terms of civility found in the documents appear to rest on imitation of Chinese style. If the writer speaks of himself as "I, the bad one" (*bdag ṅan-pa*), we are reminded of such Chinese phrases as *kua žön* 寡人, *yü* 愚, *pi* 敝, *tsien* 賤, and other 自稱之詞.

An interesting case in the history of Sino-Tibetan is mentioned by CHANDRA DAS. [5] The Chinese Buddhist priest, called in Tibetan Zan-tᶜan šan ši, [6] visited the monastery bSam-yas at the invitation

[1] *J. R. A. S.*, 1914, pp. 37—59.

[2] Such anomalies of transcription indeed occur: in *Pol. D.* (7, p. 33) we meet *bod* as equivalent of Chinese *pu* 部 ("section of a book"), which also is devoid of a final explosive dental.

[3] *Ibid.*, p. 43

[4] *Tᶜoung Pao*, 1914, p. 105.

[5] *Sacred and Ornamental Characters of Tibet* (*J. A. S B*, Vol LVII, 1888, p. 43).

[6] The title *šan ši* appears to be a transcription of Chinese *čan ši* 禪師, "master of meditation" (dhyāna).

of King K'ri-sroṅ ldeu-btsan, and was so much struck with the capacity of the Tibetan alphabet to express Chinese words that he undertook both to transliterate and to translate some Chinese works into Tibetan and certain Tibetan works into the Chinese language. In an inscription found at bSam-yas it is stated that this priest translated Chinese documents into Tibetan. On plate VI of the article of Das, a copy of this inscription in Chinese and Tibetan is reproduced. The Tibetan portion is clear and intelligible (see No. 285), but the Chinese characters are so disfigured that they defy reading.

Despite the preliminary notice of E. VON ZACH,[1] the Tibetan words hidden in a Chinese garb in the *Yüan ši* require more profound study. The Sino-Tibetan inscription from the epoch of the Mongols, published and translated by CHAVANNES,[2] furnishes several interesting examples of Tibetan transcriptions of Chinese words.[3]

A glance over Tibetan historical works — as, for instance, *Hor č'os byuṅ*, edited and translated by G. HUTH — is sufficient to convince one of the fact that this department of literature teems with transcriptions of Chinese names. Huth recognized these and their importance,[4] but did not identify them in his translation, as he planned to issue a commentary to it in a separate volume; his premature death unfortunately prevented him from carrying this plan into effect. This is most regrettable, as many of these Chinese names are not self-evident, and in their strange Tibetan garb are

[1] *China Review*, Vol. XXIV, 1900, pp. 256—257.

[2] *T'oung Pao*, 1908, pp. 418—421, and plates 28—29.

[3] See Glossary, Nos. 158, 214, 289, 306, 307. The Tibetan version is a literal rendering of the Chinese text. I hope to publish occasionally a complete transcription of the Tibetan version in comparison with the original.

[4] *Transactions Ninth Int. Congress of Orientalists*, Vol. II, p. 640 (London, 1893).

not familiar to sinologues, while Tibetan scholars unacquainted with Chinese are not in a position to understand them. The Tibetan transcriptions follow throughout the modern Northern Mandarin, and have therefore no interest from the viewpoint of Chinese phonology. They are not made according to hard and fast rules, but appear arbitrary to a high degree. It is impossible to establish any certain phonetic rules according to which these transcriptions could be identified with their Chinese equivalents. The historical or geographical point of view is the only criterion that may guide us. Nobody, for instance, could say positively what is understood by "the great palace of the park of Yai ho," if this passage were culled from the context. Only the context [1] shows us that Yai ho is intended for Chinese Je ho 熱河 (Jehol). [2] Yo mi yvan [3] seems to be the palace Yüan miṅ yüan 圓明園. Such-like transcriptions seem to be based on inexact hearing rather than on a knowledge of Chinese writing. It would be impossible to recognize in Žun šu waṅ [4] the Emperor Šön-tsuṅ 神宗 of the Ming dynasty but for the year 1616, under which he is mentioned.

A knowledge of Chinese is indispensable for the study of Tibetan numismatics and the exact reading of the legends on the Chinese-Tibetan coins. Mr. E. H. C. WALSH has issued a meritorious work under the title "The Coinage of Tibet," [5] in which he figures and

---

[1] HUTH, *B. M.*, Vol. II, p. 316.

[2] In other Tibetan works (for instance, in the Kanjur Index of Čo-ne) this name is transcribed že-hor; in the stages of the journey of the Paṇ-č'en Lama dPal-ldan Ye-šes to Peking it is written Ye-hor (*J. A. S. B.*, 1882, p. 52); *hor* answers to 河兒.

[3] HUTH, *ibid.*, p. 321.

[4] *Ibid.*, p. 244. Huth adds, "mit der Regierungsbezeichnung Tai waṅ." Such a reign period does not exist, and that of the Emperor Šön-tsuṅ was Wan-li. Tibetan *Tai waṅ* is transcription of Chinese *ta waṅ* 大王, and is merely a title.

[5] *Memoirs A. S. B.*, Vol. II, 1907, pp 11—23, with 2 plates.

describes (p. 21), among other coins, the Chinese silver *ṭaṅ-ka*, [1] minted by the Chinese for circulation in Tibet. The first of these issues bears on the reverse the Tibetan legend: "Č'an-luṅ gtsaṅ pau," translated by Mr. Walsh, "the pure money of Chhan Lung." Accordingly, he takes the word *gtsaṅ* for the adjective *gtsaṅ-ba*, "clean, pure;" but this is impossible. The Tibetan is merely a transcription of the Chinese legend on the obverse, which runs: 乾隆藏寶. This means, "Precious object (or treasure) of Tsaṅ (that is, Tibet) of the period K'ien-luṅ." The Chinese word Tsaṅ is a transcription of Tibetan gTsaṅ, one of the provinces of central Tibet, — a name extended by the Chinese to the whole country. The word gTsaṅ on this coin, therefore, is not connected with the adjective *gtsaṅ-ba*. [2] The Tibetan writing Č'an-luṅ for K'ien-luṅ approaches the modern Č'ien-luṅ, and was perhaps in vogue during the eighteenth century. ạJigs-med nam-mk'a, who wrote in 1819, transcribed the name K'yān-luṅ. [3] The *nien-hao* Kia-k'iṅ 嘉慶 is written on the coins *bča-ạč'in*, [4] by ạJigs-med nam-mk'a Čyā-č'iṅ. Tao-kuaṅ 道光 appears on the coins in the form rDau-kvoṅ. [5]

---

[1] Regarding this word see Glossary, No. 98.

[2] The correct explanation of the Tibetan legend has already been given by A. T. DE LACOUPERIE, *The Silver Coinage of Tibet*, p. 351 (*Numismatic Chronicle*, 1881),—a treatise not consulted by Walsh.

[3] HUTH, *B. M.*, Vol. II, p. 77. The addition of the subjoined letter *a* seems to be suggested by the high tone of 乾. In the *Pol. D.* (4, p. 16), this word is transcribed in Tibetan *k'yan*. Kiang-nan 江南 is written in Tibetan *kyaṅ-nan* and *čaṅ-nan* in the *Pol. D.*

[4] Not *ač'en*, as written by Walsh; nor *ats'aṅ*, as written by De Lacouperie, who, however, transcribed *htsiṅ*.

[5] The work of Mr. Walsh, by the way, is very interesting and has many merits. He is the first to give a correct explanation of the legend on the dGa-ldan ṭaṅ-ka, which runs: "dGa-ldan p'o-braṅ p'yogs-las rnam rgyal," and which is translated by Walsh, "The Ga-den Palace victorious on all sides," the Ga-den Palace being a designation for the seat of Government in Lhasa. Sanskritized, the legend would be "Tushita-prāsāda-dik-vijaya." The same interpretation as offered by Walsh was given me by a Tibetan Lama at Peking

Tibetans are very fond of providing utensils and vessels, particularly those for ritualistic purposes, with brief inscriptions alluding to the character of the object, or containing a sage dictum or sometimes even a date. In Peking and in several places of Mongolia the Chinese have developed special industries to meet the requirements of the Lama temples, and to cater to the taste of the Lamas and wealthy Mongols. Inscriptions on such pieces sometimes are wholly composed of Chinese sentences transcribed in Tibetan. A bronze wine-jar, [1] for example, of most elegant shape and execution, bears on the base the engraved legend "Tä čᶜeṅ kᶜaṅ žes ñan so." I wonder what Tibetan scholars unfamiliar with Chinese would make of this! Of course, these words yield no sense if taken as Tibetan, but are a transcript of 大淸康熙年作 *Ta Tsᶜiṅ Kᶜaṅ-hi nien tso,* "Made in the years of the period Kᶜaṅ-hi (1662—1722) of the great Tsᶜiṅ Dynasty."

In 1766 an interesting geographical dictionary in six languages (Chinese, Manchu, Mongol, Tibetan, Kalmuk, and Eastern Turkī) was published by order of the Emperor Kᶜien-luṅ in eight volumes. It bears the title *Kᶜin-tiṅ si yü tᶜuṅ wên či* 欽定西域同文志, and contains 3111 geographical names of Central Asia, with their transcriptions in Chinese and Manchu and explanations of their meaning. Klaproth has made use of it in his commentary to

---

in 1901. DE LACOUPERIE (*l. c.*, p. 345) misunderstood it thoroughly by placing the words in the wrong order, "rnam-rgyal dga-ldan pᶜo-braṅ pᶜyogs-las," and taking *rnam-rgyal* (Sanskrit vijaya) as the 27th year of the Jovian cycle; he thus arrived at dating the coin in 1771 (instead of 1773, as he dated the first year of the first cycle in 1025). In fact, none of the coins bearing this legend is dated. The reading and explanation of this legend, as given by ROCKHILL (*Notes on the Ethnology of Tibet*, p. 718), are likewise erroneous. It may be added that the floral design in the centre on the reverse, taken by Rockhill to be a lotus, was explained by my Lama informant as a *dpag bsam ljon šiṅ,* "wish-granting tree" (Sanskrit *kalpalatā*).

[1] In the collections of Field Museum, Chicago (Cat. No. 122672).

Friar Francesco Orazio's *Breve notizia del regno del Thibet* (also *Description du Tubet*, p. 46). We are indebted to E. von Zach for the publication of the portion dealing with Tibet.[1] The entire work, which was out of print long ago, would merit publication, important as it is not only for the geography, but also for the languages of Central Asia. The list of Tibetan names is carefully drawn up: the author who placed them on record was well informed and possessed a fairly good ear. As will be seen, his transcriptions of Tibetan words are made according to a uniform and logical system, and therefore allow of some inferences as to the state of Tibetan phonetics in the eighteenth century, which is confirmed by the transcriptions due to the Jesuit and Capuchin missionaries of the same period. His interpretation of the names appears to be based on local tradition, and generally inspires confidence, although misunderstandings have occasionally slipped in.[2] The following observations may be based on the transcriptions of Tibetan names:

All prefixes are silent, with the exception of two cases, *ạbroṅ* 阿博隆 (Manchu *aboruṅ*) and *lčog* ("tower") 羅爵克 (Manchu *lojiyok*), in another passage, however, sounded *čok* 爵克. The transcription *yi* 伊 for *dbyi* ("lynx") deserves mention, as it agrees with the modern pronunciation.

Prefixes are articulated in composition in certain fixed terms:[3]

[1] *Lexicographische Beiträge*, Vol. I, pp. 83—98; Vol. III, pp. 108—135.

[2] For instance, *ljim* is not artemisia, but rhubarb; *dug roṅ la* is not "pass of the black ravine," but "poisonous ravine;" that is, ravine exhaling poisonous vapors. *Duṅ la* is not "shell pass," but simply "white pass" (white like a shell). In *nags-gi Dzam-bha* (Jambhala)-*ri*, *nags* hardly signifies "ox" (Zach: "god of oxen"), but rather "mountain where Jambhala dwells in the forest" *Wa-go*, that cannot be explained according to the Chinese author, apparently means "fox-head."

[3] Compare *T'oung Pao*, 1914, pp. 86, 91, note; 1915, pp 3, 420.

la-rgan (*lar-gan*), old pass, 拉爾干.

rgya-mts'o (*rgyam-ts'o*), sea, 佳木磋.

klu mts'o (*lum-ts'o*), sea of the Nāga, 魯木磋.

glo mč'in (*lom-č'in*), liver, 羅木沁.

rta rdzi (*tar-dzi*), herdsman tending horses, 達爾子.

rta rgod (*tar-god*), wild horse, 達爾果特.

mgo mt'o (*gom-t'o*), high peak, 郭木托.

a-bkra (*ab-ṭa*) 阿布扎 (Manchu *abja*).

mts'o-ldiṅ (*ts'ol-diṅ*, not *mts'o lo ldiṅ*, as written twice by E. VON ZACH, Vol. III, p. 134) 磋羅定.

But: *rdo bzaṅ* (*do zaṅ*) 多桑.

bu bču (*bu ču*) 布珠.

mi dpon (*mi pon*) 密本.

γyu mts'o (*yu ts'o*) 裕磋.

Initial sonants and surds, aspirate and non-aspirate, with following *r* (*ra-btags*), appear to have undergone transformation into the cerebral series, with the sole exception of the sonant and surd labials:

bkra 扎 (*ča*).

k'ra 察 (*č'a*), *k'ri* 赤 (*č'i*), *kr'o* 綽 (*č'o*), *k'rus* 垂 (*č'ui*), *k'rob* 綽布 (*č'o-pu*).

[Compare Desideri's transcriptions *trò* for *k'rod*, *tri* for *k'ri*.]

grub 珠布 (*ču-pu*), *grum* 珠木 (*ču-mu*), *groṅ* 莊 (*čuaṅ*), *gri* 濟 (*či*), *ugram* 扎木 (*ča-mu*), *mgron* and *sgron* 準 (*čun*). [Compare Desideri: *trang* for *graṅ*, *trubbà* for *grub-pa*, *drovà* for *ugro-ba*.]

p'ra 察 (*č'a*), *ąp'raṅ* 昌 (*č'aṅ*).

dre 德 (*to*, same character transcribes also *bde*), *dru* 珠 (*ču*), *druṅ* 中 (*čuṅ*), *dren* 眞 (*čen*).

However: *brag* 巴喇克 (*pa-la-ku*, Manchu *barak*), *p'o-braṅ* 坡巴朗 (*p'o-pa-laṅ*, Manchu *pobaraṅ*), *ąbrog* 博羅克 (*po-lo-ko*). [Compare Desideri: *breepà* for *abras-bu*.]

sprin 必林 (*pi-lin*).

Note also *srid* 錫里特 (*si-li-tʿe*, Manchu *sirit*).

The Chinese, not having cerebrals, resorted to palatals or dentals, respectively, to render the peculiar Tibetan sounds. [1]

Final *g*, *b*, *d*, and *l* were sounded as follows: final *g* is indicated by 克 in *gog*, *lčag-s*, *stag*, *nags*, *ug*; final *b* is expressed by 布 in *skyab-s*, *skyib-s*, *rgyab*, *bya-ma-leb* ("butterfly"); final *d* is represented by 特 in *bkod*, *skyid*, *brgyad*, *stod*, *tʿod*, *dud*, *gnod*, *od*; final *l*, by 勒 in *dkyil*, *ygul*, *rgyal*, *yul*. In this case we must not generalize, as it is always likely that geographical names retain in Tibet the old, stereotyped form of pronunciation.

Final *s* was silent and affected the stem-vowel, which was lengthened, or changed into an *i*-diphthong:

γyas (*yai*) 雅衣 (*ya-i*).
yas in *bsam-yas* (*yai*) 崖 (*yai*).
sñas (*ñai*) 愛 (*ñai*).
dños (*ñoi*) 衞 (*wei*, Manchu *oi*).
čʿos (*čʿoi*) 吹 (*čʿui*, Manchu *čoi*).
dus (*dui*) 堆 (*tui*, Manchu *dui*).
ydus (*dui*) 堆 (*tui*).
mdzes-po (*dzē-po*) 澤博 (*tso-po*, Manchu *tseibo*).
smos (*möi*) 梅.

---

[1] As regards initial *my*, Friar Francesco Orazio della Penna (on p. 73 of his *Breve notizia*) speaks of two sorts of fasts styled *gnunnè* and *gnennè*, two Tibetan words not identified by his editor Klaproth. Italian *gn* denotes the palatal nasal *ñ* Thus we have *ñun-ne* and *ñen-ne*. The former represents Tibetan *smyuṅ gnas* (pronounced *ñuṅ nä*), "the act of fasting;" the latter, Tibetan *bsñen gnas* (pronounced *ñen nä*) with the same meaning. The former corresponds to Sanskrit *uposhadha*; the latter, to Sanskrit *upavāsa* (Foucaux, *Lalitavistara*, Vol. II, p. 177). It is interesting to note from the transcription of *smyuṅ* that *my* was articulated *ñ* in the beginning of the eighteenth century. Desideri transcribes the word *dmyal-ba* in the form *gnee-va*

[Compare Desideri: *Sang-ghieè* for *Saṅs-rgyas*, *kiepù* for *skyes-po*, *soo* for *sos*, *tuu* for *bsdus*, *lee* for *las*, *nee* for *gnas*: Orazio della Penna: *c'iö* for *č'os*: *Alphabetum Tibetanum*: *re* for *ras*, *sre* for *sras*.]

The dictionary in four languages (Manchu, Tibetan, Mongol, and Chinese) published by order of the Emperor K'ien-luṅ yields a goodly number of Chinese loan-words in Tibetan, but no rational system of transcription is followed there. In the glossary the work is quoted as *Pol. D.* (Polyglot Dictionary). The Tibetan section of the *Hua i yi yü* has also been utilized.

I. J. SCHMIDT and O. BÖHTLINGK [1] make mention of a List of Drugs (*sman sna ts'ogs-kyi miṅ č'ad*) in Tibetan, Mongol and Chinese, printed in Chinese style. E. BRETSCHNEIDER, [2] in 1882, gave a fairly accurate description of a list of 365 drugs, in which their names in Tibetan and Chinese are enumerated, the pronunciation of the Chinese characters being added in Tibetan letters. Bretschneider makes a Peking firm, Wan I hao, responsible for this booklet; it may be that this firm in its own commercial interests has issued a special reprint of it under its signature. I have never seen this edition, but know that two editions — one in Chinese, the other in Tibetan book style — have left the press of the Tibetan printing establishment near the temple Sun ču se, for copies of both are in my possession. [3] The same I have seen in the British Museum and in the Royal Library of Berlin. The little work first attracted my attention when in 1900 my brother and myself co-operated in

[1] *Verzeichnis der tib Handschriften und Holzdrucke im Asiatischen Museum*, Nos. 37—41, on p. 63

[2] *Botanicon Sinicum*, pt 1, p. 104.

[3] They bear the title *Fan Han yao min* 番漢藥名, "Names of Drugs in Tibetan and Chinese;" in Tibetan: *sman miṅ bod daṅ rgyai skad šan sbyar-ba*; in Mongol: *em-iin nurü tubet kitat khadamal*. The edition in Chinese style has ten folios, that in Tibetan style fourteen folios. The trilingual edition mentioned above is not known to me.

a study of Tibetan medicine.[1] Afterwards I prepared a critical text of it, and succeeded also in identifying most of the technical terms, — a task considerably facilitated by the circumstance that I had occasion to make a collection of a great number of these drugs in 1901 at Peking.[2] The list in question seemed to have some importance for Tibetan lexicography, as it contains many words not recorded in our current dictionaries, and others of hitherto dubious identification, with a Chinese gloss. In the latter respect, however, a somewhat critical attitude is necessary, as the Tibeto-Chinese equations do not always establish an absolute identity of the articles; in some cases they merely point to similarity, and in others are certainly wrong.[3]

In 1908 appeared A. Pozdn'äyev's first volume of a translation from the Mongol of one of the Four Medical Tantra (*man-ṅag-gi rgyud*).[4] On pp. 247—301 of that work we find a list of 381 medicines, with their names in Tibetan and Mongol, the Chinese equivalents being added in the footnotes. The latter have been supplied by Pozdn'äyev from the Peking edition above mentioned; and he has fully recognized that the list of this edition and his text, aside from the surplus in the latter, are identical. Pozdn'äyev's publication happily relieved me from my own plan of publishing

---

[1] H. LAUFER, *Beiträge zur Kenntnis der tibetischen Medicin*, two parts (Berlin and Leipzig, 1900).

[2] This collection is in the American Museum of Natural History of New York.

[3] He who has perused Bretschneider's notice carefully would expect this, for the firm Wan I hao itself cautions us against the belief that the Tibetan drugs named in this list are exactly the same as the original productions of Tibet bearing these names in that country; but their medical virtues are stated to be similar. A few examples may be given. Chinese *ma nao* 瑪瑙 ("agate") does not correspond to Tibetan *rdo sñiṅ*, but to *mc'oṅ*. Tibetan *,utpal* (Sanskrit *utpala*) has nothing to do with Chinese *šu k'uei-tse* 蜀葵子.

[4] Учебникъ Тибетской медицины, Vol. I (St. Petersburg, 1908).

this glossary, although little has been done by him in the matter of identification. His work demonstrated at once the source for the above Tibeto-Chinese Pharmacopœia. The firm Wan I hao, therefore, is very far from being able to claim the authorship of the work. In fact, it is derived from this Tibeto-Mongol standard work on medicine; so that the Tibetan text appears as the original, while the Chinese translation is a subsequent addition, made in the interest of the Chinese druggists of Peking trading with Tibet and Mongolia, and with the Tibetans and Mongols living in the capital. In 1913 Dr. Hübotter presented us with a volume in which, on pp. 49—147, the list of drugs mentioned appears autographed by the author's hand.[1] As a beginner, Dr. Hübotter has a just claim to attenuating circumstances; he is very enthusiastic about his subject; like many another beginner, he dates from himself a new era, and looks down with sublime contempt on everything accomplished by his predecessors. With all sympathy for the author's good intentions, however, it must be frankly said, in the interest of Tibetan studies, that he lacks the philological training necessary for such a task, and that his edition is uncritical and valueless. He is content to copy his text; and, moreover, he has copied it badly and carelessly.[2] The future lexicographer of Tibetan will thus be obliged to base his entries on the originals published in Peking, and may utilize to advantage also Pozdn'äyev's edition.[3]

---

[1] *Beiträge zur Kenntnis der chinesischen sowie der tibetisch-mongolischen Pharmakologie* (Berlin and Wien, 1913).

[2] A few examples may suffice: he writes *dak* (p. 51) instead of *dkar-po, ts'ır* (p. 53) for *ts'er, kaṅs* (p. 54) for *gaṅs, pa-bla* (p. 56) for *ba-bla, mtuṅ rtse dmaṅ-bo* (p. 59) for *mduṅ rtse dmar-po, sug-smal* (p. 69) for *sug-smel, nla-ga* (p. 71) for *nā-ga, ze-ra* (p. 72, twice) for *zi-ra, u-sus* (p. 73) for *u-su, star-bus* (ib.) for *star-bu*, etc.

[3] Further, it must be denied that Dr. Hübotter, as announced on the title-page of his book, has furnished a useful, or even a new, contribution to Tibeto-Mongol pharmacology.

Neither Pozdn'äyev nor Hübotter has endeavored to ascertain the author and the date of the little work. The well-known Sanskrit formula of blessing, *maṅgalam kurvantu*, is taken by Hübotter for the name of the author. The editor of the work was mGon-po skyabs,[1] a teacher at the Tibetan School (*bod-kyi slob-grva*) of Peking, and well known to us from his participation in the translating of Chinese-Buddhist books from Chinese into Tibetan,[2] which were embodied in the Kanjur. He seems to have lived during the K'aṅ-hi period, and probably still under the reign of Yuṅ-čeṅ. The date

---

From a modern student of the materia medica of the Chinese, Tibetans, and Hindu, we justly demand that he have an actual knowledge of the drugs under discussion, and that he be at least acquainted with the abundant literature on these subjects disseminated by orientalists, botanists, and pharmacologists. A fundamental work like Hanbury's *Science Papers*, which would have furnished numerous correct identifications, is unknown to Dr. Hübotter. H. Laufer's results are not utilized, but old errors are repeated; and matters perfectly known to every serious student are treated in an absurd manner, as though they were still mysteries to us. To quote only a few examples: Tibetan *gur-gum* (Chinese *huṅ hua* 紅花) is identified with Carthamus tinctorius (an error refuted by the writer eighteen years ago); it is the saffron (Crocus sativus). Chinese *yaṅ k'i ši* 陽起石 (Tib. *rdo rgyus*) is left without determination, being provided with an interrogation-mark, although HANBURY (*l. c.*, p. 218) explains the Chinese term as "asbestos tremolite; silicate of lime and magnesia," and H. LAUFER (*l. c.*, p. 82, note 3) interprets the corresponding Tibetan term correctly as "asbestos." Dr. Hübotter is unable to explain Chinese *ör č'a* 兒茶 (see No. 107), with the Tibetan equivalent *rdo ja* (written by him *rdo rda*), which is catechu, — on the employment of which by Tibetan women so much has been written. Such a well-known affair as *dbyar rtsa dgun abu* (see p. 445), a parasitic fungus growing on the pupa of a caterpillar (Cordyceps sinensis) is taken by Hubotter (p. 135) for "a kind of grass gathered in November and contracting in December, a grass luxuriously thriving in Tibet, notably in Kham." This absurdity is copied (with the error *dbyuṅ* instead of *dguṅ*) from the *Tibetan Dictionary* of Chandra Das; while the product is correctly described and well figured in ENGLER's *Naturliche Pflanzenfamilien* (Vol. I, pp. 368, 369), which Dr. Hübotter, as a physician, who was obliged to study natural sciences and pharmacology, certainly ought to know. Gypsum, musk, bear's gall, salt, salpetre, natrium, etc., are classified by Hubotter among the remedies of the vegetable kingdom.

[1] Styled by BRETSCHNEIDER (*l. c.*) Gonbedjan.

[2] LAUFER, *Dokumente*, I, p. 52, note. See SCHMIDT, *Index des Kanjur*, Nos. 446, 502. In the colophon of the latter, a brief Dhāraṇī, he is titled "the great translator of the present great Ts'iṅ dynasty."

of his edition is not stated in the colophon; but a re-edition of it, as it is there recorded, took place in 1734 ("twelfth year of Yuṅ-čeṅ, corresponding to a *wood tiger* year"). [1]

In this medical glossary the Chinese equivalents are transcribed in Tibetan, apparently with the intention of facilitating their reading for Tibetans. These transcriptions have little scientific interest, as they reflect the last stage in the development of Tibetan phonology; but we glean from them the one point, not unimportant for the history of the Tibetan language, that its present phonetic state existed at least as early as the beginning of the eighteenth century, — a fact, as stated previously, confirmed by the contemporaneous romanizations of Tibetan words on the part of the Catholic missionaries. Tibetan transcriptions in the above glossary, like *nas* for Chinese *nai* 奈, *gus* for Chinese *kui* 桂, *wes* for Chinese *wei* 魏, *kus* for Chinese *kui* 鬼, *šus* for *śui* 水, *hus* for *hui* 灰, and many others, show that they were made at a time when final *s* was silent, and affected the preceding vowel (compare p. 437). Very curious is the transcription *kʻy* for the Chinese palatal aspirate, as *kʻyiṅ* for *tsʻiṅ* 青, *kʻyi* for *kʻi* 起, and even *kyiṅ* for *tsiṅ* 精. The value of this glossary for our present purpose does not lie in these transcriptions, which are a mere curiosity, but in the Tibeto-Chinese concordances, which allow us to recognize Chinese loan-words in Tibetan, or Tibetan translations of Chinese terms. For instance, the equation Tibetan *ha-šig*—Chinese *hua ši(k)* 滑石 ("steatite") shows us that the Tibetan word is an old derivation from Chinese

[1] This portion of the colophon giving the date of the second edition is appended only to the print in Tibetan style; that in Chinese style lacks this part, and closes with the mention of mGon-po skyabs.

(No. 226), and that it does not mean "alabaster," as asserted by Jäschke and his copyists. [1]

There is a similar pharmacological list containing the names of 179 drugs in Tibetan, Mongol, and Chinese, the Chinese characters being accompanied by a transcription in Manchu. The title of this work is *P^cu tsi tsa fañ* 普濟雜方 (in Tibetan: *Kun p^can sna ts^cogs sman sbyor žes bya ba*), "Various Prescriptions for the Healing of All Complaints." The book is printed on Chinese paper and in Chinese style, and makes a single volume of 66 leaves, ten of which are occupied by the vocabulary. The text is written in Mongol, and is quoted as *Mong. Pharm.*, while the aforementioned book is cited as *Tib. Pharm.*

Tibetan grammarians were conscious of the presence of Chinese elements in their language. The following curious passage in the *Li-šii gur k^cañ* (fol. 23 a) relates to this matter, although it is the author's object not to discuss the Chinese loans in Tibetan, but to caution his readers against taking them for "ancient Tibetan," as had apparently been done:

"K^ca-čig-tu rgya-nag-gi skad brda rñiṅ-du ạk^crul-pa snaṅ-ste | blon-po-la ạp^cyiṅ saṅ žes-pa daṅ | dṅul rdzus-ma-la ha-yaṅ-ha žes-pa daṅ | rgya-la huṅ la daṅ | gžau yaṅ grum-rtse daṅ | čog-ts^co daṅ | lugs rgyu mts^can bsñad-pa-la | γyaṅ-rtse žes-pa daṅ | p^ciṅ č^caṅ daṅ gzau-gyi miṅ-la zin gzau daṅ | čiṅ p^ciṅ daṅ | ʾas-man gur-gum sogs maṅ-du snaṅ-la | rgyal-poi žal-la gser žal daṅ | baṅ č^ceu la gser yig-pa daṅ | gser skyems sogs kyaṅ rgya-nag-gi brdar snaṅ žiṅ." "Some words which have been mistaken for 'old language' belong to the Chinese language, thus, *ạp^cyiṅ saṅ* corresponding in meaning

[1] Neither does *t^cod-le-kor* mean "alabaster," but it means likewise "steatite" (*hua ši* Pol. D., 2, p. 61).

to *blon-po* (see No. 301), *ha-yan-ha* which means 'counterfeit money,' [1] *hun la* which means 'seal' (No. 236); further, *grum-rtse*, [2] *čog-tsʿo* (No. 225), and *g-yan-rtse* which signifies 'one propounding the foundation of customs' [in the Mongol rendering *yosun-i siltagan ügüläkü*], *pʿin-čʿan*, *zin gzan* which is the name for a *gzan* (?), *čin-pʿin*, *ʾas-man* [3] *gur-gum*, and many others. Chinese phrases, further, are *gser žal* ('gold face') which has the meaning 'face of the king,' [4] *gser-yig-pa* ('bearer of a gold document') [5] answering to *ban čʿen* ('courier'), and *gser skyems* ('libation-offering'), [6] and others."

---

[1] I do not know the Chinese prototype (*kia yin-tse* 假銀子?).

[2] *grum-rtse* (rendered *dabuskur* in the Mongol translation of the work), "rug, pillow. Jäschke has "*grum-tse*, thick woollen blanket," as occurring in Mi-la-ras-pa. I venture to doubt that this is a Chinese word.

[3] In Persian *āsmān* آسمان means "heaven." Whether Tibetan *ʾas-man* (otherwise unknown to me) is intended here as an independent word, or whether *ʾas-man gur-gum* denotes a particular variety of saffron, I do not know. It is likewise difficult to guess what the three preceding words, apparently transcriptions of Chinese, are to represent.

[4] Now it becomes intelligible also why in the document issued in favor of the Capuchins in 1724 the Emperor of China is styled *gser-gyi rgyal-po*, "golden king" (GEORGI, *Alphabetum Tibetanum*, p. 651). Hence GEORGI (p. 654) drew the erroneous inference that *gser* is the general Tibetan designation for China and the foundation for the Seres of the ancients. KLAPROTH (*Leichenstein*, p. 39), in 1811, noted Georgi's error, without explaining its source, however, and, curiously enough, asserted that "in Tibet the western part of the Chinese province of Shen-si is called Sser [in agreement with Georgi, wrongly written in Tibetan letters *gsar* or *sar*], which means 'gold,' and that the same region was formerly designated *Kin* ('gold') by the Chinese." In the *Hua i yi yü*, the term *gser mkʿar* (JÄSCHKE: "imperial castle") is explained as the capital of China, Peking (京城); *gser deb* means "imperial genealogy" (*ti hi* 帝系), and *gser dan* "genealogical record of the imperial house" (*yü tie* 玉牒, *Pol. D.*, 3, p 9); *gser lam*, "imperial road" (*yü tao* 御道; *ibid.*, 19, p. 48).

[5] Tibetan *gser yig* translates terms like Chinese *kin pʿai* 金牌, *kin tsʿe* 金册, *kin šu tse* 金書字 (regarding the last-named see PELLIOT, *Bull. de l'Ecole française*, Vol. IV, p. 241, note 5). The term is applied also to the letters of great personages in Tibet; *gser-yig-pa* was formerly the title of the imperial envoy sent from Peking to Lhasa in order to summon the Dalai Lama to Peking and to accompany him on the journey. In the *Hua i yi yü* it is equalized with *ši čʿen* 使臣.

[6] The literal meaning is "golden beverage." In our dictionaries various definitions of this term are given. JÄSCHKE interprets it as "beer together with grains of corn, as an offering to the gods for the success of an enterprise, a journey, etc." The *French*

While first transcriptions of Chinese words are enumerated, the last clause contains Sinicisms, renderings into Tibetan of Chinese phrases. Some Sinicisms in Tibetan seem to be of comparatively ancient date, as they occur in ancient translations from Sanskrit. The word *sgrog rus* ("collar-bone") has been traced in the Citralakshaṇa: it is a literal translation of Chinese *so-tse ku* 銷子骨 (literally, "chain-bones"); Tibetan *sgrog* meaning "chain," and *rus*, "bone."[1] Also *mgo nag* ("black-headed") as a designation for the common people, and *ñi og* (rendering of *tᶜien hia* 天下), are old imitations of Chinese style.

Many hundreds of Sinicisms might be gathered from the *Pol. D.*, but most of these are artificial productions, and have hardly any real life in Tibetan literature. A few examples may suffice:

gas lčags zan, tapir (literally, "eater of split iron"); translation of Chinese *nie tᶜie* 齧鐵 ("iron-eater"), epithet of the tapir *mo* 貘.

čᶜu glaṅ, buffalo: translation of *šwi niu* 水牛.

ži mig, cat's eye (precious stone); rendering of *mao tsiṅ* 猫睛. Manchu *simikte* is derived from the Tibetan word.

rdo snum, petroleum: rendering of *ši yu* 石油.

lha lag ("God's hand"), Citrus medica: based on *fu šou* 佛手.

dbyar rtsa dgun ₒbu, a parasitic fungus growing on the pupa of a caterpillar (Cordyceps sinensis), a word-for-word translation of *hia tsᶜao tuṅ čᶜuṅ* 夏草冬蟲, "plant in the summer, worm in the winter."[2]

Dictionary determines it as a "gold vase holding libations." According to CHANDRA DAS, it signifies wine offered to royalty, to the gods, and the Grand Lama. An instance of the application of the term is found in the account of the journey into Tibet of the Kalmuk Bāza-bakši (edition and translation of POZDN'ÄYEV, p. 192). According to Pozdn'äyev, the Tibetan term (in Mongol *ser-čem*) refers to offerings made to protective local genii and for the greater part consisting of holy water and tea.

[1] Compare the writer's edition, p. 158.

[2] This appropriate name for that extraordinary combination of animal and vegetable

rta klad (literally, "horse brain"), agate. Reproduction of the popular etymology underlying *ma-nao* 瑪瑙. It is further Tibetanized into *rdo klad*. The real Tibetan word for agate is *č'oṅ* or *mč'oṅ*.

rgya-mts'o bži, the four seas. Used in the same sense as *se hai* 四海.

rus byaṅ (*Pol. D.*, 19, p. 34), domino (literally, "bone cards"). Translation of Chinese *ku p'ai* 骨牌.

šog byaṅ (*ibid.*), playing-cards (literally, "paper cards"). Translation of Chinese *či p'ai* 紙牌.

Sometimes entire phrases may be rendered from Chinese into Tibetan. At the end of Tibetan works printed during the K'ien-luṅ era in the imperial palace we read the formula, goṅ-ma č'en-po dguṅ-lo k'ri p'rag k'ri p'rag-tu brtan-čiṅ, "May the great emperor live ten thousand times ten thousand years!" (the well-known *wan sui wan wan sui* formula.) Tibetan *goṅ-ma* ("the upper one") is a Sinicism based on *šaṅ* 上; *goṅ sa* = *huaṅ saṅ* 皇上 (*Pol. D.*, 3, p. 2, where *wan sui* is rendered by Tibetan *k'ri t'ub*).

For the present it is sufficient for me to convince the students of Tibetan that, besides the Indian strata, there is also a strong Chinese sphere of influence, and that this Sino-Tibetan merits careful

---

indeed corresponds to the natural facts. in the dried specimens, as they are traded in commerce, the animal and vegetal portions are still discernible, the lower larger one with its rings and joints belonging to the caterpillar, the upper one closely joined to it being the fungus consisting of a spurred filament of a grayish-brown color. This product is found in considerable quantity in the district of Li-t'aṅ. It is made up into small bundles, weighing each about ¼ ounce, bound with red thread. It is highly esteemed as a medicine throughout China, worth at Li-t'aṅ 5 to 6 rupees a catty, the annual export to Ta-tsien-lu amounting to about 2400 catties to the value of some 4000 Taels See ROCKHILL, *J.R.A.S.*, 1891, p. 271, where *dgon bu* is to be corrected into *dgun abu*; G. A. STUART, *Chinese Materia Medica*, p. 126; *List of Chinese Medicines*, p 442, No 287 (Shanghai, 1889); A. ENGLER, *Die natürlichen Pflanzenfamilien*, Vol. I, 1, pp. 368, 369, where also an illustration is given; PARENNIN, *Lettres édifiantes*, nouv. ed , Vol XIX, pp. 300—303 (Paris, 1781); A HOSIE, *Journey to the Eastern Frontier of Tibet*, p. 38 (Parliamentary Papers, China, No. 1, 1905).

attention. In the Glossary great stress has been laid upon the names of plants (in particular cultivated plants) and products of industry and commerce. In meeting with such names in Tibetan texts, it must always be ascertained whether we move in a really Tibetan, an Indo-Tibetan, or a Sino-Tibetan atmosphere: for, according to these different conditions, the name in question may assume a different signification. For instance, *wa* used with reference to Tibetan surroundings means the "fox;" in all translations from Sanskrit, however, it signifies the "jackal" (*çrgāla*).[1] Tibetan *sa-rdzi-ka* is a transcription of Sanskrit *sarjikā*, which designates soda (natron) and a soda-yielding plant. In Sino-Tibetan, however, *sa-rdzi-ka* is made to correspond to Chinese *wu mei* 烏梅 ("black-plums," that is, plums gathered half ripe and smoked), as shown by the *Mong. Pharm.* (No. 105). In the same work (No. 32) we find *sa-rtsi dkar* ("white of *sa-rtsi*") as equivalent of *kuan fön* 官粉 ("fine facial powder"). Tibetan *'ut-pal* is Sanskrit *utpala*, the blue lotus (a species of Nymphæa). In Sino-Tibetan it serves for the designation of the Chinese chrysanthemum *kü* 菊 (*Pol. D.*, 29, p. 58). In Lahūl, according to Jäschke, the name seems to be transferred to an indigenous plant (Polemonium cæruleum). Tibetan *rtsva padma* ("root lotus") renders Chinese *mu-tan* 牡丹 ("peony"). Tibetan *'a-bi-ša* transcribes Sanskrit *aviṣa* ("non-poisonous"), and, according to the medical work Vaiḍūrya snon-po, is the designation of a medicinal herb (*pw: aviṣā*, Curcuma zedoaria). In Sino-Tibetan it answers to Chinese *po-ho* 薄荷 (*Mong. Pharm.*, No. 51), Mentha arvensis, furnishing peppermint;[2] but also to *pai-ho ken* 百合根

[1] HUTH (*B. M.*) has always wrongly translated the word by "fox" in proverbial sayings based on the Jātaka.

[2] I do not know whether *'a-bi-ša* is thus used in literature. The term for the mint in the written language is *byi-rug-pa* (Jäschke says only that it is a medicinal plant),

(the root of a lily) and *šan-tou* 山荳 (Cajanus indicus). The term *zar-ma*, as correctly stated by Jäschke, means the seed of sesamum (Sesamum orientale) and this is confirmed by the Chinese equivalent (in *Pol. D.*) *hu ma* 胡麻. In the Mahāvyutpatti (section 228), however, the word is equalized with Sanskrit *atasī* (Linum usitatissimum). The Sino-Tibetan glossaries are of great value also in confirming the results yielded by Sanskrit-Tibetan lexicography. Tibetan *spaṅ spos* ("fragrant plant of the meadows"), according to the Mahāvyutpatti, is the equivalent of Sanskrit *gandhamāṁsī*, which led me in 1896 to explain it as the true spikenard. This conclusion is corroborated by the *Tib. Pharm.* (No. 221), where *spaṅ spos* is identified with Chinese *kan suṅ* 甘松, which indeed relates to spikenard.[1] Again, it is interesting to note that the Tibetan term, as observed by Jäschke, with reference to the Alpine regions of Tibet, designates an aromatic composite, Waldheimia tridactylites.

In many cases, the Tibetans have not adopted the Chinese names for things Chinese, but have coined new terms for them. Thus, porcelain is styled *kar-yol* ("white pottery"). The jujube-tree (*tsao* 棗, Zizyphus vulgaris) is termed *rgya šug* (*Pol. D.*, 28, p. 53), literally, "Chinese juniper." The so-called Chinese olive, *kan-lan* 橄欖 (Canārium album), is in Tibetan *rgya ˛a-ru-ra* (*ibid.*), literally, "Chinese myrobalan." The term *rgya gul nag* ("black Guggula of China") refers to myrrh (Chinese *mu yao* 沒藥); and *rgya tsᶜva* ("Chinese salt"), to sal-ammoniac (Chinese *nao-ša* 硇砂).

---

corresponding to Chinese *po-ho*, Mongol *jirukba*, Manchu *farsa* (*Pol. D.*, 27, p. 20). In Ladākh, peppermint is called *pᶜo-lo-hu*.

[1] G. A. Stuart, *Chinese Materia Medica*, p. 278.

II. GLOSSARY.

1. *Indian Loan-Words.*

The number of Sanskrit loan-words in Tibetan is comparatively small. The tendency to translate Sanskrit terms literally into Tibetan or to convey to Tibetan words the meaning of Sanskrit words (*k^cyuṅ* = Sanskrit *garuḍa*; *klu* = Sanskrit *nāga*, etc.) is well known. Aside from the learned and purely literary transmission of Sanskrit terminology, we find in Tibetan, partially even in the colloquial language, a comparatively large number of Indian words, which are not derived from Sanskrit, but from the vernaculars, the Prākrits. It is even possible that these loans, at least some of them, were made long before the introduction of Buddhism into Tibet (that is, prior to the age of Tibetan literature) in consequence of the contact of Tibetan with Indian tribes. An exact chronology of these words cannot be established at present. Many of them, in their Tibetan forms, may be apt to enlighten the history and phonology of Indian vernaculars, but they have not yet been studied from this point of view; in fact, they have been neglected by the students of Prākrit. It is hoped that the following list will induce them to pay some attention to this subject. The Tibetan grammarians are perfectly conscious of the fact that these words are not Sanskrit, and style them Apabhraṁça (*zur k^cag*; see *S. Bayr. A. W.*, 1898, p. 593). Some go so far as to teach that any deviations from standard Sanskrit must be regarded as Apabhraṁça: while others, like the author of the *Li-šii gur k^caṅ*, discriminate between genuine Apabhraṁça words (that is, words derived directly from Indian vernaculars) and Sanskrit words which, for some reason or other, were corrupted by the Tibetans (see Nos. 78—80, 94—97).

The following forty-four words are enumerated and explained in the *Li-šii gur k'aṅ* as Apabhraṃça:

1. *'ap* (Sanskrit *ap*), water, *č'u* being given as Tibetan synonyme (*skad dod*).
2. *sto-ka* (Sanskrit, "a drop, a little, a trifle"), explained as *mar č'uṅ* ("a drop of oil"), and synonymous with *č'uṅ-ṅu* ("a little").
3. *kol-ma* (Sanskrit?), designation for "warm food" (*gzan dron*; JÄSCHKE: *zan dron*).
4. *kulmā-sha* (Sanskrit). Not explained. According to *pw.* "sour juice of fruits."
5. *gar-ba* (Sanskrit *garva*), synonyme of *ṅa rgyal* ("pride").
6. *ạ̄-ṭo-pa* (Sanskrit *āṭopa*), swelling, abundance, pride.
7. *giṅ-pa*, servant; *gin p'o*, male servant; *giṅ mo*, maid. From Sanskrit *kiṃkara*, servant, slave (translated into Tibetan: *či bgyi-ste*); Tibetan synonymes are *las byaṅ*, *p'o-ña*, and *γyog*.
8. *t'ul č'e-ba*, gross, thick, coarse. From Sanskrit *sthūla*. Tibetan synonymes are *sbom-po*, *rags-pa*, *č'e-ba*.
9. *rdul*, dust. From Sanskrit *dhūli*. I do not believe in this derivation, but am disposed to think that *rdul* is a Tibetan word.
10. *ạp'al-ka* (CSOMA: *ạp'al-ga*, "notch, incision"). From Sanskrit *phalka* (*sic*, written *k* with superscribed *l*); that is, *phalaka* ("board for writing or painting upon"). SCHIEFNER (*Mélanges asiatiques*, Vol. I, p. 337) looked in vain for a Sanskrit word *phalka*.
11. *kalpa* or *bs-kal-pa*, "have become Tibetan words" (bod-kyi brdar mdzad-pa daṅ). From Sanskrit *kalpa* (Mongol *kalab*, Uigur *ḳalb*). The double prefix added in Tibetan may be explained on the assumption that *kalpa*, on account of the ending *pa*, was taken by the Tibetans for a verb, to which

the transitive prefix *s*- and the sign of the past tense *b*- were added, the kalpa being regarded as a period back in the past. This word certainly is not Apabhraṁça, but is plain Sanskrit.

12. *Legs smin ka-ra*, said to be Apabhraṁça for Sanskrit Lakshmīkara, which is rendered into Tibetan as *dpal mdzad-de*. In this case *legs smin* ("well ripe") is a Tibetan popular adjustment based on Sanskrit *lakshmī*.

13. *pe-ña-ba*, pronounced also *pen-da-ba* (not in our dictionaries), alms. From Sanskrit *piṇḍa*, "alms" (*bsod sñom*).

14. *dpe-har*, *dpe-dkar*, monastery (*gtsug lag k^caṅ*). From Sanskrit *vihāra*. We find also the spellings *be-har*, *bi-har*, *be-kar*, *pe-kar*, *pe-dkar*, and *pe-ha-ra*, in the sense of a "tutelary deity of temples and monasteries" (*T^coung Pao*, 1908, p. 30); *dpe-dkar* is a thoroughly Tibetanized mode of writing with prefixes, conveying the meaning "white example." The Sanskrit word appears to have had various modifications in the vernaculars at the time when Buddhism was diffused over Central Asia. In Uigur the word is met in the form *bakar* or *vaχar* (F. W. K. MÜLLER, *Uigurica*, p. 47), and in Mongol as *bukar* (only in the written language). See also GAUTHIOT, *Journ. as.*, 1911, II, p. 53.

15. *ͅanda-rñil* (thus in two prints of *Li-šii* and in Citralakshaṇa, 959; also *ͅandra-rñil* and *-sñil*), sapphire. From Sanskrit *indranīla* (literally rendered into Tibetan *dbaṅ sñon-te*). The Tibetan form allows us to conjecture an Apabhraṁça **andaṅil*. According to *Tib. Pharm.* (No. 18), the Chinese equivalent is *yiṅ ts^ciṅ* 映青.

16. *ͻA-mi-de-wa*. From Sanskrit Amitābha (Tibetan *Od dpag med*).

17. *ͻĀrya-pa-lo*. From Sanskrit *Āryāvalokita* (written *ote*). Both this and the preceding word seem to have been adopted literally from Indian vernaculars.

30

18. *po-ti*, book, volume. From Sanskrit *pusta*, *pustaka*. Other Tibetan forms are *pu-sti*, *po-sti*, *pu-ti*, *pu-di*, *po-ta*, and even *pot*, *bot* (HUTH, *B. M.*, Vol. II, pp. 335, 357). Also the prototype *pustaka* is employed in Tibetan (see, for instance, Avadānakalpalatā, prose ed., p. 383, line 7) According to R. GAUTHIOT (*Mém. soc. de linguistique*, Vol. XIX, 1915, p. 130), the Sanskrit word should be derived from Iranian *pōst* ("skin"). In the form *po-ti*, it has penetrated into the popular language, where it appears in the names of several mountain-passes (*po-ti la*, pass shaped like a book, transcribed in Chinese *po-ti* 博氏 and *po-te* 博德; see E. VON ZACH, *Lexic. Beiträge*, Vol. III, pp. 118, 121). In the colloquial language of Sikkim it is used in the sense of "small volumes" (*French Dict.*, p. 592).

19. *ḥbo-de-tsi* (not in our dictionaries). From Sanskrit *putrajīva*, Nageia putranjiva roxburghii. [1] The name is literally rendered into Tibetan as *bui ḥtsʻo-ba* ("life of the son"): the nuts of this large and fine tree are strung by parents and placed around the necks of their children to keep them in good health (W. ROXBURGH, *Flora Indica*, p. 716).

20. *ru-rag-ša* (JÄSCHKE and DAS: *ru-rakša*; *French Dict.*: *ru-rak-ša*), a nut (not "a sort of berry") used for rosaries (see L. A. WADDELL, *Lamaic Rosaries*, *J.A.S.B.*, Vol. LXI, 1892, p. 29). From Sanskrit *rudrāksha* (Tibetan rendering: *drag-poi mig*), Elæocarpus ganitrus, and other species of E.

21. *se-ḥdur-ra* (indicating an articulation *sendura*), written also *sindhura*, minium, vermilion. From Sanskrit *sindūra*, Tibetan

[1] In Boehtlingk's *Sanskrit Dictionary* the designation of the genus, *Nageia*, has been omitted by some oversight. Schiefner's *Nagelia* (*Mélanges asiatiques*, Vol. I, p. 337) is a misprint.

li k^c*ri* (below, p. 456) being given as a synonyme; Apabhraṁça *sindūru*; Hindī *sēdūr*, Gujarātī *sīdūr*.

22. *gur-kum*, saffron, alleged to be Apabhraṁça for Sanskrit *kuṅkuma* (written *guṅkuma*). This, in my opinion, is wrong (see No. 109).

23. *bug-pa*, hole, alleged to be derived from Sanskrit *bhūka* ("hole"). This derivation, however, seems very doubtful, as we have such Tibetan variants as *bu-ga*, *bu-gu*, *bi-gaṅ*, *bi-yaṅ*, *big*, which are apparently connected with the verb *ạ-big-pa* ("to pierce, to bore"). There is no example known where the Tibetan language would form a verb from a Sanskrit or Prākrit noun.

24. *bram-ze*, a Brāhmaṇa. From Sanskrit *brāhmaṇa*, correctly explained as being derived from Brahma (Tibetan *Ts*^{c}*aṅs-pa*).

25. *gu-gul*, a gum resin from Boswellia serrata. From Sanskrit *guggula* (*T*^{c}*oung Pao*, 1914, p. 6).

26. *ạbum-pa*, *ạbi-ạbi* (that is, *bimbi*), disk, globe, round parts of the body; mirror, image. From Sanskrit *bimba*, explained as *sku gzugs* ("body, form"). The form *ạbi-ạbi* is apparently based on Sanskrit *bimbī*.

27. *ạgan-ạji-ra* (that is, phonetically, *ganjira*). From Sanskrit *gañjira*, explained as *mdzod ldan-te* ("having a treasury"); hence derivation from *gañja* ("treasury").

28. *ba-dan*, flag, banner. From Sanskrit *patāka*. The equation occurs also in the Mahāvyutpatti.

29. *ts*^{c}*a-ts*^{c}*a*, sacred image of clay, clay tablet. From an alleged Sanskrit (but rather Prākrit) *sāccha*, explained as *dam-pai gzugs brñan*. Other spellings are *ts*^{c}*a-tsa*, *sa-ts*^{c}*a*, *sā-ts*^{c}*a*, *sa-tsa*; in the respectful language *p*^{c}*yag ts*^{c}*a* (POZDN'ĀYEV, *Journey of Bāzabakši*, in Russian, p. 239). The Sanskrit prototype seems to be *sañcaka* ("mould").

30. *mu-tig*, pearl. From Sanskrit *muktā*, *muktikā*, explained as *grol-ba-čan* ("being released"). Hindustānī *mōtī* موتى, from Apabhraṁça *mottiyau*. The final guttural explosive of *mu-tig* is still articulated: Chinese transcription of the eighteenth century *mu-ti-k'u* 木底克 (E. von Zach, *Lexic. Beiträge*, Vol. III, p. 133). The Tibetan form might allow us to presuppose an Apabhraṁça form **muktik*, *mutik*.

31. *ge-sar*, hair, mane. From Sanskrit *kesara*, explained as *ral-ba* ("long hair, mane").

32. *bya-na* (also *bya-na-ma*, *bya-ñan-ja*), sauce, condiment, vegetables eaten with another dish. From Sanskrit *vyañjana*, explained as *ts'od-ma am zas spags-ma* ("vegetables or ingredients, pickles with food").

33. *sbe-ka*, frog. From Sanskrit *bheka* (in the prints before me written *bhaka*, and thus transcribed also in the Mongol version).

34. *dig-pa* (Chandra Das gives the meaning "to stammer"). From Sanskrit *dhik*, interjection expressive of lamentation, discontent, reproach, or contempt (Tibetan synonyme: *ñan-pa*).

35. *Za-hor*, from Sanskrit *Sahora*, explained as the name of a royal family of India (*rgya-gar-gyi rgyal rigs yan gar ba žig-gi miṅ*).

36. *ka-to-ra*, vessel. From Sanskrit *kaṭora*. The word *snod* is given as synonyme.

37. *la-mo* and *la-gor*, swift, quick. From Sanskrit *laghu* (written *la-hu*). Synonymes are *yaṅ-ba* and *myur-ba*.

38. *ma-he*, buffalo. From Sanskrit *mahirsha*, explained as *sar ñal* ("to couch on the earth"). The word intended is Sanskrit *mahisha*. Sindhī *mēhi*.

39. *ḥdzab*, to murmur prayers. From Sanskrit *japa*, *japti* (synonyme: *bzlas-pa*). The author adds that the conception of the word

as a designation for mantra is erroneous (*ạdzab ćes snags-kyi miṅ-du ạkᶜrul-pa daṅ*). Hence Jäschke's "magic sentence" is to be rectified.

40. *tri-ka*, edge of a well. From Sanskrit *trikā* (*pw.*: "a certain contrivance in a well"), explained as *ri-mo ṅis rim* and *kᶜron-pai mu kᶜyud*.

41. *go-ra*, ball, globe, round. From Sanskrit *gola* (synonyme: *zlum-po*).

42. *zi-kᶜyim*, red gold melted together with many precious stones. From Sanskrit *jhaikshim* (?).

43. *gre-ba*, neck, alleged to be Apabhraṁça for Sanskrit *grīvā*; in all probability, however, *gre-ba* is a real Tibetan word, derived from the same root as underlies *m-grin-pa* ("neck, throat").

44. *tal-tsam*, lamp. From *talala* (Sanskrit?), *sgron-me* being added as a synonyme.

In fact, the number of these words, especially those designating plants and vegetable products, is far more extensive, as shown by the following list.

45. *śa-ka-ra* and *ka-ra*, sugar. From Sanskrit *çarkarā* (Persian *šakar*, Mongol *šiker*, *šikir*). Tibetan *kᶜa-ra*, sugar. From Prākrit *sakkharā*, Mahārāshṭrī *sākhara*. Tibetan *li ka-ra* or *li kᶜa-ra*, a sort of sugar (Das: a medicinal sugar), sugar from Khotan (Li).[1]

[1] In 1881 Chandra Das (*J.A.S.B.*, Vol. L, p. 223) asserted that Li-yul is identified with Nepal by the translators of the Kanjur, and that "he has been able to ascertain that the ancient name of Nepal in Tibetan was Li-yul, while Palpo is the modern name for the monastery of Palpa." In his edition of *Pag Sam Jon Zang* (pt. II, p. x, Calcutta, 1908) we read that Nepal was called Bal-po-Li, because during the Buddhist period there was in Nepal a considerable manufacture of bell-metal called *li* in Tibetan, and that this Li-yul (alleged to mean "country of bell-metal") was different from the Chinese Li-yul which was Kashgar (sic! read "Khotan"), that is, bell-metal land. There is, however, no

46. *bu-ram*, in composition *bur* (West-Tibetan *gu-ram*, Sikkim *kuram*), raw sugar, treacle, molasses. Jäschke refers only to Hindī *guṛ*; also Chandra Das gives no Sanskrit equivalent. The foundation, of course, is Sanskrit *guḍa, gula* (Gypsy *gūlo, gūr*); according to *Za-ma-tog* it appears in Tibetan also as *go-la* (equivalent of *hvags*). The word *bu-ram* with the Sanskrit equivalent *guḍa* is listed in the Mahāvyutpatti; we find there also Tibetan *bur šiṅ*, answering to Sanskrit *ikshu*.

47. *k'an-da, k'a-ṇḍa, k'aṇ-ḍa*, treacle, molasses, candy. From Sanskrit *khāṇḍa*. *k'aṇḍa* (*Pol. D.*, 27, p. 48), mixed-fruit cakes; answering in meaning to Chinese *tsa kuo kao* 雜果糕, Mongol *ürädäsü* (*ürä*, "fruit"), Manchu *tebse*. Tibetan *sran-mai k'aṇḍa* (*ibid.*, p. 21) even serves for the rendering of Chinese *tou-fu* 豆腐, bean-curd.

48 *k'a-zur*, wild date. From Sanskrit *kharjūra* (Phoenix sylvestris); Hindustānī *khajūr* کھجور, Hindī *khajūra*; Newārī *khajur*.

Tibetan tradition explaining the word *li* in this connection as "bell-metal," and the misconception of Chandra Das seems to be wholly based on a misunderstanding of the text in *dPag bsam ljon bzaṅ* (pt. II, p. 170, line 26), where the combination *tal-po li* indeed occurs; yet *li* cannot be connected with the preceding *bal-po*, but only with the following *mč'od-rten*: the question is of a "brass (*li*) Caitya that was restored in Nepal." — It is interesting to note that many Tibetan products are named for the countries from which they hailed, or were at least supposed to come. It is very probable that the word *li* preceding the designations of several products, as already supposed by Schiefner in 1849 (*Lebensbeschreibung*, p. 97), relates to their origin from Khotan. As they are not listed in our Tibetan dictionaries, they may be pointed out here. Aside from the above *li k'a-ra*, we have *li k'ri* (Mongol *likri*, pronounced *liti*; Sanskrit *sindūra*; Chinese *huaṅ tan* 黄丹), "red lead, minium, vermilion;" *li ši* (Mongol *lisi, liši*; Sanskrit *lavaṅga*), "cloves" (Myristica moschata); *li doṅ-ra* (Sanskrit *nāgara*), a drug (Cyperus pertenuis); and *li ga-dur*, identified with Sanskrit *kuṭannaṭa* (Calosanthes indica and Cyperus rotundus) and *bhūtriṇa* (Andropogon schoenanthus and name of a fragrant grass), while the plain *ga-dur* is identified (in *Tib. Pharm.*) with Chinese *tsao hiu* 蚤休 (Paris polyphylla). Similar compounds are formed with Mon, that refers to the Himalayan tribes south of Tibet and in a wider sense to northern India, with Bal ("Nepal") and rGya ("China"); see above, pp. 409 and 448.

49. *ga-bur*, camphor. From Sanskrit *karpūra.*

50. *go-bye-la* (*Mong. Pharm.*, No. 65), Strychnos nux-vomica (Jäschke has *go-byi-la* as name of a poisonous medicinal fruit). From Sanskrit *kupīlu* (not yet mentioned by Suçruta or in the Bower Manuscript); Flückiger and Hanbury (*Pharmacographia*, p. 429) observe that we have no evidence that it was used in India at an early period, and that Garcia da Orta is entirely silent as to nux vomica. Mongol *gojila* (KOVALEVSKI, p. 2557). The corresponding Chinese term is *fan mu pie* 番木鱉 (G. A. STUART, *Chinese Materia Medica*, p. 425). The corresponding Arabic term is *juz* ("nut") *el-kei* جوز القى; other Arabic names are *izaragi*, *khanek-ul-kella*; Persian *fulūzmāhi*, *izarakī*, *kučla.*

51. *go-yu* (Tromowa *ga-ye*, *go-he*; Sikkim *go-he*), areca-palm, areca-nut (*Pol. D.*, 19, p. 18 = Chinese *pin-lan* 梹榔), Areca catechu. From Sanskrit *guvāka* (WATT gives *gubak* as Sanskrit word); Bengālī *gavā*, *guā*: Assamese *guā* (Guā-hāti, capital of Assam, is said to derive therefrom its name: S. ENDLE, *The Kachāris*, p. 26); Kačāri *goi*. Mongol *guyuk* (*Pol. D.* and KOVALEVSKI, p. 2622); Manchu *niyaničun*; Turkī *šipara*. The common Sanskrit word for the areca-nut is *pūgaphala*. A large number of Sanskrit synonymes for the tree are enumerated by RÖDIGER and POTT, *Kurdische Studien* (*Z. K. d. Morg.*, Vol. VII, p. 92).

52. *be-ta* (*Pol. D.*, 28, p. 56; also in CHANDRA DAS) coco-nut (Cocos nucifera). It answers in meaning to Chinese *ye-tse* 椰子, Mongol *beta*, Manchu *xotoči*, apparently from *xoto*, "skull," an artificial formation based on the Chinese synonyme *Yüe wan t'ou* 越王頭 (see Bretschneider, *Chinese Recorder*, Vol. III, 1870, p. 244). The origin of the Tibetan name is obscure (*nārikela* is the common Sanskrit word for "coco-nut"). According

to Jäschke, Be-ta is a geographical name, probably referring to the Himālaya and occurring in the *Pad-ma tᶜaṅ yig* (see also Grünwedel, *Z.D.M.G.*, Vol. LII, 1898, p. 455). According to Chandra Das, Be-ta is the Tibetan designation of Vidarbha, Bedar, the birth-place of Nāgārjuna.

53. *pi-spal*, Ficus religiosa. From Sanskrit *pippala*.

54. *pᶜo-ba-ri, pᶜo-ba-ris*, black pepper (Piper nigrum). Evidently derived from some Indian vernacular, but the origin of the word is still obscure. The corresponding Chinese term is *hu tsiao* 胡椒. The *Tib. Pharm.* (No. 89) gives as Tibetan equivalent for the latter *na-li-šam*.

55. *pi-pi-liṅ, pi-liṅ*, Piper longum. From Sanskrit *pippalī*. Chinese **pit-pal* 蓽茇.

56. *ša-ka-ma*, saffron, Crocus sativus (especially *kᶜa-čᶜe ša-ka-ma*, saffron from Kashmir; according to Rockhill, *Land of the Lamas*, p. 110, a variety of saffron different from *kur-kum*); *ša-ka-ma spos*, saffron-colored incense (Rockhill, *J.R.A.S.*, 1891, p. 274). This word is in general colloquial use, together with *dri bzaṅ* ("of good fragrance," synonyme for saffron), and is well known to the Peking traders in Tibetan drugs as the Tibetan equivalent of *huṅ hua* 紅花 ("red flower,"— saffron). Although Tibetan *ša-ka-ma* would seemingly point to a Sanskrit form *çakama*, such a Sanskrit word for "saffron" has not yet been traced. Yet we possess a Chinese parallel in *čᶜa-kü-mo* (**ja-ku(gu)-ma*) 荼矩磨, given as the Sanskrit name for "saffron" in the *Fan yi miṅ i tsi* and *Pên tsᶜao kaṅ mu*. Pelliot (*Bull. de l'Ecole française*, Vol. III, p. 270, note 4) thinks that this transcription defies restoration, and proposes to emendate *kü* 菊 in the place of *čᶜa*, in order to arrive at Sanskrit *kuṅkuma*.

This hypothesis is hardly necessary. It is not difficult to recognize in the Chinese transcription Sanskrit *jāguḍa,* which is the name of a country and a designation for "saffron." S. Lévi has recently confirmed the identification, first proposed by Watters, of the country Jāguḍa, with the Tsao-kü-tʿa mentioned by Hüan Tsaṅ, where, according to the pilgrim, the soil is favorable to the growth of saffron (*Journal asiatique,* 1915, janv.-févr., pp. 83—85). The Chinese transcription certainly leads to a Sanskrit form **jāguma,* in agreement with the termination *-ma* of Tibetan *ša-ka-ma.* This formation was perhaps suggested by the ending of *kuṅkuma,* **kurkuma* (see No. 109). Tibetan *ša-ka-ma,* in my opinion, is an Apabhraṁça word derived from or related to Sanskrit **jāguma.*

57. *bā-hi-ka,* saffron. From Sanskrit *vāhlīka* (Amarakosha, ed. Bibl. Ind., p. 170, stanza 123), "originating from Balkh" (Ch. Joret, *Les plantes dans l'antiquité,* Vol. II, *L'Iran et l'Inde,* p. 272). From the text of the Mahā-māyūrī, edited by S. Lévi, it appears that vāhlīka is identical with Pahlava (Tibetan *Pa-hla-ba*), but I do not know that vāhlīka, as stated by S. Lévi (*Journal asiatique,* 1915, janv.-févr., pp. 56, 83), is a synonyme of *hiṅgu* ("asafœtida"). The Prākrit form is *bahaliya* (Boyer, *Journal asiatique,* 1915, mars-avril, p. 288).

58. *šiṅ-kun,* asafœtida. See *Tʿoung Pao,* 1915, p. 274. Also *hiṅ* (abbreviated from Sanskrit *hiṅgu*) is in use.

59. *si-la,* incense. From Sanskrit *sillakī. Li-šii gur kʿaṅ* gives the Tibetan word *sihla,* and states that it is identical in meaning with *turushka*; *sihla* is intended for *silha, silhakī* ("incense-tree," Liquidambar orientale).

60. *sug-mel, sug-rmel, sug-smel,* small cardamom (Alpinia cardamomum).

From Sanskrit *sūkshmailā*. *Li-šii gur k'an* writes *sūkshmela*, and gives as synonyme *'e-la* (Sanskrit *elā*) *p'ra-mo*. The same work imparts also a motivation for the abbreviated mode of spelling the word: it was the horror of the numerous letters that brought about the process of simplification (yi-ge maṅ-bai ạjigs-pas spros-pa bsduo). According to Li Ši-čên, the Sanskrit term for cardamom, as given in the Suvarṇaprabhāsasūtra (金光明經, section 32), is *su-ki-mi-lo-si* 蘇乞迷羅細. This may be mentioned here, as BRETSCHNEIDER (*Bot. Sin.*, pt. 3, p. 121), who quotes this term, has not added the Sanskrit equivalent.

61 *'a-p'im*, opium (Tib. Pharm.), product of Papaver somniferum, corresponding to Chinese *ya p'ien* 鴉片. None of our Tibetan dictionaries contains this word. The late lamented Dr. P. CORDIER (*Bull. de l'Ecole française*, Vol. III, p. 628) has revealed a Tibetan word for "opium" in the form *'a-p'i-ma* from two medical treatises embodied in Vol. 131 of the section Sūtra of the Tanjur, and translated from Sanskrit. This observation, due to Dr. Cordier, is of great interest. The date of the two translations is unfortunately unknown. A slight historical inference may be based on the fact that this word is absent from the Mahāvyutpatti, translated into Tibetan in the first part of the ninth century. This may either hint at the fact that the word was then not known in Tibet; or it may have been known, but, not being believed to be Sanskrit, was for this reason not incorporated in the Sanskrit dictionary. At any rate, the Tibetan form *a-p'i-ma* must belong to a time ranging from the tenth to the twelfth century, and could not be posterior to the thirteenth century. There is no doubt that P. Cordier

accurately reproduced the word as he read it in the edition that he consulted. The possibility remains that a Tibetan copyist may have punctuated it wrongly, and inserted a dot between the syllables *p*^{c}*i* and *ma*. The greater probability is that *a-p*^{c}*im* is the original and only correct form. We further find *phīm* in Kanaurī, and *ap*^{c}*im* in Nepal (RAMSAY, p. 113, gives a Ladākhī form *a-fīm*). In view of the history of the poppy (Papaver somniferum) and the product yielded by it, there are theoretically three possibilities as to how the word could have reached Tibet. It may have come from India, or from Persia, or finally from China. In India, the Arabic word *afyūn* افيون (from Greek ὄπιον) appears as late as the middle ages in such transformations as *aphūka*, *ahiphena* (explained as "foam of a snake;" see P. C. RAY, *History of Hindu Chemistry*, 2d ed., Vol. II, p. LXIX), *aphena* ("foamless"), or *aphenaka* (JOLLY, *Medicin*, p. 14). There seems to be little chance that Tibetan *ᵓa-p*^{c}*im* (or eventually *ᵓa-p*^{c}*i-ma*) is traceable to any of these; Sanskrit *aphenaka* might have resulted in a Tibetan form *ᵓa-p*^{c}*en*, but could not be made responsible for the final *m*. Neither in Persian nor in Chinese do we meet any form of the word with final *m*, for Chinese *p*^{c}*ien* (**b'en*) was never possessed of a final labial nasal. In the vernaculars of India, however, we have *afīm* (Hindī, Hindustānī, and Panjābī), and the early European writers on India likewise record forms with final *m*: thus Alboquerque, in 1513, has *oafyam*; and Garcia ab Horto, in 1563, gives as Portuguese name *amfiam*, as Spanish name *ofium*. The Khasi language has *aphim* in addition to *aphin* and *aphin* (U N. SINGH, *Khasi-English Dict.*, p. 3). Thus the evidence points to Tibetan *a-p*^{c}*im* being derived from a mediæval Indian vernacular.

The Chinese transcription *a-p'ien* or *ya-p'ien* 鴉片 is difficult to diagnose, as we are ignorant of the time when it came into use. Li Ši-cên (*Pên ts'ao kan mu*, Ch. 23, p. 8 b) does not state in what *Pên ts'ao* the term makes its first appearance, or from what language it is derived; the significance of the name was unknown to him (名義未詳). The opinion of WATTERS (*Essays on the Chinese Language*, p. 345), that it represents the Malayan word *apiun*, does not seem to me to be well founded, nor is there reason to believe that the Chinese word is directly modelled on the basis of Arabic or Persian *afyūn*. The Chinese designation shows the same traits as Burmese *a-p'in* or *b'in*, and Siamese *p'in*, and, like the latter, goes back to Sanskrit *aphena* (*a-p'en-a*); compare also Gujarātī *aphina*, Tamil *abini*, Telugu *abhini*. I am convinced that the knowledge of preparing opium from the capsules of the poppy, and its medicinal employment, reached China from India overland by way of Yün-nan and Se-č'uan, and that Arabs and Malayans had no concern with this transaction. There is no account on the part of the Arabs to the effect that they introduced the poppy or opium into China, neither do the Chinese ever assign such a rôle to the Arabs or Malayans in their traditions regarding the subject. It is a gratuitous assumption that Chinese *a-fu-yuṅ* 阿芙蓉 should represent Arabic *afyūn* (BRETSCHNEIDER in A. de Candolle, *Origin of Cultivated Plants*, p. 400; YULE and BURNELL, *Hobson-Jobson*, p. 641; GILES, *Glossary of Reference*, p. 200). Li Ši-čen, again, is our earliest authority for this word, and plainly says that this name originated from the resemblance in color, of the flowers of the poppy, to those of *fu-yuṅ*, a designation of Nelumbium speciosum

(以其花色似芙蓉而得此名也); compare the analogous formation *mu* 木 *fu yuṅ* and *ti* 地 *fu yuṅ* (Hibiscus mutabilis). The *a* in *a-fu-yuṅ* may have been suggested by *a-pʿien*. Still less does the third name given by Li Ši-čên, *ṅo* (*ngo*)-*faṅ* 阿方, bear any relation to *afyūn*; the reading *ṅo* for 阿 is especially indicated (阿方音稱我也). G. A. Stuart (*Chinese Materia Medica*, p. 307) asserts that Li Ši-čên quotes a contemporary work as saying that the preparation of opium came from Tʿien-faṅ kuo 天方國, and that for this reason it is called "o-fang." This work is the *I lin tsi yao* 醫林集要, by Waṅ Si 王璽 (he died in 1488), who merely states that in the country Tʿien-faṅ (Arabia) a red poppy-flower is cultivated; but he says nothing about an introduction of this species or opium into China. The generally accepted opinion, that the poppy was introduced into China under the influence of the Arabs, goes back to an unfounded theory propounded by J. Edkins (*Opium: Historical Note*, p. 5, Shanghai, 1889).[1] He merely quotes the description of the poppy (as preserved in the *Pên tsʿao kaṅ mu*) by Čʿên Tsʿaṅ-kʿi, who has not a word to say in this connection about the Arabs or foreign introduction of any kind whatever; he simply notes the plant under the Chinese name *yiṅ-su* 罌粟 after Suṅ Yaṅ-tse 嵩陽子. In no Tʿang or Sung author has a name for opium as yet been traced. Čao Žu-kua makes no mention of opium. — In modern times the Tibetans have adopted in their colloquial language the Chinese word in the form *ʾa-pʿin* or *ya-bin* (V. C. Henderson, *Tibetan*

[1] See also *Final Report of the Royal Commission on Opium*, Vol. VII, pt. 2, pp. 9 and 29 (London, 1895), where the baseless assertion is made, "We know from very early Chinese writers that the Arab merchants brought poppy capsules to China."

Manual, p. 81). C. A. BELL (*Manual of Colloquial Tibetan*, p. 289) gives it as *ya-pʿin*, and writes it with Tibetan letters *γya-pʿin*. Both these authors have also a native formation *ṅal tʿa-kʿa* (presumably "sleeping-tobacco"). HENDERSON (p. 48) has a native colloquial name for the poppy, — *stobs-ldan me-tog* ("strong flower"). RADLOFF (*Wörterbuch der Türk-Dialecte*, Vol. I, col. 614) derives Kirgiz and Tarančí *apīn* ("opium") from Persian; it seems to me that the Turkish word goes back rather to Chinese.

62. *ʾa-ru-ra*, myrobalan (Terminalia chebula); from Sanskrit *harītakī*. *ba-ru-ra*, Terminalia belerica; from Sanskrit *vibhītaka*. Compare *Tʿoung Pao*, 1915, p. 275.

63. *seṅ-ge*, lion. From Sanskrit *siṁha* (see above, p. 81), Apabhraṁça *siṅghu*, *siṁghu*.

64. *byi-la* (West-Tibetan *bi-la*, *bi-li*), domestic cat. From Sanskrit *biḍāla*, *viḍāla* (Amarakosha, p. 130, 6); Hindustānī *billā* بلّا. The Tibetan seems to be derived from a Prākrit form **biḷā*.[1]

65. *ne-le*, *neu-le*, ichneumon. From Sanskrit *nakula*, interpreted in Tibetan as *rigs med* ("without family;" based on the analysis *na-kula*); derived through the medium of a Prākrit form which corresponds to Hindī *nevlā*, *nevlī*.

66. *rma-bya*, peacock. From Sanskrit *mayūra* (see this volume, p. 80); *bya*, bird, hen.

67. *ne-tso* (colloquially also *nen-tso*), parrot. Presumably of Indian origin, but the prototype is not known to me. At any rate, the word does not seem to be Tibetan, and it could hardly be

---

[1] *ži-mi*, *žim-bu*, *žum-bu*, is the "wild cat" (Jäschke's "domestic cat" is erroneous). Amarakosha, p. 130, 6: *žum-bu* = Sanskrit *otu*. In the *Pol. D.* it is the equivalent of Chinese *li* 貍, Mongol *tsogontai*, Manchu *ujirxi*, all referring to a wild cat.

expected that the Tibetans were acquainted with parrots otherwise than through Sanskrit literature.

68. *byi-ru, byu-ru*, coral (see this volume, p. 79). From Sanskrit *vidruma* (through the channel of some Indian vernacular). A Chinese transcription of the Tibetan word, made in the eighteenth century, is *č'i-lu* 集魯 (E. von Zach, *Lexic. Beiträge*, Vol. III, p. 130); in the Ming edition of the *Hua i yi yü* it is transcribed *si-lu* 席盧. Kitan *širu* and *šuru*.

69. *bai-du-rya, bai-ḍur*, a precious stone belonging to the class rock-crystals. Transcribed according to Tibetan pronunciation *bendruie* by Orazio della Penna (*Breve notizia del regno del Thibet*, ed. Klaproth, p. 55). Sanskrit *vaiḍūrya*, Prākrit *veluriya* (Arabic transcription *bīruraj* بيرورج: E. Sachau, *Indo-arabische Studien*, p. 17). Mongol *binduriya*.

70. *šel* (Mongol *šil*), rock-crystal; glass. Possibly from Sanskrit *çilā*, but this is not certain (see Eitel, *Handbook of Chinese Buddhism*, p. 153).

71. *mar-gad* (written also *ma-rgad*: Tāranātha, p. 173, line 19; and *markad*: *Pol. D.*, 22, p. 66), emerald. From Sanskrit *marakata* (see the writer's *Notes on Turquois*, p. 55).

72. *man-dzi-ra* (Jäschke: "a mineral medicine; perhaps Sanskrit *man-dza-ri*, pearl"), a variety of mica. Sanskrit *mañjarī* means "bunch of flowers, bud; a certain plant; pearl," and can hardly come here into question. Chandra Das gives the word only in the sense of a mineral. My identification with mica rests on the *Tib. Pharm.* (No. 28), where Tibetan *mañdzi-ra* is explained as *lha-ma lhañ-ts'er* and equalized with Chinese *yün mu* 雲母 and *p'u-sa ši* 菩薩石. The latter has been identified with mica by Biot on the basis of actual specimens (see F. de

Mély, *Lapidaires chinois*, pp. 67, 260; also Geerts, *Produits*, p. 478).[1] Tibetan *lhañ-ts'er* denotes mica, a white and black variety being distinguished, the former being the equivalent of *yin tsiñ ši* 銀精石, the latter that of *muñ ši* 礞石.

73. *ke-ke-ru*, a precious stone, cat's eye, chrysoberyl (occurs in the Citralakshana, my ed., p. 162). From Prākrit *kakkeraa*; Sanskrit *karketaṇa*.

74. *ban-de, ban-dhe*, a clergyman (*Hua i yi yü* renders it by *ho-šañ* 和尚). According to Hodgson, from Sanskrit *vandya* ("reverend"). This derivation seems doubtful to me. Chinese *pan-ti* 班第 (Mayers, *Chinese Government*, No. 611) is apparently a transcription of this word.

75. *dzo-ki, dzvo-ki, su-gi* (Mi-la-ras-pa: also *r-dza-ki*), vulgar corruption of Sanskrit *yogin*. This or a similar form appears to have existed in mediæval India, as attested by Marco Polo's *chughi* (ed. of Yule and Cordier, Vol. II, p. 365) and Ibn Baṭūṭa's *joki* جوكى (ed. of Defrémery and Sanguinetti, Vol. IV, p. 62); see also Rockhill, *T'oung Pao*, 1915, p. 616.

76. *čataraki*, chess. From Sanskrit *caturaṅga*. In Ladākh (Ramsay) *satranj* (Bengālī *satranč*, Persian *šatranj* شطرنج and *satranj* سطرنج). The Tibetan designations of the chessmen are as follows: *dpon* (Sanskrit *rāja*, "king"), *stag* ("tiger," owing to a misunderstanding of Mongol *bers* [= Persian *fersīn*] in the sense of *bars*, "tiger"), *rña-moṅ* ("camel," corresponds to the elephant, *hastin*, of India, our "bishop"), *rta* (Sanskrit *açva*,

[1] A Japanese text translated by Geerts says that the *p'u-sa* stone was formerly brought to Japan by the inhabitants of Amakawa. This is Macao, but not in the Philippines, as annotated by Geerts. A brief text relative to the stone is contained in the *Tuñ si yañ k'ao* (Ch. 2, p. 6), which lists it as a product of Champa, and states that it was sent as tribute in the period Hien-te (954—959).

"horse," our "knight"), *šiṅ-rta* (Sanskrit *ratha*, "chariot," our "rook"), *bu* ("boy," Sanskrit *padāti*, "footman;" Mongol *kübün*, "boy;" our "pawn"). Compare A. VAN DER LINDE, *Geschichte und Litteratur des Schachspiels*, Vol. II, pp. 136, 197. Chess is known in Tibet only to the Lamas (see, for instance, G. BOGLE's *Mission* in C. R. MARKHAM, *Narratives*, pp. 92, 101, 104; and S. TURNER, *Account of an Embassy*, p. 235). A Tibetan word for "chess" is *č'o-lo*; this, however, refers to Chinese chess (*siaṅ k'i* 象棋), not to the Indian game (see *Pol. D.*, 19, p. 33). *Mig-maṅ* or *mig-maṅs* does not denote "chess," as wrongly stated by JÄSCHKE, who followed I. J. Schmidt's translation in *mDzaṅs-blun* (Vol. I, p. 100; Vol. II, p. 124),[1] but is the *ta k'i* 大碁 or *wei* 圍 *k'i* of the Chinese (see K. HIMLY, *T'oung Pao*, Vol. VII, pp. 135—146). The Tibetan term has been adopted into Mongol as *miṅmaṅ* or *miṅman* (KOVALEVSKI, p. 2022), Kalmuk *miṅma* (PALLAS, *Sammlungen*, Vol. I, p. 157), and Turki *mingmā* مينك ما (*Pentaglot Dict.* of British Museum). The chessboard is called *reu mig* (*k'i p'an* 碁盤, Manchu *tonikô*, Mongol *külkä*).

77. *ri bo-ta-la* (Mongol transcription *riotala*), popular pronunciation for *ri-bo po-ta-la*, Mount Potala. According to *Li-šii gur k'aṅ*, this abbreviation is chosen in order to avoid the double closing of the lips in the production of *b* and *p*. The popular writing occurs in GEORGI's *Alphabetum Tibetanum*, p. 475.

78. *rī-raṁ*, corporeal relics of Buddha and saints. From Sanskrit *çarīram*. According to *Li-šii gur k'aṅ* (fol. 22 a), the first syl-

[1] The term is thus carried to the first part of the ninth century, and, in view of the *mDzaṅs-blun* being translated from Chinese, might be regarded as a Sinicism. Another instance of its use occurs in *sGra sbyor* (Tanjur, Sūtra, Vol. 124, fol. 56 a), where the plan of a well-built house is compared with the outlines of a *mig-maṅs*

31

lable *ça* (*śa*) was dropped in Tibetan, because Tibetan *śa* means "flesh" and interfered with the conception of *çarīra* as "body, bones." The word *rī-raṁ* is used in the sense of *riṅ bsrel* ("relics").

79. *sendha-pa* (Mongol transcription *siwendhaba*). CHANDRA DAS gives for this word the fantastic explanation "probably Tibetanized form of the word Siddha." The correct explanation is found in *Li-śii gur k'aṅ* (fol. 21 b): "*Sendha-pa* means what has originated from the sea, and is applied to salt, swords, horses, etc. (sendha-pa žes-pa rgya-mts'o-las byuṅ-ba-ste | ts'va daṅ ral-gri daṅ rta sogs la ạjug-pa)." This definition leads to Sanskrit *sindhuja* (*sindhu*, "river, Indus, sea, ocean"), "originating from the Indus," which according to *pw.* is used with reference to horses and salt. Compare *French D.* (p. 1017): *sin-dhu* (*du*) *skyes*, "e fluvio Indus natus, epitheton pro *lan ts'va* sal."

80. *su-ra k'uṅ*, pit. According to SCHIEFNER (*Mélanges asiatiques*, Vol. VIII, p. 156), from Sanskrit *suruṅgā*, "mine" (*k'uṅ*, of course, is the Tibetan word for "pit," corresponding to Chinese *k'uṅ* 孔, Ahom *kuṅ*, Shan *k'um*). Tibetan *su-ra* presupposes an Apabhraṁça form **suraṅga*, for in Marāṭhī, Hindī, and Bihārī we have *suraṅg*, and in Bengālī *suḍaṅg*.

A selection of a few Sanskrit terms naturalized in Tibetan literature, and known also to the educated, or otherwise of especial interest, may here follow:

81. *ka-ra-bi-ra*, oleander (Nerium odorum). Sanskrit *karavīra*(*ka*).

82. *kun-du-ru*, incense, frankincense, gum olibanum, resin of Boswellia thurifera. Sanskrit *kunduru*, *kundu*, *kunda*.

83. *nim-ba*, Nim bark, Margosa (from Portuguese *amargoso*, "bitter") bark, Melia indica or azadirachta L.; Azadirachta indica Juss.

From Sanskrit *nimba, nimbaka*. Mongol *nimbaka*. Anglo-Indian *neem*, from Hindustānī *nīm* (YULE, *Hobson-Jobson*, p. 622). Regarding a related species in China see STUART, *Chinese Materia Medica*, p. 261; FLÜCKIGER and HANBURY, *Pharmacographia*, p. 154. Another Tibetan term for the same product is ₒ*ag-tse*.

84. *bim-pa*, from Sanskrit *bimba*, Momordica monadelpha (Citralakshaṇa, V. 671).

85. *bil-ba*, from Sanskrit *bilva*, Aegle marmelos, known as bela, bæl fruit, Indian bæl, Bengal quince. In Sino-Tibetan it refers to the quince (Pyrus cathayensis), *mu kua* 木 瓜 (*Pol. D.*, 28, p. 51).

86. *byi-daṅ-ga* (Avadānakalpalatā, Vol. II, p. 204 [Bibl. Indica], pallava 109, avadāna of Kuṇāla; corresponding to *viḍaṅga* of the Sanskrit text), *byi-tᶜaṅ-ka* (Tibetan prose edition of the same work, p. 230, line 13); *bye-daṅ-ka* (REHMANN, *Beschreibung einer tib. Handapotheke*, No. 28), *byi-taṅ-ga* (Jäschke: a medicine), *byi-dam-ga* (Chandra Das, in addition to *byi-taṅ-ga*). From Sanskrit *viḍaṅga*. Compare *Tᶜoung Pao*, 1915, p. 287. The transcription of the Avadānakalpalatā is the oldest and correct one, and in a satisfactory manner renders certain the identification with Sanskrit *viḍaṅga*. Chandra Das identifies the Tibeto-Indian term with Erycibe paniculata, a large climbing shrub related to Embelia ribes, and, like the latter, yielding a black berry (W. ROXBURGH, *Flora Indica*, p. 197); as it occurs in Bengāl, the home of Chandra Das, it may be that the plant is called there *viḍaṅga*. As the berry of Erycibe is not employed medicinally, however, or exported from India, it is not very likely that the Tibetan drug can be identified with it. As formerly stated, the question is of Embelia ribes.

87. *ma-ša*, pea. Sanskrit *māsha*.

88. *ma-ša-ka*, a small gold weight and coin. Sanskrit *māshaka*.

89. *mal-li-ka*, Jasminum champaca or zambac. Sanskrit *mallikā*. Chinese **mat-li* 末利. Also Sanskrit *kunda*, Jasminum pubescens, is found in Tibetan (*kun-da*) and Mongol.

90. *tsan-dan*, sandal-tree. Sanskrit *candana*, Sirium (Santalum) album and myrtifolium; τζανδανά (Cosmas Indicopleustes); Persian *čandan* and *čandal*; Chinese *čan-tᶜan* (**čan-dan*) 旃檀.

91. *dza-ti*, nutmeg (Myristica moschata). Sanskrit *jātī*; Hindī *jāyaphala*; Hindustānī *jāēphál*.

92. *ša-na*, hemp. Sanskrit *çaṇa*.

93. *bhañ-ge*, hemp, charras. From Sanskrit *bhañgā*.

94. *šim-ša-pa* (our dictionaries state, after Csoma only, "a kind of tree or wood"), Dalbergia sissoo (*pw.*). From Sanskrit *çiṁçapa*. The *Li-šii gur kᶜañ* (fol. 22 a) has an interesting comment on this word as follows: "A general designation for Agaru [besides Amyris agallocha, aloe-wood, this term refers also to Dalbergia sissoo] is *šim-ša-pa*. As the syllable *šim* is written in Sanskrit with an Anusvāra, and is accordingly read with the letter *ñ*, it is mistaken in Tibetan for the word *šiñ* ('tree'), so that people say *ša-pai šiñ* ('ša-pa tree')." — ᾿A-ga-rui miñ-gi rnam-grañs šim-ša-pa žes-pai šim-gyi klad kor ña yig-tu klog-pa | bod skad-kyi šiñ-du ạkᶜrul-nas ša-pai šiñ zer-ba dañ. An example of the application of the word occurs in Tāranātha's *bKa-babs* (Grünwedel's translation, p. 71).

95. *kᶜa-tañ-ga*, *kᶜa-ṭvam-ga*, *kᶜa-tra* (pronounced *kᶜatomga*, *kᶜato*), ceremonial trident. From Sanskrit *khāṭvañga*.

96. *ḍa-ma-ru*, *ḍa-ru*, ceremonial tambourine. From Sanskrit *ḍamaru*.

97. *dzam-bu gliṅ, dzam gliṅ, ḥdzam-bu*, from Sanskrit *Jambudvīpa* (Uigur *čamputivip*).

98. *ṭaṁ-ka, ṭaṅ-ka, ṭaṅ-kᶜa, taṅ-ka*, a silver coin. From Sanskrit *ṭaṅka*, a weight or a coin weighing four māsha or twenty-four raktikā. Compare also WATTERS, *Essays on the Chinese Language*, p. 358; and ROCKHILL, *Tᶜoung Pao*, 1915, pp. 435—436, 448. The word is found also in Persian, *tanga* تنکه ("cash, coin") and *dāng*, and in Turkish languages (RADLOFF, Vol. III, col. 1046; and YULE, *Hobson-Jobson*, p. 896).

The following words belong to the most recent phase of the Indo-Aryan languages, and to all appearances have been borrowed from Hindī or Hindustānī.

99. *sa-laṅ-gi*, a stringed instrument, with nine chords, played upon with a bow. Hindustānī *sāraṅgī* سارنگی. This instrument seems to have originated in Nepal, and is known also in China (M. COURANT, *Musique classique des Chinois*, p. 182).

100 *ḍan-ḍi, dran-dri* (Lahūlī), beam of a pair of scales; a kind of litter. Hindī *ḍanḍī*.

101. *tim-pi* (in Mi-la-ras-pa), kid leather imported from India.

102. *m-do-le, ḥdo-li* (*Pol. D.*, 26, p. 76), sedan-chair. From Hindī *ḍolī*.

103. *ti-pi* (JÄSCHKE), *tī-bi* (RAMSAY). "The black bag-like patoo cap which hangs down on one side is called *bot teep*, and the close-fitting cap, lined with lambskin, covering the ears, is called *gónda*; a gun cap is called *topee* or *tubak i topee*" (RAMSAY). Seemingly connected with Hindustānī *ṭopī*, Anglo-Indian *topee* (YULE, *Hobson-Jobson*, p. 935).

104. *tᶜan*, a piece of cloth. From Hindī *tᶜān*.

105. *be-za* (Ladākh), interest. From Hindī *vyāja*; Hindustānī *biyāj* بياج.

106. *skyes-sdoṅ*, *skyes-la-sdoṅ* (pronounced *kyedoṅ*, *kyeladoṅ*), banana (only in Sikkim). From Hindustānī *kēlā* كيلا (also in Hindī and Pañjābī), from Apabhraṁça *ka(y)alaa* = Sanskrit *kadalaka*, "plantain" (GRIERSON, *Z.D.M.G.*, Vol. XLIX, 1895, p. 418).

107. *rdo-ja*, *stod-ja*, *dwa* (*d* with *va zur*)-*já* (the latter in *Mong. Pharm.*, No. 40; Mongol *duwa*), catechu, cutch, an astringent extract obtained by boiling in caldrons chips of the brown heart-wood of several species of Acacia (catechu, suma, sundra, and probably more) and Areca until the fluid becomes sirupy, when it is taken out, cooled, and shaped into cakes, which are used as a dye for nets, sails, and other articles. The term corresponds to Chinese *ör č'a* 兒茶 (BRETSCHNEIDER, *Bot. Sin.*, pt. 3, p. 333; G. A. STUART, *Chinese Materia Medica*, p. 2). W. W. ROCKHILL (*J.R.A.S.*, 1891, p. 226) took the total Tibetan term for a transcription of the Chinese, but it is difficult to reconcile the first element in the Tibetan compound with Chinese *ör*. The vacillating spelling, leading to the same phonetic type *do*, *to*, *tö*, shows well that not a Tibetan word, but an attempt at rendering a foreign element, is involved (JÄSCHKE gives *dwa-ba* as "a plant yielding an acrid drug," and the compounds *da-ts'od*, *da-rgod*, and *da-γyuṅ*); and this apparently is identical with the *te* in *catechu*, *cate*, of Garcia and Acosta. Besides *cate* and *cacho* (the latter in BARBOSA, Hakluyt Soc. ed., 1866, p. 191), the earlier European writers offer the word also in the form *catcha* (YULE, *Hobson-Jobson*, p. 173), which would seem to have resulted from an Indian term *ca-te-ča*, — the prototype of Tibetan *tö-ja*. In this case

the Tibetan appellation would have been formed in imitation of the Indian, not of the Chinese word; and it is reasonable to conclude that the Chinese terms are also derived from India (*hai'r č^ca* 孩兒茶, from modern Indo-Aryan *khair*, the latter from Apabhraṁça *khairu* = Sanskrit *khadira*). The French spelling *cachou* is based on Tamil *kašu*. — For the tree Acacia catechu, the Tibetans have a native name *señ-ldeñ*, identified in the Mahāvyutpatti with Sanskrit *khadira* (JÄSCHKE quotes "Schiefner" to this effect; this reference occurs in *Mélanges asiatiques*, Vol. VIII, p. 12). Among the names of the mountains we find in the same Sanskrit-Tibetan dictionary Señ-ldeñ-čan, corresponding to Sanskrit Khadiraka. Regarding the employment of catechu on the part of Tibetan women see ROCKHILL (*Land of the Lamas*, p. 214) and LAUFER (*Globus*, Vol. LXX, 1896, pp. 63—65). See also ROCKHILL, *T^coung Pao*, 1915, p. 463.

108. *gi-gu-ša*, cloisonné enamel, is given in the *Pol. D.* as the equivalent of Chinese *fa-lañ* 琺瑯. According to a Lama teacher whom I consulted at Peking in 1901, the correct form is *gu-gu-ša*, to which he assigned the meaning "cloisonné enamel." JÄSCHKE says in regard to the latter word that it is peculiar to the province of Tsañ, and signifies "plate, flat dish." The *French Dict.* (p. 149) attributes to the word (I believe, wrongly) the meaning "porcelain." CHANDRA DAS (p. 219) observes that it refers to enamelled plates, cups, etc., and generally to enamels on copper. HENDERSON (*Tibetan Manual*, p. 39) has it also with the meaning "enamel ware, enamel." JÄSCHKE, further, quotes after Schmidt a word *gi-gu-ša* or *gi-gu-šel* in the sense of "having a white speck

in the eye, wall-eyed (of horses)." In letter No. 4 published by E. H. C. WALSH (*Examples of Tibetan Letters*, Calcutta, 1913) we read of a *gu-ša-i spyad-pa* ("enamelled pot").

2. *Persian Loan-Words.*

109. *kur-kum*, *gur-kum*, *gur-gum*, saffron (product of Crocus sativus). From Persian *kurkum*, *'karkam*, *karkum* کرکم, derived on its part from Semitic (Assyrian *karkuma* [OPPERT, *L'ambre jaune chez les Assyriens*, pp. 6, 15; and E. BLOCHET, *Sources orientales de la Divine Comédie*, p. 130]; Hebrew *karkōm* כַּרְכֹּם: SOLOMON, *Canticles*, IV, 14; Aramaic *kurkāma*, Arabic *kurkum*), on which in all likelihood are based also Greek *κρόκος* and Armenian *kᶜrkᶜum*. The preservation of *r* is sufficient to place the Tibetan word with the Persian-Semitic forms, not with Sanskrit *kuṅkuma*; still less is there any reason to conjecture a Tibetan form *guṅ-gum*, as has been done by L. FEER (*Journal asiatique*, 1865, p. 504). *Kuṅkuma* is peculiar only to Sanskrit, and is doubtless developed from **kurkuma*. According to CH. JORET (*Les plantes dans l'antiquité*, Vol. II, p. 272), this word has never penetrated into India proper. Saffron was never cultivated in India, but solely in Kashmir, where it appears to have been transplanted from Iran.[1] The *Liaṅ šu* (Ch. 54, p. 7b) states that saffron (*yü-kin*) is solely produced in Kashmir (see also *Kiu Tᶜaṅ šu*, Ch. 221 B, p. 6 = CHAVANNES, *Documents*, p. 166; Hüan Tsaṅ's account of Kashmir: JULIEN, Vol. II, pp. 40, 131; TAKAKUSU, *I Tsing*, p. 128). The cultivation of saffron in Persia is attested by Iṣṭakhri and Edrisi (A. JAUBERT, *Géographie*,

[1] The Tibetan synonyme *kᶜa-čᶜe skyes* does not mean, as stated by WATTERS (*Essays*, p. 348), "gift of Kashmir" (he evidently thought of the noun *skyes*, "present, gift"), but "produced by, grown in, Kashmir." It is a literal rendering of Sanskrit *kaçmīrajanman*.

pp. 168, 192); also by the *Sui šu* (Ch. 83, p. 7 b). HIRTH's opinion (*Chau Ju-kua*, p. 91), that Chinese *yü-kin* (anciently **hat-kam*) should go back to Persian *karkam*, is not entirely convincing, owing to the diverse initials. YULE (*Hobson-Jobson*, p. 780) quotes from Sontheimer's translation of Ibn al-Baiṭār that the Persians call the root of the saffron *al-hard*. This statement is not found in LECLERC's translation of the work (*Traité des simples*, Vol. II, p. 208). If such a Persian word should exist, Chinese **hat* might possibly correspond to it.

110. *zi-ra*, cummin-seeds (Cuminum cyminum). The word is not derived from Sanskrit and Hindī *jīra*, as stated by JÄSCHKE (Sanskrit *jīra* would be transcribed in Tibetan *dzi-ra*), but from Persian *zīra* زيره ("cummin-seed"), also *žīra* ژيره, Grusinian and Ossetian *zira*. Cummin was cultivated by the ancient Persians (CH. JORET, *Les plantes dans l'antiquité*, Vol. II, p. 66), and penetrated at a remote period from Iran to Egypt on the one hand and to India on the other (*ibid.*, p. 258):[1] Sanskrit *jīra*, *jīraka*, is based on the Persian word, as is likewise Chinese **ži-ra* (*ši-lo*) 時蘿. It is not certain, however, that the plant was brought to China directly from Persia; it seems, rather, that it was introduced from the Malayan Archipelago. Cᶜên Tsᶜan-kᶜi of the Tᶜang period reports that it grows in Bhoja (Sumatra, 佛誓), and Li Sün 李珣 quotes the *Kuań čou ki* 廣州記 to the effect that it grows in the country Po-se 波斯. If Su Suṅ 蘇頌 tells us that the plant was cultivated in his time in Liṅ-nan 嶺南 and the adjoining

[1] According to FLÜCKIGER and HANBURY (*Pharmacographia*, p. 331), the plant is indigenous to the upper regions of the Nile, but was carried at an early period by cultivation to Arabia, India, and China, as well as to the countries bordering on the Mediterranean. See also V. LORET, *La flore pharaonique*, p. 72.

regions, it is not likely that Po-se is intended for Persia, but rather for the Malayan Po-se (*Chinese Clay Figures*, I, p. 145).

111. *ba-dam*, almond (Amygdalus communis). From Persian-Hindustānī *bādām* بادام. Jäschke's note, derived from Shakespear and reproduced by Watters (*Essays*, p. 349), to the effect that the Persian word is derived from Sanskrit *vātāmra* ("wind-mango"), must be discarded as a baseless conjecture. The Sanskrit word is *vātāma* or *bādāma*, which, as also indicated by Boehtlingk, represents a Persian loan (or ultimately from Pahlavī *vadam*). The home of the almond-tree is not in India, but in western Asia (De Candolle, *Origin of Cultivated Plants*, pp. 219—220). The Tibetan term *p'a-tiṅ* ("dried apricots") bears no relation to the Persian designation of the almond, as wrongly asserted by Watters (*Essays*, p. 348).

112. *se-rag dur-sman* (West-Tibetan, pronounced *sĕrak turman*), carrot (Daucus carota). From Persian *zardak turma* زردك ترمه (*zardak*, "carrot;" *turma*, "radish").

113. *dal čini* (Ladākh: Roero, Vol. III, p. 230; Ramsay, p. 21), cinnamon, bark of Cinnamomum cassia (in common Tibetan *šiṅ ts'a*, "wooden salt"). From Persian-Hindustānī *dār-čīnī* دار چینی (Arabic *dār-ṣīnī*); that is, wood of China (*dār* = Sanskrit *dāru*, wood, tree). Puštu *dāl čīnī*. Hindī *dāra-čīnī*.

114. *kram*, cabbage. From Persian *karamb* کرنب or *kalam* کلم (see this volume, p. 87).

115. *'alu-bakara* (Ladākh: Roero, Vol. III, p. 245), also *bokhara* (Vigne, Vol. II, p. 457), dried plums and apricots from Bokhāra. Persian *ālū-bokhāra*. Yule, *Hobson-Jobson*, p. 16. In the Tibetan written language Bokhāra is transcribed *Bho-har* (Vasilyev, *Geography of Minčul Chutuktu*, p. 5).

116. ʾā-lu (Ladākh), potato. From Hindustānī *ālu* آلو, which itself comes from Persian.

117. ʾa-lu-ča (Ladākh), plum. From Persian *āluča* آلوچه.

118. *čob-čīnī* (Ladākh), root of Smilax pseudo-china, Chinese *tᶜu-fu-liṅ* 土茯苓 (see G. A. Stuart, *Chinese Materia Medica*, p. 410), known in commerce as China root, and imported from Turkistan to Le, Ladākh (Roero, Vol. III, p. 12). From Persian *čūbi čīnī* چوبی چینی ("China root").

> "In some cases, I was told, that they obtain the benefits of an alterative medicine by the persevering administration of a decoction of Chob-i-Chini,—the strength of which is increased at intervals of two or three days, until the cure is effected." — G. T. Vigne, Vol. II, p. 123.

119. *zar-babs* (Jäschke), *zar-baft* (Ramsay), gold brocade. From Persian *zar-baft* زربفت (also *zar-bāf* زرباف, *zar-bāfta*; *bāf*, "weaving"). Under another entry Jäschke has *sal-báb* (West-Tibetan), "gold ornament, gold lace, and the like;" this, of course, is a transcription of the same Persian word. Compare J. Fryer, *New Account of East India and Persia*, Vol. II, p. 167 (ed. of Hakluyt Soc.).

120. *sag-lad*, fine cloth of goat-wool; European broadcloth. From Persian *saqlāt* سقلات (see Yule, *Hobson-Jobson*, p. 861; *J.A.S.B.*, 1910, pp. 263—266; W. Heyd, *Histoire du commerce du levant*, Vol. II, p. 700; Watters, *Essays*, pp. 341—342).

> "All Tibetans admire and value English broadcloth beyond one of our products." — J. D. Hooker, *Himalayan Journals*, p. 405.

121. *kimkhāb* (Ladākh: Ramsay); *kincob*, *čincob*, gold brocade. The latter is given as a Tibetan term by W. W. Rockhill (*J.R.A.S.*, 1891, p. 125), and said by him to be intended for Chinese *kin kia* 金叚 (see also his *Land of the Lamas*, p. 282, note 1); more

probably, the Chinese prototype is *kin hua* (**kim hwa*) 錦花, as, for instance, used by Čao Žu-kua with reference to the brocades of Ta Tsʿin. In the History of Ladākh Tibetanized in the form *rkyen-kʿab* (K. Marx, *J.A.S.B.*, Vol. LX, 1891, p. 135). The term must have reached the Tibetans from Persia or India, where we find the forms with final *b* (Persian *kimkhāb* كمخاب; Hindustānī *kamkhāb*, *hamkhwāb*). The earliest Arabic reference to the term seems to occur in Ibn Khordādbeh, who wrote between 844 and 848, and recorded it in the form *kīmkhāw* كيمخاو (G. Ferrand, *Textes relatifs à l'Extrême-Orient*, Vol. I, p. 31). See Yule, *Cathay* (new ed. by Cordier, Vol. III, p. 155), and *Hobson-Jobson*, p. 484.

"From Hindustan to Yarkand are carried madder, pearls, English calicoes, Dacca muslins, chintzes, kimkab or golden cloth of Benares" etc.—G. T. Vigne, Vol. II, p. 345. "Exclusive of tea, the following are among the principal exports from Ta-chien-lu to Tibet: cotton, silks, satins, gold brocades or chinkobs."—Rockhill, *Land of the Lamas*, p. 281.

122. *tsa-dar*, *tsa-sar*, *tsa-zar*, *tsa-dir* (Ladākh), shawl, plaid. According to Jäschke, from Persian-Hindustānī *čaddar* چدّر. The *Persian-English Dictionary* of Steingass does not contain this word, but only *čādar* with the meanings "tent, pavilion; mantle, scarf, veil, shroud; table-cloth" (see No. 142).

123. *sag-ri*, *sags-ri*, shagreen. From Persian *sagrī* سغرى, shagreen. The Persian word denotes properly the croup of a horse or donkey, from which the peculiar granulated leather was originally made (Yule, *Hobson-Jobson*, p. 818; also his edition of *Marco Polo*, Vol. I, p. 395). Mongol *sagari*, *sarisu*; Manchu *sarin*.

124. *pi-ši*, cat (Jäschke: "perhaps from the Persian"). In the first line, this word seems to be connected with such Piçāca forms as *puši*, *pīšī-r*, *piš*, *pišō*, *pušak*, as enumerated by Grierson

(*Piśāca Languages*, p. 66), and with corresponding Dardu words like *būši, puši, pušak* (LEITNER, *Languages of Dardistan*, p. 2); and these point to Persian *pušek, pōšek* پوشك (also *pūšank*), Afgan *pišō* (compare F. JUSTI, *Les noms d'animaux en kurde*, p. 5, Paris, 1878).

125. *p^co-lad*, steel. From Persian *pūlād* پولاد (see this volume, p. 82, note 4).

126. *ta-ba, tao*, iron pan. From Persian *tāwā* تاوا.

127. *dig*, large kettle. From Persian *dig* ديگ (found also in Šugnan: Восточныя Замѣтки, p. 290).

128. *ta-ra-tse* (Ladākh), a small pair of scales, gold weights. From Persian *tarāzū* ترازو (also current in Turkistan: A. VON LE COQ, *Turfan*, p. 86), a balance, scale, weight.

129. *nal* (given in our dictionaries only as the name of a precious stone), balas ruby. From Persian *lāl* لعل (Mongol *nal ürdüni*, Manchu *langča*). The Chinese equivalent (*Pol. D.*, 22, p. 66) is *pi-ya-se* 碧琊包, derived from Persian-Arabic *piyāzakī* پيازكى, balas ruby from Piyāzak (see the writer's *Notes on Turquois*, pp. 45—48). CHANDRA DAS has a term *pi-tsu lā-la* as "name of a gem," and designated as Sanskrit, but not identified. I do not know for what Sanskrit word it could be intended; maybe *lā-la* stands for *lāl*, and the whole term refers to a certain variety of balas ruby.

130. *saṅ-gi-ka* (JÄSCHKE with ?), greenish stone of which knife-handles and similar articles are said to be made. It might be permissible to connect this word with Persian *sang* سنك ("stone"). On the other hand, there is also a West-Tibetan word of uncertain spelling, *yaṅ-gi-k^ca*, said to be the name of a green stone wrought into handles of knives. Another form of this word

is *yaṅ-tri*, which to all appearances is identical with *g-yaṅ-ṭi*, the common Tibetan term for "nephrite," as attested by the *Pol. D.* (22, p. 64), where it is equalized with Chinese *yü* 玉.

131. *dambu-ra, tambu-ra*, a stringed instrument (*Pol. D.*, 7, p. 18, = Chinese *čeṅ* 箏, Mongol *yaduga* or *yadaga*, Manchu *yatuxan*). From Persian *tambūr* تنبور ("lute, lyre, guitar; a drum"). We find the same instrument and word in Chinese *tan-pu-la* 丹布拉 (see COURANT, *Essai sur la musique classique des Chinois*, p. 178); in Eastern Turkī *dambura*, Osmanli *tambur* and *tambura*; Kirgiz *domra, dombra, dunbura*, and *dumbra*: Kazan Tatar *dumbra*; Russian *domra* and *dombra* (the latter referring to the Jew's-harp).

132. *sur-na* (Chinese *so-na* 鎖吶), hautboy, flageolet. From Persian *surnā* سرنا. The word seems to be known only in western Tibet, for the compilers of the *Pol. D.* (7, p. 16) were ignorant of it, and rendered Chinese *so-na* by *rgya gliṅ* ("Chinese flute").

133. *kab-ša*, shoe, boot (the ordinary and general term for shoe in Ladākh). From Persian *kafš* كفش, in all probability adopted into Ladākhī from Hindustānī, as the word refers to the ordinary shoe of Hindustan (RAMSAY, p. 12). *ba-bu*, soft shoe. According to Jäschke, from Persian *pāpōš* پاپوش.

134. *dur-bin*, telescope. From Persian *durbīn* دوربين (literally, "far-seeing").

135. *saṅ-gin*, bayonet. From Persian *sangīn* سنگين.

136. *pʿugs-ta, pʿogs-ta, pogs-ta* (Ladākh), firm, strong, durable. From Persian *pukhta* پخته.

137. *pe-ban*, graft, scion. From Persian *pewand, paiwand* پيوند, relations, allies.

138. *po-la*, *p*^{c}*u-la*, *p*^{c}*o-la* (Jäschke: Ladākh; but known also in Central Tibet), pilau, pilaw, a dish consisting of boiled rice with fowl or mutton, and spices, the Tibetans adding also butter and dried apricots (*p*^{c}*a-tiṅ*); in Ladākh, sweet rice prepared with butter, sugar, and apricots. From Persian and Hindustānī *pulāo*, *pilāv* پلو or پلاو; [1] Radloff (Vol. IV, col. 1337) derives Osmanli *pilaw* from Persian. The word has penetrated also into Slavic: Russian *pilav* (пилавъ), *plav* (плавъ), and *plov* (пловъ). Perhaps also the (Niüči?) term *pi-lo* 畢羅 ("stuffed pastry") belongs to this type (Chavannes, *T*^{c}*oung Pao*, 1904, p. 168).

"On arriving at a halting place, the traders opened their bazaar, in which the soldiers could buy rolls, meat pasties, pilau and tea."—Kuropatkin, *Kashgaria*, p. 231.

139. *pai-k*^{c}*am-pa* (Tāranātha, translation of Schiefner, p. 80), "the Ṛishi of the Mleccha" (Mohammedans). From Persian *paigāmbar* پيغمبر, prophet. This identification is due to the ingenuity of Ernst Kuhn, *Barlaam und Joasaph* (*A.B.A.W.*, 1893, p. 85).

140. *deb-t*^{c}*er*, *deb-gter*, *deb-ster*, document, record, book. From Persian *däbtär* دفتر (said to be connected with Old-Persian *dipi*, "writing;" see P. Horn, *Neupersische Etymologie*, No. 540; according to others, derived from Greek *διφθέρα*, "leather, parchment for writing upon," see E. Blochet, *Inscriptions turques de l'Orkhon*, p. 46). Sogdian *dipêr-ṭ* ("one versed in the Scripture;" Müller, *Soghdische Texte*, I, p. 17). The same word has been adopted into Mongol *däbtär* and Manchu *debtelin*

---

[1] Yule (*Hobson-Jobson*, p. 710) refers also to Sanskrit *pulāka* ("ball of boiled rice;" properly, a certain cereal), but I do not believe that it should be connected with our word. Regarding the preparation of pilau in Persia see J. Fryer, *New Account of East India and Persia*, Vol. III, p. 147 (Hakluyt Soc., 1915); F. Bernier, *Travels in the Mogul Empire* (2d ed. by V. A. Smith), p. 121.

("chapter, volume"). Compare W. Schott, *Zur Uigurenfrage*, II, *A.Be.A.W.*, 1875, p. 54; and Rémusat, *Recherches*, p. 137.

141. *pʻe-raṅ*, *pʻa-raṅ*, *pʻi-liṅ* (*rgya pʻi-liṅ*), *pʻo-raṅ* (Georgi, *Alphabetum Tibetanum*, p. 654), at present refers to British India, Englishmen, Europe and Europeans (see Jäschke, p. 106, who cites also the derivation from "Feringhi." The opinion that "*pʻi-liṅ* represents only the more vulgar pronunciation of the genuine Tibetan word *pʻyi gliṅ*, a foreign country and especially Europe," is untenable). The four forms evidently are transcriptions of the same foreign term; and since Tibetan lacks *f* and renders it by a labial surd or sonant, aspirated or non-aspirated, we arrive at *fe-raṅ*, *fa-raṅ* (or *fa-laṅ*), *fi-liṅ*, *fo-laṅ*. These forms are on a par with Persian *farangī*, *firingī* فرنگی; Arabic *al-faranj*, *ifranjī*, *firanjī*; Tamil *pʻarangi*, Singalese *parangi* (Yule, *Hobson-Jobson*, p. 352; Dalgado, *Vocab. português em línguas asiáticas*, p. 10), Neo-Sanskrit *phiraṅga* (*phiraṅgaroga*, "Morbus americanus"); Chinese *fu-laṅ* 佛郎, *fa-lan* 發藍, *fa-laṅ* 琺瑯 (Manchu *falasu*), *fu-laṅ-gi* 佛郎機 (compare Bretschneider, *Mediæval Researches*, Vol. I, pp. 142—144; Watters, *Essays*, p. 334), — all connected with the term "Frank." It cannot be positively asserted, of course, that the Tibetan terms are derived from Persian, as long as we have not secured their exact history (and this is not possible in our present state of knowledge); but there is a great deal of probability in this assumption.

The Tibetan term Pʻi-liṅ has passed into the Chinese History of the Gorkha War: "En ce temps, les Gorkha étaient en mauvais termès avec Pʻi leṅ, pays de l'Inde qui est situé au sud des frontières de leur pays, et qui, depuis longtemps, était sous la domination des Yiṅ-ki-li (Anglais)."—M. C. Imbault-Huart, *Histoire de la conquête du Népâl par les Chinois*, p. 24 (Paris, 1879, extrait du *Journal asiatique*). In another passage of this work, the same name is applied to Calcutta (*ibid.*, p. 29).

142. *p'ya-t'er* (KOVALEVSKI, p. 2104), tent. From Persian *čāder* چادر (Uigur *čadir*, Mongol *tsatsar*, Manchu *tsatsari*, Russian *čadra* and *šater*, Polish *szatra*, Magyar *sátor*, etc.).

3. *Arabic Loan-Words.*

143. *ʾa-rág* (RAMSAY: *árak*), alcoholic beverage made from barley; recorded also in *Pol. D.*, 27, p. 35 (Chinese *nai-tse tsiu* 奶子酒). From Arabic *ʾáraq* عرق ("perspiration, exudation or sap drawn from the date palm"), that has become a universal word in Asia. In China this word first appears as *a-la-ki* (*a-ra-ki*) 阿剌吉 in the writers of the Yüan dynasty (see J. DUDGEON, *Beverages of the Chinese*, p. 23, Tientsin, 1895; and WATTERS, *Essays*, p. 354). Turkish *rāḳi*; Mongol *araki*, *ariki*, *arki*; *arača* (*Pol. D.*); Manchu *arčan*. The forms ending in *-ki* are evidently based on Arabic *ʾaraqī* عرقى.

"Among the Horba I saw more drinking than farther north: nä ch'ang and a strong spirit distilled from it called *arrak* being the only two liquors in use. The latter is imbibed either cold or warm in Chinese fashion."—ROCKHILL, *Land of the Lamas*, p. 248. "On leaving, in place of tea, we drank some 'arrack,' an extremely potent liquor slightly resembling Kummel in flavor, after which we were able to bid a cheerful goodbye to our rather gruesome hosts."—G. G. RAWLING, *The Great Plateau, being an Account of Explorations in Central Tibet*, p. 187.—From my own experience I can confirm Mr. Rockhill's impression: it is rather difficult to have an opportunity of meeting Horba Tibetans in a sober state. I met Horba Lamas of the Red Sect in a hopeless condition of intoxication.

144. *sa-heb* (colloquially *sāb*, *sab*), from Arabic *sāheb* صاحب, formerly title of Moslems of high rank, now address of every white man in India. In the History of Ladākh it is used in connection with the names of the first Englishmen who came to

32

Le (K. Marx, *J.A.S.B.*, Vol. LXIII, 1894, p. 104). The Tibetan adaptation *sa-yab* ("earth-father") is interesting in showing how the Tibetans adjust foreign words to their own language. *Sab*, *salām*, and *bakšīš* (a *salām*, in fact, means or calls for *bakšiš*) are the three most important words which are hurled at the European in India daily and hourly, and also from the lips of Tibetans.

145. *bag-šis*, tip, pourboire. From Arabic *bakhšiš* بخشيش, now a universal word in all countries of the Indian ocean.

146. *kab-za*, hilt of a sword. According to Jäschke, from Arabic *kabṣa* قبضة.

147. *bi-ṣli-mli*, "the great demon from the clan of the Asura" (Tāranātha, p. 63, line 10; Schiefner's translation, p. 79). As Islam is involved in this passage (see E. Kuhn, *A.B. A.W.*, 1893, p. 85), Schiefner may be right in deriving this word from the Arabic *bismillāh* بسم الله ("in the name of God"), or rather the abbreviated formula بسملة. In a commentary of the Tanjur, the god of the Kla-klo (Mleccha; that is, Mohammedans) is styled *bi-ši-mil-la* (Schiefner, *ibid.*, p. 304).

148. *ma-si-ta*, a temple of the Mohammedans. Schiefner (*Tāranātha*, p. 128) derives the word from Arabic *masjid* مسجد. It may have come to Tibet from Hindustānī. — *mā-žia*, *ma-zīd* (Ladākh: Ramsay), mosque.

149. *gyiṅ*, name of a deity (Jäschke), the djinn of the Arabs (in the literature relating to Padmasambhava; see Grünwedel, *Bastian Festschrift*, p. 465). From Arabic *jinn* جن. Perhaps also *giṅ bdud* ("*giṅ* demon," noted in the *French D.*) is the same word.

150. *da-fan* (Ladākh: Ramsay), Mohammedan burial. From Arabic *dafan* دفن.

151. *ka-bar* (Ladākh: RAMSAY), Mohammedan grave. From Arabic *qaber* قبر.

152. *ma-zār* (Ladākh: RAMSAY), Mohammedan graveyard. From Arabic *mezār* مزار.

153. *kasāba* (RAMSAY), the red turban worn by Argon women.

154. *dāōn* (RAMSAY), the white veil worn over the *kasāba*.

155. *ha-lál* (*čo-če*), to slaughter an animal according to Mohammedan rites. From Arabic (through Hindustānī) *halal* حلال (see JÄSCHKE).

156. *hu-ka*, the hookah pipe. From Arabic *huka* حقّة.

4. *Uigur Loan-Words.*

After discussing the occurrence of Chinese words in Tibetan, the *Li-šii gur k'an* (fol. 23 a) refers to the Uigur language thus: la-la Hor-gyi brdar yan snan-ste | btsun-pa-la pag-ši dan | dban č'e-ba-la dar-k'a-č'e dan | dar rgan byin sogs man la, "Some are words of the Uigur: for instance, *pag-ši*, corresponding to *btsun-pa*; and *dar-k'a-č'e* and *dar-rgan*, which have the meaning 'endowed with great power,' and many others."

157. *pag-ši* (not *pa-ši*, as written by JÄSCHKE), teacher. CHANDRA DAS (p. 777) has justly combined this word with Mongol *bakši*. The "Tibetan Lama of the Karma-pa sect who visited China to preach Buddhism," cited by him, is the so-called Karma Bakši (1204—83), whose life has been narrated by ạJigs-med nam-mk'a (HUTH's translation, p. 136). As observed by O. KOVALEVSKI (*Mongol Chrestomathy*, in Russian, Vol. I, p. 350; Vol. II, p. 247), the Mongol word has the meaning "teacher," and is synonymous with Sanskrit *guru* and *ācārya*. In *T'oung Pao*, 1914 (p. 411), I have given some indications on the word, disconnecting it

from Sanskrit *bhikshu*, and stating that, according to the Tibetan source above cited, it is derived from the language of the Hor; that is, the Uigur. Indeed, W. RADLOFF (in GRÜNWEDEL, *Bericht über arch. Arbeiten in Idikutschari*, pp. 193, 194; and *Wört. d. Türk-Dialecte*, Vol. IV, col. 1445) read the word in the sense of "Buddhist scholar" in an Uigur inscription from Turfan, written on a wooden pillar (see also F. W. K. MÜLLER, *Uigurica*, pp. 47, 49; and *Zwei Pfahlinschriften*, *A.P.A.W.*, 1915). The word, accordingly, was familiar to the Uigur language; and the Tibetan tradition regarding the Uigur origin of the word, as far as Tibetan literature is concerned, may well be correct. This being the case, there is every reason to assume also that the Mongols received the word, like many others, from the Uigur, and that the Manchu in their turn adopted it from the Mongols (Manchu *baksi*, *bakši*, "teacher, sage, scholar, one who knows books, master of any profession," according to SACHAROV; but Manchu *faksi*, "artist, artisan, clever," in all probability, is an independent, indigenous word). The question now arises, What is the origin of the Uigur word? P. P. SCHMIDT (*Essay of a Mandarin Grammar*, in Russian, p. 50, Vladivostok, 1915) has proposed to derive the word *bakši* from Chinese *po ši* 博士 ("a wide-read scholar, professor"), and this appears to me a happy solution of the problem: *po* was anciently **bak*, and is still *pok* in Cantonese and Hakka (Korean *pak*, Annamese *bak*). Only it seems to me that the Uigur transcription *ši* would rather go back to Chinese *ši* (**š'i*) 師 than to *ši* (**j'i*) 士 (compare PELLIOT, *T'oung Pao*, 1911, p. 668); the analogous case of *fa-ši* (**fap-š'i*) 法師 presumably may have exerted some influence on the shaping of *bak-ši*. The Chinese terms given by WATTERS (*Essays*, p. 371) are merely modern transcriptions of the Mongol word *bakši*; but in the *Yüan ši* we meet a transcription *pa-ha-ši* (*baχši*) 八哈失 (PELLIOT, *Journal asiatique*, 1913, mars-avril, p. 456). It is well known that our word has had many vicissitudes and re-interpretations in various languages. Its fate is ably expounded by H. YULE (*Hobson-Jobson*, p. 134). The derivation from Sanskrit *bhikshu* (still repeated by VLADIMIRTSOV, *Turkish Elements in the Mongol Language*, in Russian, *Zap.*, Vol. XX, 1911, p. 12; and by J. NÉMETH, *Z.D.M.G.*, 1912, p. 553) must definitely be abandoned. This conclusion was based on the observation that, according to a passage in the *Ain-i Akbarī* (translation of H. S. JARRETT, Vol. III, p. 212), the religious of the Buddhist order are named *bakhši* by the learned among the Persians and Arabs; but this recent text can only prove that in the India of the Moguls an adjustment or confusion between *bhikshu* and *bakši* had taken effect. In Central Asia, *bakši* never had the Buddhistic significance "religious mendicant" (*bhikshu*), for which notion specific terms are in use; first of all, it is applied to persons able to read and write, and hence it passed into the sense of a clerk (in Comanian *bacsi*, "scribe, secretary"). It remains an open question whether *bakši*, denoting a surgeon among the Moguls, a bard in western Turkistan, and a medicine-man among the Sart, is actually the same word as that under consideration. The Sart word is perhaps connected with or assimilated to the Turkish stem *bag*, "charm, sorcery" (Z. GOMBOCZ, *Bulg.-türk. Lehnwörter*, p. 39). *Bakši* as a military office ("Master of the Horse") under the Mohammedan emperors of India, and the Anglo-Indian *buxee* ("paymaster"), are independent words, to be dissociated from the Uigur-Chinese term.

158. *dar-k'a-č'e* (thus transcribed in the Sino-Tibetan inscription of 1341, published by CHAVANNES in *T'oung Pao*, 1908, p. 418,

and plate 28, line 4), *dar-rgan*, empowered with authority, official (see above). The interlinear Mongol translation of the *Li-šii gur kʿaṅ* renders the former expression by *darkači*, the latter by *darkan*. In the Orkhon inscriptions, *tarkan* (Djagatai and Osmanli *tarchan*) appears as the designation of a dignity. A. VON LE COQ (*Turfan*, p. 91) says that *dŏroγa* or *dăroγa* (compare Persian *dārogā*) has in Turfan the meaning "mayor," but that in the western part of Turkistan, from Kuča to Kašgar, it is the title of the lower Turkish officials (soldier or policeman) in the Yamen. Compare, further, Djagatai *darga* ("chief, officer"); and Mongol, Djagatai, and Osmanli *daruga* (see also CHAVANNES, *Tʿoung Pao*, 1904, p. 389). Our word appears in two official Tibetan documents of 1724 and 1729 as *rda-ro-kʿa*, transcribed *ta-ro-kha* (GEORGI, *Alphabetum Tibetanum*, pp. 652, 660), again written *da-ro-ga* (p. 653), and explained as "Tartarorum, idem ac Ciu-phon [*bču dpon*] Tibetensium." For further information see YULE, *Hobson-Jobson*, p. 297; J. FRYER, *Account of East India*, Vol. III, p. 23 (Hakluyt Soc., 1915).

159. *dam-ga*, *dam-kʿa* (in *rGyal rabs*), *tʿam-ga*, *tʿam-ka*, seal. From Uigur *tamγa*, Orkhon inscriptions *tamka*; Mongol *tamaga*. Originally the word designated among the nomadic tribes a property-mark that was branded on the skin of the animals (see illustrations in POTANIN, *Sketches of North-Western Mongolia*, Vol. IV, plate I); hence seal, mark, stamp (RADLOFF, Vol. III, col. 1003; YULE, *Cathay*, new ed. by CORDIER, Vol. III, p. 143). In composition it is also abbreviated into *tʿam*; for instance, *rgyal tʿam*, "royal seal" (HUTH, *B.M.*, Vol. I, p. 10, line 10), which shows that the Tibetans learned to regard the word as one of their own, and the element *ka* (*kʿa*, *ga*) as a suffix on a par with their suffix *ka*.

160. *t'em yig* (*Pol. D.*), memoirs, corresponding in meaning to Chinese *ki* 記. Connected with Mongol *tämdägläl* (KOVALEVSKI, p. 1732; "journal, agenda, mémorandum"), derived from *tämdäk* ("sign, mark, proof, document, seal"), that seems to be associated with the preceding word *tamaga*. It is therefore possible that also in this case there is direct connection of Tibetan with Uigur.

161. *dar*, silk. Perhaps anciently connected with Uigur *torγu*, Djagatai *torka* (RADLOFF, Vol. III, col. 1183, 1185); Koibal *tórga* (CASTRÉN), Mongol *torgan*, *torga*, *torgon*, *torgo* (KOVALEVSKI, p. 1891). A. VON LE COQ (*Turfan*, p. 91) notes a Qara-Khōja word *dūryä* with the meaning "kind of silk stuff from Kašgar," and *dūrdü'n* meaning "inferior silk stuff."

162. *'am-č'i*, *'em-č'i* (Tromowa and Sikkim *am-ji*), physician. According to Jäschke, a Turkish word. From Uigur *ämči* (F. W. K. MÜLLER, *Uigurica*, I, p. 7), Mongol *ämči*. The word has no connection with Tibetan *sman-byed*, as suggested by W. W. ROCKHILL (*J.R.A.S.*, 1891, p. 234). HODGSON (*The Phœnix*, Vol. III, p. 46) notes a Chepań word *či-me* ("physician"), which apparently is based on Tibetan *'em-či*, the two members being inverted, and *em* being changed into *me*.

163. *g-yer-ma*, the black pepper-like seeds of Zanthoxylum bungei and piperitum. In all probability, from Uigur *yarma* (KLAPROTH, *Sprache und Schrift der Uiguren*, p. 14), that corresponds in meaning to Chinese *hua tsiao* 花椒. Tibetan *g-yer-ma* is identified with the latter term in the *Hua i yi yü* and in *Pol. D.* (29, p. 25), where Mongol *irma* is added. JÄSCHKE wrongly attributes to the Tibetan word the meaning "Guinea-pepper" (Capsicum annuum), likewise his faithful copyist CHANDRA DAS, while the *French Dict.* gives the correct definition. "Guinea-pepper" is out of the question, being an American plant that was introduced into Asia in post-Columbian times. Cf. No. 237.

5. *Turkī Loan-Words.*

164. *yam-bu* (Central Tibet *yam-bᶜu*), Chinese ingot of silver. From Chinese *yüan pao* 圓寶. The Tibetan form, however, indicates that the word did not come directly from China, but was adopted from Turkī *yam-bō* or *yam-bū* of Central Asia (see YULE, *Hobson-Jobson*, p. 830; N. ELIAS and E. D. ROSS, *The Tarikh-i-Rashidi*, p. 256). The Turkī word *kurus* or *kūrs* is used in Ladākh beside *yam-bu* (RAMSAY). The Tibetan *yam-bu*, like the Chinese *yüan pao*, has fifty ounces or taels (*sraṅ*, *šraṅ*) silver (about 125—156¼ rupees). It is known in Tibetanized garb as *r̂do tsᶜad* (literally, "weight-measure") or *rta rmig-ma* (*ta mig-ma*, Ladākhī *star-mik-ma*, "horse's hoof"). There are, further, a "sheep's hoof" (*lug rmig-ma*), an ingot of silver worth 15 rupees, 10 annas; and a "yak's hoof" (*g-yag rmig-ma*), equal to 31 rupees, 4 annas.

"Argento puro in yambou in thibetano, *Kurus* in turki, dalla Kashgaria."—O. ROERO, Vol. III, p. 71; enumerated among the import-articles of Le, Ladākh.

"Forty ducats are equal to a *yamb*. *Yambs* are bars of silver of three different sizes, used by the Chinese in their monetary system. The largest size is about 4¾ lbs. in weight.... In Russian money, the exchange being at the rate of 10 per cent on silver, a *yamb* would cost about 130 rubles."—KUROPATKIN, *Kashgaria*, p. 65.

165. *čᶜu-ba*, *čᶜu-pa*, *čᶜo-pa*, a long loose gown (see ROCKHILL, *J.R.A.S.*, 1891, p. 122; and *Notes on the Ethnology of Tibet*, pp. 684—685). Turkī *juba* (Russian *šuba*, *šubka*; Polish *czuba*; Czeš *čuba*, *šuba*; German *schaube*). Compare also Eastern Turkī *čapan* ("wadded coat with long sleeves") in A. VON LE COQ (*Turfan*, p. 89), and Persian *čapkan* چپکن ("a sort of short coat").

166. *bol-gar* and *bul-ha-ri* (JÄSCHKE: West-Tibetan *bul-gar*, Russian leather, yufts), *bu-la-ha-ri* (CHANDRA DAS). ROCKHILL (*Notes*

on the Ethnology of Tibet, p. 712) gives *bulgar* as general Tibetan term for Russian leather. From Turkī-Persian *bulgār*, also in Hindustānī. Compare YULE, *Hobson-Jobson*, p. 125; and *Marco Polo*, Vol. I, pp. 6—7, 395. The Swedish captain Ph. J. VON STRAHLENBERG (1676—1747), in his remarkable work *Das nord- und östliche Theil von Europa und Asia* (p. 381, Stockholm, 1730), states in his notice on yufts that the Tatars designate this leather *bulgarie*, perhaps as such leather was first introduced to them from that region (Bulgaria); that is, the Bulgar kingdom on the Volga.

"Pelli o cuoi russi, d'una conica peculiare, chiamati Bulgaro."— O. ROERO, Vol. III, p. 71.

The Russians imported this leather into China during the eighteenth century (G. CAHEN, *Le livre de comptes de la caravane russe à Pékin*, p. 104). According to P. COUVREUR (*Dict. français-chinois*, p. 364), the Chinese term for yufts is *hiań p'i* 香皮 ("fragrant leather"): according to the English-Chinese Dictionary of the Commercial Press, *o-lo-se ko* 我羅斯革 ("Russian leather").

167. *l-čags-mag*, *b-ča-mag*, steel for striking fire, flint-stone, tinder-box. From Turkī *čakmak* چاقماق.

168. *l-čags-p'ra* (Central Tibet), "a kind of musket imported from Rūm (Turkey)," see JÄSCHKE, *Dictionary*, p. 148. [1] The first element is not *l-čags* ("iron"), but is the abbreviation of the preceding Turkī word.

---

[1] *Rum* is given by Jäschke as the name of Turkey, the Ottoman Empire, the site of which is but vaguely known to the Tibetans, though some commodities from thence find their way to Lhasa. He further cites the formations *rum-pa* ("a Turk") and *Rum-šam* ("Syria"). The authors of the *French Dictionary* are inclined to think that *Rum* designates rather the Roman Empire, and query the rendering "Turkey." An instance of the application of the name occurs in the Geography of the Minčul Chutuktu (VASILYEV's translation, p. 5), where it follows from the context that the Turkish Empire is understood. It is used also with respect to the Byzantine Empire in a conclusive passage of *B. M.* translated by HUTH (p. 29), where Kod-k'ar, the fourth son of Č'agatai, is established as king of the country Rom (*Rom yul*), with the capital Stambhola (that is, Stambul).

169. *top* (Ladākh: RAMSAY), cannon. Djagatai and Osmanli *top*; Turkī *tōp* (A. VON LE COQ, *Turfan*, p. 91); Persian *tōb*, *tufañ*.

170. *tu-pag* (*tu-bak*), gun. From Turkī *tupak* توپك (Buriški *tūmak*).

171. *pi-č'ag*, large butcher-knife (Ramsay: *pīč'ak*, a long dagger-like knife). According to Jäschke, from Turkī *čākū* چاكو. The same word occurs also in Persian with the meaning "clasp-knife, pen = knife."

172. *u-lag* (in eastern Tibet pronounced *úla*), socage service, compulsory post-service, beasts of burden requisitioned for government purposes, courier. From Turkish *ulak* اولاق (also *ulañ*, *ulau*, *ulā*), Mongol *ulaga*, Persian (Rashid-eddīn) *ulag* اولاغ (RADLOFF, *Wörterbuch der Türk-Dialecte*, Vol. I, col. 1679; E. BLOCHET, *Djami el-Tévarikh*, Vol. II, p. 312; KLAPROTH, *Description du Tubet*, p. 75). Manchu *ula*. Regarding *ula* in Tibet see ROCKHILL, *J.R.A.S.*, 1891, p. 218; as to the Chinese transcriptions of the word, see WATTERS, *Essays*, p. 374.

173. *'ar-gon* (Ladākh), "an offspring of parents not having the same rank, or the same religion, and not belonging to the same nation" (JÄSCHKE). The proper significance of the word is "half-breed, hybrid," with reference to man and animals. In Le, Ladākh, it is the designation for the offspring of Mohammedan traders from Central Asia and Tibetan women temporarily married by them after the *nikah* ceremony. The condition of these Argon, who speak Tibetan and affect Islam, has been best described by Captain H. RAMSAY in his very useful dictionary *Western Tibet* (pp. 56—57). He renders justice to their character in contradistinction to the grotesque generalizations indulged in by other writers (as cited, for instance, by YULE, *Marco Polo*, Vol. I, p. 290): "A good Argon is the best man in Ladākh; he has more intelligence, more courage, and more enterprise than the Ladākh Buddhist, and so far as I know,

he is not a whit less trustworthy.... They are far better traders than the Buddhists.... In physique they are certainly equal to them, and in personal appearance, more particularly the women, are far superior to them." There is no doubt that this Mohammedan term of Ladākh is identical with the one applied by Marco Polo to the Argon in the realm of Prester John (YULE, *l. c.*, p. 284), who have "sprung from two different races, to wit, of the race of the Idolaters of Tenduc and of that of the worshippers of Mahomet; they are handsomer men than the other natives of the country, and having more ability, they come to have authority, and they are also capital merchants" (which strikingly agrees with Ramsay's judgment). Certainly Polo's Argon has nothing to do with the Arkawūn اركاوون of Rashid-eddīn, who are not half-breeds, but Christians, the Ye-li-ko-wen 也里可溫 of the Chinese, and the arkaχun of the Mongols (CHAVANNES, *T'oung Pao*, 1904, p. 420). While Yule, in his *Marco Polo*, rejected Klaproth's unfounded opinion connecting the words "Argon" and "Arkawūn," yet he referred to Argon in his translation of Rashid-eddīn (*Cathay*, Vol. III, p. 120, new ed., where the additional note on Argon from Ladākh is not justifiable). It is strange that E. BLOCHET (*Djami el-Tévarikh*, Vol. II, p. 470), commenting on Rashid-eddīn's Arkawūn, is able to say that Marco Polo writes this word "Argon." According to Yule, Argon is of Turkī origin, which seems plausible. Radloff's Dictionary does not contain the term.

6. *Mongol Loan-Words.*

174. *a-ja-sa*, *a-ja-mo*, edict, diploma (stated by CSOMA to be Chinese). From Mongol *dsasak*, *jasak* ("law, government, administration;"

from *dsasa-*, "to govern"). As to the prefix *ạ* in Tibetan, see *J.R.A.S.*, 1915, p. 784, note). The *Hua i yi yü* gives *tsʿi šu* 勅書 ("credentials, letters patent") as rendering of the Tibetan term, likewise the *Pol. D.* (3, p. 9), where Manchu *ecexe* and Mongol *jiguku* are added.

175. *mu-či*, province. From Mongol *moji* (see HUTH, *B.M.*, Vol. II, p. 22).

176. *ạ-jam-mo*, *ạ-ja-mo* (*Pol. D.*, 19, p. 46), *rta-zam*, post-station; *rta-zam-pa*, courier, express, estafet. The element *jam*, *zam*, is derived from Uigur and Mongol *jam* (*dzam*, *yam*), post-stage, post-horses (RADLOFF, *Wörterbuch der Türk-Dialecte*, Vol. II, col. 298); *rta* ("horse") is Tibetan. The Tibetan compound certainly does not mean "horse-bridge," as asserted by Chandra Das. It is more likely that the Tibetan term hails straight from Mongol rather than from Uigur, as it was the Mongol rulers who introduced the relay service into Tibet. Moreover, under the Mongols, Tibet was divided into twenty-seven *jam* ("departments"), a chief officer (*jam dpon*) being appointed in each. I do not share the opinion of E. BLOCHET (*Djami El-Tévarikh*, Vol. II, p. 311) that the Uigur-Mongol word is based on Chinese *yi-ma* 驛馬 ("post-horse"). See particularly PELLIOT, *Journal asiatique*, 1913, mars-avril, p. 458.

177. *kʿa-tʿun*, princess, in connection with a female name (SCHLAGINTWEIT, *Könige von Tibet*, p. 867; MARX, *J.A.S.B.*, Vol. LX, 1891, p. 128). From Mongol *khatun* (Orkon Inscriptions *katun*, Turkish *katin*).

178. *tʿob-čʿu*, *tʿob-čʿi*, *tʿob-čʿe*, *tʿeb-čʿu*, *tob-čʿi*, and *tob-či*, button. From Mongol *tobči* (Teleutic and Altaic *tobčy*, Osmanli *toplu*).

179. *ʾo-mo-su* (RAMSAY: *omosuk*), stocking. JÄSCHKE states that this is a Mongol word; but neither Kovalevski nor Golstunski has such a Mongol word.

180. *tālai* (*talai*) *bla-ma*, Dalai Lama, the Lama of the sea (who is as big as the sea). Mongol *dalai* ("sea"), translating Tibetan *rgya-mts'o*. See WATTERS, *Essays*, p. 370; MAYERS, *Chinese Government*, p. 113.

181. *paṇ č'en' er-te-ni* (instead of the Tibetan form *paṇ č'en rin-po-č'e*; see, for instance, HUTH, *B.M.*, Vol. II, p. 317), title of the hierarch of bKra-šis lhun-po. Mongol *ärdäni*, precious object, precious stone. As *paṇ* is abbreviated for Sanskrit *paṇḍita*, and as *č'en* ("great") is Tibetan, words from three languages are welded in this compound. It is known also in Chinese (WATTERS, *Essays*, p. 370; MAYERS, *l. c.*, p. 115).

182. *bog-to*, *pog-ta*, holy, venerable, divine, lord. From Mongol *bogda*, first used in the title of Chinggis Khan (HUTH, *B.M.*, Vol. II, pp. 16, 18). The Mongol title *khān* (*khagān*) is usually translated into Tibetan as *rgyal-po*. E. BLOCHET's derivation of the Mongol word from Persian *bokhta* (*J.R.A.S.*, 1915, p. 308) is not convincing to me,[1] any more than W. SCHOTT's etymology based on Sanskrit *bhakta* (*A.Be.A.W.*, 1877, p. 5). I see no reason why *bogdo* should not be simply a Mongol word.

183. *se-mo-do*, *se-mo-to* (in Mi-la-ras-pa), kind of ornament, for instance, made of pearls. In the *Za-ma-tog*, written in 1514 by Rin-č'en č'os-skyoṅ bzaṅ-po of Ža-lu (1440—1526), the term is explained as a neck lace of precious stones (*rin-po-č'ei do-šal*). CHANDRA DAS assigns to it the specific and probably correct meaning "amber." In my opinion, this word is a hybrid formation composed of Chinese *si* 瑿 ("jet, gagate,

[1] There is a strange misconception on his part with respect to Mongol phonetics; he transcribes the word *bokhta*, and thinks that *kh* represents an aspirated guttural. Mongol writing, as is well known, has no signs for expressing aspirates, and the sound in question is a spirant.

black amber")[1] and Mongol *modo* ("wood"), gagate being justly regarded as a woody formation, and being obtained by the Chinese from Yarkhoto and Turfan (see *Pên tsʿao kaṅ mu*, Ch. 37; also the writer's *Jottings on Amber*, pp. 222, 231). So far as I know, this presupposed Mongol (or eventually Turkish) word **simodo* or *semodo* has not yet been traced in Mongol (or Turkish). The Mongol term given by Kovalevski as corresponding to Tibetan *se-mo-do* is *sugub-čilagur*.

184. *has ᵓer-te-ni*, nephrite. From Mongol *khas ärdäni*. See HUTH, *B.M.*, Vol. II, p. 21 (not "jaspis," as translated by him); *ibid.*, p. 16, we meet the transcription *has pāu tʿa-ma-ga ᵓer-te-ni* ("precious seal of nephrite"), where *pāu* stands for Chinese *pao* 寶.

185. *sku bde rigs* (pronounced *ku-de-ri*), a species of musk (Moschus sifanicus). From Mongol *kudäri*. The Tibetanized form of writing is notable; the genuine Tibetan word is *g-la-ba*.

186. *bu-gu* (Amdo), a species of stag (Cervus eustephanus). From Mongol *bugu* (Manchu *puku*); according to PALLAS (*Zoographia Rosso-Asiatica*, p. 216), Cervus elaphus; see also F. JUSTI, *Les noms d'animaux en kurde*, p. 19.

187. *ạdzeg-ran*, a species of antelope (Procapra subgutturosa). From Mongol *dsägārän*, *dsārän*. This word has been adopted into Russian зеренъ, fem. зерена. It is the *huaṅ yaṅ* 黄羊 of the Chinese (Capra flava of Du Halde; see PALLAS, *l. c.*, p. 251).

188. *rdza-ra* (*za-ra*), a species of hedgehog (Erinaceus amurensis, occurring in the Kukunōr region). From Mongol *dsarāga*, *dsarā*. The Tibetan word for "hedgehog" is *rgaṅ*.

---

[1] I am inclined to regard this word as an imitation of Arabic *sīḥ* سيح, the oldest form of the name for "gagate;" otherwise *sabaj* سبج or *šabaj* شبج, Persian *šaba* شبه (see RUSKA, *Steinbuch*, p. 55). Spanish *azabache* and Portuguese *azeviche* come from the same Arabic word.

rdza-bra (*za-bra*), another species of Erinaceus (Central Tibet). In all probability, this is only another way of writing the same Mongol word, with reference to Tibetan *bra* ("marmot").

189. *sbu-la-kʿa*, sable. From Mongol *bulaga*, *bulagan*. According to PALLAS (*l. c.*, p. 83) and JÄSCHKE, Mustela zibellina; according to G. SANDBERG (*Tibet and the Tibetans*, p. 294), Putorius auriventer.

190. *o-god-no* (*wo-gʿö-no*), Lagomys ogotona (occurs in Tsaidam and Amdo). From Mongol *ogotona* (Lepus ogotona, see PALLAS, *l. c.*, p. 151);[1] Manchu *oxotono* (according to Sacharov, Lepus dauricus and alpinus).

191. *tu-la*, Lepus tolai (Kuku-nor region, Mongolia, and Transbaikalia). From Mongol *taolai*, *tulai*. PALLAS (*l. c.*, p. 149) adds a "Tangutan" word *rangwo*.

192. *'i-man šiṅ* (*Pol. D.*, 29, p. 26), a kind of Sophora (Chinese *šan huai* 山槐). From Mongol *iman* (*boro*).

193. *'ag-lig* (or *šiṅ wa:*[2] *Pol. D.*, 29, p. 41). From Mongol *agli*, Manchu *ageli* (Chinese *šu pao* 樹包), excrescences or knots on the stems of trees, from which is made a decoction for treating the backs of animals galled by riding.

194. *had* (*Pol. D.*, 28, p. 57), a species of wild pear or crabapple (Pyrus betulæfolia). From Mongol *khat* (KOVALEVSKI: name of a fruit). The Chinese equivalent, according to *Pol. D.*, is *tu-li* 杜棣, which appears to be identical with the *tu-li* 杜梨 discussed by BRETSCHNEIDER (*Bot. Sin.*, pt. II, p. 304). The Manchu equivalent is *eikte*.

195. *na-ra-su* (*Pol. D.*, *ibid.*), a wild blue berry, the *golubitsa* of the Russians (Vaccinium uliginosum). From Mongol *narasu*. The Chinese equivalent is *tsao li-tse* 皂李子 (Manchu *duksi*).

---

[1] KOVALEVSKI's identification with the striped squirrel (Tamias striatus) is not correct.

[2] JÄSCHKE writes this word *lba*. The combination of the letters *lb* is very similar to *w*.

196. *ya-šel abru dmar* (*Pol. D.*, *ibid.*), a species of wild cherry (?). From Mongol *yašil* (Kovalevski: Rhamnus frangula). The Chinese equivalent is *huṅ yiṅ* 紅櫻 (Manchu *fafaxa* or *fafuxa*; according to Sacharov, the fruit of a tree, looking like a peach pressed flat, red of color, and of acid flavor).

197. *šid* (*Pol. D.*, 28, p. 55), hazel-nut (Corylus heterophylla and mandshurica). From Mongol *šit* (Chinese *čön-tse* 榛子, Manchu *sisi*). Regarding *čön-tse* in Tibetan, see above, p. 409. In the *Hua i yi yü*, a curious word for "hazel-nut" is noted as *pe-tog*, which I have not seen elsewhere.

198. *pir-tᶜi* (*Pol. D.*, 9, p. 43), quick-match. From Mongol *bilta*? (Chinese *huo šön* 火繩).

199. *beg-tse*, a hidden shirt of mail; name of the God of War. Probably from Mongol *begder* (Kovalevski, p. 1125), "cotte de mailles cachées;" but on the following page, Kovalevski gives *begji* side by side with *begder*, and *begji* doubtless transcribes Tibetan *beg-tse*. Cf. Persian *bagtar*, Djag. *bäktär* ("armor").

200. *hom*, felt saddle for a camel. From Mongol *khom* (Manchu *komo*).

201. *Änätkäk*, the Mongol designation for India, being a transcription of Chinese *Yin-tu kuo* (*In-du kwok). According to Rockhill (*J.R.A.S.*, 1891, p. 132), this word is frequently used by Tibetans who have travelled in Mongolia or China.

202. *'O-ro-su*, *'O-lo-su*, *'U-ru-su*, from Mongol *Oros*, *Urus*, Russia. *'O-ro-su sram*, Russian (Siberian) sable. *Orosu* means also a "foreigner" in Central Asia. "Olosu Amban," appellation of Przewalski (Rockhill, *Land of the Lamas*, p. 163). In *dPag bsam ljon bzaṅ* (p. 74), Russia is enumerated among the prosperous "barbarous border-countries" (*mtᶜa ạkᶜob-pa*) along with Tokhāra, Persia, Turushka, Sog-po (Sogd) and Ho-tᶜon (Khotan). The *Pol. D.* mentions the *'O-ro-sui ra-ba* ("hotel 館 of the Russian Mission in Peking") and the *'O-ro-sui yi-gei grva*

("Russian School in Peking;" see J. DUDGEON, *Chinese Recorder*, Vol. IV, 1871, p. 37). A real Tibetan designation for Russia is *rgya ser* ("yellow country"), in accordance with the Tibetan color-scheme system referring to the surrounding countries (*rgya gar*, *rgya nag*). Originally, however, the term *rgya ser* in general denoted the regions to the north of Tibet, particularly Sartagol (SCHIEFNER, *Mélanges asiatiques*, Vol. I, p. 426, where some data on Russia are extracted from the Geography of the Jambudvīpa written by Sum-pa mkʻan-po in 1717; in this work *rgya ser* is identified with the empire of the 'O-ro-su). Hence it occurs that Sartagol is styled also 'O-ro-swo by ạJigs-med nam-mkʻa (HUTH's translation, pp. 23, 28). In the Geography of the Minčul Chutuktu (VASILYEV's translation, p. 93), Russia is likewise styled *rgya ser*. In eastern Tibet, a popular designation for Russia is Hor-sog or Hor-ro-sog (*French Dict.*, p. 222), composed of the well-known tribal names Hor and Sog, to which the name 'O-ro-su is wittily adapted.

203. *hor-du*. This term is contained in the *Tibetan Dictionary* of CHANDRA DAS, who explains it thus: "A Mongolian encampment, from *Hor* 'Tartar or Mongolian' and *du* 'smoke.' Each nomad's tent represents a fire-place and chimney, i.e., a family." This would not be so bad if the term in question were really applied to the tents of Mongols only. In fact, however, when Tibet formed a province of China, it was applied by the Mongols with reference to Tibetan families, and served as the unit in drawing up a census of the population of Tibet (compare CHANDRA DAS, *Narrative of a Journey round Lake Yamdo*, p. 63). It is therefore clear that Tibetan *hor-du* is based on Mongol *ordu* ("camp, encampment, tent of the Khan"), and was assimilated by the Tibetans to the tribal name Hor. The Mongol word,

on its part, appears to be derived from the Uigur. For further information see YULE, *Hobson-Jobson*, p. 639; BRETSCHNEIDER, *Mediæval Researches*, Vol. I, pp. 18, 58.

7. *Manchu Loan-Words.*

A few of these occur in Tibetan in consequence of the Manchu organization extended over Tibet by the Ts'in dynasty.

204. *'am-ban*, from Manchu *amban* ("great, official, minister"), appellation of the imperial Resident at Lhasa under the Manchu (see WATTERS, *Essays*, p. 366). The common Tibetan pronunciation of this word is *ampa*.

205. *ho-šo* (HUTH, *B.M.*, Vol. II, p. 329), explained by Huth from Mongol *khošoi*, that "according to Kovalevski should be the equivalent of Chinese *kho-šo*, province." There is, of course, no Chinese word with this meaning. The word intended appears to be Manchu *xošo* (transcribed in Chinese 和碩), which signifies "corner, region; appanage of a prince," and is connected with the titles of some Manchu princes and princesses (see MAYERS, *Chinese Government*, Nos. 14, 19; BRUNNERT and HAGELSTROM, *Political Organization of China*, Nos. 14, 15, 16, 35, 873). In the passage translated by Huth, the word indeed means "prince." In the *Pol. D.* (3, p. 6) we find Tibetan *ho-šoi koṅ jo* (= 和碩公主).

8. *Chinese Loan-Words.*

A. Old Chinese.

206. *bug-sug* (*French Dict.*, p. 667), alfalfa, lucerne (Medicago sativa). In Ladākh (Jäschke), bird's-foot trefoil (Melilotus). This is a good and ancient transcription of Chinese **buk-suk* (*mu-su*)

苜蓿, denoting the same plant, the original habitat of which was Iran, and which was introduced from Fergana into China by Čaṅ K'ien. The Iranian equivalent has not yet been discovered. The Chinese word bears no relation to Pahlavī *aspast* (explained as *asp-ast*, "horse-fodder" [NÖLDEKE, *Z.D.M.G.*, Vol. XXXII, 1878, p. 408], that is, lucerne), Persian *uspust*, *aspist*, *ispist* (*supust*, "trefoil, clover"), Balūčī *aspust*, Puštu *spastu*. In my opinion, the Pamir languages will some day yield the word required: in Wakhī we have *wujerk* as designation for "lucerne" (R. B. SHAW, *On the Ghalchah languages*, *J.A.S.B.*, 1876, pp. 221, 231), and in Sariqolī *vurj* for "horse." The etymology *μεδικὴ βοτάνη* noted in the dictionary of Giles is impossible for phonetic and historical reasons, for the Greek designation was certainly not current in Iranian regions. Russian *burkún*, *burkúnčik*, *burundúk*, seem likewise to go back to Iranian. A Tibetan word for "clover" is *ol* (JÄSCHKE: snail-clover, medic, Medicago; RAMSAY: a high-growing kind of clover or lucerne). The *Pol. D.* (29, p. 11) gives two artificial words, *rtswa bsuṅ* ("fragrant grass") and *rgya spos*,[1] as equivalents of Chinese *mu-su*; Manchu *morxo* (from *morin*, "horse").

207. *tsi-tse*, *či-tse*, *či-ts'e*[2] (Kunawar *tse-tse*; on the tablets from

[1] The element *rgya*, accordingly, in this case means "Chinese" (*rgya spos*, "Chinese fragrant plant"). Otherwise, however, this compound means "Indian fragrant plant," for we meet it twice in the Mahāvyutpatti; first, in the list of aromatics, where it corresponds to Sanskrit *vāyana*; second, in the list of drugs, where it is identified with Sanskrit *tagara* (Tabernaemontana coronaria), a tree indigenous in Atali from the wood of which incense is made (EITEL, *Handbook of Chinese Buddhism*, p. 168). As *tagara* (Chinese 多伽羅 *ta-g'a-ra) is explained in the Chinese glossaries as "putchuk" (根香 or 木香), this meaning also may adhere to the Tibetan term. The usual Tibetan designation for "putchuk" is *ru-rta* (Sanskrit *kuṣṭha*, Costus speciosus). Tibetan *rgya spos dkar* is identified with Chinese *žu hiaṅ* 乳香, the *olibanum* produced in India.

[2] While the ending *tse*, *tsi*, or *ts'e* (transcribing Chinese *tse* 子) in some cases is a criterion indicating Chinese loan-words, we must not generalize on this rule, since there is also a genuine Tibetan suffix *tse* (written also *rtse*). Thus we have *k're-tse* (Central

Turkistan: *či-tsʻe* [Francke, *J.R.A.S.*, 1914, p. 54]), millet. From Chinese *tsi-tse* 稷子 (Panicum miliaceum, panicled millet). The Tibetan word is listed in the Mahāvyutpatti (section 228) as corresponding to Sanskrit *anuphala*; further, *či-tsʻe nod* (or *kʻre rgod*) answering to Sanskrit *çyāmāka* (Panicum frumentaceum), and *či-tsʻe* being the equivalent of Sanskrit *kodrava* (Paspalum scrobiculatum).

208. *ske-tsʻe* (*Pol. D.*, 27, p. 19), black mustard (Sinapis ramosa). From Chinese *kai* (*kiai*) *tsʻai* 芥菜. The fact that the Tibetan loan is old and goes back at least to the ninth century, becomes evident from this word being on record in the Mahāvyutpatti (section 228), where we find *ske-tsʻe-i ạbru* = Sanskrit *rājikā* ("black mustard"). It thus appears that the Tibetans received name and object from the Chinese prior to the contact with India. The *Pol. D.* (*ibid.*, p. 30) has *sga-rgod* = *ye kiai tsʻai* 野芥菜. Tibetan *sga* is possibly also a transcription of Chinese *kiai*.

209. *bim-pa* (*Pol. D.*, 28, p. 49), apple (Pyrus malus). Transcription of Chinese *pʻin kuo* 蘋(𬃊)果 (Manchu *pʻingguri*, Mongol *almurat*), or rather **bim-ba* (*pʻin-pʻo*) 頻婆, which itself represents a transcription of Sanskrit *bimba*. Li Ši-čên says that the apple (*nai* 柰) is thus styled in Sanskrit literature, and that this term is employed also by the northerners.

210. *čag šiṅ* (not in our dictionaries), čag tree (Scirpus tuberosus), a black eatable tuber, resembling the chestnut in appearance (hence called "water-chestnut"), Chinese *šao* 芍. In the *Pol. D.* (Ch. 29, p. 29) we meet the following equation: Manchu *jak moo*, Tib. *čag šiṅ*, Mongol *jak modo*, Chinese *šao mu* 灼木 (灼 being substituted for 芍). The latter is in

Tibetan *ṭʻe-tse*), "Chinese vermicelli" (from *kʻre*, "millet"); *tsem-tse* ("scissors"); *keu-rtse* ("jacket"); *kuṅ-rtse* ("cuckoo"); *bag-tse* ("basket"); *pil-tse* ("sieve"); *čʻag-tse* ("a small grain"); *rag-tse* ("stone in fruits"); *dze-tse* ("vent-hole for smoke").

Cantonese *čök*, Hakka *čok*, Korean *čak*, Japanese *jaku* and *šaku*. Hence Tibetan *čug* is an ancient transcription of Old Chinese **čak* 芍. In the *T'ai p'iṅ huan yü ki* (Ch. 31, p. 2 b), the plant is mentioned as a product of Yao čou 耀州 in Shen-si.

211. See p. 542.

212. *ram-s*, *ram*, indigo. From Chinese **lam* (*lan*) 藍. E. HUBER (*Bull. de l'Ecole française*, Vol. IX, p. 397) conceived the two words as anciently related, adding Siamese *k'ram* and Dioi *thom*; but it seems preferable to me to regard Tibetan *ram-s* as a loan-word.

213. *lá-buk* (Ladākh: RAMSAY); *lá-p'ug* (occurs in the wooden documents from Turkistan: FRANCKE, *J.R.A.S.*, 1914, p. 54), radish (Raphanus sativus). From Chinese **la-buk* (*lo-po*) 蘿蔔 (see this volume, p. 83). There can be no doubt that *la-bug* must have been the original Tibetan form, as *bug*, in the same manner as Chinese **buk*, is deep-toned, while Tibetan *p'ug* is high-toned. For this reason I am not yet convinced that the form *la-p'ug*, as asserted by Francke, really occurs in the wooden documents which have not yet been published. If it does occur there, the change from the labial sonant to the aspirate surd must have been brought about at an early date, or *la-p'ug* was derived by the Tibetans from another language than Chinese. In Bunan we have the curious form *am-p'aṅ*, which, as far as the final guttural nasal is concerned, agrees with Mongol *lobaṅ*.

214. *guṅ la-p'ug* (Tromowa *goṅ la-p'up*, Sikkim *goṅ-la-pup*), carrot (Daucus carota). M. PELLIOT (*T'oung Pao*, 1915, p. 11) has proposed to regard this as a transcription of Chinese **γuṅ la-buk* 紅蘿蔔. This would be possible if the Tibetans had adopted the Chinese term in the T'ang period, when the word *huṅ* 紅 ("red") was indeed sounded *γuṅ*, but for historical reasons it is out of the question to admit this possibility.

The carrot was not known in China under the T'ang; it is not mentioned in the herbals of either the T'ang or the Sung. Li Ši-čên states in regard to the carrot (*hu lo-po* 胡蘿蔔) that it began to make its appearance at the time of the Yüan, from the land of the Hu (元時始自胡地來, see also BRETSCHNEIDER, *Chinese Recorder*, Vol. III, 1871, p. 223), and this opinion appears to be well founded; for in the *Cêṅ lei pên ts'ao* (Ch. 27, fol. 16 b *et seq.*), the herbal of the Sung period, no mention is made of the carrot. Under the Yüan, 紅 was sounded *huṅ*, which the Tibetans, in case they should then have received the term from China, would doubtless have transcribed *huṅ*: thus, indeed, they transcribe the Chinese word (see Nos. 236, 281). In the Sino-Tibetan inscription of 1341 (*T'oung Pao*, 1908, plate 28, line 12), Chinese *ho* 河 is written in Tibetan *ho* (not *kho*, as transcribed by S. Lévi). Hence M. Pelliot's opinion can hardly be upheld. It is not plausible also for another reason: the term *huṅ lo-po* is merely presupposed by him, but in fact does not exist in the Chinese language. The *Pol. D.* of K'ien-luṅ enumerates three varieties of *lo-po*, — *ta* 大 *lo-po* = Tibetan *la-p'ug dmar-po* ("red l."), *šui* 水 *lo-po* ("a kind of sweet radish") = Tibetan *la-p'ug dkar-po* ("white l."), and *hu* 胡 *lo-po* = Tibetan *la-p'ug mṅar-po* ("sweet l."). These examples show that the Tibetan nomenclature is not based on Chinese. On the other hand, JÄSCHKE (p. 540) tells us that the carrot is colloquially called in Central Tibet also *la-p'ug ser-po* ("yellow l."), and he cites (p. 69) another Central-Tibetan term, *guṅ dmar la-p'ug*. As *dmar* is the Tibetan word for "red," it is out of the question that *guṅ* should represent the Chinese word for "red:" the Tibetans would certainly not have produced such a monstrosity, for which there is no precedent in their language. At the outset there is no valid reason to assume that Tibetan *guṅ* must transcribe a Chinese word. We know nothing, either, to the effect that the Tibetans received the carrot from China. It may have reached them as well from another region. Indeed, it is evident from the presence of a Persian loan-word (No. 112) in Ladākh that the carrot became known there from Irān. If, further, the Tibetans discriminate between *bod la-pʻug* ("Tibetan radish;" that is, the common black radish) and *rgya la-pʻug* ("Chinese radish," a red species of an acidulent taste), the chances are that the word *guṅ* in *guṅ la-pʻug* likewise hints at a geographical or ethnical name connected with the introduction of the vegetable. It may bear no relation to Chinese *hu* 胡, but still less has it anything to do with Chinese *huṅ* 紅.—The term *lĕa-ba*, which according to Jäschke denotes a sort of carrot, is identified with *yü ču* 玉竹 (Polygonatum officinale) in the *Tib. Pharm.*

215. *ja* (*dža*), tea. Exact reproduction of ancient Chinese **dža*, *da* (modern *čʻa*) 茶, the tea-shrub (Camelia japonica). The Chinese word has the even lower tone, and is therefore correctly rendered by the Tibetan deep-toned *ja*, while Tibetan *čʻa* is high-toned. It is notable that of the many Asiatic languages which have adopted with the product the Chinese designation for it, Tibetan is the only one that has preserved the ancient initial sonant. Hence the conclusion is justified that the acquaintance of the Tibetans with tea goes as far back as the Tʻang period. In fact, tea was then exported to the Tibetans and Taṅ-hiaṅ from Kʻiuṅ-čou 邛州 in Se-čʻuan, being made up into cakes or bricks of forty ounces each (*Tʻai pʻiṅ huan yü ki*, Ch. 75, p. 3).

216. *ja-ạbiṅ* (pronounced *jʻam-biṅ* in Central Tibet), tea-kettle, tea-pot (respectful language: *gsol ạbiṅ*, pronounced *sol-biṅ*).

Tibetan *ja* = Chinese *č'a*, tea; Tibetan *abiṅ* = Chinese **biṅ* (*p'iṅ*) 瓶, jar, jug, pot. The loan therefore appears to be old. It is interesting that the Tibetans have received the word also in the modern form *p'iṅ* ("earthenware pitcher, cup"), so that a double loan has taken place at different times. In my opinion, Tibetan *p'iṅ* cannot be correlated with Burmese *p'yaṅ*, as proposed by B. HOUGHTON (*J.R.A.S.*, 1896, p. 45), who believes that this equation testifies to the existence of earthenware before the separation of the Burmese and Tibetans.[1]

217. *ben*, *bön*, a large pitcher, jug. According to Jäschke, this word is attested in *rGyal-rabs*. It is apparently derived from **bŏn*, *bun* (*p'ŏn*) 盆, "basin, bowl, pot." The Tibetan transcription exactly reproduces Chinese **bŏn*, as the Tibetan letter *e* conveys also the timbre *ö*. Both the Chinese and the Tibetan have the even lower tone. In the same manner as Chinese *p'iṅ*, this word also has been re-imported into Tibet in recent times. JÄSCHKE (p. 487 b) cites *zi-liṅ pan-tse* or *baṅ-tse*, "basin of tutenague, white copper." HENDERSON (*Tibetan Manual*, p. 64) has *č'u-ban* (pronounced *b'en*), "water-jug, pitcher" (*ibid.*, p. 87, he says that *ban* is from Chinese *p'en*).

218. *doṅ-tse* (the reading *doṅ-rtse* is rejected by SCHIEFNER, *Ergänzungen*, p. 27), copper coin, especially of Chinese coinage, money (*mDzaṅs blun*). Transcription of Chinese **duṅ-tse* (*t'uṅ-tse*) 銅子. Although Tibetan *doṅ* ("pit, ditch") is anciently related to Chinese **duṅ* (*t'uṅ*) 洞, and Tibetan

---

[1] The word *k'o-ti* (Central Tibetan), "tea-kettle," is accompanied by JÄSCHKE with the note: "Chinese?". WALSH (*Vocabulary of the Tromowa Dialect*, p. 32) spells the word *k'og-tib* (but Tromowa pronunciation *k'o-ti*). The same word *tib* is found in *tib-ril* (respectful language: *gsol-tib*), K'ams *teb-rel*, "tea-pot." In this case *ril* is the Tibetan word *ril*, "globular, globe, ball" (compare *rdza tib*, "earthen tea-pot"), so that *tib* itself has the meaning "tea-pot." As to the word *k'o*, it is almost certain that it merely presents an abbreviation of *k'og*, and that the latter is a Tibetan word (*k'og-ma*, "pot, earthen vessel"). B. HOUGHTON (*J.R.A.S.*, 1896, p. 45) has compared it with Burmese *ō:*.

doṅ-po, *l-doṅ-po* ("tube") to Chinese **duṅ* (*t^cuṅ*) 筒, the word *doṅ-tse* represents only a transcription, as outwardly indicated by the Chinese ending *tse*; above all, the Tibetan equivalent of Chinese **duṅ* (*t^cuṅ*) 銅, evolved from **džuṅ*, *dzuṅ*, is represented by *zaṅ-s* (see this volume, p. 83). The Mongol word corresponding in meaning is *dsoγos*; the Manchu word, *jiχa*. In the *Hua i yi yü*, the Tibetan word is translated by *ts^cien* 錢.

219. *kim* (*mDzaṅs blun*; SCHMIDT's translation, p. 274), guitar. The reading in the unreliable text of Schmidt is *ki-ma*. It was rather tempting to a Tibetan scribe to place the dot (*ts^ceg*) wrongly, as the word thus adapted itself to a Tibetan formation with suffix *ma*. SCHIEFNER (*l. c.*, p. 49) recognized correctly that the word is based on Chinese *k^cin*, the Mongol translation offering *khuγur*. From Chinese **gim* (*k^cin*) 琴, "guitar" (this, in fact, is the word used in the Chinese text, Ch. 10, p. 17 b). In all probability, the original Tibetan reading was likewise *gim*. See also PELLIOT, *T^coung Pao*, 1914, p. 258, note 2. The Chinese instrument is usually styled in Tibetan *rgyud bdun-ma* ("seven-stringed").

220. *keu-le*, *-i rgya*, seal (*mDzaṅs-blun*, p. 109, line 1). The Chinese version (Ch. 4, p. 17) offers *fuṅ yin* 封印, which does not help us in explaining the Tibetan term. The Mongol translation made from the Tibetan work renders this term by *kaoli tamaγa*, "customary seal" (SCHIEFNER, *l. c.*, p. 27, who thought that *keu-le* might be a corrupted foreign word); we have the same word in Manchu *kōli* ("law, rule, custom"). The second element, *li*, in all probability, is Chinese *li* 例.

221. *li*, league, mile (in the wooden documents; see above, p. 430). From Chinese *li* 里.

le-bar (*Hua i yi yü*: *sa le-bar*), *le-war*, mile; *le*, from Chinese *li* 里. In the description of Jambudvīpa (ạDsam-gliṅ spyi bšad ṅo mtsᶜar gtam sñan), written by Sum-pa mkᶜan-po in 1717, it is stated, "India is semi-circular in shape, and its circumference is said to amount to 90,000 *le-war* of Chinese style" (rgya-gar-gyi sa dbyibs zla gam lta-bu mtᶜa kᶜor rgya-nag lugs-kyi le-war' dgu kᶜri yod zer). Here, accordingly, *le-war* is characterized as a Chinese word. I cannot explain the second element, *war*. Colloquially, a Chinese *li* is called in Tibet *le-bor*.

222. *gi-lin*, listed by JÄSCHKE as "a fabulous animal" after Klaproth's *Description du Tubet* (p. 157). We further have the transcriptions *gi-liṅ* ("strong-bodied, durable horse") and *gyi-liṅ*, mentioned in *rGyal-rabs* as an excellent breed of horses. In the *Pol. D.* the names of the eight famous steeds of Mu Wang and other designations of horses are rendered into Tibetan by means of the word *gyi-liṅ*. The *French Dict.* (p. 152) notes also a variant *ger-liṅ*. Transcription of Chinese **gi-lin* (*kᶜi-lin*) 麒麟. In the *Hua i yi yü*, Tibetan *gi-lin* is identified with this Chinese term.

223. *gi-waṅ*, *gi-bam*, *giu*, *giu-waṅ*, *gi-wam* (Mahāvyutpatti *ạgi-haṅ* = Sanskrit *gorocanā*), *ghi-dbaṅ* (*Hua i yi yü*), *gi-haṅ* (*Pol. D.*, 19, p. 19), bezoar. Derived from Chinese **giu-waṅ* (*niu hwaṅ*) 牛黄 (Japanese *giū-kwō*). Mongol *giwaṅ* (KOVALEVSKI, p. 2554).

224. *ạ-jab-tse*, nippers, tweezers. From **č'ap*(*k'ap*)-*tse* 鋏子.

225. *čog-tsᶜe*, *l-čog-tse*, *l-čog-rtse* (*Li-šii gur kᶜaṅ*, fol. 23: *čog-tsᶜo*),[1] table. From Chinese **čok-tse* 桌子. Also *rgya lčog*, Chinese table; *rgya jog* or *ạ-jog*, long table, bench; and *ạ-joṅ-tse*, *ạ-joṅ-tsᶜe*.

---

[1] See above, p. **444**. The fact that this word means "table" becomes evident from the Mongol translation *širdgün*.

226. *ha-šig* (*Tib. Pharm.*, No. 54), steatite. From Chinese **hwa šik* (*hua ši*) 滑石. JÄSCHKE defines the word as "a mineral medicine, used as a remedy for the stone."

227. *yol* (*French Dict.*, pp. 877, 918), month. From Chinese **yüet, yüt* 月; anciently **ń'wiet*. The Tibetan form is curious, inasmuch as it shows the recent initial *y* in combination with the ancient final consonant, which, in like manner as in Korean (*wöl*), is transformed into a liquida.

228. *gur*, dynasty (not in our dictionaries), occurs only in connection with the names of the Chinese dynasties; for instance, *T'ań-gur* (the "T'ang dynasty"), *Č'iń-gur* (the "Ts'ing dynasty"); compare *T'oung Pao*, 1907, p. 394, note 2. In *Pol. D.* (Appendix, 1, p. 33) the term *Čou i* 周易 is rendered into Tibetan *Ṭeu gur-gyi rtsis-kyi rig-byed* (literally, "the science of the calculations of the Čou dynasty"). In all probability, this word is a reproduction of Chinese **kwok, *kwor, *gwor* 國 ("kingdom, state, dynasty"), from which also Manchu *gurun* is derived.

229. *pir*, writing-brush, pencil. From Chinese **bit, bir* (*pit*) 筆 (Mongol *bir*). King Sroń-btsan sgam-po asked Emperor Kao-tsuń (650—683) for workmen to manufacture paper and writing-brushes (*T'ań hui yao*, Ch. 97, p. 3 b). The Tibetans usually employ for writing a wooden or bamboo stylus (*smyug-gu*), in the same manner as the ancient Chinese did prior to the invention of the brush.

230. The Tibetans have adopted from the Chinese the system of *pa kua* 八卦 for purposes of divination. These are styled in Tibetan *spar k'a*.[1] The *Pol. D.* gives *p'yag rgya* as Tibetan rendering of Chinese *kua*. The transcription *spar*

[1] *T'oung Pao*, 1914, p. 79, note.

(in the Sūtra of the Eight Phenomena *par*) with final liquida, again, is notable, and suggests an Old-Chinese pronunciation *par* beside *pat*. A comparative list of the Tibetan and Mongol names has been given by O. KOVALEVSKI [1] and A. POZDN'ÄYEV;[2] a Tibetan list, by CHANDRA DAS.[3] None of these authors has compared them with their Chinese equivalents. In order to establish the correct spelling of the Tibetan names, I availed myself of the *Pol. D.* and several wood-engravings executed in Peking, which give the designs of the *pa kua* with their Tibetan equivalents.[4] The figures are sometimes wrongly identified; thus, for instance, the figure ☴ is called *dva*, while its Chinese designation is *sun* 巽. The Tibetan equivalent of the latter character, of course, is *zon*, while *dva* is the transcription of *tui* 兌. No regard need be taken here of this confusion of the trigrams; it is merely our task to study the Tibetan terms in their relation to the Chinese.

1. 乾 *k'ien* (Japanese *ken*, Sino-Annamese *kien*) = Tibetan *kin* or *gin*,[5] Mongol *kin*, *gin*.
2. 兌 *tui* (Japanese *de*, Sino-Annamese *dwai*) = Tibetan *dva* (the letter *d* with *va zur*, accordingly *d̤a*),[6] Mongol

---

[1] *Dict. mongol-russe-français*, p. 2599. The Mongols employ for Chinese *kua* their native word *külil* ("knot"), also the transcription *guwa*.

[2] *Sketches of the Life in Buddhist Monasteries*, in Russian, pp. 414—415. There are two misprints in the Tibetan: *mi* for *li*, and *gir* for *giṅ*.

[3] *Tibetan Zodiac* (*Proc. A. S. B.*, 1890, p 4), where the "eight *parkha*, or factors of luck" are given as follows: *k'en*, *k'on*, *gin*, *dva*, *k'am*, *sson* (intended for *zon*), *li*, *ssin* (*zin*). In his *Dictionary* they are enumerated thus: *li*, *k'on*, *dva*, *k'en*, *k'am*, *gin*, *zin*, *zan* (for *zon*).

[4] A similar divination-chart has been published by Satis Chandra Vidyabhusana under the title *Srid-pa-ho: a Tibeto-Chinese Tortoise Chart of Divination* (*Mem. As. Soc. Bengal*, Vol. V, No. 1, pp. 1—11, 1913).

[5] Chandra Das, Vidyabhusana (*l. c*, p. 2), and the aforementioned Sūtra of the Eight Phenomena, transcribe *gin*; *Pol. D*, *k'en*.

[6] Vidyabhusana (*l. c*, p. 2) transcribes *dvo*.

da which Pozdn'äyev transcribes *dva*. This transcription must be old, having been made at a time when the initial sonant was still preserved in 兌 **dui* (compare Annamese *dwai*).

3. 離 *li* = Tibetan and Mongol *li*.
4. 震 *čön*, *tsön*, *čin* = Tibetan *tsin*, *zin* [1] (Pozdn'äyev: *jen*, likewise in Mongol; Kovalevski: *dsen*).
5. 巽 *sun* (from **dzun*, *zun*) = Tibetan *zon*, [1] Mongol *sün*.
6. 坎 *k'an* (**k'am*) = Tibetan *k'am*, Mongol *kam*. The final *m* shows that the transcription is old and was effected at a time when 坎 was still sounded *k'am*.
7. 艮 *ken* = Tibetan *gin* (*Pol. D.*), *k'en*, [2] Mongol *ken*.
8. 坤 *k'un* = Tibetan *k'on*, Mongol *kon*.

As the Tibetan transcriptions have partially preserved the ancient initial sonants and the ancient finals of Chinese, they are thus well attested as coming down from the T'ang period. In this connection we might remember the Tibetan tradition recorded by Vidyabhusana, [3] that the Chinese method of divination by means of the tortoise, in which the system of the *pa kua* was employed, is said to have been imported into Tibet by the Chinese princess who was married to King Sroṅ-btsan sgam-po in 641.

231. The Ten Celestial Stems of the Chinese are transcribed by the Tibetans as follows: 甲 *gya* (*g'a*), 乙 *yi*, 丙 *biṅ*, 丁 *tiṅ*, 戊 *wu*, 己 *kyi* (*k'i*), 庚 *giṅ*, 辛 *zin*, 壬 *žim*, *žiṅ*, 癸 *gui*. The Twelve Branches are transcribed as follows: 子 *tsi*, *či*, 丑 *ts'iu*, 寅 *yin*, 卯 *mau*, 辰 *šin*, 巳 *zi*, 午 *wu*, 未 *wui*, 申 *šin*, 酉 *yiu*, 戌 *zui*, 亥 *hai*.

---

[1] Thus likewise in the Sûtra of the Eight Phenomena mentioned above.

[2] The Sûtra of the Eight Phenomena writes *k'yen*.

[3] *L. c.*, p. 1.

232. *par* (written also *s-par* and *d-par*),[1] printing-block, printed book; form, mould. Presumably derived from Chinese *pan* 板 ("printing-block"). Also the compounds *par rkyen* and *č'a rkyen* (the latter in HUTH, *B.M.*, Vol. II, p. 165) seem to be reproductions of Chinese terms.

233. *pi-wan̄*, *pi-ban̄* (*Pol. D.*: *pi-wam̄*), guitar. The word is on a par with Chinese **bi-ba* (*p'i-p'a*) 琵琶 (Japanese *biwa*, Mongol *biba*, Manchu *fifan*), which in all probability is not a Chinese word. The instrument of this name, written *p'i-pa* 批把, is first mentioned in the dictionary *Ši min̄* 釋名 of Liu Hi 劉熙, and in the *Fun̄ su t'un̄* 風俗通 of Yin̄ Šao 應劭 of the second century, who says that it was made by musicians of recent times, but that it is unknown by whom, and that it received its name from being struck with the hand (compare also M. COURANT, *Essai hist. sur la musique classique des Chinois*, p. 177). This explanation, of course, is a fantastic afterthought; and it is reasonable to regard **bi-ba* as the transcription of the foreign appellation for this instrument that seems to have reached China from Central Asia. It is doubtful whether the Tibetan form is borrowed directly from Chinese. If this were the case, we should expect an initial labial sonant, for the Tibetan word is of ancient date, being listed in the Mahāvyutpatti and Amarakosha; it could by no means be a more recent transcription of Chinese, for in that case we should have the initial labial aspirate. Further, the final guttural nasal is curious, and cannot be justified from the standpoint of the Chinese. It is therefore more likely that the Tibetan word is not based on the Chinese, but is derived from another language,

[1] Over the entrance to the library of the monastery sKu-ạbum (Kumbum) I read the legend dpar-gyi lha k'an̄ n̄o-mts'ar rin-č'en glin̄, "the library, the temple-hall of wonderful treasures."

probably from Khotanese.[1] I am led to this supposition by the presence of the final guttural nasal which seems to be a peculiar feature of the language of Khotan (see PELLIOT, *Journal asiatique*, 1914, sept.-oct., p. 400, note).

234. *šiṅ-tse*, lion (SCHIEFNER, *Çâkyamuni*, p. 96; the form *šaṅ-tsi* must be regarded as a misprint). Transcription of *ši-tse* 獅子 (see this volume, p. 81).

B. Modern Chinese.

235. *la* (also with the Tibetan suffix *la-ba*), wax-light, wax-candle, taper. Jäschke already annotated that it is from the Chinese *la* 蠟. The Tibetan loan cannot be very old, as in the T'ang period the Chinese word was still possessed of a final labial explosive, which appears at present in Amoy *lap*, Korean *rap*, and Mongol *lab*.

la-č'a, also *la-ja*, sealing-wax. JÄSCHKE (*Z.D.M.G.*, Vol. XXX, p. 112) sought the etymology of this word in Sanskrit *lākshā*; it is not likely, however, that Sanskrit *lākshā* would yield a Tibetan form *la-č'a*, or we must presuppose without reason a Prākrit form **lāchā*. It is more probable that Tibetan *la-č'a* is a transcription of Chinese *la-ča* 蠟渣 ("wax-sediment"). I believe that also KOVALEVSKI (p. 1960) was of this opinion when he identified Mongol *latsa* with Chinese *la-tcha* (characters not given) and Tibetan *la-č'a*, translating it by "sédiment de

[1] Any corresponding words of Central-Asiatic languages are not known to me. It is not correct, as stated by me in *T'oung Pao* (1914, p. 89), that *pišik* is the Djagatai word. The derivation of the Tibetan word *pi-waṅ* from Sanskrit *vīṇā* must positively be rejected.—Chinese *pa-waṅ* 巴汪 is considered by M. COURANT (*Musique classique des Chinois*, p. 179) as a reproduction of Tibetan *pi-waṅ*.—The *Pol. D.* (7, p. 18) gives Tibetan *waṁ-še* as equivalent of Chinese *yüe k'in* 月琴 ("moon guitar"). In Mongol it is *biwalik* ("an instrument like the biwa"), in Manchu *fiतuxan* (*fi-* from *fifan*); hence also the element *waṁ* of Tibetan *waṁ-še* may be identified with *pi-waṁ*.

cire." Chinese *ča*, as indicated by the Korean reading *č'a*, may formerly have had, at least in certain dialects, an initial surd aspirate.

236\. *huṅ la*, according to *Li-šii gur k'aṅ* (above, p. 444), a Chinese word with the meaning "seal." From *huṅ la* 紅蠟, "red wax."

237\. *la-tse* (generally in eastern Tibet), Guinea pepper, red pepper (Capsicum annuum). From *la-tse* 辣子 (*la tsiao* 辣椒). According to G. A. Stuart (*Chinese Materia Medica*, p. 92), this solanaceous plant is not mentioned in the *Pên ts'ao*. In the opinion of A. de Candolle (*Origin of Cultivated Plants*, p. 290), the original home of the plant is probably Brazil. According to Watt (*Economic Products of India*, II, p. 138), it was recently (comparatively speaking) introduced into India from South America. Chinese *la tsiao* means "acrid pepper," but the designation *la* was presumably suggested by the Hindustānī name *lāl-mirč* لال مرچ (*mirč* = Sanskrit *marica*, "pepper"); see also Bretschneider, *Chinese Recorder*, Vol. II, 1870, p. 224. The extensive cultivation of red pepper in Yün-nan and Se-č'uan suggests an overland transplantation from Indian regions. The word is found also in Turkistan: A. von Le Coq (*Turfan*, p. 97) notes it in the form *lá-zā* (the derivation from Chinese is certain, and need not be queried, as was done by the author). There is another Tibetan word for "red pepper," *su-ru-p'an-ts'a* or *sur-p'an*, the origin of which is unknown to me. The word *mar-rtsi*, indicated by Chandra Das (p. 1270), surely is a transcription of Sanskrit *marica*.

238\. *u-su*, coriander (Coriandrum sativum). Apparently transcription of *hu-sui* 胡荽. The tradition reported by Li Ši-čên, to the effect that this plant was introduced by Čaṅ K'ien, is unfounded; it is contained only in the *Po wu či*, but not in the Han Annals. The plant is first mentioned in the *Ts'i min yao šu* (fifth century), and was introduced into the pharmacopœia under the Sung, when it appeared in the *Pên ts'ao* of the Kia-yu period (1056—64). For more information see my *Sino-Iranica*.

239. *t'au*, peach, Prunus persica (recorded in the *Hua i yi yü*). From *t'ao* 桃.

240. *čaṅ-bkwa-tsi* (*Pol. D.*, 27, p. 33), transcription of *tsiaṅ kwa-tse* 醬瓜子 (Manchu *čanggôwa*, Mongol *jaguwasa*), melons (Citrullus vulgaris) seasoned in soy. The transcription *bkwa* appears to be of modern origin; the usual mode of transliteration is *kau*, *gau*, *ku*, *ku-ba*. Likewise *čaṅ* for *tsiaṅ* is a recent affair, for in the *Hua i yi yü* the Chinese word is transcribed *tsaṅ*.

241. *pe-tse*, *pi-tsi*, Chinese cabbage (Brassica chinensis, cultivated in the U. S. under the name "celery cabbage"). From *pai ts'ai* 白菜 (see *T'oung Pao*, 1916, p. 88).

242. *pi-tsi* (*Pol. D.*, 28, p. 58), Scirpus tuberosus. From *pi-ts'i* 荸薺.

243. *taṅ-kun* (*Tib. Pharm.*), transcription of *taṅ kuei* 當歸 (Cryptotaenia canadensis; see STUART, *Chinese Materia Medica*, p. 133). An older form of transcription is *daṅ-gu*, noted in the *Hua i yi yü*. The plant is mentioned as a product of Mao čou 茂州, Wei čou 維州, and other localities of Se-č'uan (*T'aṅ šu*, Ch. 42, p. 3), and may have been exported as a medicine from there to Tibet.

244. *hor-len* (*Tib. Pharm.*), a species of lotus (?). Transcription of *hu lien* 胡蓮; *hor* is not part of the transcription, but the corresponding translation of *hu*. In the edition of Pozdn'äyev, the Tibetan word is written *hoṅ len* (Mongol *khoṅlen*), and, according to Rehmann, the plant is identified with Scutellaria baicalensis. A species of Scutellaria is termed in Chinese *huaṅ k'in* 黃芩 (STUART, *Chinese Materia Medica*, p. 400),

while *huaṅ lien* 黄連 refers to Coptis teeta (*ibid.*, pp. 125, 401). Some confusion has obviously crept in here between *huaṅ lien* and *hu* 胡 *huaṅ lien* (Barkhausia repens).

245. *mdzo-mo šiṅ* is explained by CHANDRA DAS as "a tree the wood of which resembles the red sandal wood, and being largely imported into Tibet from China is used in dyeing the garments of the Lamas of Amdo." The word *mdzo-mo* is Tibetan, and denotes the female of that cattle breed which is a cross between a yak bull and the ordinary cow, and is known to the Chinese under the name *p'ien niu* 犏牛. "Wood of the female yak mongrel" certainly yields no sense; and the wood, we are told, is imported from China. It is therefore reasonable to regard the word *mdzo-mo* as the adaptation of a Chinese term. The *Tib. Pharm.* (No. 170) given as equivalent of the above Tibetan term, Chinese *su mu* 蘇木 ("sappan-wood," Caesalpinia sappan) and Tibetan *mdso-mo* (pronounced *dzu-mu*), seems to be a jocular imitation of the Chinese sounds.

246. *li-tsi* (*Pol. D.*, 28, p. 50), plum (Prunus domestica). From *li-tse* 李子.

247. *se-yab*, *b-se-yab*, according to JÄSCHKE, "(in medical literature), fig." This, however, rests on an erroneous identification. The *Pol. D.* (28, p. 49) has *se-yab* = Chinese *ši piṅ* 柿餅, dried Diospyros kaki (Mongol *šikir šabtala*, "sugar-cake"); see G. A. STUART, *Chinese Materia Medica*, p. 152. Tibetan *se*, therefore, represents a transcription of Chinese *ši*. In the same manner we have Tibetan *se-pad* = Chinese *ši šuaṅ* 柿霜, "crystallized Diospyros, prepared by taking off the skin of the fruits, and then exposing them to the sunlight and night-dew until they are dry, when a whitish powder gathers upon the surface" (STUART).

248. *bo-ti-tse*, rosary, given in *Pol. D.*, 22, p. 68: Chinese *p'u-*

tʻi-tse 菩提子, Manchu *bodisu*. CHANDRA DAS (p. 877) writes the Tibetan word *bo-dhi-rtsi* and (p. 922) *ạbo-dhi-tsi*. The Tibetan term is not based on Sanskrit, as shown by the transcription *ti* and the last element *tse*, but transcribes the Chinese name (not *vice versâ*, as suggested by WATTERS, *Essays*, p. 377). Chinese *pʻu-tʻi* transcribes Sanskrit *bodhi*, and the whole term signifies "seeds or fruits of the Bodhi-tree" (Ficus religiosa). The Tibetan form *bo-de*, cited by Watters from Jäschke, does not represent a transcription, but is a loan-word orally received from an Indian vernacular, while Tibetan *bo-dhi* is the correct and learned reproduction of the Sanskrit word.

249. *pʻan šu* (*Pol. D.*, 28, p. 59), potato (Solanum tuberosum). From *fan šu* 蕃薯.

250. *bin-tsi* (*Pol. D.*, 28, p. 50), a sour crabapple. From *pin-tse* 檳子 (Manchu *merseri*).

251. *po-so-i siu* (*ibid.*, Appendix 3, p. 9), date (Phoenix dactylifera). From *po-se tsao* 波斯棗 (Manchu *bosoro*), see BRETSCHNEIDER, *Chinese Recorder*, Vol. III, 1870, p. 265; and my *Sino-Iranica*.

252. *'u-bi-tsi* (*Mong. Pharm.*, No. 87), transcription of *wu pei tse* 五倍子, denoting the galls that are produced by an insect upon the leaves or leaf-stalks of Rhus semialata.

253. *pʻu-guṅ-yiṅ* (*ibid.*, No. 139), transcription of *pʻu kuṅ yin* 浦公陰 (*sic*, for *yiṅ* 英), Taraxacum officinalis.

254. *čʻau-tse* (*ibid.*, No. 143), transcription of *čʻie-tse* 茄子, Solanum melongena.

255. *ti-gu-pʻi* (*ibid.*, No. 62), transcription of *ti ku pʻi* 地骨皮, Lycium chinense, a solanaceous plant, root and seeds of which are used medicinally. It is mentioned in the *Tʻaṅ šu* (Ch. 38, p. 2 b) among the taxes sent from Kuo čou 虢州 in Ho-nan.

256. *kʻru-lan-šuṅ* (*ibid.*, No. 61) answers to *čʻuan hiuṅ* 川芎,

Conioselinum univittatum (G. A. STUART, *Chinese Materia Medica*, p. 123), an umbelliferous plant. Tibetan *šuṅ* is a bad transcription of Chinese *hiuṅ*, *hsiuṅ*.

257. *su-mi*, a cereal on which the Black Lo-lo (kLo nag) subsist. VASILYEV (*Geography of Tibet by the Minčul Chutuktu*, p. 86) tentatively proposed that it should be a transcription of Chinese *su mi* 粟米 ("millet"); but it rather seems to be a Lo-lo word.

258. *tā-ts'ra*, *tā-ts'raṅ* (CHANDRA DAS), "two kinds of Chinese tea greatly used in Tibet." Evidently *ta čuan* 大磚 ("great brick"), brick-tea.

259. *sin-'i me-tog* (*Pol. D.*, 29, p. 51). Transcription of *sin i hua* 辛夷花, Magnolia conspicua.

260. *ts'uu*, vinegar. From *ts'u* 醋 (see this volume, p. 77).

261. *ts'ai* (*ts'e*) *yaṅ-tse* (written *ts'al yaṅ-ts'e*; JÄSCHKE, p. 447), vegetable-garden. According to HENDERSON (*Tibetan Manual*, p. 51), from Chinese *ts'ai yüan-tse*. More probably, however, it is *ts'ai yaṅ-tse* 菜樣子. Chinese *ts'ai*, in the form *ts'e* of the Se-č'uan dialect (written *ts'al* and *ts'as*), is much used in colloquial Tibetan (p. 122); but it is incorrect to say, as assumed by Henderson, that there is no word in Tibetan for "vegetables" (compare *sṅo*, *sṅo-ts'od*, *ts'od-ma*, *ldum*).

262. *Zi-liṅ*, the Chinese city and prefecture of *Si-niṅ* in Kan-su Province; especially in *Zi-liṅ ja*, "tea from Si-niṅ." YULE (*Marco Polo*, Vol. I, p. 276) observed that "Si-ning is called by the Tibetans Ziling or Jiling, and by the Mongols Seling Khoto." The town is frequently referred to by mediæval writers as Seling. DESIDERI, who heard the name of the city in Tibet, spelled it Sinim or Silingh (C. PUINI, *Il Tibet*, p. 25); ORAZIO DELLA PENNA, Sciling, Scilingh, and Silin. According to ROCKHILL (*Land of the Lamas*, p. 49), the Tibetans call the

place at present Seling k'ar (*mk'ar*, "walled town"); the Mongols, Seling kutun.

Zi-liṅ t'a k'ug (*t'a* is the abbreviation for *t'a-ma-k'a*), an embroidered Chinese tobacco-pouch, as illustrated on plate 20, fig. 1, of ROCKHILL's *Notes on the Ethnology of Tibet*. This designation was given me by a Lama on being shown this illustration; the pouch in fig. 2 he styled *bod-kyi t'a k'ug* ("Tibetan pouch"), and that in fig. 3 *rgya-mii t'a k'ug* ("Chinese pouch").

Si-liṅ, sliṅ, name applied in Ladākh and the Panjāb to a stuff of goat's wool made at Si-niṅ and used for shawls (YULE, *Hobson-Jobson*, p. 847).

"A cloth called 'Siling' is manufactured from the shawl wool in Yarkand and China; it somewhat resembles a coarse English kerseymere in texture."—G. T. VIGNE, Vol. II, p. 129.

"Siling, stoffa proveniente dalla China, tessuto di pashum, generalmente di colore cenericcio; la stoffa più soffice al mondo, molto cara e rara."—O. ROERO, Vol. III, p. 71.

263. *zi-liṅ, zi-lim, zi-laṅ*. JÄSCHKE states that these words are derived from the Chinese and denote a composition metal similar to German silver. There is no doubt, accordingly, as to the meaning of the word, which designates the "white copper" (*pai t'uṅ* 白銅) of the Chinese, our "tootenague." The words *zi-liṅ* and *zi-lim* may refer to the city of Si-niṅ; *zi-laṅ*, possibly, is an adaptation from *si la* 錫鑞 ("spelter, pewter, solder").

264. *yu ši*, jade. From *yü ši* 玉石. Frequently in the biography of the lČaṅs-skya Rol-pai rdo-rje; for instance, *yu ši-i p'or-ba gčig*, a jade bowl; *yu ši-i ąk'or-lo*, a jade prayer-wheel; *yu ši-i yid bžin lčags-kyu*, a jade žu-i sceptre.

265. *gruṅ-ru-ši* (*Mong. Pharm.*, No. 27), transcription of *čuṅ žu ši* 鐘乳石 (the text in question writes 中), carbonate of lime

in stalactitic masses (D. HANBURY, *Science Papers*, p. 218).

266. *k'yen-hun* (*Mong. Pharm.*, No. 25), transcription of *kien fön* 鉛粉, carbonate of lead, white lead (D. HANBURY, *l. c.*, p. 223).

267. *big-ban* (*Tib. Pharm.*, No. 68; *Mong. Pharm.*, No. 16) [not *big-pan*, as written by JÄSCHKE, or *big-pag*, as in the *French Dict.*] corresponds to *tan fan* 膽礬, sulphate of copper; accordingly Tibetan *ban* = Chinese *fan*.

268. *k'am-pa*, porcelain clay, kaolin. Perhaps from *kaṅ* (*t'u*) 崗(土), kaolin. The transcription is irregular, though the Tibetan aspirate may be explained from the tendency to preserve the high tone of the Chinese word; the change of the guttural into the labial nasal might be due to assimilation to the ending *pa*. It is difficult to realize that the Tibetans, unfamiliair with porcelain and its manufacture, should have a word of their own for the designation of kaolin. "Porcelain" is expressed by *kar* (or *dkar*) *yol* ("white pottery"). [1]

269. *hai* (CSOMA), shoe (compare *ho lham*, "espèce de chaussures:" *French Dict.*, p. 1065). From *hiai* 鞋.

---

[1] In Mongol, porcelain is called *tsagājaṅ*, *tsājaṅ*, *šagādsaṅ*, *šādsaṅ*, explained by KOVALEVSKI as being developed from Chinese *č'a čuṅ* 茶鍾 ("tea-cup"). This derivation was justly called into doubt by A. SCHIEFNER (in the preface to Castrén, *Burjät. Sprachlehre*, p. XIII), for in Buryat we have *šāzaṅ*, *šāzen*. In other Mongol dialects we find *šāǯin* (A. RUDNEV, Матеріалы по говорамь восточной Монголіи, p. 156, who adheres also to Kovalevski's wrong etymology). "Tea" in Mongol, however, is *čai*, in dialects *ts'ai*, *č'ä* (RAMSTEDT, Срав. фонетика монг. письменнаго языка,' Russian translation by Rudnev, p. 53), in Buryat *čai*, *sai*, but never with initial *š*; and there is no reason why a Chinese *čuṅ* should be transformed by the Mongols into *dsaṅ*, *zaṅ*, or *zen*; it would simply be retained by them as *čuṅ* or *dsuṅ*, as proved indeed by Mongol *dsuṅdsa* ("goblet"), which is the equivalent of Chinese *čuṅ-tse* 鍾子 (KOVALEVSKI, p. 2409). The variants *tsagā-*, *tsā-*, *šagā-*, *šā-*, plainly show that this is the Mongol word *tsagān*, *tsān*; in dialects *šagān*, *šaṅān*, Buryat *sagan*, *sagaṅ*, which means "white." The second element, *dsaṅ* or *jaṅ* (the alternation of initial *ds*, *j*, and *y* in Mongol is well known), means "cup, bowl, pot, pottery," and may have been derived from Persian *jām* ("glass, cup"), Pahlavi *jām*, *yām*, Avestan *yāma*. The Mongol term *tsagājaṅ*, accordingly, presents an analogon to Tibetan *dkar yol*.

270. *yan ljiṅ*, dulcimer, a musical instrument employed in gTsaṅ (Central Tibet). From *yaṅ č'in* (*k'in*) 洋琴.[1] The Tibetan word apparently is a recent loan based on oral communication of a Se-č'uan dialect, in which the word is sounded *yaṅ jin* (the confusion of final *n* and *ṅ* being prevalent in Se-č'uan as in other dialects, for instance, in Nankiṅ). The addition of the prefixed *l* is merely prompted by the modern tendency to lend foreign words a native appearance; for it cannot refer to the tone, as 琴 has the even lower tone, while *ljiṅ* is high-toned; the correct transcription, therefore, would be *jiṅ* in the deep tone. As stated by M. Courant, the instrument in question was introduced into China from abroad, in all probability, not earlier than the end of the seventeenth century or the beginning of the eighteenth.

271. *čoṅ*, bell. From Chinese *čuṅ* 鐘.

272. *mog-mog* (Jäschke, p. 419: "meat-pie, meat-balls in a cover of paste," where "paste" is meant for "dough;" Henderson, *Tibetan Manual*, p. 84: "pastry-puff, Chinese word"). From *mo-mo* 饝饝. The two final *g*'s in the Tibetan transcription are not articulated.

273. *ts'ai-tau* (Jäschke, p. 447: "Chinese"), chopping-knife. Transcription of *ts'ai tao* 裁刀.

274. *liṅ-ts'e*, gratings, lattice. From *liṅ-tse* 櫺子, lattice of a window, sill.

275. *lan-gan*, *lan-kan*, balustrade, railing, barrier. Transcription of *lan kan* 欄杆 [Mongol *käräsgä*, Manchu *jerguwen*] (*Pol. D.*).

276. *lu-kaṅ*, crucible for gold and silver. Jäschke is inclined to regard the word as a "misspelling for *lugs-koṅ*." It seems

---

[1] Figured and described by J. A. van Aalst, *Chinese Music*, p. 67; A. C. Moule, *Chinese Musical Instruments*, p. 118; M. Courant, *Essai hist. sur la musique classique des Chinois*, p. 180.

more reasonable to look upon it as the transcription of a Chinese word, the element *lu* being apparently identical with *lu* 鑪 ("stove;" *tan* 丹 *lu*, "crucible"), and the element *kaṅ* presumably being *kaṅ* 釭 ("earthen vat").

277. *hp'uṅ hwa-aṅ*, phœnix. Transcription of *fuṅ-huaṅ* 鳳凰 (see HUTH, *B.M.*, Vol. II, p. 54). In the same work we meet *hp'u* for *fu* 府, and *hp'u-saṅ* as transcription of Fu-saṅ 扶桑, while in *Pol. D.* (29, p. 53) the Fu-saṅ flower is styled in Tibetan *bu-ts'aṅ me-tog*.

278. *yā-tse*, duck. From *ya-tse* 鴨子.

279. *k'ra-rtse* is given in the *French Dict.* (p. 114) as a Chinese word with the meaning "balance, scale." CHANDRA DAS ascribes to the same word the significance "a kind of biscuit or pastry made in the shape of a grating." The Chinese scale is usually designated in Tibetan *rgya-ma* and *rgya-t'ur*; the latter name was given me by a learned Lama in explanation of plate 27 (figs. 1 and 2) of ROCKHILL's *Notes on the Ethnology of Tibet*, while he termed the native scale on plate 28 *ña-ga* (or *ñag*, literally "notch," from the scale of notches cut in the beam). As *k'ra-rtse* is pronounced *ṭ'ar-tse*, it would seem that this spelling is identical with *t'ur* of *rgya-t'ur*; but I do not know what Chinese word it should transcribe. The word for the scale used by the Chinese traders in Tibet is *töṅ-tse* 戥子.

280. *man-tsi*, a kind of silk cloth; *men-tsi* (DAS), a colored silk handkerchief. It occurs in the History of Ladākh, where K. MARX (*J.A.S.B.*, Vol. LX, 1891, p. 135) explains it as "silkgauze with dots," while *g-liṅ-zi* (綾子) should be the same without dots. From *man-tse* 縵子, thin, plain silk, sarsenet. DAS gives also *s-man-rtse* as an incorrect spelling of *maṅ-tse*, with the meaning "yellow silk scarf with red spots impressed on it."

281. *ja-hoṅ* (Jäschke, who erroneously writes *ja-hod*: "yellowish red"), the color of boiled tea, red brown; according to *Li-šii gur k'aṅ*, the color of madder (*btsod mdog*, Sanskrit *mañjishṭā*, Chinese *k'ien* 茜, Rubia cordifolia). Transcription of Chinese *č'a huṅ* 茶紅, tea-red. Regarding *hoṅ* = Chinese *huṅ*, see also *French Dict.*, pp. 111, 1066.

282. *dar liṅ*, fine silk material. This term is given in the Ming edition of the *Hua i yi yü* as equivalent of *liṅ* 綾, so that Tibetan *liṅ* appears as a transcription of this word. Regarding *dar* ("silk"), see No. 161.

283. *č'aṅ čan*, wine-cup (*Hua i yi yü*); *čan*, from Chinese *čan* 盞.

284. *k'rau*, paper money (*Hua i yi yü*). Transcription of *č'ao* 鈔. The Tibetan transcription is only intelligible if we assume that at the time when it was made (under the Ming), *k'rau* was sounded *ṭ'au*. In this case the Tibetans employed a cerebral in order to reproduce a Chinese palatal; *vice versâ*, the Chinese reproduce Tibetan cerebrals by means of their palatals (above, p. 409).

285. *zan yaṅ*, transcription of *san yaṅ* 三樣, "three styles (of art)," used with reference to the monastery bSam-yas (compare *T'oung Pao*, 1908, pp. 20, 24). Hence the latter is styled Zan yaṅ mi-ḡgyur lhun-gyis grub gtsug-lag (Vasilyev, *Geography of Tibet by the Minčul Hutuktu*, in Russian, p. 33; Vasilyev could not explain the term *zan yaṅ*, and took it for a copyist's error). In an inscription of this temple, copied by Chandra Das (*Sacred and Ornamental Characters of Tibet*, *J.A.S.B.*, Vol. LVII, 1888, plate VI and p. 43), this title is written *gzaṇ yaṅ*. The transcription *gzan* is justified, inasmuch as the word, by the addition of the prefix *g*, is high-toned (while *zan* is deep-toned), thus rendering the high tone of Chinese *san*. An older Tibetan transcription of the same word

is *zam* = Chinese **sam* (*T'oung Pao*, 1907, p. 396, note 2).

286. *ha-šaṅ* (also *hva-* and *hā-*), a Chinese Buddhist priest. From *ho-šaṅ* = Sanskrit *upādhyāya* (compare EITEL, *Handbook*, p. 186; LEGGE, *Fa Hien*, p. 58; J. BLOCH, *I.F.*, Vol. XXV, 1909, p. 239; or *T'oung Pao*, 1909, p. 719; and PELLIOT, *Journal asiatique*, 1914, sept.-oct., p. 400).

287. *kiṅ-kaṅ*, *kaṅ daṅ kiṅ*, mentioned as a terrifying deity in *rGyal-rabs*. From *kin kaṅ* 金鋼 (see *T'oung Pao*, 1908, p. 23).

288. *jiṅ*, sūtra. From *kiṅ*, *čiṅ* 經 (CHANDRA DAS [*Tib. Dict.*, p. 449] erroneously states that *jiṅ* is the "Chinese term for Buddhism"). The older Tibetan form is *giṅ* (see, for instance, *T'oung Pao*, 1907, p. 392; and above, p. 423).

289. *gu ši*, *gu šrī*, *ko ši*, *ko šrī* (in the *Hua i yi yü*: *gui šrī*), state preceptor, royal teacher; a title conferred on the Buddhist clergy. From *kuo ši* 國師. Mongol *guši*. The writing *šrī* seems to have been prompted by adaptation to Sanskrit *çrī*. In the Sino-Tibetan inscription of 1341 (*T'oung Pao*, 1908, plate 28, line 2) we find the form *gui šri*. In the *Hua i yi yü* we meet also *č'en šri* as transcription of *č'an ši* 禪師 ("master of contemplation").

290. *r-luṅ rta*, *k-luṅ rta*, dragon-horse. The first element is a transcription of Chinese *luṅ* in the compound *luṅ ma* 龍馬 ("dragon-horse"). The Tibetans possess a large variety of charms printed on paper, cotton, or hemp cloth. These contain manifold designs accompanied by invocations, stereotyped prayer formulas, Dhāraṇī, etc. One of the most frequent designs to be met with is the figure of a running horse, usually carrying on its back the flamed jewel, which is composed of three individual precious stones encircled by a line. This jewel trinity has a double significance: first, it is an illustration of the Sanskrit term *triratna*, "the Three Jewels" (Buddha, his doctrine,

and the clergy); and, second, it represents the cintāmaṇi, the fabulous jewel granting every wish (Tibetan *yid-bžin nor-bu*, Chinese *žu i pao ču* 如意寶珠). F. W. K. MÜLLER, who published a Japanese *ema* 繪馬 (*Z. f. Ethnologie*, 1899, Verh., p. 529), referred only to the latter symbolism. The Japanese votive picture described by Müller differs from the dragon-horse of the Chinese and the *luṅ rta* of the Tibetans: it presents a syncretism of Shintō and Buddhist ideas, horse and jewels being separated from each other, the horse in the upper panel being tethered to two stakes, the three jewels in the lower panel being placed on a dish and enclosed in a shrine. The Tibetan term *r-luṅ rta* (pronounced *luṅ ta*), literally translated, means "wind-horse" (JÄSCHKE, p. 538, "the airy horse." His statement, derived from Schlagintweit, that the figure of the horse signifies the deity *rta mč'og*, is erroneous: the latter is not a deity at all, but merely means "excellent horse"). In view of the fact that strips of cloth imprinted with figures of such horses and attached to poles are made into flags merrily fluttering in the wind from the roofs of houses or from the top of an *obo*, the etymology "wind-horse," which is indeed advanced by the people, would seem to have a certain degree of plausibility. But two objections to this theory present themselves immediately. In lieu of the above spelling, we find another orthography of the word *luṅ* in the form *k-luṅ*, likewise articulated *luṅ*; and, as this mode of writing occurs as early as 1514 in the *Za-ma-tog*, there is reason to believe that it is even the older of the two.[1] Further, from an iconographic point of view, the coincidence of the Tibetan

[1] The expression quoted in *Za-ma-tog* is ***kluṅ rta dar*** (*dar*, "flag"); and besides ***rluṅ rta***, the term ***rluṅ** dar* is still in use (the difference between the two is established in the ***French Dict***, p. 479).

designs with the Chinese is obvious. Also L. A. WADDELL (*Buddhism of Tibet*, p. 412) rightly identifies *r-luṅ* with Chinese *luṅ* ("dragon"). He translates *luṅ-ma* by "horse-dragon;" but what is figured on p. 410 under this title is not the Chinese dragon-horse, but the *lin* 麟, as indicated also by the Chinese legend that accompanies the illustration. In the *Gazetteer of Sikhim* (p. 347), where the *luṅ-rta* is styled "the pegasus-horse of luck," Mr. Waddell derived it from "the jewel horse of the universal monarch, such as Buddha was to have been had he cared for wordly grandeur." This theory had already been advanced by E. SCHLAGINTWEIT (*Annales du Musée Guimet*, Vol. III, p. 164), who likewise fell back on the horse of the Cakravartin. The seven treasures (saptaratna) of the Cakravartin are the wheel, the wishing-jewel (cintāmaṇi), wife, minister, elephant, horse, and general. In this series the jewel and the horse are two distinct affairs; their combination into a jewel-carrying horse, it seems to me, was brought about, not in India, but in China, and the Tibetans received this conception from the Chinese. All this leads to the conclusion that Tibetan *r-luṅ* or *k-luṅ* is merely a transcription of Chinese *luṅ* ("dragon").

291. *zin-šiṅ*, the Taoists. From Chinese *sien šöṅ* 先生. In the Tibetan history of Buddhism in Mongolia (*Hor č'os byuṅ*, ed. of HUTH, p. 97) it is narrated that under Kubilai the followers of the sect Zin-šiṅ were very numerous, and adhered to the doctrine of T'ai šaṅ la gyin; that is, Chinese T'ai šaṅ lao kiün 太上老君, designation of Lao-tse. [1] The Tibetan transcription, however much it may have been disfigured in this work of recent date, corresponds to Marco Polo's *sensin*,

---

[1] HUTH, in his translation (p. 153), has misunderstood the entire passage.

Rašid-eddīn's *šinšin* (see YULE in his edition of *Marco Polo*, Vol. II, p. 322), and Mongol *senč'ing-ud* (DEVÉRIA, *Notes d'épigraphie mongole-chinoise*, p. 41). Semedo's *shien-sien*, cited by Yule, however, is a different word, answering to Chinese *šön sien* 神仙. The above Tibetan text, in close concordance with Rašid-eddīn, irrevocably proves that Marco Polo's Sensin are nothing but Taoists, and most assuredly have no connection with the Tibetan Bon sect. See also CHAVANNES, *T'oung Pao*, 1904, p. 377.

292. *t'e-se* (*Pol. D.*, Appendix, 2, p. 3), the planet Jupiter. Transcription of *t'ai sui* 太歲.

293. *t'e-an*, *t'ian*, heaven; Allah (see, for instance, *J.A.S.B.*, Vol. XLI, 1882, p. 114). Transcription of *t'ien* 天.

294. *hu-aṅ* (*hvaṅ*)-*dhi*, (also *dī* and *tī*), title of the Chinese emperor. Transcription of *huaṅ ti* 皇帝.

295. *k'ā* (*Pol. D.*, 4, p. 16), guardian, adjutant, body-guard (*ši wei* 侍衛). The Manchu equivalent is *xiya* (*x'a*), the Mongol *kiya* (*k'a*), of which Kovalevski says that it represents Chinese *hia*. I do not know for which character this is intended. The Tibetan transcription appears to be based on the Mongol form.

296. *tā žin*, excellency. Transcription of *ta žön* 大人 (see, for instance, HUTH, *B.M.*, Vol. II, p. 191). Chinese *žön* is even transcribed in Tibetan *bžin*, as already observed by SCHIEFNER (*Mélanges asiatiques*, Vol. I, p. 337).

297. *Dá-lo-ye*, aide-de-camp of the Amban of Lhasa. From *ta lao ye* 大老爺.

298. *guṅ*, or *kuṅ* (the latter in *Pol. D.*, 4, p. 6). JÄSCHKE (p. 69) observes, "(Chinese?) title of a magistrate in Lhasa, something like Privy Counsellor." CHANDRA DAS (p. 220) gives a more complete definition of the title. Transcription of *kuṅ* 公.

The *Pol. D.*, further, has Tibetan *hiu* (from *hou* 侯) and *pe* (from *po* 伯).

299. *t'ai rje* (*French Dict.*, "title of a provincial governor"). Seemingly modelled after *t'ai ši* 太師; *rje* ("lord"), of course, is a Tibetan word.

300. *tai hu*, title of certain officials. From *tai fu* 大夫 (see *T'oung Pao*, 1907, p. 397) or 大傅. The latter is transcribed in the *Pol. D.* (3, p. 13) *p'u*, while 大夫 is translated as *mi drags*.

301. *ạ-p'yiṅ-saṅ* is indicated as a Chinese word with the meaning "minister" (*blon-po*) in the *Li-šii gur k'aṅ* (see above, p. 444), the Mongol translation offering *čiṅsaṅ*. Although pointed out by SCHIEFNER (*Mélanges asiatiques*, Vol. I, p. 341), it is not listed in JÄSCHKE's Dictionary. CHANDRA DAS (p. 852) gives it with the explanation, "the designation in the older writings of a minister of state of Tibet = the modern *bka-blon*." That the term occurs in older writings may well be doubted; for the Tibetan transcription, apparently based on Chinese *č'öṅ* (*č'iṅ*) *siaṅ* 承相 ("minister of state"), must be of recent date when *p'* had changed into *č'*; moreover, the addition of the prefix *ạ* is meaningless: Chinese *č'öṅ* has the even lower tone, while both Tibetan *p'yiṅ* and *ạp'yiṅ* are high-toned. The Chinese word has passed into Mongol as *čiṅsaṅ*, and is written by Rašid-eddīn in the same manner (KLAPROTH, *Description de la Chine sous le règne de la dynastie mongole* trad. du persan de Rachid-eddin, p. 21 of the reprint from *Nouveau Journal asiatique*, 1833; and E. BLOCHET, *Djami el-Tévarikh*, Vol. II, p. 472; YULE, *Marco Polo*, Vol. I, p. 432, Vol. II, p. 145 [Polo: Chincsan]; YULE, *Cathay*, new ed. by CORDIER, Vol. III, p. 119). The Tibetan transcription is perhaps based on the Mongol form, not on the Chinese; compare

p'yam-ts'a, transcribing Mongol *tsamtsa* (KOVALEVSKI, p. 2104).

302. *tai se du*, president of the Board of Revenue. Transcription of *ta se t'u* 大司徒 (*T'oung Pao*, 1907, p. 397).

303. *zuṅ-č'u* or *-ču*. DAS explains this as Chinese *zuṅ* ("province") and *č'u* or *ču* ("local governor"). Apparently it is intended for *tsuṅ tu* 總督 ("governor-general").

304. *žiṅ sa*, province. The first element, *žiṅ*, is a transcription of *šöṅ* 省; the latter is given as the equivalent of *žiṅ sa* in *Pol. D.* (19, p. 42), where we find also Tibetan *ṭi-li žiṅ č'en* ("great province of Chi-li").

305. *ṅp'u* (older mode of writing *hu*, see *T'oung Pao*, 1907, p. 397, note), department. Transcription of *fu* 府.

306. *čeu*, *ču*, department. Transcription of *čou* 州. Other transcriptions of the same word are *jo* (see HUTH, *B.M.*, Vol. II, p. 416) and *ju* (for instance, *T'ou ju* = *T'ao čou* 洮州). In the Sino-Tibetan inscription of 1341 (*T'oung Pao*, 1908, plate 28, line 6) it is written *jiu*.

307. *ži-an*, district. Transcription of *hien* 縣. In the aforementioned inscription it is written *hyẹn*.

308. *'ui*, fort, military station (defined as *mk'ar č'uṅ*, see HUTH, *B.M.*, Vol. II, p. 34). Transcription of *wei* 衞.

309. *šo-gam*, customs duty. The first element presumably from *šui* (Hakka *šoi*) 税.

310. *t'uṅ-ši*, interpreter (generally used in eastern Tibet. From *t'uṅ-ši* 通事, which has passed also into Mongol, likewise into Turkī (*tuṅči bäk*: A. VON LE COQ, *Turfan*, p. 88). HENDERSON (*Tibetan Manual*, p. 63) writes the word in Tibetan letters *t'uṅ-sri*, transcribing it *t'ung-si*. Manchu *tuṅse*, Golde *tuṅsiko*.

311. In an extensive work on Tibetan chronology printed in Peking, and entitled *rTsis-kyi man-ṅag ṅin-mor byed-pai snaṅ-ba*, is contained a synoptical table of the names of the months

according to Chinese, Tibetan, Hor, and Sanskrit fashion. The transcriptions of the Chinese names, accompanied by a translation into Tibetan, prove to be the Twelve Animals of the duodenary cycle. They are enumerated as follows:

Number of Hor or Turkish Month	Name of Hor Month	Name of Chinese Month	Identification
3	*gsan yol*	1. *mā yol*	*ma yue*, horse month
4	*gsi yol*	2. *yaṅ yol*	*yaṅ yüe*, goat month
5	*u yol*	3. *huu yol*	*hou yüe*, monkey month
6	*blu yol*	4. *kyi yol*	*ki yüe*, rooster month
7	*ts'i yol*	5. *gau yol*	*kou yue*, dog month
8	*bā yol*	6. *t'ui yol*	*ču yüe*, swine month
9	*sgyeu yol*	7. *žui yol*	*šu yüe*, rat month
10	*šiṅ yol*	8. *neu yol*	*niu yue*, ox month
11	*dbyi yol*	9. *lau-hu yol*	*lao hu* 老虎 *yüe*, tiger month
12	*šiṅ-dbri yol*	10. *t'ur yol*	*tu yüe*, hare month
1	*byi yol*	11. *luṅ yol*	*luṅ yüe*, dragon month
2	*dbri yol*	12. *že yol*	*šo yue*, serpent month

These transcriptions are of comparatively recent date, as shown above all by *neu*, while the Old-Tibetan transcription for *niu* 牛 is *gi* (No. 223). The Sino-Tibetan names are given also in the *French Dict.* (p. 877), with the exception of the fourth month, and with a few errors: the name of the serpent is written *sgeu* instead of *že*; the name of the tiger is given as *lau*, and that of the hare as *ši-dgri*, apparently due to a confusion with the corresponding Hor month.

9. *Portuguese, Anglo-Indian, English, and Russian Loan-Words.*

312. *kor-do-pa, ko-do-ba* (Ladākh, Jäschke: "boot," with query), cordwain, Spanish leather. From Portuguese *cordovão, cordão*; Spanish *cordoban*, from Cordoba in Spain. Hindustānī *kardhanī*, Laskari-Hindustānī *kurdam*, Tamil *kordan*, Malayālam *koḍudam* (see S. R. Dalgado, *Influência do vocabulário português em línguas asiáticas*, p. 63, Coimbra, 1913).

313. *go-bi* (West Tibet), cabbage. Hindustānī *kobī*, from Portuguese *couve* (see this volume, p. 87).

314. *li-lam*, auction. From Portuguese *leilão*, through the medium of modern Indian languages: Gujarātī *lilám, nilám*, Hindī and Hindustānī *nīlám*, Nepalese *līlám*, etc. (Dalgado, *l. c.*, p. 97; Yule, *Hobson-Jobson*, p. 621). Burmese *lay-lan*, Siamese *leláṅ*, Malayan *lelan, lelon, leloṅ*. Cantonese *yeloṅ*, Amoy *lelaṅ*, Swatow *loilaṅ*.

315. *sa-bon* (West Tibet), soap. Hindustānī *sābún, sábun, saban*; Gujarātī *sabu, sābú*; Bengālī *sābán*; Singalese *saban, sabuṅ*; Telegu *sabbu*; Siamese *sa-bu*; Malayan *sabon, sabun*; Chinese *sa-pūn, sa-būn* (Watters, *Essays*, p. 346); Japanese *sabon, šabon*. The word was spread over Asia by the Portuguese (Portuguese *sabão*, Spanish *jabon*, from Latin *sapo*). The Arabic word *ṣabon, ṣabun* صابون is derived from Latin; and in view of the fact that the Arabs made little use of soap, it is not probable that they introduced the term into Malaysia (Dalgado, *l. c.*, pp. 138—139). The Latin word in all likelihood is of Celtic origin, since Pliny (xxviii, 51, § 191) ascribes the invention

35

of soap to the Galls (prodest et sapo, Galliarum hoc inventum rutilandis capillis).[1]

316. *'an-rgar-ajig* (pronounced *'aṅgarji*), English. In imitation of Hindī and Hindustānī *angrezí*, that is based on Portuguese *inglés*, *ingrês* (DALGADO, *l. c.*, p. 89).

317. *pa-tir*, padre; address to English missionaries and clergymen.

318. *kartūs* (Ladākh: RAMSAY), cartridge. Anglo-Indian *cartooxe.*

319. *čurut* (Ladākh: RAMSAY), cigar. From Anglo-Indian *cheroot*, which is said to come from Malayālam *čuruṭṭu*, Tamil *šuruṭṭu* (YULE, *Hobson-Jobson*, p. 188). The word was probably diffused by the Portuguese (*charuto*).

320. *t'a-ma-k'a*, West-Tibetan *t'a-mag*, *tamak*, and *da-mag*, tobacco. In all Indian languages the word for "tobacco" terminates in *ku*, *khu*, or *k* (DALGADO, *l. c.*, p. 147); *-k'a* is peculiar to Tibetan, and it seems that it owes its existence to an adaptation to the Tibetan suffix *k'a*. The *Pol. D.* (29, p. 12) has the Tibetan form *t'a-ma-k'i* (Mongol *tamaga* and *tamaki*, Manchu *dambaku*, Chinese *yü* 菸).

321. *mulmul* (RAMSAY), muslin. Anglo-Indian *mulmull*, Hindustānī *malmal* (see YULE, *Hobson-Jobson*, p. 595).

322. *ka-fī*, coffee (Ladākh: RAMSAY, p. 22). The missionaries in Lahūl have given to *k'a-ba* the signification of "coffee," which is otherwise unknown in Tibet (JÄSCHKE, p. 37). There is no reason, however, for the initial aspirate, and, according to Tibetan practice, *f* would preferably be represented by *p* or *p'* (compare in Indian languages *kaphi*, *kāphi*, *kopi*, *kappi*, *kapi*, etc.).

---

[1] The Tibetan word *pa-le* ("bread"), however, which DALGADO (*l. c.*, p. 120) derived from BELL's *Manual of Colloquial Tibetan* and with an interrogation-mark placed among the derivatives from Portuguese *pão*, does not belong to the Romance languages. It is written *bag-leb*, both elements being genuine Tibetan words, *bag* meaning "flour, pap, porridge," and *leb* "flat."

323. *ṭi-ked*, *tri-ked* (Jäschke), card, postage-stamp. From English *ticket*.

324. *ra-sid ṭi-ked*, money-stamp (*ra-sid* from Persian *rasīd*, "receipt").

325. *rafal* (Ramsay), rifle. From English *rifle*.

326. *kōt* (Ramsay), European coat. From English *coat*.

327. *ma-ni 'or-da* (written *gra*) = money-order.

328. *mim*, mistress. From *Ma'm*, *Madam* (through Hindustānī *mēm*).

329. *samāwār* (Ramsay, p. 157), *samavár* (Roero, Vol. III, p. 252), tea-boiler. From Russian *samovar* (many of these being brought to Le, Ladākh, by traders from Central Asia).

APPENDIX I. TIBETAN LOAN-WORDS IN CHINESE.

Although the number of Tibetan words, which are transliterated in books dealing with Tibet and Lamaism, is very large, only a few Tibetan loan-words have penetrated into Chinese. The word *la-ma*, which does not yet appear in the T'ang Annals, is as familiar to the average Chinese as to any of us. Most Tibetan words, however, are familiar only to those Chinese who have come in contact with Tibetans, live among them or along the frontier, or are in commercial relations with Tibet. Words like *p'ön ša* ("carbonate of soda") from Tibetan *bul* (*T'oung Pao*, 1914, p. 88),[1] *p'u-lu* from Tibetan *p'rug* (*ibid.*, p. 91; 1915, p. 22), *ka-ta* from Tibetan *k'a-btags* (Watters, *Essays*, p. 377), may be regarded as genuine loan-words

[1] The term *ta p'ön ša* 大鵬砂 occurs in the *Ta T'an leu tien* 大唐六典 (Ch. 22, p. 8; ed. of *Kuan ya šu ku*, 1895), where it is remarked that this product comes from Persia 波斯 and Lian čou 涼州. Persia, in fact, is one of the great centres for the supply of borax, and this very word is derived from Persian *būra* بوره (Russian *burá*), Arabic *būraq* بورق. Likewise *tincal*, *tincar* (crude borax found in lake-deposits of Persia and Tibet) comes from Persian *tinkār*, *tankār* تنكار (Sanskrit *ṭaṅkaṇa*; Yule, *Hobson-Jobson*, p. 923). Besides the passage of the *Wu tai ši*, formerly noted, the *Pien tse lei pien* (Ch. 209, p. 1 b) points out a text in the *Sun ši* and another in the *Mon hua lu* 夢華錄. For more information see my *Sino-Iranica*

naturalized in the Chinese language. Words, however, like *ńuan (*yüan*) 羱 (*T'oung Pao*, 1914, p. 71; provided this word be a reproduction of Tibetan *gñan*), and *fu-lu* in the T'ang Annals, are mere transcriptions of isolated occurrence, which left no imprint on the Chinese language. The term *p'u-t'i-tse* noted by Watters is not Tibetan (see No. 248). His *tie-lie* (from *t'er*) is not in general use, but is merely a bookish transcription; in the Tibetan-Chinese documents of the *Hua i yi yü* (Ch. 20), Tibetan *t'er-ma* is transcribed *t'ie-li-ma* 鐵哩麻. A few more interesting examples of Tibetan words in Chinese may follow.

1. *tsan-pa* 糌粑, transcription of Tibetan *r-tsam-pa*, roasted barley-flour, the staple-food of the Tibetans. This word is well known to all Chinese living in Tibetan regions.
2. *kien'r* 繭耳 is the designation, on the part of the people of the West (西人), of the wild yellow goat 黃羊 of Tibet (*Pên ts'ao kań mu*, Ch. 50 A, p. 18). The corresponding Tibetan name is *r-gya-ra*, *r-gya-ru*, or *r-gya* (*French Dict.*: *r-gya-ba*, in certain places also *r-gya-ka-ra*), identified with Procapra gutturosa (family Capridae), that occurs in the Kuku-nōr district and Amdo (G. Sandberg, *Tibet and the Tibetans*, p. 298; the rendering "Saiga-antelope," given by Jäschke after Schmidt, is erroneous). The Chinese transcription, answering to *gyer, gyar (g'er, g'ar), may well be based on the Tibetan term. Li Ši-čên observes that the animal's ears are very small, which may have given rise to the choice of the characters ('cocoon ears'). The same animal is called also *fan* 羳, an artificial formation meaning the "goat of the Fan (Tibetans)," and defined in the *Šuo wên* as a "goat with yellow abdomen" (黃腹羊); Giles explains it as "a small-sized deer found among the mountains of Kuku-nōr" (for "deer" insert "goat").

3. *kio-ma* 脚嗎, Potentilla anserina (compare ROCKHILL, *J.R.A.S.*, 1891, p. 284; the Chinese name is *žön šou kuo* 仁壽菓), is derived from Tibetan *gro-ma* (*ḍo-ma*), referring to the same plant. In the *Pol. D.* (29, p. 14) the latter is identified with Chinese *kou šê ts'ao* 狗舌草, which is said to refer to Senecio campestris (STUART, *Chinese Materia Medica*, p. 403).
4. *k'ü-ma ts'ai* 曲麻菜 and *k'u-ma'r* 苦麻兒 (*Pol. D.*, 27, p. 27) appear to be transcriptions of Tibetan *k'ur-ma* (also *k'ur-mań*, *k'ur-mańs*), dandelion. The Mongol equivalent *idara* is explained by KOVALEVSKI as "chiccory," a plant that does not occur in China, either.
5. Chinese *kan-pu* 甘卜 does not render Tibetan *mk'an-po*, as stated by E. BLOCHET (*Djami el-Tévarikh*, Vol. II, p. 544), but Tibetan *s-gam-po*, 甘 being formerly **kam*, **gam*, and, like *sgam*, possessing the high tone. The Tibetan aspirate would be reproduced also in Chinese by the corresponding aspirate. The Mongol writing *mkanpo* for Tibetan *mk'an-po* proves nothing, as Mongol is devoid of aspirates, at least as far as the script is concerned. As already stated by T. WATTERS (*Essays*, p. 376), *mk'an-po* is transcribed in Chinese *k'an pu* 堪布.

The official organization of Tibet has been dealt with from the Chinese angle by F. W. MAYERS (*Chinese Government*, 3d ed., 1896, pp. 105—122), T. WATTERS (*Essays*, pp. 375—377), W. W. ROCKHILL (*J.R.A.S.*, 1891, pp. 219—221), and BRUNNERT and HAGELSTROM (*Present Day Political Organization of China*, pp. 465—477; Russian edition, pp. 389—399). None of these expositions is complete or entirely clear, and the identification of the Chinese transcriptions with their Tibetan equivalents leaves much to be desired. I have no intention of canvassing the same ground again, but restrict myself to a few identifications or observations.

1. *dza-pe, ža-pe, ša-pe*, the colloquial designation for the *bka-blon* (*ka-lon*, "minister"), is not written *gšags dpe* ("model of justice;" this compound indeed does not exist), as suggested by ROCKHILL, but is *žabs pad* ("foot lotus;" that is, a lotus placed beneath the feet, as in the images of Buddhas and saints).

2. The Chinese transcription *ka-pu-lun* 噶布倫 (MAYERS, No. 567) of Tibetan *b-ka-blon* ("prime minister") is based on the Tibetan pronunciation *kab-lun*.

3. *ka-hia* (*zia*) 噶厦, Council Chamber. Transcription of Tibetan *bka g-šag-s* (pronounced *ka ša*), abbreviated for *bka-blon g-šag-s lhan rgyas* ("union of the court of ministers"). In the Ming edition of the *Hua i yi yü*, 厦 serves for the transcription of Tibetan *žag, šag*, and *ša*. In the *Pol. D.*, Tibetan *gšags* is employed for the rendering of *se* 司.

4. *tsai puṅ* 仔琫 (MAYERS, No. 569), Councillor of the Treasury, of the first class. Transcription of Tibetan *m-dzod d-pon* (pronounced *dzö pun*).

5. *šaṅ-čo-t'ö-pa* 商卓特巴 (MAYERS, No. 570), Councillor of the Treasury, of the second class. Transcription of Tibetan *p'yag mdzod-pa*, pronounced *č'an* or *č'aṅ dzot-pa*. The *ts'aṅ-* or *č'aṅ-ču-pa*, noted by WATTERS (*Essays*, p. 376), represents the same Tibetan word; this, however, is not the title of a Lama of rank, as stated by Watters, but an appellation of the secular governor or regent (*sde-srid*) of Tibet, who is styled *sa skyoṅ-bai p'yag mdzod* ("treasure of the government").

6. *yer-ts'aṅ-pa* 業爾倉巴 (MAYERS, No. 571), Controller of the Revenue, fifth rank. Presumably transcription of

Tibetan *g-ñer-ts'an-pa* ("one in charge of storehouses"), pronounced *ñer-ts'an-pa*.

7. *lan-tsai-hia* 郎仔轄 (MAYERS, No. 572), Controller of Streets and Roads. Presumably transcription of Tibetan *lam-mdzad* (pronounced *dzai*, *dzä*) *g-šag-s* ("road-making court").

8. *hie'r pan* 協爾幫 (MAYERS, No. 573), Commissioner of Justice. It is not clear on what Tibetan term this transcription is based. The usual term for this office is *žal gčod-pa* or *žu len-pa*. As to *pan*, I should be inclined to see in it *dban* ("power").

9. *šo-ti-pa* 碩第巴 (MAYERS, No. 574), Superintendent of Police. The element *ti-pa* represents the transcription of Tibetan *s-de-pa* ("chief, governor"), which occurs again in No. 578. The Tibetan word represented by Chinese *šo* is not known to me.

10. *ta-pun* 達琫 (MAYERS, No. 575), Controller of the Stud. Transcription of Tibetan *r-ta d-pon* ("horse official").

11. *čun yi* 中譯 (MAYERS, No. 576), Secretary of the Council. Transcription of Tibetan *drun yig* (pronounced *ḍun yi*; in regard to Chinese palatals representing Tibetan cerebrals, see above, p. 409), secretary. The usual designation is *bka drun*, abridged from *bka blon drun*, secretary of a minister.

12. *tun k'o'r* 東科爾 (MAYERS, No. 583), "the ancient native nobility." Transcription of Tibetan *g-dun ą-k'or* (pronounced *dun k'or*, "circle of families"). The men thus entitled form the nucleus of civil officers. They receive their training in a school of Lhasa (*gYu-t'og slob-grwa*), are then assigned for five years as apprentices to the Office of Accounts (*rtsi k'an*), may be detailed on various duties with the executive

or revenue, and may finally land the post of a *rDzoṅ dpon* (*Joṅ pon*), corresponding to that of district magistrate in China.

13. Titles of the military officers. These are enumerated, but not identified by Mayers and Brunnert. Watters has only identified the first title. Chinese *tai puṅ* 戴琫 = Tibetan *mda dpon*, general. Chinese *žu puṅ* 如琫 = Tibetan *ru dpon*, captain, commander of 200 men. Chinese *kia puṅ* 甲琫 = Tibetan *brgya dpon*, centurion, commander of 100 men. Chinese *tiṅ puṅ* 定琫 = Tibetan *ldiṅ dpon*, officer set over 45 or 50 men. Add: *bču dpon*, corporal, set over ten.

14. *čao* 召 does not mean in Tibetan "monastery or shrine," as asserted by MAYERS (No. 585), but is the transcription of Tibetan *jo-bo*, *jo-wo*, *jo-o* ("lord") with reference to the celebrated Jo-k'aṅ temple of Lhasa (see ROCKHILL, *J.R.A.S.*, 1891, p. 74).

15. What MAYERS and BRUNNERT write *geleng* and *gylong* is in written Tibetan *dge sloṅ* (pronounced *ge-loṅ*). Brunnert's *gheneng* is Tibetan *dge-bsñen* (pronounced *ge-ñen*). In regard to Chinese *pan-ti*, see above, No. 74.

16. *ko-se-kuei* 格思規, according to Brunnert and Hagelstrom, is Tibetan *gisk-hui*. It corresponds to Tibetan *dge-bskos*, the Chinese transcription being based on a Tibetan pronunciation *ge-s-kö*.

APPENDIX II. TIBETAN LOAN-WORDS IN ENGLISH.

Tibetan words like Lama, Dalai Lama, Teshoo Lama, Kanjur, Tanjur, yak and other names of animals, have obtained naturalization in the English language. Books of travel in Tibet, as might be expected, swarm with native words. Thus we read of *gelong* or *gylong* (*dge-sloṅ*, "monk"), *kahlon* (*bka-blon*, "minister"), *gompa* (*dgon-pa*, "monastery"), *chörten*, *chorten* or *chürten* (*mč'od-rten*, "tope"), *tsamba* (*rtsam-pa*, "roasted barley-flour"), and others.[1] The following list contains our zoölogical terms borrowed from Tibetan and a note on the word "polo."

1. YAK, Bos grunniens. From Tibetan *g-yag*.
2. DZO, cross between yak-bull and Indian cow. From Tibetan *m-dzo*.
3. KYANG, KIANG, the wild horse inhabiting the table-lands of Central Asia (Equus kyang, or Equus hemionus kiang). From Tibetan *r-kyaṅ*.
4. SEROW (spelled also saraw, sarau, sarao, surow, serou), an antelope (Nemorrhœdus bubalinus). From Tibetan *b-se-ru*.
5. CHIRU (R. LYDEKKER, *Game Animals of India*, p. 184, gives also CHUHU as a Tibetan name), an antelope (Pantholops hodgsoni). Presumably based on the same Tibetan word as the preceding one.
6. GOA, RAGAO, a gazelle (Gazella picticaudata). From Tibetan *r-go-ba*, Central Tibetan *go-a*.
7. TAKIN, a horned ruminant allied to both the goats and the antelopes (Budorcas taxicolor), occurring in south-eastern Tibet and on the northern frontier of Assam (first described by J. RICHARDSON, *On the Tákin of the Eastern Himalaya*, *J.A.S.B.*, Vol. XIX, 1850, p. 65; see also R. LYDEKKER, *Game Animals of India*, p. 157). The word *takin* (omitted

[1] In his article "Cooly" (*Hobson-Jobson*, p. 250), YULE, after mentioning the Turkish word *kol* ("slave"), has inserted this observation: "*Khol* is in Tibetan also a word for a servant or slave (note from A. Schiefner; see also Jäschke's Tibetan Dict, 1881, p. 59). But with this the Indian term seems to have no connection." I do not share Yule's opinion (*ibid.*, p. 463) that the Tibetan verb *dpyaṅ* ("to hang") can be sought in Anglo-Indian *jampan* ("kind of palanquin"), although we have the compound *ạk'yogs-dpyaṅ* with this meaning.

in Yule's *Hobson-Jobson*) is usually regarded (thus in the new *Oxford English Dictionary*) as Mišmi, on the authority of T. T. Cooper (*The Mishmee Hills*, p. 183, London, 1873). E. H. Parker (*As. Qu. Rev.*, 1913, p. 429) says that it is presumably a Tibetan or Assamese word. This is justified, inasmuch as G. Sandberg (*Tibet and the Tibetans*, p. 297) gives a Tibetan word *rta skyin* (pronounced *ta kyin*) with the identification "Budorcas taxicolor," and this item thus found its way into the *Tibetan Dictionary* of Chandra Das, of which the late G. Sandberg was one of the editors. But is *rta skyin* really a Tibetan term for the animal in question? It is given neither by Jäschke, nor by the *French Dictionary*, nor by any Tibetan or Chinese-Tibetan dictionary known to me, nor does it occur in Tibetan literature. The formation itself is highly suspicious: *rta* means "horse," and *skyin* is a wild mountain-goat (Capra sibirica; see M. Dauvergne, *Bull. Mus. d'hist. nat.*, Vol. IV, 1898, p. 217). It seems to me impossible that the Tibetans should regard Budorcas as the "horse-goat," as this animal does not bear an atom of resemblance to a horse. In the words of Cooper, it resembles somewhat a cross between the deer and the bull. G. Sandberg, interested in Tibetan and in zoölogy, resided in Darjeeling, where he made his inquiries about the fauna of Tibet. His informant, whom he interrogated in regard to the takin, may have coined for his benefit the above Tibetan mode of writing. I do not believe that this is a Tibetan term at all, but simply the fancy of an individual who attempted to make the best of writing it in Tibetan. Takin is indeed a Mišmi word, and *ta* does not mean "horse" in Mišmi.

8. The most interesting of our Tibetan loan-words is polo,

the equestrian ball-game. Both the *Century Dictionary* and the *Encyclopædia Britannica* (Vol. XXII, p. 11) derive it from Tibetan *pulu* ("ball"), the former referring also to Balti *polo*. The form *pulu* as a Tibetan word is given by G. T. VIGNE (Vol. II, p. 289). JÄSCHKE writes the word *bo-lo* ("ball, for playing"), but we know positively that the Ladākhī form is *po-lo* (O. ROERO, Vol. III, p. 236, who has *po-lo* as Tibetan and *pu-lo* as Indian; RAMSAY, p. 123; A. H. FRANCKE, *Ladākhī Songs*, p. 12). The *Polyglot Dictionary* of K'ien-luṅ writes the word *p'o-loṅ* (as likewise KOVALEVSKI, see *bumbuge*); in the unique Pentaglot edition of the British Museum we find the same Tibetan spelling, which, moreover, is confirmed by the pronunciation added there in Manchu characters, *p'oloṅ*. An analogon to the alternating forms *po-lo* and *p'o-loṅ* is presented by *šo-lo* and *šo-loṅ* ("dice"), and by *č'o-lo* ("dice, chess;" see No. 76) in comparison with Lepcha *č'o-loṅ*. The Chinese equivalent given is *hiṅ t'ou* 行頭, apparently a colloquialism (see K. HIMLY, *T'oung Pao*, Vol. VI, 1895, p. 272), corresponding to Manchu *mumuhu*, Mongol *bumbuge* (the pronunciation is thus fixed by the Manchu transcription in the Pentaglot edition), and eastern Turkı *tob* توپ (transcribed *tob* in Manchu). The latter word is listed by RADLOFF (Vol. III, col. 1220) as Djagatai and Osmanli *top*, meaning "any round object, ball, globe" (hence also *topči*, "button," see No. 178). First of all, therefore, the term *po-lo* or *p'o-loṅ* relates to the ball, which, according to the Manchu description given in the Mirror of the Manchu Language, was a large ball sewed together from pieces of leather; the balls used in Ladākh are made from willow-wood. In a wider sense, the word *po-lo* denotes also the game itself and the polo ground, the latter, in addition, being styled in Ladākh *ša-ga-ran* (ROERO: *šagara*). The game is by no means cultivated throughout Tibet; in the central and eastern portions of the country it is now wholly unknown. In fact, it is restricted to Baltistān and Ladākh, and there is good reason to believe that it was introduced into Ladākh from Baltistān, or from Gilgit and Chitral, where the game also is cultivated (BIDDULPH, *Tribes of the Hindu Kush*, p. 84). Some authors regard polo as the national game of Ladākh (A. CUNNINGHAM, *Ladák*, p. 311; RAMSAY, *Western Tibet*, p. 123); see also KLAPROTH's *Magasin asiatique*, Vol. II, p. 17.

The word *p'o-loṅ* is of ancient date, for it occurs in the Mahāvyutpatti. The Emperor T'ai-tsuṅ (627—649) knew that the Tibetans made excellent polo-players. In A.D. 709 Chinese were defeated in a polo-match by a Tibetan envoy in the Pear-Garden Pavilion (see *Fuṅ ši wen kien ki* 封氏聞見記 by Fuṅ Yen 封演 of the T'aṅ period, Ch. 6, p. 2 b, ed. of *Ki fu ts'uṅ šu*; and *Kiu T'aṅ šu*, Ch. 196 A, p. 4 b). I hope to publish some day a detailed history of the game.

Through an oversight of the compositor the following item was omitted on p. 503.

211. *'an-dur* (*Hua i yi yü*), cherry (*yiṅ t'ao* 櫻桃, Prunus pseudo-cerasus, so-called Chinese cherry). The Chinese term is transcribed in Mongol *ingdōr*, in Manchu *ingduri*; hence it is probable that the Tibetan form (presumably written inexactly for *'aṅ-dur*) also represents a transcription of Chinese.

INDICES.

(Figures introduced by p. refer to pages; the others, to the numbers in the text.)

1. Tibetan.

2. Sanskrit.

3. **Prākrit.**

4. **Persian.**

9. **Manchu.**

10. **Chinese.**

11. Portuguese.

12. Anglo-Indian.

13. Russian.

14. Botanical.

ADDENDA.

The following Tibetan Sūtra printed in Peking has been translated from the Chinese. The Tibetan title runs thus: Šin-tu rgyas-pai sde snod rtogs-pa yoṅs-su rdsogs-pai sñiṅ-poi ṅeṣ don bstan-pai žes bya-bai mdo. It is accompanied by the Chinese title in Tibetan transcription: Tā hpaṅ bkwaṅ ywan kyau šeu tau (written with double *o*, presumably intended for *ō*) lo liu yi kyiṅ. This corresponds to Chinese 大方廣圓覺修多羅了義經 *Ta faṅ kwaṅ yüan kiao siu to lo liao yi kiṅ* (Bunyiu Nanjio, No. 427). According to the Tibetan colophon, this Sūtra had never previously been translated into Tibetan, but existed in the Chinese Tripiṭaka (rgyai bka-ạgyur). The translation was made by the dka-bču [title] Subhagaçreyadhvaja and the dka-bču Dhyānārishṭaṁvyāsa; a Manchu translation was utilized for this work, and a dānapati (sbyin bdag) Hiṅ līn, apparently a Chinese, contributed toward it a sum of three hundred Taels.

67. *ne-tso*, after all, may be a Tibetan word. Cf. Lo-lo-pʻo *a-jö* ("parrot").

157. The Uigur title *bakši* appears in the transcription 帛師 *bak-š'i in the *Yu yaṅ tsa tsu* (Ch. 17, p. 11) as the title of Fa-tʻuṅ 法通, a Parthian 安西人, who in his youth had lived in southern India, and had there entered the clergy. This transcription shows that *bakši* conveyed to the Chinese the impression of a foreign word. Simultaneously it demonstrates that my identification of the second element, *ši*, with 師, is correct. To Manchu *faksi* add Jurči *faši* (Grube, *Jučen*, No. 315) and Golde *paksi* or *paxsi*.

183. Mongol *bogdo* is possibly connected with Uigur *poktas*.

To the Tibetan loan-words in English, add *tangun*, *tanyan*, according to Yule (*Hobson-Jobson*, p 898), "Hindustānī *ṭānghan*, *ṭāngan*; apparently from Tibetan *rtaṅāṅ*, the vernacular name of this kind of horse (*rta*, 'horse'), the strong little pony of Bhutān and Tibet." Jaschke has noted this term after Hooker, writing it *rta-ṅaṅ*; but whether it is really written this way, and whether the term actually exists in Tibetan, is not known to me.

134

不儿罕

JOURNAL

OF THE

AMERICAN ORIENTAL SOCIETY

EDITED BY

JAMES A. MONTGOMERY
Professor in the
University of Pennsylvania

GEORGE C. O. HAAS
Sometime Instructor in the
College of the City of New York

CHARLES C. TORREY
Professor in Yale University

VOLUME 36

In Memoriam

WILLIAM HAYES WARD

AMERICAN ORIENTAL SOCIETY
NEW HAVEN, CONN., U. S. A.
1917

Burkhan.—By BERTHOLD LAUFER, Curator at the Field Museum of Natural History, Chicago, Ill.

As is well known, the word 'Burkhan' serves in the Buddhist literature of the Mongols for the designation of the Buddha. It has likewise been traced in Buddhist texts of the Uigur language,[1] and in Manichaean literature is the name given to the incarnate messengers of the God of Light to man.[2] The etymology proposed in 1866 by A. Schiefner[3] to the effect that the Turkish form Purkan, as noted by Radloff, has been derived from Mongol Burkhan, and in its origin seems to be a corruption of Indian Brahman, may now be dismissed without discussion. A more tempting explanation has been advanced by Baron A. von Staël-Holstein,[4] who believes himself to be justified in tracing Uigur *purkhan* (read *burkhan*) to Chinese 佛 **pur* (read **bur*) + Turkish *khan*. At first sight this hypothesis would seem convincing to the uninitiated, nevertheless it is fallacious and indefensible. As will be shown, the term *burkhan* does not represent a transcription, but is an ancient and indigenous word of the Altaic languages. The proposition of Baron von Staël-Holstein is by no means novel, but has been forestalled by his countryman P. Schmidt,[5] who has already given in substance the same etymology for the 'mysterious' Mongol word Burkhan. 'A similar root does not exist in the allied languages,' he remarks, 'and since the present notion conveyed by it is not of Mongol origin, I am inclined to regard it as a loan-word. This being the case, it must be derived either from Chinese or Tibetan. As regards the period when the loan took place, Buddha may have been known in Mongolia long before the introduction of Buddhism. In literary documents I have been able to trace it back to Marco Polo.[6] The Tibetan name of Buddha is

---

[1] F. W. K. Müller, *Uigurica* 2, p. 77.

[2] Chavannes and Pelliot, *Traité manichéen retrouvé en Chine*, p. 76; F. Legge, *Forerunners and Rivals of Christianity*, 2, p. 336.

[3] In the introduction to W. Radloff's *Proben der Volkslitteratur der türkischen Stämme*, 2, p. xi.

[4] In Radloff's *Tišastvustik*, p. 142 (*Bibl. Buddhica* 12, 1910).

[5] 'Der Lautwandel im Mandschu und Mongolischen,' *J. Peking Or. Soc.* 4. 63.

[6] Polo's spelling is *borcan* or *borcam*.

said to resemble neither the Indian nor the Mongol one. We do not either get much farther with the present Chinese 佛 *Fo*. We may therefore presuppose as the root of Burkhan either the syllable *bud* in accordance with the Indian name, or the Old-Chinese **Fut*. The latter hypothesis seems to be the more probable one. The Mongol syllable *bur* contradicts neither of these suppositions. The second syllable *khan* is here either a suffix (cf. Manchu *Fucich'i* [Buddha], from **Futich'i*), or even the well-known word *khan* (*Chan*), accordingly *Buddha Chan* = *Bud* (*Fut*) *chan* = *Burchan*.'

A serious objection must be raised to this dissection of the word. There is no analogy to such a hybrid combination of a Chinese and Turkish element; and if the second component *khan* really were this alleged Turkish word, why do we never meet the fuller form *Bur-khagan*? In order to anticipate this objection, Baron von Staël-Holstein assures us that according to an oral communication of Radloff the word *khan* is frequently attached to Turkish names of gods and idols but hastens to add that he knows of no examples for this phenomenon in Turkish-Buddhistic documents. For phonetic reasons the conception of the ending *-khan* in *burkhan* as the word for 'king, sovereign' is out of the question. As is well known, the vowel of *khān*, being contracted from *khagān*, is long, whereas the *a* in *burkhan* is short. This is clearly evidenced by the writing of Kalmuk in which the long vowels are marked by the addition of a small dash: while *burkhan* (plural *burkhat*) is written in Kalmuk with a short *a*, the word *khān* (plural *khāt*) is expressly fixed with a long *a*.[7] In the Tungusian, Mongol, and Turkish languages we find a suffix *-khan*, *-kan*, *-gan*, with such vowel changes as are conditioned by the laws of vowel-harmony, usually having the meaning of a diminutive.[8] Whether this suffix may be recognized in *burkhan* cannot be decided. Further, we are entitled to raise the question, what authority could have induced the Uigur to style Buddha (either the one Buddha or any other Buddhas) a king or sovereign? Every one knows that Buddha never was a king, and is not so designated in any passage of Sanskrit or Chinese literature of Buddhism[9]; he was, however,

[7] See A. Popov, *Grammar of the Kalmuk Language* (in Russian), § 62, 66.

[8] See particularly W. Schott, *Altaische Studien I* (*Abh. Berl. Akad.* 1860), p. 591-594; *IV* (*ibid.* 1870), p. 275 *et seq.*

[9] The epithet *dharmarāja* ('king of the law'), in Mongol *nom-un khagān*, is of course a seeming exception only, being a metaphorical expression.

the son of a king, and is therefore styled the 'crown-prince' (*kumārarāja, rājaputra*). The insinuation that the Uigur should have been guilty of such a gross violation of sacred tradition, as would crop out of this fantastic dismemberment *burkhan,* is an absurdity on the very face of it. That the element *khan* bears no relation to the word for 'king' becomes clear also from the compound *purkan kān* ('the king purkan') noted by Radloff[10] with the meaning 'a spirit worshiped by the shamans.' Among the Turkmen of Khiwa, according to Radloff, the word *porkhan* even designates the shaman.[11] In fact, *burkhan* is a term peculiar to the ancient shamanism of Siberia, and was diffused there over an extensive area long before the introduction of Buddhism.

Among the Tungusians of Nerčinsk, M. A. Castrén[12] noted a word *burkan* with the significance 'God,' and derived it from Buryat *burkhan.* The same word he recorded also among the Karagas in the Altai, who speak a Turkish dialect, and there also concluded that it was adopted by them from the Buryat.[13] True it is that the two Tungusian dialects studied by Castrén, as emphasized by Schiefner in the introduction to his work, have been strongly affected by Buryat influence both lexical and grammatical; but the word *burkan* can prove nothing along this line, as it occurs also in other Tungusian languages, particularly in that of the Gold on the lower Amur. As I spent a whole summer among this people, particularly studying its religious concepts, the word *burkhan,* as used by the Gold, is deeply stamped on my mind, for my conversations with them turned on this subject frequently, and I had a large collection of *burkhans* made for me. The best study of this subject thus far is contained in the book of P. P. Shimkevič,[14] where we read as follows (p. 38): 'With their notions concerning the life beyond

---

[10] *Wörterbuch,* 4, col. 1368.

[11] As regards this double significance, compare the observation of Hubert and Mauss (*L'Année sociologique,* 7. 87): 'L'esprit que possède le sorcier, ou qui possède le sorcier, se confond avec son âme et sa force magique: sorcier et esprit portent souvent le même nom.'

[12] *Grundzüge einer tungusischen Sprachlehre,* p. 95.

[13] M. A. Castrén, *Koibalische und karagassische Sprachlehre,* p. xiii, 144.

[14] 'Materials for the Study of Shamanism among the Gold (Materialy dla izucheniya šamanstva u Goldov),' in the *Zapiski of the Amur Section of the Imp. Russian Geogr. Soc.* 2. 1 (Chabarovsk, 1896). L. Sternberg will deal with the same subject in the Publications of the Jesup Expedition.

and the existence of various spirits bringing to man luck or calamity the Gold combine a great number of most diverse gods (*burkhan*) personifying a certain spirit. Whatever work a Gold may commence, it is incumbent upon him to resort to the *burkhan* for help. The shaman appears as the mediator between him and the spirit, and has supernatural power to communicate with the spirits. According to the circumstances, the shaman orders the people to make such and such a *burkhan* and to appeal to him in accordance with established precepts, but occasionally when the *burkhan* thus made does not bring the expected advantage, he is destroyed by the shaman or exchanged for another *burkhan*. For every kind of disease, on every special occasion of life, the *burkhans* are invariably made after the direction of the shaman in a strictly prescribed order. They consist of representations of men, animals, birds, fish, and reptiles; sometimes also amulets are made in the shape of joints, palms, soles, heart, etc. As to material, they are made of wood, metal, fish-skin, paper, cloth, grass, or clods from marshes. *Burkhans* are delineated on wood, cloth, or paper, or are carved from wood, cast from tin or silver, and skilfully forged from iron.' The author then proceeds to give a classification and detailed description of the *burkhans* (p. 39-60), and in the following chapter records some legends concerned with them, many of which are figured on the plates attached to the volume. Every one will recognize that this sort of *burkhan* has not a flavor of Buddhism, but is a genuine and original shamanistic element. In fact, I did not discover among the Gold any trace of Buddhism, which has never reached the Amur. The word *burkhan* is foreign to the Manchu language. Buddha is called in Manchu *Fučihi*, *Fu-* being a transcription of the corresponding Chinese designation of Buddha, the second element *-čihi* being as yet unexplained.

The word *burkhan* may be traced also in ancient Chinese records; at least, this is the opinion of the Japanese scholar K. Shiratori.[15] In discussing the name of the mountain T'u-t'ai 徒太 in the country of the Mo-ho or Wu-ki, Shiratori states that this mountain is also styled Pu-hien 仆險 (anciently But-kan). He refers to a passage in the *Shan hai king* (*Ta huang pei king* 大荒北經) to the effect that 'in the desert there is

[15] *Über die Sprache des Hiung-nu Stammes und der Tung-hu Stämme*, p. 60 (Tokyo, 1900).

a mountain Pu-hien, where there is also a country styled Su-shen 肅愼'; and further to *Tsin shu* (Ch. 97, p. 2 b), where it is said: 'The Su-shen tribe is also called Yi-lou, and its habitat is north of Mount Pu-hien.' Shiratori adds: 'In Mongol God is styled *tägri* or *burkhan*. Pu-hien is assumed to be a transcription of the word *burkhan.*' From a phonetic viewpoint this identification is possible, and it is equally possible that the said mountain was personified as a deity and worshiped under the title Burkhan. As is known, mountains and rivers were (and partially still are) the object of worship among all Tungusian, Mongol, and Turkish tribes (as well as in ancient China). One of the sacred mountains revered by the Mongols is the Burkhan-khaldun in northern Mongolia, on which Tchinggis Khan is said to have been interred.[16] A Mongol book, dealing with sacrifices to the deity of fire, and according to the well-founded testimony of Banzarov, devoid of any Buddhistic influence, begins thus: 'Mother Ut, mistress of the fire, created from the elm-tree, growing on the summits of the mountains Khangai-khan and Burkhatu-khan!'[17] These mountains are entitled 'sovereigns' (*khan*), and *burkhatu* is apparently a derivation from *burkhan* by means of the possessive suffix *-tu,* meaning as much as 'having a deity' or 'deified.' Potanin[18] mentions a pass under the name Burkhan-boksin-daban, and argues that this name presumably designated a pre-Buddhistic Mongol deity, while at present it is referred to Buddha.

In the same manner as among the Gold, so also among the Mongol, *burkhan* is a fixed term of their ancient shamanistic religion which still flourishes among the Buryat. Generally speaking, *burkhan* is a synonym of *tengeri* (or *tengerin*) or *zayan,* the chief deities of the Buryat, to the number of ninety-nine, each known under his proper name.[19] A special group

[16] Dordji Banzarov, *The Black Faith* (in Russian), p. 21; I. J. Schmidt, *Sanang Setsen,* p. 57, 59; H. Yule, *Marco Polo,* 1. 247; G. N. Potanin, *Tanguto-Tibetan Borderland* (in Russian), 2. 303.

[17] Banzarov, *op. cit.* p. 25.

[18] *Op. cit.* 2. 337.

[19] M. N. Khangalov, 'New Materials Relating to the Shamanism among the Buryat,' p. 1, in *Zapiski East-Sib. Section of the Russian Geogr. Soc.* vol. 2, no. 1, Irkutsk, 1890. *Burkun* and *burkhan* are dialectic variations of the word in Buryat (M. A. Castrén, *Burjätische Sprachlehre,* p. 171); *burkhyn* also occurs.

among these is formed by the *satini burkhat* (*burkha-t* being the plural of *burkhan*), who belong to the western, white gods, especially worshiped by the Buryat of Kudinsk.[20] Likewise in the tales and traditions of the Buryat the term *burkhan* is referred to their own gods.[21] In several Mongol dialects the Dipper is styled *Dolon burkhyn,* in Kalmuk *Dolon burkhut.*[22] Among the Turkish tribes of the Altai, as previously stated, *purkan kān* denotes a spirit worshiped by the shamans; and among the Turkmen of Khiwa the word (in the form *porkhan*) designates the shaman himself.

Burkhan in Mongol by no means conveys exclusively the limited notion of Buddha, but, first of all, signifies 'deity, god, gods,' and secondly 'representation or image of a god.' This general significance neither inheres in the term Buddha nor in Chinese Fo; neither do the latter signify 'image of Buddha'; only Mongol *burkhan* has this force, because originally it conveyed the meaning of a shamanistic image. From what has been observed on the use of the word *burkhan* in the shamanistic or pre-Buddhistic religions of the Tungusians, Mongols, and Turks, it is manifest that the word well existed there before the arrival of Buddhism, fixed in its form and meaning, and was but subsequently transferred to the name of Buddha. This being the case, it cannot represent a transcription, and the theories of P. Schmidt and Baron von Staël-Holstein should be discarded. A single concession may be made, and this is that the indigenous word *burkhan* for the designation of Buddha may have been chosen as a more or less conscious adaptation in sound to the latter.

---

[20] Khangalov, *op. cit.* p. 30.

[21] See for instance the collection of Tales of the Buryat edited by D. G. Gomboyev, p. 24, 63, 69 (*l. c.,* vol. 1, no. 2, 1890); A. D. Rudnev, *The Khori-Buryat Dialect* (in Russian), pt. 3, p. 039.

[22] Potanin, *Tanguto-Tibetan Borderland,* 2. 318, 319.

JOURNAL

OF THE

AMERICAN ORIENTAL SOCIETY

EDITED BY

JAMES A. MONTGOMERY
Professor in the
University of Pennsylvania

GEORGE C. O. HAAS
Sometime Instructor in the
College of the City of New York

VOLUME 37

PUBLISHED FOR THE AMERICAN ORIENTAL SOCIETY
YALE UNIVERSITY PRESS
NEW HAVEN, CONNECTICUT, U. S. A.
1917

in the much discussed כלה מקללוני, Jer. 15. 10. In a letter dated December 9, 1836, he writes:

"כלה מקללוני, נראה לי לקרוא מקללוני, מקללים אותי, כמו בתלמוד ירושלמי (דמאי פ"ז) דאינן מחשדונך "שהוא כמו מחשדין יתך".[6]

In the Aramaic parts of the Talmud *û* as plural ending of the participle occurs very often, cf. Margolis, *Lehrb. d. Aram. Spr. d. Talmuds,* p. 40 ff. Margolis, it is true, considers it as a later form developed by analogy of the perf., but may we not assume that it represents the old plural ending *û?*

M. SEIDEL

Baltimore, Md.

Burkhan

With reference to my note on the word *Burkhan* (in the JOURNAL, 36. 390—395) I now note that R. Gauthiot (*Mélanges Sylvain Lévi,* Paris, 1911, p. 112) had already opposed the theory of Baron A. von Staël-Holstein of Petrograd. Gauthiot regarded that etymology as 'very doubtful,' and remarked (in the same manner as I did) that compounds of this kind do not exist in Turkish. Moreover, he justly emphasized that the historical facts run counter to such a conception of the term, and that the history of the expansion of Buddhism in the Iranian regions toward the northwest of India and the fluctuations of Chinese influence in Central Asia render that theory rather improbable. While regretting that I overlooked Gauthiot's comment, I am glad to find myself in full accord with the opinion of that eminent philologist, whose premature death we have every reason to deplore.

In regard to the Manchu term *Fučihi,* Professor P. Schmidt, now president of the Oriental Institute of Vladivostok, has been good enough to write me that he regards *-i-hi* as a suffix added to

---

[6] אגרות שד"ל, p. 361. So also Steinberg in his Hebrew Grammar מערכי לשון עבר, p. 142, but in his lexicon משפטי האורים he suggests to read כִּלַּהֵם קִלְלוּנִי (Baer reads כֻּלֹּהֶם) but we would rather expect the imperfect instead of the perfect, cf. also Kittel *ad loc.*—The use of the participle with the plural ending *û* when combined with a suffix occurs also in the פיוט for Shebu'oth: הַמַּלְבִּישׁוּךְ=המלבישים אותך לבוש עדנים.

the stem *Fut,* pointing to such analogous formations as *guč-i-hi, ginč-i-hi, sol-o-hi, tarb-a-hi, tarb-i-hi.* This explanation is quite satisfactory.

B. Laufer

Field Museum of Natural History, Chicago.

PERSONALIA

Professor John Williams White, of Harvard University, died May 9, 1917. He was professor of Greek at Harvard from 1884 to 1909, and was one of the founders of the American School at Athens. He became a member of the American Oriental Society in 1877.

Morton William Easton, Ph.D., Emeritus Professor of English and Comparative Philology at the University of Pennsylvania, died Aug. 21, 1917. He was born in 1841 in Hartford, Conn., and completed the course in medicine at Columbia in 1865, but returned to philology, taking his degree in Sanskrit at Yale.

He was called to the classical chair at the University of Tennessee in 1873, and came to the University of Pennsylvania in 1880. His subjects ranged from Sanskrit to English, in all of which he was a profound student and a most distinguished teacher. He directed the presentation of the first Greek comedy to be given in this country, the *Acharnians,* presented in 1886 by students of the University of Pennsylvania.

135

关于按指印的历史

SCIENCE

NEW SERIES
VOL. XLV. No. 1169 | FRIDAY, MAY 25, 1917 | SINGLE COPIES, 15 CTS.
ANNUAL SUBSCRIPTION, $5.00

simplification of its constituent elements. The main lesson that he learned by his journey is that France can no longer remain stationary in these matters and that it should make efforts to organize biology as applied to agriculture upon a large and solid basis, and he proceeds with practical suggestions in this direction. He praises the Federal Horticultural Board, the Federal Insecticide Board, and the Horticultural Commission of California, and thinks that all of these should be imitated in France. He especially points out the necessity for the introduction into France of such education as our young men get in applied biology in the agricultural colleges and universities like Cornell and Illinois. There is, he points out, in France at the present time no way of getting a scientific education in biological studies as applied to agriculture.

After pointing out some of the great examples of monetary saving in this country as the result of work in applied biology, he closes with the sentence, "These are great examples which it is well to recall, for they establish with the most complete evidence the fact that there is no other sure way than that of scientific organization of work to get full value from the national soil and to give back to agriculture the greatest possible part of the riches which are lost to it annually from pests."

L. O. Howard

CONCERNING THE HISTORY OF FINGER-PRINTS

Sir William J. Herschel published recently a brief pamphlet of 41 pages under the title "The Origin of Finger-Printing" (Oxford University Press, 1916). This is mainly an autobiographical sketch, giving in detail the story of how the author during the time of his useful service in India (1853–78) conceived the notion of finger-prints and elaborated this system, which was subsequently developed and placed on a truly scientific basis by Sir Francis Galton. We are indebted to Sir W. Herschel for his interesting document: it is always valuable when one who has played a prominent rôle in inaugurating a new movement presents us with a record of what he believes was his share in bringing about this innovation or invention. The inventor, however, will seldom be able to write impartially the history of his own invention; no one, in fact, whether statesman, artist, poet or scholar, while recording his own history, has the faculty (I should even say, the right) of clearly determining his own place in the long chain of historical development. This judgment must be left to the historian of the future. The principal purpose by which Sir W. Herschel was guided in writing his account is to demonstrate that he was the real "discoverer" of finger-prints in Bengal in 1858, entirely from his own resources, and to discredit all other claims to priority in this matter, especially those on the part of the Chinese. I regret that the author has failed to take notice of the "History of the Finger-Print System" published by me in the Smithsonian Report for 1912 (pp. 631–652, Washington, 1913). Not only are Sir W. Herschel's great merits and his share in the history of the invention, if invention it may be called, duly acknowledged and objectively expounded there, but he would also have found there all the available evidence in favor of the Chinese, Japanese and Tibetans, all of whom applied ages ago with full consciousness the system of finger-prints for the purpose of identifying individuals. The few modern traces of evidence known to Sir W. Herschel are treated by him slightly, and he wonders that "a system so practically useful as this could have been known in the great lands of the East for generations past, without arresting the notice of western statesmen, merchants, travelers and students." The Mohammedan authors who visited China did not fail to describe this system. Rashid-eddin, the famous Persian historian, who wrote in 1303, reports as follows:

> When matters have passed the six boards of the Chinese, they are remitted to the Council of State, where they are discussed, and the decision is issued after being verified by the *khat angusht* or "finger-signature" of all who have a right to a voice in

the council. This "finger-signature" indicates that the act, to which it is attached in attestation, has been discussed and definitively approved by those whose mark has thus been put upon it. It is usual in Cathay [China], when any contract is entered into, for the outline of the fingers of the parties to be traced upon the document. For experience shows that no two individuals have fingers precisely alike. The hand of the contracting party is set upon the back of the paper containing the deed, and lines are then traced round his fingers up to the knuckles, in order that if ever one of them should deny his obligation this tracing may be compared with his fingers and he may thus be convicted.[1]

Professor Henri Cordier of Paris, the editor of Yule's famous work, adds to this passage a footnote relative to the history of finger-prints, and commenting on the claim of Sir W. Herschel, tersely remarks:

Sir W. Herschel was entirely wrong; Mr. Faulds protested against the claim of Sir W. Herschel, and finally a Japanese gentlemen, Kumagusu Minakata, proved the case for the Japanese and the Chinese. None of these writers quoted the passage of Rashid-eddin which is a peremptory proof of the antiquity of the use of finger-prints by the Chinese.

Indeed it is, and the observation that no two individuals have finger-marks precisely alike is thoroughly Galtonian. There is the earlier testimony of the Arabic merchant Soleiman, who wrote in A.D. 851, and who states that in China creditor's bills were marked by the debtor with his middle finger and index united (see my History, p. 643). But we have more. E. Chavannes, in reviewing my article in the *T'oung Pao* (1913, p. 490), has pointed out three contracts of the T'ang period, dated A.D. 782 and 786 and discovered in Turkestan (two by Sir Aurel Stein), which were provided with the finger-marks of both parties, and contain at the end the typical formula:

The two parties have found this just and clear, and have affixed the impressions of their fingers as a distinctive mark.[2]

A clay seal for which no later date than the third century B.C. can be assumed, and which bears on its reverse a very deeply and clearly cut impression of the owner's thumb-mark, has been brought back by me from China, and is illustrated and described in the above paper. I have also shown how the system was developed in ancient China from magical beliefs in the power of bodily parts, the individual, as it were, sacrificing his finger in good faith of his promises; in its origin, the finger-print had a magical and ritualistic character.

Sir W. Herschel states that he fails to see the definite force of the word "identification" in the Chinese finger-print system. In his opinion, there must be two impressions at least, that will bear comparison, to constitute "identification." He thinks, of course, one-sidedly of the detection of criminals to which the process has been applied by us, but never in the East (for what reason, I have stated elsewhere). Most certainly, the idea underlying Chinese finger-prints was principally that of identification, as expressly stated by Rashid-eddin and all Chinese informants. If a doubt or litigation arose, all that was necessary was to repeat the finger impression of the contractor who had formerly signed the deed.

B. LAUFER

FIELD MUSEUM,
CHICAGO, ILL.

[1] See H. Yule, "Cathay," new ed., Vol. III., p. 123, London, 1914, Hakluyt Society.

[2] See A. Stein, "Ancient Khotan," Vol. I., pp. 525–529, Oxford, 1907, where the three documents are published and translated by Chavannes.

SPECIAL ARTICLES

ON THE COLLOID CHEMISTRY OF FEHLING'S TEST

I

As familiarly known, when Fehling's solution is treated with a reducing substance, it is generally expected that a bright red precipitate will be obtained. Frequently, however, an orange or yellow precipitate is obtained and in certain instances nothing but a yellowish-green or bluish-green discoloration results. The attempts to account for these differences are, for the most part, chemical in nature; it is held that the red reaction represents a precipitate of cuprous oxide, the orange or yellow ones more doubtful suboxides or hydrated forms of the oxide, while the character of the greenish discolorations is left doubtful. It is often be-

136

鹿皮靴

AMERICAN ANTHROPOLOGIST

NEW SERIES

ORGAN OF THE AMERICAN ANTHROPOLOGICAL ASSOCIATION,
THE ANTHROPOLOGICAL SOCIETY OF WASHINGTON,
AND THE AMERICAN ETHNOLOGICAL
SOCIETY OF NEW YORK

VOLUME 19

LANCASTER, PA., U. S. A.
PUBLISHED FOR
THE AMERICAN ANTHROPOLOGICAL ASSOCIATION

1917

tell Verendrye of the Panani and Panana above them. Considering that they were the Panana themselves the surprise at this would be rather diminished. If Verendrye traveled from the country of the upper Platte, as Dr. Libby contends, he would have traveled rather east and northeast than north-northeast and northwest. The above course, however, would just fit the route from the Arikara villages at the mouth of Bad river to the mouth of Heart river.

It is contended also that the party could not have followed the course of the Missouri because they did not meet the Panana whom we have shown that they had already left behind. Dr. Libby overlooks the fact, however, that they did meet one tribe, and this meeting is very valuable evidence. The account tells rather circumstantially of meeting with the Cheyenne, to whom, fortunately, they give the French name by which they were generally known. The Cheyenne were camped temporarily, and pointed out the spot where they intended to build their permanent village. We have corroborative historical and traditional evidence showing that this was just the period when the Cheyenne crossed the Missouri and built their last earth lodge village between the villages of the Arikara and the Mandan.

There is no question of the value of traditional evidence secured from the Indians, and much of such evidence has been used in the preceding pages. It becomes infinitely more valuable, however, when coupled with the careful reading of all written accounts.

GEORGE F. WILL

BISMARCK, N. D.

MOCCASINS

DR. HATT became favorably known to many of us when during 1915 he visited the museums of this country for the study of North American and North Asiatic costume, for which task he was well prepared by his book *Arktiske Skinddragter* and an article "Mokkasiner" published in the *Geografisk Tidskrift*. We are grateful to the author for his courtesy in giving us the benefit of his present researches in the English language.[1] Hatt's method is excellent, and his enthusiasm for his subject is supported by a wealth of solid information, a thoroughness and keenness of analysis that is deserving of high praise. His study is divided into two parts,—a descriptive one in which all available types of moccasins and related foot-gear are described, illustrated, and classified, and a theoretical discussion of the origin of the moccasin group. As the

[1] "Moccasins and Their Relation to Arctic Footwear," *Memoirs of the American Anthropological Association*, vol. III, no. 3 (1916), pp. 149–250.

material and the theory based upon it cover both North America and northern Asia, it becomes incumbent both upon Americanists and students of Asiatic ethnography to define their attitude toward Hatt's results. In order to preserve full impartiality, I shall not await the Americanistic verdict, but will briefly set forth my opinion solely from the Asiatic point of view. It may be taken for granted that Hatt's treatise is in the hands of all readers of the *Anthropologist*, and that a detailed analysis of its contents is therefore superfluous.[1] Dr. Hatt says;

> The material at hand enables us to say that the moccasin group has *not* furnished the prototypes for all the forms of footwear, which we find in Central Asia and Asia Minor, it is even quite probable that *none* of them were derived from a moccasin type (p. 236).

While Dr. Hatt is very keen at recognizing moccasin types, it is striking that he fails to grasp the origin and culture-historical position of other types of footgear. Not only has that of Central Asia positively nothing to do with the moccasin, but also it has had a quite different development. What this history was we recognize clearly from ancient Chinese records. King Wu-ling (325–299 B.C.) of Chao introduced into China the tactics of mounted infantry, as it was then practised by the nomadic tribes (Turkish and Tungusian) of inner Asia. This innovation was a military necessity, as the infantry of the Chinese was powerless against the inroads of those swift horsemen. This measure, further, necessitated on the part of the Chinese the adoption of the tight-fitting riding-costume of their enemies, together with narrow, long-legged riding-boots, while their own national garment had always been spacious, loose, and flowing, and only sandals and shoes (but no boots) had previously been known to them. This case is instructive in many respects: it affords a good example of how the attire of one group of peoples may suddenly become the property of another, and it shows that even a higher civilized nation may take lessons in clothing itself from peoples regarded as barbarians. Above all we see that at this early date the long-legged boot existed among the roving horsemen of central Asia, and that without any doubt the equestrian practice had called it into being. The wide diffusion of mounted infantry and subsequently

[1] It is regrettable that the author has not added a map illustrating the geographical distribution of the various types of moccasin, and that he has not tabulated in convenient form his twenty-one American series which it is not quite easy to remember in detail. Another omission is that the publications of the Jesup Expedition, L. von Schrenck's classical work on the Amur peoples, and Russian literature on the ethnography of Siberia, have not been utilized.

cavalry, together with the constant migrations, shiftings, expeditions, and conquests of these ever restless nomads, ultimately led to the distribution of the riding-boot all over Asia. For this reason I have always been convinced that the long boots of the present Chinese, as worn by officials and soldiers, whether made of leather or in connection with ceremonial costume of satin or velvet, and further the boots of the Tibetans, Mongols, and Amur tribes, are traceable to that one primeval type. I am therefore somewhat suspicious of what Hatt styles in chapter III "transformed moccasins." If, as he states (p. 229), "the moccasin-boot may undergo further transformation and thereby lose all of its moccasin character," the question may justly be raised whether this theory is necessary, and whether it would not be more plausible to dissociate this type from the moccasin group altogether and simply to regard it as a descendant of the riding-boot. The association of the latter with the horse which I am inclined to assume meets its parallel in the correlation of the moccasin with the snowshoe disclosed by Hatt, and this impresses me as the most important point brought out by his investigation. Unfortunately we still know too little about the history of the snowshoe in order to reach a positive result (some historical data on snowshoes will be found in my article on the Reindeer which is soon to appear). The rationalistic evolution traced by Hatt (p. 243) does not satisfy me.

In the general theory summed up by the author at the conclusion of his study we miss again to some degree historical insight. He attempts at reconstructing historical movements from the data of the geographical distribution of a single ethnographical phenomenon, and while this effort may be justifiable for North America, it is apt to invite disaster in Asia where history exists and must be consulted. The author ought to have pursued the problem as to how his deductions stand in relation with the ethnic conditions of northern Asia in ancient times. At all events it is manifest that we cannot (and need not) reconstruct from present-day observations a history for those Asiatic peoples whose history is perfectly well known. Hatt's ultimate theory culminates in the assumption of two large cultural waves which "in prehistoric times" a (rather vague determination) swept over the northern regions.

The oldest of these, now most fully represented and highest developed in the culture of the Eskimo tribes, did not have snowshoes and therefore could not conquer the vast inland areas; it must have followed and taken into possession the rivers and coasts, and we would call this first great wave the coast culture. The younger culture wave is found fullest and most unmixed in the culture of the Tungusians, although its influence is felt from Lapland to Labrador; it still has the character of an inland culture and must have originated as such. Its most valuable possession is the snowshoe, which has carried it over the greater part of the Arctic.

At the outset this reconstruction appears tempting, and as the author claims, may prove of some value as a working hypothesis. Yet it seems to me to be subject to some caution. Hatt does not define what he understands by Tungusians. The term "Tungusian," however, comprises two different things. There is a group of scattered clans, widely distributed over northeastern Siberia from the upper Yenisei as far as the Okhotsk Sea and claiming for themselves the name Tongús. These are Tungusians proper. This designation, further, has been extended by our science to a large number of compact tribes massed in the Amur region and Manchuria. These are Tungusians in the wider sense, and their grouping with the Tongús of northeastern Siberia, first of all, is a scientific theory. The history of the entire stock is well known and can be traced in detail from the Chinese annals for several millenniums.[1] Certain it is that the original home of the family was located in Manchuria, and that Tungusians are newcomers and late arrivals in Siberia. From oldest times they were tribes of horsemen, and did not possess any such characteristics that we find at present among the Tungusians of Siberia. These adopted the reindeer culture with all its appurtenances from those Siberian peoples whose territory they occupied. The culture of Siberian Tungusians is very far from exhibiting any originality, and does not occupy the place which Hatt is inclined to assign to it. At any rate I am under the impression that in Hatt's "prehistoric times" there were no Tungusians in Siberia, and that his so-called Tungusian culture did not then exist. There is no reason whatever to argue that the ethnical conditions of northern Asia were the same anciently as at present, while Hatt evidently takes this as a foregone conclusion. Not only the Tungusians, but also the Yakut and Samoyed are but new and recent guests of Siberia, and some of the so-called Palaeo-Asiatic peoples formerly extended much farther to the south. It is hoped that Dr. Hatt will see fit to revise his theories somewhat in the light of historical facts; also if he will take up the history of the Lapp in connection with that of the Samoyed, northern and southern, he will doubtless recognize the interrelations of the Lapp with Siberian groups and be able to place on a more solid basis his assumption that "the Lapp moccasin and the Amur moccasin have sprung from the same source" (p. 238). Whether the

[1] The reader may be referred for a general account to E. H. Parker's *A Thousand Years of the Tartars* (Shanghai, 1895).

oldest moccasin forms originated "somewhere" in northern Asia and spread thence to America is a theory which must be left for the discussion of the Americanists. I must confess, however, that it does not convince me.[1]

B. LAUFER

FIELD MUSEUM OF NATURAL HISTORY,
CHICAGO, ILL.

COMMENTS ON "THE PLACE OF COILED WARE IN SOUTHWESTERN POTTERY"

IN his article on "The Place of Coiled Ware in Southwestern Pottery"[2] Mr. Morris suggests certain points in the technique which are deserving of somewhat more extended notice. The question of how the vessel was held and how manipulated, arises, because if the pottery as a feature is worth study, surely the way in which it was made may have importance, and is of interest. The "spiral coil which begins at the bottom and proceeds continuously to the rim" is so often laid in one direction as to make the other seem exceptional. From the examination of many jars of various shapes, and from many fragments, it seems probable that the clay ribbon was somewhat flattened, that after the start of the coil on the center of the bottom of the jar, the jar was held in the left hand, fingers inside, thumb outside, the manipulation of the coil done with the right hand, the pressure with the right hand thumb making the pattern as the pressure was applied to make the coil adhere. The jar must have been held with the outside toward the worker, the coil being applied from left to right on the near side of the jar, with the pattern always in view of the worker. This direction of the spiral may

[1] The spelling of tribal and geographical names in Hatt's paper calls for some criticism. Aino is wrong for Ainu—an error already rectified a generation ago. Instead of Golds (p. 211) read Gold or Golde; instead of Giljaks (p. 212 and *passim*) read Gilyak; instead of Wiljuisk (p. 231) read Wilui or Wil'ui. The plural form of tribal names is arbitrarily formed with or without *s*: on p. 232, for instance, we have "the Ostyak, Vogul, and Samoyed," but eleven lines below "the Lapps." He who spells Sakhalin, as has been done here, must be consistent in writing Okhotsk (not Ochotsk, p. 232) and Turukhansk (not Turuchansk, p. 231). Orotchon (p. 229) is redundant; phonetically the question is merely of a palatal explosive surd (Oročon), so that in English spelling Orochon is sufficient, or at best Orotshon would be admissible. Finally, the spelling Chukchee for Chukchi is a barbarism; then, why not spell Asqueemoe?—Is it unknown to the editors of our Memoirs that the name Field Columbian Museum (pp. 149 and 185), for which the abbreviation F. C. M. is especially invented, has been abolished since November 1905?

[2] *American Anthropologist* (N. S.), vol. 19, no. 1 (Jan.-Mar., 1917), p. 24.

137

“萨满”一词的来源

ORIGIN OF THE WORD SHAMAN

By BERTHOLD LAUFER

IN *Chinese Clay Figures* (p. 198) the writer had occasion to discuss briefly the history of our word shaman, and to refer at the close of his naturally succinct note to the new theory of J. Németh, according to which the term should be an ancient constituent of the Turkish-Tungusian languages. The latter reference was an addition inserted in the proofs at the moment when Mr. Németh's article reached me (summer 1914), but since then it has been possible to scrutinize his theory at close range, with the result that it can be confirmed and even be supported with new data and arguments. As ethnologists may be expected to take an interest in the history of a term which has become part and parcel of anthropological nomenclature, and the origin of which has given rise to numerous discussions and speculations, a brief examination of the case might not be unwelcome to the readers of the *Anthropologist.*

There could never be a serious doubt of the real source from which the word shaman is derived. We received it from the Russians: it was the Russian (chiefly Cossack) explorers and conquerors of eastern Siberia in the second half of the seventeenth century, who heard and recorded the term among Tungusian tribes. It was first brought to Europe by the Hollander, E. Ysbrants Ides, and by Adam Brand, who from 1692 to 1695 accompanied a Russian embassy sent by Peter the Great to China. Some examples from the former's *Driejaarige Reize naar China*, first published at Amsterdam, 1698 (again 1704 and 1710), may be cited:

Eenige dagen reizens van hier, is de groote steenachtige waterval, Schammanskoi, of Toverval geheten, om dat aldaar een beroemde Schamam, of Tungusche duiveldienaar woont (p. 34).

Eenige mylen van hier opwaarts woonen veele Tunguzers, waar onder ook hun beroemde Schaman, of Duivelskonstenaar (p. 35; follows a lengthy description of the shaman and his costume).

Zy [the Tungus] weeten van geen andere Priesteren, dan hunne Schamannen, of Duivelbanners (p. 39).

It is thus shown that our "shaman" had its origin in the Tungusian languages, in all of which the word indeed is known as *šaman* or *saman*, and with several other variations to be noted hereafter. It would be reasonable to regard the term as of native origin, for it is not likely that the name of a religious institution and function so characteristic of all the tribes of northern Asia should be borrowed from an outside quarter. Nevertheless it was possible that at a time when little was known about Siberian shamanism, and when rigid philological principles were not yet established, the theory could be advanced that Tungusian *šaman* or *saman* were to be derived from Chinese *ša-men*, itself a transcription of Pāli *samaṇa* corresponding to Sanskrit *çramaṇa*, a technical term for the designation of a Buddhist monk or ascetic. The germ of this idea was well prepared as early as the eighteenth century. The Augustinian Pater A. Georgi[1] states correctly that the Samanaei mentioned by Clemens Alexandrinus[2] are the adherents of Buddha, and in this connection refers to La Croze, who already hazarded the combination of the Samanaei with the shamans of the Tungusians ("le nom de schaman est un nom de religion d'un peuple tartare, voisin des frontières de la Chine"). Sanskrit not yet being well known at the time of La Croze and Georgi, the origin of the Greek term Samanaioi escaped them; this is what the philologists of the nineteenth century supplied.[3]

Which of them was the first to assert the connection of the Siberian *šaman* with the Indian term I am unable to ferret out; this point is immaterial, as the question is merely that of the history of an error. D. Banzarov[4] was inclined, without giving a specific reference, to fix the responsibility on F. Schlegel, according to whom

[1] *Alphabetum Tibetanum* (Rome, 1762), p. 222.

[2] In fact Clemens avails himself of the form Sarmanai.

[3] Maturin Veyssiere La Croze was born at Nantes in 1661, joined the clergy, suffered persecution on account of his liberal views, fled to Germany, where he turned a Protestant in 1696, and was made a member of the Academy of Sciences of Berlin, where he died in 1739. His work *Histoire du Christianisme des Indes*, to which Georgi refers, unfortunately is not accessible to me. La Croze was a scholar of wide knowledge and linguistic ability: he was the first who had a presentiment of the relationship of Indian with Persian.

[4] *Černaya v'ära*, p. 34 (see p. 366).

the *çramana* among the crude tribes of Central Asia transformed themselves from hermits into shamans. I believe that Banzarov had in mind Friedrich von Schlegel (1772–1829), romantic poet and Sanskrit scholar, who published in Heidelberg, 1808, an essay under the title *Über die Sprache und Weisheit der Indier, ein Beitrag zur Begründung der Alterthumskunde*.[1] This is the only one of his works in which such a statement might be supposed to occur; but on a cursory perusal of this book I am unable to trace it. The idea itself being romantic, there would be no reason to wonder that it took its root in the Romantic School. At all events it was alive in the first part of the nineteenth century. The French sinologue J.-P. Abel-Rémusat (1788–1832) became a champion of this theory.[2] An ingenious scholar of the type of J. Klaproth[3] was capable of saying, "Les peuples tartares ont sans doute reçu le mot *chaman* de l'Inde avec le bouddhisme, car il est indien d'origine, et signifie un homme qui a vaincu toutes ses passions." In 1857 Max Müller[4] of Oxford wrote,

> Shamanism found its way from India to Siberia *viâ* Tibet, China, and Mongolia. Rules on the formation of magic figures, on the treatment of diseases by charms, on the worship of evil spirits, on the acquisition of the supernatural powers, on charms, incantations, and other branches of Shaman witchcraft, are found in the Tanjur, or the second part of the Tibetan canon, and in some of the late Tantras of the Nepalese collection.

Needless to add that the magical writings of the Tanjur are unknown to the shamans of Siberia and have nothing to do with Siberian shamanism. Still less is it intelligible how Müller could say that "the only trace of the influence of Buddhism among the Chudic races, the Finn, Lapp, etc., is found in the name of their priests and sorcerers, the shamans." The word shaman is foreign to any Finno-Ugrian language; it found its way into Finnish (*shamani*) and Hungarian as a scientific term only, in the same

[1] An English translation was edited in London, 1889, in *F. v. Schlegel's Aesthetic and Miscellaneous Works*.

[2] *Recherches sur les langues tartares* (Paris, 1820), p. 133, and *Observations sur la doctrine samanéenne et la triade suprême* (Paris, 1831).

[3] *Mémoires relatifs à l'Asie* (Paris, 1828), vol. III, p. 67.

[4] *Chips from a German Workshop*, vol. I, pp. 233–234.

manner as into the Indo-European languages. A. H. Sayce[1] has the following:

> In shamanism, so called from the Shaman or Siberian sorcerer, who is himself but a transformed *çramaṇa*, or Buddhist missionary priest, we rise to a higher conception of religion.

These superstitions were tenaciously upheld under the influence of the romantic movement of pan-Indianism, which held the minds of scholars enthralled in the first part of the nineteenth century, and the germs of which are not yet entirely exterminated.

True it is that the Sanskrit word *çramaṇa* with the Prākrit form *samaṇa* has become widely known outside of the pale of India. It reached the ears of the Greek writers on India, who described the Buddhist monastics under such names as Sarmanes, Sarmanai, or Samanaioi. The first of these names was recorded by Megasthenes,[2] though the single text of Strabo from which all the existing codices have been copied offers the erroneous transcription Garmanes.[3] The word has also passed into early Arabic records for the designation of the Indian and Bactrian Buddhists (for instance, in Masūdi, Gurdezi, and Abu'l Faradj) in the form *samaniyya*, while in later literature the form *šaman* was adopted.[4] In Firdausī's Shāhnāme, completed in A.D. 1010, we find *šaman* in the sense of a worshipper of idols, and this Persian word likewise is traced to Sanskrit *çramaṇa* and its congeners.[5] Again, it bears no relation to "our" or the

[1] *Introduction to the Science of Language*, vol. II, p. 293.

[2] Strabo XV, 1, 59.

[3] The texts in question are easily accessible in J. W. McCrindle, *Ancient India as described in Classical Literature*, pp. 65, 168, 170. See further C. Lassen, "De nominibus quibus a veteribus appellantur Indorum philosophi," in *Rhein. Museum für Philol.*, vol. I, 1833, pp. 178–180, and *Indische Alterthumskunde*, vol. II, p. 700; E. Burnouf, *Introduction à l'histoire du Buddhisme indien*, p. 245; T. Watters, *Essays on the Chinese Language*, p. 403; S. Lévi, *Journal asiatique* (1911), I, p. 445.

[4] Geza Kuun, *Keleti Szemle*, vol. IV, 1903, p. 133; G. Ferrand, *Textes géographiques arabes relatifs à l'Extrême-Orient*, vol. I, p. 131. The Mohammedan writers were perfectly conscious of the fact that the term hailed from India and denoted the Buddhists of India and China. See particularly D. Chwolsohn, *Die Ssabier und der Ssabismus* (St. Petersburg, 1856), vol. I, pp. 165, 214, 217, etc. As far as I know, the shamanistic religion of Siberian tribes is not mentioned by any early Arabic or Persian author.

[5] P. Horn in *Grundriss der iranischen Philologie*, vol. I, pt. 2, p. 7; K. Vullers (*Lexicon persico-latinum*, vol. I, p. 466) renders the term by *idololatra* and *idolum*.

Siberian shaman, but evidently refers to Indian Buddhists, for these šamans say, "Our religion is one of peace and quiet, and fighting and slaying is prohibited, as well as all kinds of shedding of blood."[1] Finally the Indian word has been traced in the newly discovered Indo-European language styled Tokharian in the form *samāne;* and this induced A. Meillet to formulate the opinion that the Tokharian form (not a Prākrit form of Sanskrit *çramaṇa*) would account for the Tungus word *šaman.*[2] With all respect due to the scholarship of this eminent philologist it must be admitted, however, that there is a far cry from Tokharian to Tungusian, that there is no evidence for any contact of these two groups, and that the Tungus notion of a shaman is radically different from the conception of a Buddhist monk. If two words in geographically remote languages are physically alike or similar, but fundamentally diverse in meaning, it is safe to assume that the face resemblance is purely accidental.

Indeed this theory was brilliantly antagonized as early as 1842 by W. Schott; and H. Yule,[3] accepting his verdict, reached the conclusion, "Whether the Tungus word is in any way connected with this or adopted from it [that is, the Sanskrit-Pāli term], is a doubtful question." Schott has devoted no less than three studies to this problem,[4] in which he endeavored to dissociate *šaman* from *samaṇa*. His arguments are briefly as follows: the Chinese strictly discriminate between Manchu *saman* (shaman) and Sanskrit *çramaṇa*, and never confound a shaman and a Buddhist monk, the two having totally diverse functions; a direct or indirect connection of Tungusian tribes with India is unproved; Buddhist missionaries never advanced into regions inhabited by Tungus;[5]

[1] Compare the quotation in H. Yule, *Hobson-Jobson*, p. 820.

[2] A. Meillet, "Le Tokharien," *Indogermanisches Jahrbuch* (Strassburg, 1914), vol. I, p. 19.

[3] *Hobson-Jobson*, p. 820.

[4] "Über den Doppelsinn des Wortes Schamane und über den tungusischen Schamanen-Cultus am Hofe der Mandju-Kaiser," *Abhandlungen Berliner Akademie*, 1842, pp. 1–8; "Wohin gehört das Wort Schamane?" in his Altaische Studien, no. 3, *ibid.*, 1867, pp. 138–141; and "Das Wort Schamane," in *Erman's Archiv für wissenschaftliche Kunde von Russland*, vol. XXIII, 1865, pp. 207–210.

[5] As emphasized in my article "Burkhan," *Journal American Oriental Society* (1917), pp. 390–395, there is no trace of Buddhism in the religion of the Tungusian peoples on the Amur.

it would be strange if they should have borrowed the only word in their language for the designation of their national priests; the forms of shamanism are uniform in a vast territory extending from Lappland to the Bering Sea and far beyond into America; Vasilyev's hypothesis that shamanism should be a caricature of ancient Buddhist sorcery and jugglery and have been brought into existence by such influences is a mere phantom; the outward resemblance of the words *saman* and *samaṇa* is purely fortuitous. Schott's etymological explanations of the word from Samoyed are not convincing, but his remark that Tungusian *sam* possibly is anciently related to Turkish *kam* testifies to great foresight and ingenuity; and this thesis, as will be seen, can now really be proved. Also Ch. de Harlez[1] joined the ranks of the adversaries of the Indian hypothesis, reiterating in the main Schott's arguments. In Russia, it had been severely criticized as far back as 1846 by Dordji Banzarov, a Buryat and an excellent Mongol scholar.[2] Nevertheless in Russian dictionaries also the nightmare of the *çramaṇa* is perpetuated.[3] As it happens, twilight reveries appeal more to the multitude than plain reason and continue to live even if abandoned by the thinking minority; it is not always the best and fittest that survives. The ghost of the Indian *çramaṇa* still haunts the poor shaman in our standard dictionaries and cyclopaedias (see Webster, Nouveau Larousse, and Littré, *Dictionnaire de la langue française;* Schaff-Herzog *Religious Encyclopaedia*, vol. x, p. 385; Funk and Wagnalls' Dictionary offers even a "Persian-Hindustani shaman, pagan"!).[4] It is a relief, therefore, to read in the new Oxford Dictionary,

> The Persian *shemen*, idol, idol temple, sometimes cited as the source, is unconnected. Evidence seems to be wanting for the plausible suggestion that the Mongolian word is an adoption of Chinese *sha men*, an ordained member of a Buddhist fraternity.

Eliminating "plausible" and substituting "Tungusian" for

[1] *La Religion nationale des Tartares orientaux* (Bruxelles, 1887), p. 28.

[2] His works were collected and edited by Potanin in 1891 under the title *Černaya v'ära ili šamanstvo u Mongolov i drugiya stat'i* ("The Black Faith or Shamanism among the Mongols and Other Articles"). Regarding the word shaman see p. 34.

[3] For instance, in Gor'aev, *Sravnitelny etimologičeski slovar' russkago yazyka*, p. 418.

[4] Even in the 7th edition of O. Peschel's *Völkerkunde* (p. 274) the *cramana* still holds sway.

"Mongolian," this statement marks a progress in lexicographical literature. So there is hope that our great-grandchildren will perhaps be treated to a correct definition in the dictionaries of the following century.

The Hungarian scholar J. Németh[1] has attacked the problem in the right spirit, and ably solved it by means of a sane philological method. A schooled phonetician, he has successfully worked for several years on the phonetic history of the Turkish and Mongol languages and established very interesting correspondences of phonetic phenomena in these two linguistic families. In the study cited he advances the law that primeval Turkish initial *k'* (palatalized *k*) has developed into Tartar *k*, Čuvaš *j*, Yakut *x* (spirant surd as in German *ach*), Mongol *ts*, *č*, and Manchu-Tungusian *s*, *ś*, or *š*. He gives seven examples to this effect, the first of which is:

> Turkish (Uigur and in the dialects of southern Siberia) *kam* or anciently *qam* (shaman), in some dialects *küär*, *kum* (with such derivations as *kamdi*, to prophesy; *kamla*, to heal; *kamna*, to shamanize; etc.). Čuvaš *jum*, *jumeś* (shaman); *im-jum* (curative charms). Manchu *saman*, Golde *šama*, *š'ama*, Olča *šam*, Oročon *s'ama*, Tungus on the middle Amur *samañ*, Tungus of Udsk *saman* and *šaman*, Kondogir and Wilui Tungus *xaman*.

Németh concludes that the origin of the much disputed word "shaman" is thus definitely solved, and that it is an ancient property of the Turkish-Mongol languages. The phonetic alternation determined by him is perfectly correct; and Manchu-Tungusian *sam*, *s'am*, or *šam*[2] is indeed the exact phonetic equivalent of Turkish *kam*.[3] Apart from the Tungusian series noted by Németh

[1] "Über den Ursprung des Wortes Šaman und einige Bemerkungen zur türkisch-mongolischen Lautgeschichte," *Keleti Szemle*, vol. XIV, 1913–14, pp. 240–249.

[2] This in fact is the stem of the word, not as formerly believed by some scholars, *sa-*, which was explained from Manchu *sa-* (to know). Németh justly rejects this derivation. By the way, Schott, the father of this etymology, disavowed it himself in his second treatise on the word shaman, quoted above. D. Banzarov (*l. c.*, p. 35) justly recognized *sam* as the stem.

[3] It was hitherto believed that the word for shaman was different in all languages of Siberia, while there is a fairly uniform designation for the female shaman (*udagăn*, *utygan*, *ubakxan*, etc.). Hence V. F. Troščanski (Evolutsiya černoi v'äry [šamanstva] u Yakutov = "Evolution of the Black Faith among the Yakut," Kazan, 1902, p. 118) was prompted to the theory that the Siberian tribes originally had only female shamans, and after the time of their separation instituted also male shamans. This reconstruction is contestable on many grounds; for instance, we have male and female shamans

after Grube's *Goldisches Wörterverzeichnis*, several other forms of the word have become known. De la Brunière[1] recorded it in 1846 among the Golde as *tsama* or *tsamo*. I heard only *s'ama* from the Golde, in the same manner as Shimkevič,[2] but in view of the existence of several Golde dialects such a form may be regarded as possible. A. Rudnev[3] noted from a Manchu, who was still conversant with his mother tongue, the following: *sáma*, *sáman*, *sámą* (nasalized *a*), and *sāmăn*.

The Gilyak style the shaman *čam* (*tšam*). This certainly is a loan-word received from Tungusian,[4] but it is not necessary to assume that the change of the initial *š* or *s* into the palatal explosive surd is due to the Gilyak; for this alternation is common within Tungusian, so that a Tungusian *čam* borrowed by the Gilyak may be presupposed for some Tungusian dialect spoken in the neighborhood of the Gilyak. Compare the following examples: Manchu *nesuken* and *nečin* (equal)—Jučen *nušin;* Manchu *ničuxe* (pearl)—Old Jučen *inšuko* or *inšuxa*, modern Jučen *ninčuxe;* or *vice versâ*, Jučen *ančun* (gold)—Manchu *aisin*.

It has been said that the word *šaman* is found only in the eastern part of Siberia and in Manchuria. There is, however, one well-authenticated instance on record to the effect that it was known also in northwestern Siberia. In the earliest account that we have of the Irtysh-Ostyak, written in 1715 by G. Novitski, their priests were styled *šamančik*.[5] In view of the early date of this record we cannot well assume that this is a loan-word received through intercourse with the Russians. Whether the term still exists I

co-existent in the most ancient days of China. The supposition on which the conclusion is based is now no longer valid: the word for the male shaman is in fact identical throughout from the Turkish to the Tungusian and Gilyak tribes.

[1] In H. E. M. James, *The Long White Mountain or a Journey in Manchuria* (London, 1888), pp. 433, 434.

[2] *Materialy dl'a izučeniya šamanstva u Goldov*, p. 8.

[3] *Novyia dannyia po živoi mandžurskoi r'äči i šamanstvu* (*New Data on the Live Manchu Speech and Shamanism*), p. 9.

[4] There is no doubt that the shamanism of the Gilyak is derived from and a weak echo of that of the neighboring Amur tribes. There are but a few shamans among the Gilyak, and shamanizing is rarely practised by them (see for the present L. von Schrenck, *Reisen und Forschungen im Amur-Lande*, vol. III, p. 752).

[5] B. Munkácsi, *Keleti Szemle*, vol. IV, 1903, p. 88.

am unable to say from the sources at my disposal. S. Patkanow[1] gives for the shamans of the Ostyak only the name *tonx-ort*.[2]

The Turkish word *kam* is of ancient date.[3] It is attested (and this is its earliest occurrence) in the Chinese Annals of the T'ang dynasty (618–906), where it is said with respect to the Kirgiz[4] that they designate their sorcerers or shamans (Chinese *wu*) with the word *kam* 呼巫爲甘. The last character now reads *kan*, but was anciently possessed of a final labial nasal, which is still preserved in the dialects of southern China. Chinese *wu* is a very ancient term for the native medicine-men.[5] The word *kam*, further, is found in Uigur[6] and in the *Codex cumanicus* of 1303.[7] It is also used by the Persian historian Rashid-eddīn (1247–1318) in 1302, with reference to the shamans of the Mongols.[8] The existence of the Tungusian word *šaman* can now be traced to the first part of the twelfth century, for Pelliot[9] has found it as a Jučen word in a Chinese gloss, where it is transcribed in the form *šan-man* and explained as a sorceress. The Chinese transliteration would presuppose a Jučen form *šamman*. Indeed we still meet forms with double *m* in the dialect of the Wilui Tungus,—*samman* and *xamman*.[10] The latter with initial spirant represents the missing link between Turkish *kam* (in the language of the Soyot or Urang-khai *xam*) and the Tungusian forms with initial sibilant. The

[1] *Die Irtysch-Ostjaken*, pt. I, p. 120.

[2] Consulting Castrén's vocabulary in his *Versuch einer ostjakischen Sprachlehre* I find that *tonx* means idol, and *ort*, workman, slave. If accordingly *tonx-ort* should mean servant of the idols, the term might hardly be claimed as a primeval designation of the shaman.

[3] The Russians have derived from it the verb *kamlat'* (to shamanize, to conjure, to spell, to prophesy) and the noun *kamlanie* (the act of shamanizing).

[4] Regarding this name and its ancient Chinese transcriptions see Pelliot and Laufer, *T'oung Pao*, 1916, p. 295.

[5] Regarding their activity see *Chinese Clay Figures*, pp. 198–200.

[6] In the Kudatku Bilik, written in A. D. 1068 (H. Vámbéry, *Uigurische Sprachmonumente*, p. 221).

[7] W. Radloff, *Aus Sibirien*, vol. II, p. 67. Németh overlooked these data, but it is important to note the wide distribution of the word in time and space.

[8] F. von Erdmann, *Temudschin*, p. 553; E. Blochet, *Introduction à l'histoire des Mongols de Rashid Ed-Din*, p. 165.

[9] *Journal asiatique*, 1913, mars-avril, pp. 466–469.

[10] Recorded by R. Maack, *Wil'uski okrug* (*The District of Wilui*), pt. 3, p. 116.

alternation of *k* and *x* is well known;[1] likewise *x*, *h*, and *s*, alternate in Tungusian dialects.[2] For this reason it is justifiable to correlate the forms *kam* and *xam* with Yakut *xamnā*, *xamsā* (to move, to be agitated; *xām*, to step, to stride),[3] and to interpret the significance of the word from the shaman's peculiar behavior, his solemn and pompous pacing, his ecstasy, his convulsions, his frantic leaps and dancing.[4]

Németh gives no form of the word for Mongol, and explains the absence of a corresponding word in this language through a characteristic trait of the Mongols to whom the use of many words was tabooed; owing to such a taboo, the word *šaman* should have been lost. This is an arbitrary, far-fetched, and unnecessary theory. First of all, the word *šaman* does occur in Mongol, at least it is listed as a Mongol word in Kovalevski's and Golstunski's Mongol dictionaries. Whether it is frequent is another question. I never found it in a Mongol text. This is accounted for by the fact that the common term for a shaman in Mongol is *bügä*, *bögä*, Buryat *buge* and *bȫ*, corresponding to Turkish *bȫgü*, Djagatai *bügi* (sorcery), *bügiči* (sorcerer), Osmanli *büi* (sorcery).[5] This series seems to bear some kind of relation to Chinese *bu*, *wu* (shaman), *buk*, *puk* (to divine), and Tibetan *ạba* (pronounced *ba*, sorcerer);[6] but how

[1] W. Radloff, *Phonetik der nördlichen Türksprachen*, p. 108.

[2] A. Schiefner, *Mélanges asiatiques*, vol. VIII, 1877, pp. 338–339.

[3] O. Böhthingk, *Jakutisches Wörterbuch*, p. 79.

[4] This explanation was first given by V. F. Troščanski, *Evolutsiya černoi v'äry (šamanstva) u Yakutov* (*Evolution of the Black Faith among the Yakut*), (Kazan, 1902), p. 117. D. Banzarov (*Černaya v'ära*, p. 35) held a similar view, but arrived at it through a different process by connecting the word with Manchu *samarambi* and Mongol *samoromoi*, *samagu;* aside from the fact, however, that these words do not have the meaning required, or forced into them by Banzarov, they can by no means be reduced to the stem *sam.*—*Kam* is not a Yakut word for the shaman, as wrongly stated by A. van Gennep ("De l'emploi du mot 'chamanisme,'" *Revue de l'histoire des religions*, vol. 47, 1903, p. 52); the Yakut term is *oyun*.

[5] According to I. J. Schmidt (*Geschichte der Ost-Mongolen*, St. Petersburg, 1826, p. 416), *kami* should be the Mongol name of the shamans. It would hardly be necessary to point out this blunder, had it not been immortalized in Yule's *Hobson-Jobson* (p. 820). There is no Mongol word *kami* with such a meaning. Schmidt found it in the Mongol chronicle of Sanang Setsen as the proper name of a Tibetan Bon-po, and this name reads correctly Khami or Xami. Schmidt's interpretation, evidently suggested by Turkish *kam*, is baseless and purely imaginary.

[6] Compare my concordance in *T'oung Pao*, 1916, p. 68.

this relation is to be explained is beyond our present knowledge. The type *kam—sam* is possibly preserved in Mongol in a somewhat disguised form. According to the phonetic law established by Németh, the Mongol initial corresponding to Turkish *k* and Tungusian *s* should be *ts*, *č*. Thus we get at Mongol *tsam*, *čam*, that means a dance, and at present refers in particular to the religious pantomimic dance-performances of the Lamas in which, as is well known, many ancient shamanistic notions and rites of pre-Buddhistic times have survived. The same word in the same sense is encountered in Tibetan as *č'am* or *ts'am* (written *ạč'am*), and it would be justifiable to regard the Mongol as a loan-word adopted from Tibetan;[1] but considering the fact that the Tibetan word is isolated (that is, not found in any other Indo-Chinese language), it is likewise permissible to stamp the Tibetan as a Mongol loan. For the rest, I give with all reserve this theory of a possible connection of Mongol *tsam* with Turkish-Tungusian *kam—sam* (religious dancer).

At any rate it is obvious that the word *šaman* has now legitimately secured an absolute and irrevocable decree of divorce from its pseudo-mate *çramaṇa*, *samaṇa*, or *ša-men*, and that this mismated couple cannot live together any longer. Tungusian *saman*, *šaman*, *xaman*, etc., Mongol *šaman*, Turkish *kam* and *xam*, are close and inseparable allies grown and nourished on the soil of northern Asia,—live witnesses for the great antiquity of the shamanistic form of religion.

FIELD MUSEUM OF NATURAL HISTORY,
CHICAGO, ILLINOIS

[1] In a forthcoming study on *Loan-Words in Tibetan*, where an almost complete list of Mongol loan-words occurring in Tibetan is given, the writer has set forth the difficulties in recognizing the mutual loan-words of the two languages. Mongol and Tibetan have a certain number of words in common, partially due to a prehistoric contact, partially due to a lively intellectual interchange in historical periods; and it is not always possible to decide in each and every case which side is the recipient, and which the borrower.

138

驯鹿及其驯养

MEMOIRS

OF THE

AMERICAN ANTHROPOLOGICAL ASSOCIATION

VOLUME IV

1917

PUBLISHED FOR
THE AMERICAN ANTHROPOLOGICAL ASSOCIATION
LANCASTER, PA., U. S. A.

THE REINDEER AND ITS DOMESTICATION

By BERTHOLD LAUFER

THE domestication of the reindeer has not yet been satisfactorily expounded. Some interesting though brief essays on the subject have been contributed by scholars engaged in the research of animal domestication,—first of all, by E. Hahn,[1] in his admirable work *Die Haustiere*, whose chapter on the reindeer is the best hitherto written; then follow C. Keller,[2] R. Müller,[3] L. Reinhardt,[4] and M. Hilzheimer.[5] These various contributions are useful as far as they go; but what we miss in them, above all, are the historical and ethnographical points of view, and the exploitation of the abundant material accumulated by ethnographers who have had occasion to study reindeer-breeding tribes at close quarters. The Russian explorers of Siberia occupy here the first place; and it was one of the writer's chief aims to avail himself of their data, as far as this literature is accessible to him. While the observations of ethnographers working in the field are of prime importance, the interpretations of their data must occasionally be subjected to certain modifications, not all ethnographers being sufficiently schooled in the problems of domestication, or familiar with the methods and results of that science. The novel feature of the present investigation lies in the fact that here for the first time early Chinese sources relative to the domesti-

[1] *Die Haustiere und ihre Beziehungen zur Wirtschaft des Menschen, eine geographische Studie* (Leipzig, 1896), pp. 262–267. Compare the same author's "Die Transporttiere in ihrer Verbreitung und in ihrer Abhängigkeit von geographischen Bedingungen," *Verhandlungen des XII. Deutschen Geographentages in Jena* (1897), pp. 186–187.

[2] *Naturgeschichte der Haustiere* (Berlin, 1905), pp. 198–202; *Stammesgeschichte unserer Haustiere* (Leipzig, 1909), p. 93; also in Kraemer's *Der Mensch und die Erde*, vol. I, p. 257.

[3] *Die geographische Verbreitung der Wirtschaftstiere* (Leipzig, 1903), pp. 137–148.

[4] *Kulturgeschichte der Nutztiere* (München, 1912), pp. 228–237.

[5] *Die Haustiere in Abstammung und Entwicklung*, pp. 72–73.

cated reindeer are laid under contribution, and that an effort has been made to determine the origin of the domestication more precisely as to time and space. The writer attempts to answer three questions, as far as this is possible in the present state of science:— When did the primeval domestication originate? Where was the center of it, and how did it propagate from this center to other culture areas? What was the process that brought the primeval domestication about?

At the outset two current popular notions connected with the Old-World reindeer should be banished,—that the reindeer is exclusively an inhabitant of the tundra of northern Europe and Asia, and that it is employed exclusively by the peoples inhabiting the northern littorals of Europe and Asia. The reindeer haunts the woods of high mountainous districts as well, and thrives in the forests of the Ural and Baikal regions. The records referring to the woodland reindeer are much older than those pertaining to the tundra reindeer of the maritime coasts. It will be seen that in all likelihood we have to assume an historical relation between the two varieties; that is to say, the woodland reindeer is the first in point of time that was domesticated, and spread from southern into northern regions, gradually developing into the tundra reindeer through infusion with the blood of wild forms of the tundra. The wild reindeer has the same southern expansion: it abounds in the extensive woods of the governments Vyatka and Perm and in the adjoining northern portion of Kazan, in Russia. Entire herds formerly migrated from the Ural into the afforested region between the Kama and Ufa (56° N. lat.), even as far as the southern woodland boundary line, almost as far as 52° N. lat.[1] The Bashkir hunt the animal along the Ufa under 55° N. lat.

[1] J. F. Brandt, *Zoogeographische und palaeontol. Beiträge* (St. Petersburg, 1867), p. 65. See also A. Nehring, *Ueber Tundren und Steppen der Jetzt- und Vorzeit* (Berlin, 1890), pp. 31, 108. P. S. Pallas (*Reise durch verschiedene Provinzen des russischen Reichs*, vol. III, p. 470) reported in 1773, "In the fir-tree woods on the Ufa and throughout the woodlands as far as the Kama, there are, aside from other deer, still many wild reindeer (in Bashkir *yušá*), frequently wandering in large herds, and, judging from the antlers I saw, somewhat smaller than the northern ones."

Historical Notes

The first and most prominent fact about the domesticated reindeer is that it is entirely lacking in aboriginal America (the artificial introduction into Alaska is of very recent date), and represents an exclusive cultural property of the Old World. North America abounds in wild reindeer (known as caribou) and elk or moose; but the native population only hunted these animals, and never made any endeavor to domesticate them. Consequently the Old-World domestication cannot be *a priori* of very ancient date, but was accomplished only at a late time, when the population of America was settled. This consideration will be amply confirmed by the history of the domestication.

Certain it is that the classical authors have left us no account whatever of the domesticated reindeer. The Danish archaeologist G. F. L. Sarauw[1] has made a very interesting study of the information contained in the writings of the ancients in regard to elk and wild reindeer, but there is complete silence as to tamed forms. Hahn[2] is quite right in maintaining that the Greeks were not so unfamiliar with the north of eastern Europe that such a striking phenomenon as the tamed harts should not have been known among them in one form or another, had they existed at the time; but all observations of the ancients strictly refer to wild forms. This state of affairs meets its parallel among the Chinese. They were well acquainted with the host of tribes living in the north and northwest of their country, but in no Chinese author of the pre-Christian era do we meet with a single notice of the reindeer. Only at the end of the fifth century A.D. did tidings of a tame stag, used for drawing sledges and for milking, reach the ears of the Chinese. It is well known that the wild reindeer was among the game hunted by paleolithic man of western Europe. There is no evidence that he ever attempted to domesticate this animal. Its domestication manifestly falls within historical times; and, if so, there must be some way of calculating by historical methods more

[1] "Das Rentier in Europa zu den Zeiten Alexanders und Caesars," published in *Mindeskrift for Japetus Steenstrup* (København, 1913), 34 p., 4°.

[2] *Haustiere*, p. 263.

exactly the *terminus a quo*. The students of domestication have usually regarded that of the reindeer as a comparatively recent event, and as the most recent of all domestications; but their impressions naturally have remained of a somewhat vague character. C. Keller[1] remarks:

> The passage into the state of domesticity cannot have taken place at an early date, since neither specific races have as yet been formed, nor is the submissiveness to man much developed. The herds graze wherever it suits them; and the business of milking is very complex, as the cows behave stubbornly.

L. Reinhardt[2] has expressed the following opinion:

> The reindeer was elevated by man into a domesticated animal at a very late period, and generally is still domesticated very deficiently. The time when this happened can no longer be determined; however, it cannot have taken place much earlier than five hundred years ago.

This figure is far too low, and must be multiplied at least by three, as we have Chinese allusions to the domestic reindeer dating in the fifth century A.D. Even without such historical data, Reinhardt's calculation would hardly be acceptable, as the wide geographical distribution of the reindeer would argue in favor of a much earlier domestication. M. Wilcken's assertion[3] that the domestication of the reindeer took place in prehistoric times misses the mark entirely.

The earliest reference to tame reindeer in western sources is contained in the famous narrative of the Norseman Ohthere, who "said to his lord, King Alfred, that he dwelt farthest north of all Northmen." Ohthere, of whom we unfortunately know very little, was born in Haloga (Helge)-land in Norway, and undertook in A.D. 890 several voyages, one of which was from Norway toward the extreme northern coasts. In the course of his travelings he rounded the North Cape, discovered the White Sea, where he reached the south coast of the Kola Peninsula, and became acquainted with the Finn and Biarmians (Beormas) or Permians in the northeast of European Russia. The memorable account of

[1] H. Kraemer, *Der Mensch und die Erde*, vol. 1, p. 257.

[2] *Kulturgeschichte der Nutztiere*, p. 232.

[3] *Grundzüge der Naturgeschichte der Haustiere* (2d ed. by J. U. Duerst), (Leipzig, 1905), p. 172.

his expeditions was included by Alfred the Great in his Anglo-Saxon translation of the Hormista of Paulus Orosius.[1] Here we read as follows:

He [Ohthere] was a very rich man in those possessions in which their wealth consists, that is, in wild animals. He still had when he came to the king six hundred tame deer unsold. These deer they call 'reindeer;' six of them were decoy-deer; these are much prized among the Finn, because they capture the wild deer with them. He ranked with the foremost men in the land, though he had not more than twenty cattle, twenty sheep, and twenty swine; and the little that he ploughed he ploughed with horses.[2]

Schlözer[3] and I. A. Sjögren,[4] taking the term "Finn" in Ohthere's narrative in the sense of "Lapp," have advanced the theory that he lived among Lapp and spoke their language,[5] and that it was Lapplanders, who cared for his reindeer purchased from them. This theory is baseless, and we gain nothing from it. Whether Ohthere had obtained his reindeer from Lapp or Finn or Scandinavians, or had captured them himself, his story can prove little or nothing along the line of domestication; at best, it shows the very first stage necessary in reaching this object. All members of the family *Cervidae* may easily be driven into enclosures and kept there indefinitely, for which many examples will be cited hereafter. Ohthere does not state that he made any practical

[1] The original manuscript of Alfred's work, beautifully written, is preserved in the Cottonian collection of manuscripts in the British Museum. It was first published by Daines Barrington under the title, *The Anglo-Saxon Version from the Historian Orosius, by Alfred the Great. Together with an English translation from the Anglo-Saxon* (London, 1773).

[2] J. McCubbin and D. T. Holmes, *Orosian Geography*, p. 8. J. Bosworth, *Description of Europe and the Voyages of Ohthere and Wulfstan, written in Anglo-Saxon by King Alfred the Great*, p. 12, translates: "He had, moreover, when he came to the King, six hundred tame deer of his own breeding." The Anglo-Saxon text of the above passage runs as follows: "þā dēor hī hātaδ 'hrānas;' þāra wǣron syx stælhrānas; δā bēoδ swȳδe dȳre mid Finnum, for δǣm hȳ fōδ þā wildan hrānas mid."

[3] *Allgemeine nordische Geschichte*, p. 445.

[4] *Gesammelte Schriften*, vol. 1, p. 314.

[5] This point is rather doubtful. All that Ohthere himself tells us in point of language amounts to this: "The Permians told him many stories both of their own land and of the lands which were around them, but he did not know how much was truth as he did not see it himself. It seemed to him that the Finn and the Permians spoke nearly the same language." This observation does not lend itself to far-reaching conclusions.

use of his deer. In all probability, it was merely the venture of a sportsman, who had an aesthetic pleasure in the animals, like a park-owner in fallow deer. Ohthere's account stands perfectly isolated: we read no more about tame reindeer during or after his time. Only as late as the fifteenth century do we hear for the first time about domesticated reindeer from Russian sources. If at Ohthere's time the Finn or the Lapp had really possessed the reindeer, we should justly expect to find it mentioned in the *Kalewala;* but this is not the case. The songs of the *Kalewala* know only of the elk and the wild Tarandus.

It is stated by Hahn[1] that, according to Lehrberg, in 1499 the Samoyed, besides dog-sleighs, had reindeer on the backs of which they used to ride. C. Keller[2] has adopted this from him, and the "fact" has finally been popularized in H. Kraemer's *Der Mensch und die Erde.*[3] It is striking, of course, that the Samoyed should have mounted reindeer in 1499, while they never did so at any later time, nor do so at present. In fact, the reindeer is ridden only by the Soyot and Tungus, not, however, by any western tribes.[4] Thus suspicion is ripe that there may be some misunderstanding of the original Russian source on which this deduction is based. Lehrberg's work in the original German is not within my reach,[5] but I have access to a Russian translation of it and to the Russian document on which his data are based. This is reprinted in Shčšeglov's *Chronological Review of Important Data from the History of Siberia,*[6] and relates to the year 1499. In order to

[1] *Haustiere,* p. 265.

[2] *Naturgeschichte der Haustiere,* p. 201.

[3] Vol. 1, p. 257. Here we even read the absurdity that "the oldest accounts of tame reindeer come from Lehrberg, who in 1499 observes that the Samoyed ride on them,"—a complete misunderstanding.

[4] Hahn himself was struck by this anomaly, stating farther on (p. 266) that "this exception would seem doubtful to him until further confirmation were received."

[5] The work of A. C. Lehrberg bears the title *Untersuchungen zur Erläuterung der aelteren Geschichte Russlands* (St. Petersburg, 1816). An interesting analysis of his researches has been given by Klaproth, *Mémoires relatifs à l'Asie,* vol. 1 (Paris, 1824), pp. 116–146.

[6] I. V. Shčšeglov, *Xronologičeski perežen' važn'äišix dannyx iz istorii Sibiri 1032–1882* (Irkutsk, 1883), p. 12, published by the East-Siberian Section of the Imperial Russian Geographical Society.

understand these events, it is necessary to premise that Ivan the Great (1462–1505), after destroying the liberty of Novgorod, began the conquest of northern Russia, and in that year the Russians completed the subjugation of what was called by them Yugra; that is, the territory of the Ural Mountains, inhabited by Wogul and other Ugrian tribes, and formerly under the jurisdiction of the Republic of Novgorod, in the documents of which Yugra is mentioned as early as 1264. The expedition of 1499 was conducted under the command of the Prince Semyón Fedorovič Kurbski, Prince Pyotr (Peter) Fedorovič Ušati and Vasili Ivanovič Zabolotski-Bražnik. This enterprise is described in detail in the synchronous Russian documents, the result being given thus:

> The military chiefs (*voyevody*) slew fifty men of the Samoyed[1] on the rock,[2] and captured two hundred reindeer. From this Rock they marched for a week as far as the first town, L'apino,[3] covering altogether 465 *verst* over this territory. Proceeding from L'apino, they met the Yugor princes who came on reindeer from Obdor;[4] but from L'apino the [Russian] military chiefs (*voyevody*) traveled on reindeer; the army, however, on dogs.

This is a literal translation, and in the spirit of the Russian language means that they traveled on sledges drawn by reindeer and dogs respectively. The same verb, *šl'i* ("they went"), is used with the reindeer and with the dogs (*na olen'ax, a rat na sobakax*); and,

[1] The land of the Samoyed, under the name Samoyad', is mentioned as early as 1096 in the chronicle of Nestor as being situated north of Yugra. In 1246 their name is mentioned by Plano Carpini, who styles them "Samogedes," and ascribes to them dog-heads, as the ancient legend of the κυνοκέφαλοι was alive in his day. The name may be related to Sameyadna, which the Lapp (in Russian Lop', Lopari) confer on their country.

[2] The Rock (Kamen'), also Rocky Girdle (Kamennyi Poyas), is a designation of the Ural, in accordance with the Ostyak term *keu, kev* ("stone, mountain, Ural"). See B. Munkácsi, *Keleti Szemle*, vol. III (1902), p. 276.

[3] Small place (also L'apina) on the banks of the Sygwa in the district (*okrug*) Berezov, now called Vorulsk. The Sygwa is a side-river of the northern Soswa, which falls into the Ob not far from Berezov.

[4] The original document has the misprint Odor. The question is of Obdor province (*Obdorskaya oblast'*) on the lower Ob. The settlement Obdor is situated not far from the mouth of the Ob. According to A. Castrén, *Reiseerinnerungen aus den Jahren 1838–1844*, p. 279, who has given a very interesting description of the place, this name should be of Syryän origin, meaning "mouth of the Ob." An account of Berezov and Obdorsk is found also in P. S. Pallas, *Reise durch verschiedene Provinzen des russischen Reichs*, vol. III, pp. 17–24. Reindeer are still kept in this region.

since it cannot be supposed that the soldiers rode astride dogs, it is equally out of the question that riding on reindeer is understood.[1] The Samoyed have nothing whatever to do with this affair; the Russian documents of that period clearly distinguish between Yugra and Samoyed, and the situation is perfectly clear. It was the Yugor (Yugrian, Ugrian) princes (*Yugorskie kn'azi*) who were in possession of reindeer-sledges, in the same manner as their Wogul descendants are at the present time. These were duly captured by their Russian conquerors and placed at the disposal of the commanders on their further inroads into the Ugrian territory, while the soldiers were transported on dog-

[1] The Russian text is by no means ambiguous. If the Russian writer meant to express "riding," he would have used the verb *yäxat' verxom.* The usual question addressed to the winter traveler in Siberia on his arrival is, "In what way did you come?" which is answered by such phrases as, "On horses" (*na lošad'ax* or *kon'ax*), "On dogs" (*na sobakax*), "On reindeer" (*na olen'ax*); and it is perfectly understood that he traveled on a sledge drawn by horses, dogs, or reindeer. In the same manner Avril, *Travels into Divers Parts of Europe and Asia* (London, 1693), p. 161, says in regard to the Samoyed that "they travel upon harts and dogs." The text of Lehrberg (in the Russian translation, p. 14) is quite clear. "Iz L'apina na vstr'ăču Russk'im v'äxali Yugorskiye kn'az'ya na olen'ax. L'apino zavoyevano, i ot s'uda voisko pošlo dal'äye, voyevody na olen'ax, a pročie na san'ax, zapr'ažennyx sobakam'i." Lehrberg comments in a note that traveling with dogs was in full swing on the Irtysh as early as 1580, and is still practised in northwestern Siberia, horses not being kept under 62° N. lat.; that formerly also west of the Ural in Perm dogs were employed for transportation, in more ancient times even farther west along the Baltic Sea, as shown by the Esthonian and Finnish phrase for "mile," *penni koorm, penicuorma* (literally, "dog-load"). Karamzín (*Istoriya gosudarstva rossiskago*, St. Petersburg, 1819, vol. VI, p. 286), the eminent Russian historian, has interpreted the document in the same manner by saying, "Each of these princes sat in a long sledge drawn by reindeer. The voyevody of John likewise drove on reindeer (*yäxali na olen'ax*), but the soldiers on dogs (*na sobakax*), holding in their hands fire and sword for the annihilation of the poor inhabitants." Regarding the Russian expedition of 1499 see also Sjögren, *Gesammelte Schriften*, vol. I, p. 309; and Aleksandra Dmitrieva, "Pokorenie ugorskix zemel'i Sibiri," pp. 87 *et seq.*, *Permskaya Starina* (Perm, 1894), no. V. In A. Rambaud's *History of Russia* (Boston, 1886), vol. I, p. 221, this event is thus narrated: "In 1499 the voyevodi of Ustiug, of the Dwina, and of Viatka advanced as far as the Petchora, and built a fortress on the banks of the river. In the depth of winter, in sledges drawn by dogs, they passed the defiles of the Urals, in the teeth of the wind and snow, slew fifty of the Samoyedi, and captured two hundred reindeer; invaded the territory of the Voguli and Ugrians, the Finnish brethren of the Magyars; took forty enclosures of palisades, made fifty princes prisoners, and returned to Moscow, after having reduced this unknown country." Here the transportation on reindeer-sleighs as too unimportant or troublesome to the historian has been passed over in silence,—a curious example of history-writing.

sleighs. Let us hope that "the reindeer-riding Samoyed of 1499" will thus remain buried never to rise again. The document quoted is of importance, for it shows us that the Uralic Ugrians were acquainted with the domesticated reindeer as a draught animal toward the end of the fifteenth century. In regard to the Samoyed, we can assert on the basis of this account only that reindeer were kept by them.

When Baron Sigismund von Herberstein was ambassador from the Emperor Maximilian to the Grand Prince Vasili Ivanovič of Muscovy in the years 1517 and 1526, he met at the Court of this Prince in Moscow his interpreter, Gregory Istoma, who in 1496 had been sent by the Prince to the Court of King John of Denmark, where he acquired the Latin language. He gave Herberstein an account of his journey, which had taken him over Great Novgorod to the mouths of the Dvina and Potivlo. There the party embarked in four boats, and sailed along the right-hand shore of the ocean; and after accomplishing sixteen miles and crossing a certain gulf, they sailed along the left shore. Leaving the open sea to their right, they came to the people of Finlapeia. Although these people dwell in low cottages, scattered here and there along the seacoast, and lead an almost savage life, Istoma reported, yet they are more gentle in their manners than the wild Laplanders. He stated that they were tributary to the Prince of Muscovy. A voyage of eighty miles, after leaving the land of the Laplanders, brought them to the country of Nortpoden, which was subject to the King of Sweden. The Russians call the country Kaienska Semla; and the people, Kaiemai. After having passed two perilous promontories, they sailed up to the country of the Ditciloppi, who are wild Laplanders, to a place named Dront [Drontheim], two hundred miles north of the Dvina.

They then left their boats and performed the rest of their journey by land, in sledges. He further related that there are herds of deer there, as plentiful as oxen are with us, which are called in the Norwegian language 'rhen.' They are somewhat larger than our stags, and are used by the Laplanders instead of oxen, and in the following manner: they yoke the deer to a carriage made in the form of a fishing-boat, in which the man is bound by his feet lest he should fall out while the deer is at full speed; in his left hand he holds a bridle, to guide the course of the deer, and in his right a staff, with which to prevent the upsetting of the carriage, if it should happen to lean too much on either side. He stated that, by this mode of travelling, he himself had accomplished twenty miles in one day, and had then let loose the deer; which returned of its own accord to its own master and its accustomed home. Having at length accomplished this journey, they came to Berges [Bergen], a city of Norway, quite in the north, amongst the mountains, and then reached Denmark on horseback.[1]

As Herberstein's narrative is based on the report of Gregory Istoma, whose experience dates back to 1496, we are entitled to say that the Lapp were in the possession of sleigh-drawing reindeer in the latter part of the fifteenth century.

Olaus Magnus, Archbishop of Upsala and Metropolitan of Sweden, who died in 1568, published in Rome, 1555, his famous work *Historia de gentium septentrionalium variis conditionibus*,[2] where a somewhat lengthy and fairly correct description of the reindeer of Lapland is given. Certainly he is not the first author, as asserted by Hahn, who told Europeans about the tame reindeer, as Baron von Herberstein preceded him by a generation. Olaus' account is not based on personal experience, but evidently draughted from hearsay. The English naturalist E. Topsell[3] then gave a description based on Olaus, and justly emphasized that the beast was altogether unknown to the ancient Greeks and Romans.

It is thus shown that the documentary evidence presented by European history does not mention the domestic reindeer before the latter part of the fifteenth century. I regret not having access to ancient Russian chronicles, especially those of Novgorod and Archangelsk, which might contain facts bearing upon the problem. There is a noteworthy negative evidence presented by the *Kalewala*, the national epic poem of the Finn. Here we have a true picture of the primeval cultural conditions in which the Finn lived prior to their christianization (A.D. 1151), also a description of their

[1] *Notes upon Russia: being a Translation of the Earliest Account of that Country, entitled Rerum Moscoviticarum Commentarii by the Baron Sigismund von Herberstein*, translated by R. H. Major, vol. II, pp. 105–108, Hakluyt Society.

[2] An English translation appeared in 1658 under the title *Compendious History of the Goths, Swedes, and Vandals, and Other Northern Nations*. His description of the reindeer is on p. 176. Those who have not access to this edition may be referred to E. Phipson, *Animal-Lore of Shakespeare's Time*, p. 123, where the passage is extracted.

[3] *Historie of Foure-Footed Beastes* (1607), p. 592.

relations to the Lapp. Sledge-driving is most frequently mentioned, but the sledges are always drawn by horses. The wild reindeer was an object of the hunt, but there is not the faintest allusion to reindeer kept in captivity under the control of man. The period of this ancient Finnish culture is difficult to gauge by exact dates, but it is generally admitted that the beginning of this national poetry falls between A.D. 800 and 1000.[1] If we assume that the Lapp adopted the domesticated reindeer from the Samoyed during the eleventh or twelfth century, we shall probably not commit too great an error of calculation.

Before leaving the European field, it should be remembered that the theory of a Scandinavian origin of reindeer domestication has also been propounded. Its main champion was a Norwegian scholar, A. Frijs.[2] According to him, the Lapp of the ninth century were not yet reindeer-nomads, but merely hunters and fishermen, whose only domesticated animal was the dog. The domestication of the reindeer they learned from the Scandinavians. The evidence for this bold statement is based on philological arguments: it is proved by the language of the Lapp, for only the dog has a genuine Lapp name; with the reception of the other domestic animals, the Lapp adopted also their designations; the Lapp has no word for "taming," and has therefore accepted the Scandinavian word for it. It is generally known how fallacious such play with alleged linguistic evidence is; in fact, no serious scholar any longer derives historical conclusions from conditions of language. Frijs evidently traced Lapp *raingo* ("reindeer") to Scandinavian *hreinn*, but there is as good reason to believe that the latter is based on the former. In fact, no lesser scholar than Jacob Grimm[3] regards the Lapp word as the foundation of the Germanic forms (Anglo-Saxon *hrān*, Old Norse *hreinn*, Swedish *ren*, Danish *rensdyr*, German *rein*, *reiner*, *renn*). Be this as it may, neither the one nor the

[1] D. Comparetti, *Kalewala*, p. 280 (authorized translation from the Italian). It is noteworthy also that Tacitus (*Germania*, 46), in his notice of the Fenni, the oldest account of some Finno-Ugrian tribe, makes no mention whatever of deer.

[2] *Globus*, vol. XXII (1872), p. 2, translation of his work *En Sommer i Finmarken, Russisk Lappland og Nordkarelen* (Kristiania, 1871).

[3] *Deutsches Wörterbuch*, vol. VII, col. 2007.

other supposition could prove that the domestication is due to Scandinavians, or to any other nation. It is merely indicative of a fact of language, and nothing else.

In others, the theory of the Scandinavian origin of reindeer domestication may have been inspired by certain efforts in Sweden to tame the elk (*Alcės alces* or *Cervus alces*). These, however, belong to recent times, and stories relative to them are not well substantiated by historical records. Although Louis Figuier, in his *Mammalia*, asserts that in Sweden for two or three centuries the elk was used in the harness, but that the custom is now given up, the objection has justly been raised by J. D. Caton[1] that it is difficult to understand why this alleged domestication was abandoned in a country so well adapted to its use. Sporadic cases of training elks to harness may formerly have occurred in Sweden; but no general attempt to tame the animal, and certainly no "domestication" of it, has ever taken place.

As a consequence of geographical conditions, the Chinese were far removed from reindeer-breeding localities; and for this reason we cannot expect to find in their records any coherent and comprehensive accounts, which would permit us to elaborate an intelligent history of the domestication. The expansion of their political power and the extension of their influence over neighboring tribes, however, enabled the Chinese occasionally to get a glimpse of the curious animal; and for lack of any other sources, their casual mentions of it are of capital importance, and at the same time represent the oldest extant references to the reindeer.

A very curious allusion to reindeer occurs in the Annals of the Liang dynasty in the description of the mythical country Fu-sang.[2] In A.D. 499 the Buddhist monk Huei Shen returned to King-chou, the capital of the Liang, and gave a fabulous account of Fu-sang, alleged to have been situated far off in the northeastern ocean. As to means of conveyance, he reported, the people there have vehicles drawn by horses, oxen, and stags; they raise deer in the

[1] *The Antelope and Deer in America*, p. 278.

[2] *Liang shu*, ch. 54, p. 12. This work was compiled by Yao Se-lien in the first half of the seventh century from documents of the Liang dynasty, which ruled from A.D. 502 to 556.

manner as oxen are reared in China, and make cream[1] from their milk. The allusion to the reindeer is unmistakable: they are plainly described as being kept in the state of domesticity for the purpose of drawing vehicles (that is, sledges) and for milk-consumption. Such an economic condition, as described in this text—the simultaneous breeding of horse, cattle, and reindeer—is not found, however, in any region of the northern Pacific; and if Fu-sang has been identified with America by some fantasists, the fact remains that neither the domestic horse nor cattle nor reindeer ever existed in pre-Columbian America. Nor are these conditions applicable to the Island of Saghalin, which Schlegel put on a par with the Fu-sang country of the Chinese account: horse and cattle were introduced there only by the Russian settlers in the latter part of the nineteenth century; and the reindeer, as already shown by L. von Schrenck, is there likewise a recent introduction going back to a few centuries. We do not even know whether Saghalin was populated at all in the fifth century. Neither can any Tungusian tribe come into question, since the Tungus employ the reindeer only as a beast of burden and for riding-purposes, but rarely for drawing sledges. The Fu-sang account is a fantastic concoction, devoid of any geographical value, pieced together from heterogeneous elements emanating from different sources and quarters. While each of these elements bears a germ of truth, their combination makes an unreal picture. The breeding of horse, cattle, and reindeer combined, in reality, occurs only in the Baikal region, particularly among the present Soyot; and Huei Shen's account of the reindeer in connection with horse and cattle has doubtless hailed from that quarter. The ethnic and economic

[1] The Chinese term *lo* 酪 denotes any dairy products, as cream, butter, cheese, sour or fermented milk. The former translators of this text have made a liberal choice without being concerned about what products are actually made of reindeer-milk. Bretschneider had butter made from reindeer-milk, but butter is never produced from it by any East-Siberian tribe. Schlegel (*T'oung Pao*, vol. III (1892), p. 123) decided on a fermented liquor, but such is never made. In fact, reindeer-milk is not made into any product in northern Asia, but is consumed as it is, in its natural state, as a fatty, creamy substance. S. W. Williams (*Journal American Oriental Society*, vol. XI, (1882), p. 93) therefore was quite right in translating, "and make cream of their milk."

condition of this locality, which is of paramount importance for the history of reindeer domestication, will be fully discussed hereafter.

Before mentioning the three kinds of vehicles used in Fu-sang, Huei Shen speaks also of a peculiar breed of oxen with very long horns. According to Williams, the horns were so long that they would hold things—the biggest as much as five pecks. According to Schlegel, the oxen could carry on their horns loads weighing up to twenty quintals.[1] Schlegel[2] thinks also that the reindeer is intended by this ox, but it is improbable that Huei Shen would first designate the reindeer as an ox and in the following sentence describe it as a deer. Further, loads are never placed on the reindeer's antlers; and it is equally inconceivable that loads were ever packed on the horns of an ox.[3]

[1] The passage is not very clearly worded, and the text presumably is corrupted. In all probability, it means that the people used these horns for carrying loads in them, the horns holding up to twenty corns (*hu* 斛, a measure of capacity).

[2] *L. c.*, p. 142.

[3] There are several other misconceptions in Schlegel's discussion of the subject. The Manchu term *kandahan* refers to the elk only, not to the reindeer. The Tungusian name for the " reindeer," *oron*, has no connection with Russian *olen'*, or *vice versâ*, as asserted by Schlegel. Russian *olen'* is an old Indo-European word connected with Lithuanian *élnis, alnis;* Lettic *alnis*, Old Prussian *alne*, German *elen*, Greek ἔλαφος (from **elṇ̥-bhos*) and ἐλλός (from **elnos*, " young hart "); Armenian *eln* (doe); Cymric *elain* (doe). The Russians, according to Schlegel, do not discriminate between " stag " and " reindeer," and call both indifferently *olen'*. Russian *olen'*, however, is the general term for *cervus*, and the reindeer is properly *s'äverni olen'* (northern deer), abbreviated into *olen'* when so understood from the context.—Neither H. C. von der Gabelentz nor Zacharov, in their Manchu dictionaries, have noted a word for the reindeer. Such, nevertheless, exists, though it is doubtless a loan-word from Tungusian. The *Ts'ing wen hui shu* (in Manchu: *Manju isabuha bithe*, ch. 1, p. 46), a Manchu-Chinese dictionary published in 1751 by Li Yen-ki, records the well-known Tungusian term *oron* as the Manchu designation for the reindeer, giving in Chinese the definition, " name of a cervine animal, antlers growing on the heads of both male and female, subsisting on moss, and raised by the deer-hunters." The same lexicographer notes *oronggo* in the sense of " deer-hunter " (the same word signifies otherwise a wild sheep with long and flat horns, resembling the " yellow sheep," *Antilope gutturosa;* in this sense the word appears also in Mongol) and the tribal name *Orončo-i niyalma*. It is not probable that Manchu *oron* (domesticated reindeer) and *iren* (wild reindeer) are interrelated words, as proposed by W. Schott (" Ueber das altaische Sprachengeschlecht," *Abhandlungen Berliner Akademie* (1847), p. 366); for Manchu *iren* is related to the Tungusian forms *hiru* and *siru* (below the Ussuri) and *iru* (above the Ussuri), in other Tungusian dialects also *hirun, ira* (W. Grube, *Goldisches Wörterverzeichnis*, p. 54). Neither is there any likelihood that, as supposed by Schott, there is interrelation of Manchu *oron* and Lapp *rončo* or *ronča* (male reindeer), in which the initial vowel should have been eliminated.

The Annals of the T'ang Dynasty (618–906) contain an interesting notice of a reindeer-breeding tribe, the Wu-huan, then settled in a region east or southeast of Lake Baikal. This notice runs as follows:

> The Wu-huan 烏丸 or Ku-huan 古丸, also styled Kü 鞠 or Kiai 祴, live in the northeast of the Pa-ye-ku 拔野古 (Bayirku). In their country there are trees, but grass is lacking, while there is plenty of moss. The inhabitants have neither sheep nor horses, but keep reindeer (stags) in the manner of cattle or horses. These animals subsist only on moss. They are trained to drawing sledges (carts). Reindeer-skins, moreover, are utilized as material for clothing.[1]

In the *T'ang hui yao*[2] this text is worded as follows:

> Traveling for six days in a north-easterly direction from this country (Pa-ye-ku), one arrives in the country Kü, where there are trees, but no grass. While sheep and horses are absent, there are reindeer. In like manner as cattle and horses are employed in China, the reindeer are used there for drawing sledges, which are capable of carrying three or four persons. The people clothe themselves with reindeer-skin. The reindeer subsists on the moss of the soil.

The fact that in these texts the reindeer is spoken of as a domestic animal is well attested by the use of the verb *huan* 豢 (" to feed domestic animals with grain ") and by the peculiar employment of the animals in the service of man. The *T'ung tien*,[3] written by Tu Yu (A.D. 735–812) between the years 766 and 801, with reference to the Wu-huan, employs straightway the term " domesticated stag " (*kia ch'u lu* 家畜鹿).

The history of the Wu-huan is well known from the Chinese Annals.[4] In the time of their early history we hear nothing to the effect that they kept reindeer. Their domestic animals were cattle, horses, sheep, and dogs. In the beginning of the Han dynasty (about 200 B.C.) they were broken up by the Hiung-nu, who are usually regarded as identical with the Huns, and, while subject to the latter, paid their annual tribute in cattle, horses, and sheep.

[1] *T'ang shu*, ch. 217 B, pp. 7a–b.

[2] Ch. 98, p. 16b. This work was written by Wang P'u, and completed in A.D. 961.

[3] Ch. 199, p. 18b.

[4] See Visdelou in D'Herbelot, *Bibliothèque orientale* (La Haye, 1779), vol. IV, pp. 79–86; E. H. Parker, " History of the Wu-wan or Wu-hwan Tunguses of the First Century," *China Review*, vol. XX, pp. 71–100, and *A Thousand Years of the Tartars*, pp. 117–125.

They were pastoral nomads, roaming about with their herds wherever there was grass and water; tents, always faced toward the east, formed their habitations. Each man, from the chieftain downward, possessed his own flocks and managed his own property, nobody serving another. They were skilful horsemen and archers, given to hunting. Flesh and dairy products were their chief means of subsistence; in a small measure they also grew millet. Their garments were made from bird's down, though they understood the preparation of leather and felt. According to the Chinese system of classification, the Wu-huan were counted among the Tung Hu (Eastern Hu or Barbarians), which term has without any reason been identified with "Tungus."[1]

There is no wonder that the early Wu-huan had no reindeer, for in their habitat this animal did not occur. They were first settled in southern Manchuria, and after the Chinese victories over the Hiung-nu in 120 B.C., were transplanted by the Emperor Wu into what is now the northern part of Chi-li Province and Liao-tung, in order to serve as a sort of buffer-state between China and the Hiung-nu. Skilled horsemen, they were organized into cavalry squadrons. From this time onward that tribe did not play any important rôle in history. In A.D. 207 they were decisively defeated by Ts'ao Ts'ao at Liu-ch'eng. It is somewhat surprising to meet them in the T'ang period (618–906) in a new geographical environment as northeastern neighbors of the Bayirku, a branch of the Turkish Uigur, and in the entirely new economic condition

[1] This theory belongs to the category of paper etymologies. Phonetically there is not a shadow of a coincidence between Chinese Tung Hu and Tungus, except the initial consonants. The word "Tungus," about the antiquity of which nothing is known, would never have been transcribed by the Chinese in that manner. First, it is used at present only by a few clans of Tungusian tribes, and by just those who are so remote from China, that it may well be doubted that they were ever in contact with her. Secondly, the word written by us "Tungus" is pronounced by the natives claiming this name To-ṅús, as noted by myself in Siberia. There is neither a *g* nor an *h* in it, and the guttural nasal opens the second syllable bearing the accent. The Chinese, accordingly, should they have had occasion to hear this name, would have transcribed it To-ṅu (ngu), or To-ṅu-se. The Chinese term Hu is applied to many other peoples also, especially the Iranians of Central Asia, and even to India. Klaproth (*Tableaux historiques de l'Asie*, p. 83) has already observed with correct instinct that it appears not very probable that the name "Tungus" is derived from Chinese Tung Hu.

of reindeer-breeders, as which they do not appear in any earlier period. This branch of the Wu-huan appears to have been a scattered horde, which had remained in its ancient seats, and was driven thence farther to the west, presumably as far as the country east or southeast from Lake Baikal, where the natural conditions for the maintenance of reindeer prevail. In all probability, they struck there a tribe which had already domesticated the reindeer; for Huei Shen's report has shown us that the domestication must have been an accomplished fact in the fifth century. At any rate, there is no valid reason for crediting the Wu-huan with the initiative or with any originality in this enterprise. They were originally horsemen and cattle-breeders; and when they drifted into their new domicile, they adopted what they found, adapting themselves to this novel economy.

Marco Polo[1] speaks of a tribe called by him Mescript, and identified by Yule with the Merkit in the country of Bargu, near Lake Baikal:

> They are a very wild race, and live by their cattle, the most of which are stags and these stags, I assure you, they used to ride upon.

Certainly this is the reindeer. Yule is inclined to think that Marco embraces under this tribal name in question characteristics belonging to tribes extending far beyond the Mekrit, and which in fact are appropriate to the Tungus; and continues that Rashid-eddīn seems to describe the latter under the name of Uriangkut of the Woods, a people dwelling beyond the frontier of Barguchin, and in connection with whom he speaks of their reindeer obscurely, as well as of their tents of birchbark, and their hunting on snow-shoes. As W. Radloff[2] has endeavored to show, the Woodland Uryangkit, in this form mentioned by Rashid-eddīn, should be looked upon as the forefathers of the present Yakut. Rashid-eddīn, further, speaks of other Uryangkit, who are genuine Mongols, and live close together in the territory Barguchin Tukum, where the clans Khori, Bargut, and Tumat, are settled. This region is

[1] Yule and Cordier, *The Book of Ser Marco Polo*, vol. I, p. 269.

[2] "Die jakutische Sprache," *Mémoires de l'Académie des Sciences de St.-Pétersbourg* (1908), pp. 54–56.

east of Lake Baikal, which receives the river Barguchin flowing out of Lake Bargu in an easterly direction. The tribal name Bargut (*-t* being the termination of the plural) is surely connected with the name of the said river. The Persian historian Rashid-eddīn, in his history of the Mongols written in 1302, speaks of a tribe styled Woodland Uryangkit living in forests northeast of Lake Baikal.[1] Their clothing consisted of animal skins. Cattle and sheep were not reared by them, but in place of sheep and cattle they kept mountain-oxen (*gāwi kohī*), mountain-sheep (*mīš*), and *jur* (Saiga antelope). They tamed these animals, milked them, and consumed this milk. During their peregrinations they loaded the mountain-oxen, but never quitted their forests. Wherever they stopped, they made huts and yurts of birchbark. Rashid-eddīn, further, narrates how they bored the birches and drank the birch-juice, and how they hunted in the winter on snowshoes, employing snow-sticks, and dragging along the spoils of the chase on sleighs. This text is very interesting,[2] but the Persian author's description of the domestic animals is by no means clear. Radloff infers that he alludes to a reindeer-breeding hunting-tribe, but he fails to inform us by which of the three animals named in the text he wishes to have the reindeer understood. It might not be impossible that the latter may be hidden under the mountain-ox; the Scandinavians and Lapp, for instance, apply terms like "ox," "cow," and "calf" to the reindeer.[3] On the other hand, however, as the tame yak occurs in the Baikal region, and particularly among the Uryankhai, the descendants of Rashid-eddīn's Uryankit,[4]

[1] In another passage of his work, Rashid-eddīn states that the designation "woodland peoples" is meant in contradistinction to peoples inhabiting the steppe; but there are many kinds of forest peoples, because one or another yurt of almost every tribe is in the vicinity of a forest, and because some tribes are distant from forests a month's journey, others two months' journey, others again only a day's journey.

[2] It has frequently been translated: d'Ohsson, *Histoire des Mongols*, vol. I, pp. 9, 421; F. von Erdmann, *Uebersicht der Völkerstämme nach Raschid-ud-din* (Kazan, 1841), p. 124, and *Temudschin*, p. 191; Berezin, *Istoriya Mongolov, sočinenie Rashid-eddina*, pp. 90, 141; Radloff, *l. c.*, p. 54 (revised edition of the text reprinted by Salemann, p. 84).

[3] Finnish *härkä* and Lapp *herke* mean "ox," and are applied to the tame reindeer.

[4] In the high mountainous portions of the eastern Sayan, cattle are reared in a few specimens by the Soyot up to an altitude of from five to six thousand feet, but

it may be permissible to think of the yak as well. There remain the sheep and the *jur*. The latter is a Mongol word (*dsür*) referring to the Saiga antelope. That this animal *might* be tamed and kept in captivity I do not doubt, but that it has actually been done in the region in question is not known to me. It is not plausible, either, that Rashid-eddīn should avail himself, for the designation of the domestic reindeer, of a Mongol term, which strictly denotes a wild beast. Thus the word " sheep " (*mīš*) would be the last resort for the reindeer interpretation.[1] The fact that it is not an ordinary sheep becomes evident from the assertion that this people does not rear sheep. Rashid-eddīn obviously speaks from hearsay, without entertaining correct notions of the matter, and his terms are evidently chosen in a state of embarrassment. If we are allowed to read from his text that he describes a reindeer-breeding people, it is less his obscure nomenclature that justifies us in this conclusion than the facts that also contemporaneous Chinese records and Marco Polo know of reindeer in this region, and that these still exist there at the present time, together with yak and horse.

There is another, ethnographical reason, which for a long time caused me to hesitate to believe in Rashid-eddīn's reindeer. Radloff regards this writer's Woodland Uryangkit as the ancestors of the modern Yakut, chiefly on the ground that a former appellative of the Yakut was Urāngkhai (Urāngxai) Sakha; and he looks upon Urāngkhai as the original tribal name of the Yakut. Now, according to Radloff, the Woodland Uryangkit were a typically reindeer tribe; the Yakut, however, are not.

If, accordingly, Radloff's theory of a connection of Rashid-

southward among the Uryankhai and Darkhat they are frequently replaced by *Bos grunniens* and its bastard forms with the domestic cattle, the so-called *khaĭluk*. The Buryat prefer the latter to *Bos grunniens*, which is known to them through the Uryankhai. Among these Soyot, as it hardly occurs otherwise in the south of eastern Siberia, the domestic ox is found, together with reindeer and horse; the reindeer, however, remains for them the most important of the three domesticated animals. G. Radde, *Reise im Süden von Ost-Sibirien*, vol. 1, *Säugetier-fauna* (St. Petersburg, 1862), p. 270.

[1] In the language of the Koibal, the reindeer is styled " white goat " (*ak kīk*), according to A. Castrén, *Koibalische Sprachlehre*, p. 75.

eddīn's Uryangkit with the present Yakut be correct, we are confronted with a fundamental contrast between the cultures of the two peoples. The Uryangkit are supposed to have been active reindeer-breeders, milking the animals, and subsisting on their milk; while the Yakut do not milk them at all, and look upon the whole business as an incidental affair of their life and as a foreign invasion. This contradiction has escaped Radloff, but attention should be called to this anomaly. In their present condition, the Yakut have lived at least since the seventeenth century, when the Russians first became acquainted with them. Rashid-eddīn wrote in 1302, so that the transition, if it took place, must have been the outcome of some three centuries; but this would be difficult to accept. In all probability, we shall have to interpret the events somewhat differently. While part of the Uryangkit may have been absorbed by the Yakut, this process need not be invoked to explain the entire ethnic composition of the Yakut. It was merely one of the political events that tended to contribute to the formation of this now powerful tribe, but currents from other directions as well have had their share in its ultimate organization.

Among the Yakut, now numbering about two hundred thousand, the reindeer represents a secondary acquisition, which they received from the Tungus. This borrowing is upheld by the traditions of the Yakut themselves, who assert that the Tungus are acquainted with no other domestic beast than the reindeer, and that the latter is the truly Tungusian cattle, which for this reason they style "foreign cattle."[1] This fact is brought out by the very conditions obtaining among the Yakut in regard to the reindeer. The Yakut are not a people of nomadic habits, but lead a sedentary life, based chiefly on the maintenance of cattle and horses, on agriculture and fishery. Reindeer take only an insignificant share in their culture, and are kept but reluctantly, mainly in the northern districts of the province of Yakutsk. Reindeer-breeders, as are found among the Tungus, Chukchi, and Samoyed, do not exist in their midst. They merely keep small herds, mainly utilized for driving or as pack-animals. Solely among the Dolgan of Turukhansk, who have

[1] V. L. S'ároševski, *The Yakut* (in Russian), vol. I, pp. 146, 307.

adopted the Tungus practice of nomadism, is the reindeer the exclusive domestic animal. The most curious fact is that the Yakut do not milk their reindeer at all, and slaughter it on rare occasions only, so that no reindeer meat is for sale among them. Their aversion toward nomadic life, and their habit of living in blockhouses, impose many restrictions on the keeping of reindeer, which without any doubt they adopted from Tungusian tribes.[1]

There can be no doubt, however, that during the Mongol period (thirteenth century) the reindeer was kept in a state of domesticity in the Baikal region. We have excellent testimony to this effect in the Chinese Annals of the Mongol Dynasty.[2] Here mention is made of the Kirgiz on the upper Yenisei, and, in connection with them, of five smaller territories, apparently inhabited likewise by Kirgiz. One of these is styled Han-ho-na, situated at the source of the Yenisei and east of the River Wu-se (Us), an affluent of the Yenisei.

> This region is accessible only over two mountain-passes and abounds in wild game, while domestic animals are scarce. The poor have no regular means of livelihood and erect hovels from birch-bark. They transport their chattels on white deer and consume the milk of this deer.

This certainly is the reindeer. It is worthy of note that in the same period we have a report from the Chinese traveler Ch'ang Te (1259) to the effect that the Kirgiz used dogs instead of horses for drawing sledges.[3] Accordingly, we are here confronted with the curious fact that a people in the central and southern part of Siberia was familiar with two specific methods of transportation, which we are wont to connect with the cultures of the peoples in the high north and northeast of Asia. Klaproth[4] thinks that the Han-ho-na were of Samoyed stock, presumably because they kept reindeer; and there is certainly a basis for this assumption. It must be considered, however, that the reindeer is not restricted to

[1] The Yakut's power of assimilation is well characterized by A. v. Middendorff (*Die Eingeborenen Sibiriens*, p. 1561), who says that among Tungusians and Samoyed the Yakut turns a Tungusian or Samoyed within the briefest space of time.

[2] *Yüan shi*, ch. 42; 63, p. 32 b (K'ien-lung edition).

[3] E. Bretschneider, *Mediæval Researches*, vol. 1, p. 129.

[4] *Mémoires relatifs à l'Asie*, vol. 1, p. 113.

certain ethnic groups, but is first of all bound to certain localities of specific floristic environment. When a tribal movement took place in the Baikal region, it could well happen that the ownership of the reindeer changed hands. The Kirgiz, taken in their entirety, were neither reindeer-breeders nor keepers of sleigh-dogs: neither the T'ang Annals, which have preserved for us the oldest account of this nation, nor Rashid-eddīn[1] or Abulgazi, state that it ever maintained herds of reindeer.

During the eighteenth century the Chinese noticed the reindeer also in the possession of Tungusian tribes like the Oroči,[2] but these recent references are hardly of historical interest. The news of the occurrence of domestic reindeer on Saghalin was then received in Peking as a novelty.[3]

Reindeer have been traced by sinologues in Chinese records where reindeer cannot be discovered by an unbiased mind. The term Ti 狄 is one of those general designations under which the ancient Chinese comprised a certain group of barbarous or semi-barbarous tribes occupying the southern part of present Mongolia. Klaproth[4] argued that the word *ti* signifies also a large wild stag, and concluded that in ancient times the hordes in question availed themselves of reindeer, like their eastern neighbors, and that for this reason they received the name Ti.[5] This argumentation is open to several objections: true it is, *ti* may denote a wild stag, but it is nowhere explained as a tamed deer or reindeer. There is no such interpretation, as intimated by Klaproth, of the ethnic term Ti on the part of the Chinese, neither is there any record that the alleged eastern neighbors of those Ti ever kept reindeer.

[1] His account of the Kirgiz has been translated by Klaproth, *Mémoires relatifs à l'Asie*, vol. III, p. 366.

[2] *Huang ts'ing chi kung t'u*, ch. 3. In the memoirs of the Manchu Tulishen's embassy to the Kalmuk (1712–15) the reindeer among the Tungus in the region of Irkutsk is briefly described. See G. T. Staunton, *Narrative of the Chinese Embassy to the Khan of the Tourgouth Tartars* (London, 1821), p. 70.

[3] Compare Du Halde, *Description of the Empire of China*, vol. II, p. 247. The Japanese traveler Mamia Rinsō, who visited Saghalin in 1808, brought the first account of the reindeer to Japan. Ph. von Siebold, *Nippon*, vol. II, pp. 229–230.

[4] *Tableaux historiques de l'Asie*, p. 102.

[5] According to Klaproth, *Mémoires relatifs à l'Asie*, vol. I, p. 188, the term Pei Ti (Northern Ti) would date only from the T'ang period. It is found, however, at an earlier date; for instance, in the *Nan shi* (ch. 79, p. 8 a).

It is asserted also, after Chinese sources, that the northern Shi-wei should have tamed the reindeer.[1] The text here referred to, however, contains nothing to this effect, but merely says that the country of this people abounded in wild deer.[2] According to the Chinese account, this tribe raised cattle, swine, and dogs as domestic animals, and fish-skin formed their clothing; reindeer nomads certainly wear reindeer-skins. Another group of this people, plainly called Shi-wei, who lived a thousand *li* north of the Mo-ki or Wu-ki, the center of their territory being in the basin of Kerulen river, subsisted on pork and fish, reared cattle and horses, but lacked sheep; they clothed themselves in the skins of white deer.[3] This "white deer" may have been elk or wild reindeer. Theophrastus[4] already mentions that the skin of the wild reindeer (*tarandus*), which according to him occurs in the territories of the Scythians and Sarmatians, is of the thickness of a finger, and is so durable that it is made into thoraxes; and the lexicographer Hesychius (fifth century A.D.) says that the Scythians employed the furs of the tarandus as clothing.[5]

Archaeological monuments do not shed much light on the ques-

[1] J. H. Plath, *Die Mandschurey*, p. 82, who accepted the translations of the Jesuit missionaries of the eighteenth century.

[2] *T'ang hui yao*, ch. 96, p. 7. Compare Vasilyev in *Trudy* of the *Oriental Section of the Imperial Archæological Society*, vol. IV (1859), p. 32.

[3] *Wei shu*, ch. 100, p. 4 b. According to the T'ang Annals (*T'ang shu*, ch. 219, p. 7), the Shi-wei raised a large breed of swine, the tanned skin of which was used for garments. The so-called Northern Annals give the following notice of this tribe: "The Shi-wei lived a thousand *li* north of the Mu-ki, subsequently styled Mo-hö, six thousand *li* from the capital Lo-yang. In speech they were related to the Kitan. They raised cattle and horses, but not sheep, and also kept swine, subsisting on pork and fish. In the summer they led a sedentary life; in the winter they roamed along the river-courses, catching sables. They used the composite horn bows and long arrows. White-deer skins formed their clothing. Corpses were buried in the trees [as still practised by Tungusian tribes and often observed by myself]. They used coracles; and their primeval forests and pasture-lands teemed with a rich fauna, and [unfortunately, as at present] also with mosquitoes" (*Pei shi*, ch. 94, p. 9 b). Deer-skin clothing is ascribed by the Chinese annalists to several other tribes of Siberia; thus, for instance, the women of the Liu-kuei, a tribe to be located in Kamchatka (see *T'oung Pao*, 1916, p. 368), employed for their winter costume skins of swine and stag, and fish-skins for summer-dress.

[4] *Fragments*, 172 (*opera*, ed. Wimmer, p. 458).

[5] Sarauw, *Rentier in Europa*, p. 10.

tion. Wild deer, particularly the elk, are frequently represented on so-called Scythian and Siberian antiquities of the bronze age.[1] In Mongolia many sepulchral stones with figures of stags have been found.[2] A representation of domestic reindeer accompanied by men, of ancient date, has not yet been traced.

From the preceding notes it becomes manifest that the domestication of the reindeer does not go back to times of a dim antiquity, but is of a comparatively recent date, falling within the historical era. The Chinese account of A.D. 499, as far as we know at present, is the earliest in existence. The reindeer was then milked and employed as a draught-animal; in other words, its domestication was then an accomplished fact. By calculating several centuries upward of that date, we thus arrive at the primeval period when the initial steps leading to the domestication were taken. The interval required for the process of domestication in its various stages will naturally remain a matter of conjecture and speculation, but a fair compromise may be reached by the formula that the incipient stage may belong to the beginnings of our era. It is obvious also, from a purely historical standpoint, that the domestication is far older in Asia than in Europe, and that consequently the center from which the domestication has taken its starting-point must be sought for on Asiatic soil.

Center of Domestication

All observers agree in regarding the domestication of the reindeer as an imitative process leaning toward that of horse and cattle. In fact, the reindeer is utilized by man in exactly the same manner as those two breeds,—as a draught, pack, and riding animal. The recent date of the domestication also brings out its secondary character. One of the most peculiar and uniform features which is apt to illustrate the imitative tendency is the castration of the stags, practised alike throughout the zone of reindeer occurrence.

[1] See Aspelin, *Antiquités du nord finno-ougrien*, p. 68, no. 307; p. 69, nos. 311, 313–315; p. 71, no. 323.

[2] *Inscriptions de l'Iénissei*, p. 16. I. G. Granö, *Archäol. Beobachtungen in Südsibirien und Nordwest-Mongolei* (Helsingfors, 1910), pp. 49, 53; and *Geogr. Verbreitung der Altertümer in der Nordwest-Mongolei* (*ibid.*, 1910), pp. 37, 45.

In the eighteenth century Knud Leems[1] reported, "Taurum rangiferinum castraturus Lappo, testiculos non, ut alias fieri solet, dissecta cute, eruit, sed, admoto ore, dentibus contundit." In the same manner the process is described in modern times by J. D. Caton,[2]

> The Lapp perform the operation with their teeth; the glands are bruised or crushed without breaking the skin. No other mode of castration has ever been known among the Lapp. This imperfect operation is probably sufficient for their purposes, for it so subdues the natural ferocity of the animal as to subject him to control, while it leaves enough of spirit to make his services highly sufficient. Were it carried as far as with us, it might so destroy his energy as to leave him practically useless.[3]

The Ostyak designate the gelded reindeer *xatri*, which, according to S. Patkanov,[4] is a loan-word received from Samoyed. Whether the Ostyak adopted the process from this people remains an open question; but this is more than probable, in view of the fact that the Samoyed are the most skilful and successful reindeer-breeders, and are doubtless responsible for the transportation of the animal from Asia to Europe.[5] The Chukchi, according to Bogoras,[6] in order to geld the bucks, bite with their teeth either through the dowcets or through the spermatic ducts. The operation is said

[1] *Beskrivelse over Finmarkens Lapper* (Kiöbenhavn, 1767; in Danish and Latin), p. 152. About a century earlier we have the same observation recorded by J. Scheffer, *Lappland* (Franckfurt, 1675), p. 374.

[2] *A Summer in Norway* (Chicago, 1880), p. 228.

[3] See also E. Demant, *Das Buch des Lappen Johan Turi*, p. 40. This book contains the autobiography of a Lapp, and is one of the finest documents of primitive life and thought that we possess.

[4] *Die Irtysch-Ostjaken*, vol. I, p. 18. See also A. Ahlqvist, *Journal de la Société finno-ougrienne*, vol. VIII, 1890, p. 6. There are many more Samoyed loan-words in Ostyak relative to reindeer-culture: hence Ahlqvist (*ibid.*, p. 21) concluded that the Ostyak appear to have adopted from the Samoyed certain important features of reindeer-breeding, or perhaps even this entire industry.

[5] Among the Samoyed, a very specialized nomenclature of the reindeer and the equipment relating to it obtains, as shown by a glance at A. Castrén's *Wörterverzeichnisse aus den samojedischen Sprachen*, pp. 262–263. Terms denoting the wild and domesticated animal, the gelded and ungelded male, are strictly differentiated; and there are peculiar words for the female, the calf in its various stages of growth, the old and the hornless animal, with many variations in the dialects.

[6] *Jesup North Pacific Expedition*, vol. VII, p. 84.

not to affect the reindeer much, for immediately afterward it continues to graze. Sometimes the scrotum is tied very tightly with a sinew thread, and after a while becomes atrophied and drops off.

The milking of the reindeer is another practice which demonstrates the dependence of the domestication. There can be no doubt that it came into existence in imitation of milking cows, mares, and sheep. The fact that this economy is comparatively old is attested by the Chinese account of the fifth century. Even the Tungus, who, with a few exceptions, use the reindeer solely for riding, milk the calving females. Four teacupfuls of milk within twenty-four hours make the whole produce. The Chukchi even try to suck milk from the doe's udder.[1] The reindeer is plainly not a milk-furnishing animal, and has been forced by man into assuming a rôle which is denied to it by nature.[2]

Property-marks for the purpose of recognizing their animals are utilized by all reindeer-breeding tribes. The Chukchi again betray their fondness of biting likewise in this case; for they mark their property by biting a piece out of the fawns' ears in late summer, or the next spring during the separation of bucks from pregnant dams. The Lapp,[3] Samoyed, Tungus, and other reindeer peoples, cut marks in the ears of their animals. Thirteen such marks from the Tungus of Ayan have been illustrated by Pekarski and Tsv'ätkov.[4] One or two cuts, in straight lines, angular, or rounded, are made in one ear or in both. This practice has been perpetuated by our Government in Alaska.

> Every local superintendent must take careful oversight of the annual marking of the reindeer and see that all reindeer are correctly marked according to ownership. He shall keep a complete list of such marks in the records of the station.[5]

[1] Bogoras, *l. c.*

[2] In regard to peculiar methods of milking on the part of the Lapp, see E. Demant, *Buch des Lappen Johan Turi*, pp. 30, 39; on the part of the Soyot Ø. Olsen, *Et primitivt folk* (Kristiania, 1914), p. 67.

[3] J. Scheffer, *Lappland*, p. 379.

[4] "Očerki byta Priayanskix Tungusov," *Publication du Musée d'Anthropologie*, vol. II, p. 37.

[5] *Rules and Regulations regarding the U. S. Reindeer Service in Alaska, approved June 10, 1907, and December 7, 1908* (Washington, 1911).

Aluminum button markers are employed for this purpose.[1] The reindeer-breeders of Siberia are not the originators of this custom, but it was doubtless transmitted to them by Turkish-Mongol tribes. The term *tamaga, tamga, tamka*, denoting a property-mark on cattle and subsequently a seal, is common to all of these; it is diffused all over Siberia, and is even known in China and Tibet (*dam-k'a, t'am-ga*).[2]

The uniformity of reindeer-breeding is characterized also by the universal method of lassoing the animals. Everywhere a long lasso, either plaited from horse-hair or from thin seal-skin straps, is used for catching the deer after pasturing in the morning, when its services are required. The Tungus are very skilful in throwing the lasso from a respectable distance; and most animals will patiently halt, or even run to their master's side, as soon as merely touched by the rope. A classical description of this procedure is given by the Yakut Uvarovski in his autobiography.[3]

The reindeer-breeders cannot lay claim, either, to any original thought or invention as to the entire apparatus utilized by them in connection with the reindeer. Above all, the pack-saddle and the method of loading, riding-saddle, harness, sledge, and snowshoes, are all borrowed institutions.[4] The geographical distribution of sledge and snowshoe by no means coincides with the area of reindeer domestication. On the one hand, we encounter the two implements among the primitive dog-breeding tribes of northern and northeastern Asia, inclusive of the Amur and Ussuri regions, where the reindeer is unknown; and, on the other hand, they extend far into the south of Siberia, even into Mongolia and Turkistan, where they are associated neither with the dog nor with the reindeer. Sledge and snowshoe, accordingly, cover an infinitely wider territory than the domestic reindeer, and obviously were in existence in

[1] S. Jackson, *Fourteenth Annual Report on Introduction of Reindeer into Alaska, 1904* (Washington, 1905), p. 108. On plates 33 and 34 of this report will be found illustrations of several such marks.

[2] W. Radloff, *Wörterbuch der Türk-Dialecte*, vol. III, col. 1003; T. Watters, *Essays on the Chinese Language*, p. 374.

[3] O. Böhtlingk, *Ueber die Sprache der Jakuten*, text, p. 45.

[4] Bogoras (*l. c.*, p. 88) has called attention to the uniform character of the collar for the sledge-reindeer among Chukchi, Tungus, Samoyed, and Lapp.

times prior to its domestication. As we learn from the early Chinese account relating to the year A.D. 499, the reindeer must have been trained to the sledge at that date (and certain it is that this utilization of the animal preceded its breaking-in for the saddle); and, since the same people had also horses and oxen for drawing vehicles, it is manifest that this older method was simply transferred to the reindeer. The Chinese annals furnish several classical examples of the early employment of snowshoe and sledge on the part of tribes which never availed themselves of the service of the reindeer.

According to the Annals of the T'ang dynasty (618–906), there was east of the Kirgiz, on the Yenisei, a tribe styled "Snowshoe Turks" (Mu ma T'u-küe, literally, "wooden-horse T'u-küe"),[1] consisting of three hordes.

> They covered their habitations with birch-bark and owned numerous horses. They used to cross the ice on snowshoes ('wooden horses') which they tied to their feet, taking curved branches as supports for the shoulders (snow-sticks), and thus swiftly pushing ahead.

In regard to the Pa-ye-ku (Bayirku), it is said that all people put wooden boards under their feet and pursue deer over the ice.[2] The Liu-kuei, a tribe to be located in Kamchatka and mentioned on page 113, note 3, according to the T'ang Annals,[3] "fastened to their feet wooden boards six inches wide and seven feet long, and thus hunted the game over the ice." Likewise the Kirgiz on the upper Yenisei, of whom we have a description in the Annals of the T'ang, pursued the game on snowshoes.[4] A description of the snowshoe and the mode of using it is given also by Rashid-eddīn in connection

[1] In Tibetan, *šiṅ-rta* (wooden horse) means any vehicle or carriage. Compare also Russian *konki* (skates; literally, little horses), from *kon'ok*, diminutive of *kon'* (horse). Chinese T'u-küe represents a transcription of the name Türk, more exactly of the plural form Türküt (see Pelliot, *T'oung Pao*, 1915, p. 687).

[2] *T'ang hui yao*, ch. 98, p. 16. The Pa-ye-ku are mentioned under the name Bayirku in the Turkish inscriptions of Kül-tegin and Bilgä-kagan; they were a Turkish tribe living in the north of the Gobi. See also above, p. 105.

[3] Ch. 220, p. 11 b.

[4] Some authors, like Klaproth and Ritter, thought in this connection of sledges; but it has been correctly observed by W. Schott, in "Ueber die ächten Kirgisen," *Abhandlungen Berliner Akademie* (1865), p. 447; and his additional notes in *Monatsberichte Berliner Akademie* (1874), pp. 1–8, that snowshoes solely are involved.

with the Uryangkit (above, p. 108). He adds that the snowshoe is known in a large part of Mongolia and Turkistan, and that ski-running is particularly practised by the Barguchim Tukum, Khori, Kirgiz, Urasut, Telengut, and Tumat. The word used by the Persian annalist is *čane* or *čana*, which, as is well known, is found in all Turkish and Mongol languages with both significances, "snowshoe" and "sledge:" Mongol *tsana* and *čana*, Buryat *sana*, Altaic *čanak*, *čana*, etc.; Finnish *saani*, Esthonian *sāń*, Lettish *sāńus*, *sańas*, Magyar *szán*, *szány*, or *szánka*, *szánko* (diminutive); Russian *sani* (plural), *sanki* or *sanočki* (diminutive).[1]

A profound study of all types of sledge and snowshoe will doubtless yield promising results.[2] Here it may be emphasized only that the reindeer-breeders adopted ready-made what they found, merely changing some of the material: thus they preferred reindeer-skin for snowshoes, while the Turks used horse-skin and the Gilyak seal-skin. L. von Schrenck[3] has shown in particular how the Oročon (Schrenck: Oroki), scattered over a few spots of Saghalin Island, adapted the dog-sledge of the Gilyak to reindeer-transportation.[4]

From a negative viewpoint, we might say that neither the

[1] J. Kalima (*Wörter und Sachen* (1910), vol. II, p. 183) has studied to some extent the distribution of this word from the Slavistic standpoint, and arrives at the conclusion that it is a very ancient word, which Slavic, Finno-Ugrian, and Turkish languages have in common. In my opinion, the word is of Turkish-Mongol origin, and a loan-word in Finno-Ugrian and Slavic. There can be no doubt that the term has migrated jointly with the object which it denotes. The investigation of Kalima is obscured by the fact that he adds Lapp *čiöinne*, Russian *čuni*, *čunki* (in the northern dialects), and Vogul *šun*, which must be dissociated from the above series, and in fact are independent words.

[2] Compare the preliminary remarks on snowshoes by G. Hatt, "Moccasins and Their Relation to Arctic Footwear," *Memoirs American Anthropological Association*, vol. III, 1916, p. 240.

[3] *Reisen und Forschungen im Amur-Lande*, vol. III, p. 494.

[4] It is not correct, however, to say with Schrenck that the Saghalin Oročon are the only Tungusians to make use of sledges in connection with the reindeer. The practice is not generally Tungusian, as wrongly asserted by C. Hiekisch, *Die Tungusen*, p. 78, but is an exception, which, however, occurs sporadically wherever Tungusians come in contact with Palaeo-Asiatic dog-breeders. The illustration of a Tundra Tungus in the Kolyma district, driving on a reindeer-sledge, may be seen in V. Jochelson, *Očerk zv'äropromyšlennosti i torgovli m'äxami v Kolymskom okrug'ä* (Sketch of the Animal Industry and Fur Trade in the District of Kolyma), p. 36.

Lapp nor the Ugrians in the west, nor the Yakut (p. 110), nor the Chukchi and Koryak in the northeast, can come into question as the original reindeer-tamers. Among the Chukchi the introduction of the reindeer appears to be an affair of comparatively recent date, as shown, if by nought else, by the imperfect degree of domestication. It is difficult, however, to accept Bogoras' opinion that "they did not introduce the tame reindeer from their neighbors, but that, in imitation of them, they attempted to domesticate the race of reindeer inhabiting their own country." Such an expenditure of energy cannot be attributed to the Chukchi; and, as a matter of fact, such an instance of waste of energy is beyond our experience in the life of peoples. Man in general is not inclined toward work, unless compelled by sheer necessity or some inducement; still less does he try to do over again what has been accomplished by his neighbor. Bogoras believes his theory to be plausible, since the Chukchi reindeer is quite different from that of the Tungus. This fact, however, can be simply explained from the constant crossings between tame and wild reindeer, emphasized by Bogoras farther on. It is inconceivable that any Palaeo-Asiatic tribe ever undertook to domesticate the reindeer, as the maintenance of sleigh-dogs excludes the reindeer. L. von Schrenck[1] has already made the appropriate remark that

> the ancestors of the migrating Chukchi and Koryak themselves surely did not domesticate the reindeer, but received it in the domesticated state from a nomadic tribe, presumably the Tungus.

Tungusians, however, cannot be claimed to be the originators of reindeer-domestication, as L. von Schrenck maintains they are. The first Russian discoverers of eastern Siberia, who came in contact with the Tungus, speak of Reindeer, Horse, Dog, Steppe, and Woodland Tungus.[2] These divisions have no ethnographical sig-

[1] *Reisen und Forschungen im Amur-Lande*, vol. III, p. 489.

[2] P. J. v. Strahlenberg, *Das nord- und östliche Theil von Europa und Asia* (Stockholm, 1730), p. 423. Regarding the distribution and economy of the Tungus, see S. Patkanow, "Geographie und Statistik der Tungusen-Stämme Sibiriens" *Keleti Szemle*, vol. IV, pp. 141–171, 287–316; vol. V, pp. 36–56, 185–203; vol. VI, pp. 130–174, 222–283; and the same author's *O prirost'ä inorodčeskago naseleniya Sibiri* (S.-Peterburg, 1911), pp. 87–115.

nificance, but merely allude to the economic conditions under which the people were encountered at a certain time. Even this mode of life is by no means a stable characteristic, for the economy of these tribes is subject to sudden and fundamental changes. Cases have occurred where reindeer-owners lost their herds and turned to the rearing of horses or only dogs, or where woodland people were transformed into inhabitants of the steppe.[1] The Birar, settled in the river system of the Bureya and on both banks of the Amur above and below the mouth of that side-river, according to the Cossack Poyarkov, who came in touch with them in 1646, were engaged in reindeer-breeding; only thirty-five years later they are described as horse-nomads.[2] The Tungusians, accordingly, are shifting opportunists, and, in the course of their constant peregrinations, simply adopt that mode of life best suited to the geographical and economic environment of the respective places. Originally they were mere hunters and fishermen; but, being possessed of an adaptable spirit and a quick grasp of changeable conditions, they were capable of appropriating any industry offered by their neighbors. Historical considerations show us that the Tungusian tribes, in former periods of their life, were never given to reindeer-breeding. In fact, they are late arrivals in Siberia, while their original home is to be sought for in Manchuria. We can trace their history almost completely from very early times by means of the Chinese annals; but in these no mention of reindeer is made with reference to any Tungusian people, with the sole exception of a branch of the Wu-huan (p. 105). Only when they were pushed into Siberian regions did they become acquainted with the reindeer. It is even doubtful whether the Tungusians were the first to use the reindeer as a riding-beast. The Soyot, as will be seen, still ride the reindeer; and the reindeer-riding tribe alluded to by Marco Polo (p. 107) was doubtless related to the Soyot or their group.

If it is true that the reindeer represents a mere repetition of cattle and horse domestication on a smaller scale, it is logical to

[1] Examples are cited by C. Hiekisch, *Die Tungusen*, p. 47; and L. v. Schrenck, *Reisen und Forschungen im Amur-Lande*, vol. III, p. 144.

[2] Patkanow, *Keleti Szemle* (1904), vol. V, p. 41.

conclude that the reindeer can have been domesticated only in a locality where it occurred in close association with cattle and horse. In the northern regions, where the wild tundra reindeer prevails, we meet at present as domestic animals the reindeer and the dog; in the southern belt, occupied by the wild woodland reindeer, we find the domestic reindeer in company with other large domestic stocks. It is therefore clear that the original center of domestication is to be looked for in the southern belt. The fact that Ugrian peoples were in possession of reindeer-herds employed as draught-animals toward the end of the fifteenth century, has been established from Russian documents (pp. 96–99). At present the well-to-do Wogul living in Beresov (in the western part of Tobolsk government) keep cows, horses, and reindeer. They are so reduced to poverty, however, that few own more than several tens. A Wogul on the upper Tapsya River, who has a couple of hundred, is regarded as very rich in this region; whereas, compared with well-to-do Samoyed in Obdorsk, he would only be a wretched beggar, for these count their reindeer by the thousands.[1] In the beginning of the eighteenth century the wealthy among the Ugrian Ostyak still kept a large number of reindeer, together with cattle, horses, and dogs; but many of them were so poor that they had to be content with reindeer. This is the account of G. Novitski, who wrote in 1715,—the earliest historian of this tribe.[2] At the present time, only the Ostyak of the north, being neighbors of the Samoyed, still have reindeer;[3] but it lost ground among the Irtysh-Ostyak farther south. In the epic traditions of this people, ably collected and translated by S. Patkanov and traced with good reason to a period from the fourteenth to the fifteenth century, reindeer and

[1] A. Ahlquist, in *Erman's Archiv für wissensch. Kunde von Russland* (1860), vol. XX, p. 157. Regarding reindeer among the Wogul, see also A. Erman, *Reise um die Welt*, vol. I, pt. I, p. 384. Reindeer-sledges of the Wogul are illustrated by K. D. Nosilov, *U Vogulov očerki i nabroski* (1904), pp. 183, 189.

[2] G. Novitski, *Kratkoe opisanie o narod'ä Ost'atskom*, ed. of L. Maikov (St. Petersburg, 1884), p. 37. An interesting contribution to the history of this people is the article of A. van Gennep, "Origine et fortune du nom de peuple 'ostiak'": *Keleti Szemle* (1902), vol. III, pp. 13–32; reprinted in his *Religions, mœurs et légendes*, pp. 94–109.

[3] M. A. Castrén, *Reiseerinnerungen aus den Jahren 1838–1844*, p. 300.

dog are mentioned as domestic animals. At that time, also the inhabitants of the northern part of the district of Tobolsk kept herds of reindeer; while at present half-domesticated reindeer are encountered only farther northward, beneath Beresov. The domestic reindeer supplied the Ostyak with meat, skins, and sinews; served as most important draught-animal in those snow-abounding regions; and was slaughtered in honor of the gods on the occasion of the sacrificial holidays. When the breeding of reindeer was still thriving among them, this animal was exclusively chosen for the sacrifice, which is still customary in the north, among the Ostyak and Samoyed living there.[1] Patkanov holds the opinion that reindeer-breeding is only a secondary industry among the Ostyak and Wogul; that is to say, when these tribes were pushed from southern regions into their present northern domicile, they were compelled to abandon the larger domestic breeds in consequence of unfavorable geographical conditions, and to take to the reindeer. I would not subscribe to this theory unconditionally; but what interests us in this connection is merely the coexistence of reindeer, cattle, and horse among Wogul and Ostyak, neither of whom, notwithstanding, can be regarded as the original domesticator of the reindeer.

There is but one territory where all the necessary postulates for reindeer-breeding are given, and which may come into question as the original center of the domestication, and this is the region of Lake Baikal. There we meet the reindeer, wild and domesticated, and, as has been shown, from ancient times. There we meet a host of tribes partially engaged in horse and cattle rearing, and partially depending on the reindeer; there, accordingly, the contact of reindeer-breeders with horse and cattle raisers is virtually established. The ancient Chinese records, as we have seen, likewise point to the same center. In the Baikal territory we find at the present time three large and distinct stocks of peoples,—the Buryat, a branch of the Mongol family; Tungusians; and a large number of tribes, originally of Samoyed and Yenisei-Ostyak stock, but now either Turkicized (*otatarilis*, "Tatarized," as the Russians say)

[1] S. Patkanov, *Die Irtysch-Ostjaken*, vol. I, p. 109; vol. II, p. 017.

or Mongolized, and for the most part speaking a Turkish language. The Buryat occupy the area in the governments of Irkutsk and Transbaikalia from the Chinese frontier as far as the Lena system northward, and from the rivers Onon to Oka, the side-river of the Angara, westward, and still farther west into the region of Nižne-Udinsk. The Buryat element is strongest beyond the Baikal, in the valleys of the Uda, Onon, and Selenga. Those on this side of the Baikal are to some extent Russianized, even practising agriculture. The others are herdsmen and owners of horses, cattle, sheep, and goats. The reindeer is entirely foreign to them, and never was in the hands of any tribe of the Mongol family. Tungusians are scattered in the governments of Irkutsk, Yenisei, and Transbaikalia, chiefly subsisting on fishing and hunting, but also on agriculture and cattle-breeding. In Irkutsk government only a few clans on the upper Lena keep reindeer; in Yenisei government the latter are owned only by the well-to-do. In Transbaikalia we encounter among the Tungusians hunters, agriculturists, cattle-breeders, and reindeer people. Especially those inhabiting the districts of Tshitin and Barguzin keep reindeer.[1]

It seems certain that the Samoyed are not autochthonous in their present habitats, but migrated there from southern regions, in all probability from the territory of the Sayan mountains or the upper courses of the Yenisei basin, where there are still many scattered tribes of them enclosed by Mongols and Turks. Most of these split Samoyed adopted the language and customs of their superior neighbors, yet they remain conscious of the fact of their original nationality. I designate this group as Sayan tribes or southern Samoyed. Among the Soyot within the boundaries of China there are family-names that also occur among the Samoyed roving along the Arctic littorals.[2] The Woodland Kamasin still spoke Samoyed at the time of Castrén's travels (about 1840–50);

[1] S. Patkanov, *Keleti Szemle*, vol. VI (1905), pp. 278, 279. Concerning the Barguzin Tungus, see an article by N. M. Dobromyslov, "Zam'ätki po etnografii Barguzinskix Oročen," in *Trudy* of the Troitskosavsk-Kiachta Section of the Imperial Geogr. Soc., vol. V (1902), pp 78–87.

[2] M. A. Castrén, *Kleinere Schriften*, pp. 116–117; W. Crahmer, *Zeitschrift für Ethnologie* (1912), p. 110.

but fifteen years later, when visited by W. Radloff,[1] they had adopted a Turkish form of speech. Two groups of these peoples are still active reindeer-breeders,—the Karagas and the Soyot. The former roam in the territory between the rivers Oka, Uda, Biryusa, and Kan (the boundary district of the governments of Yenisei and Irkutsk), numbering about 550 individuals. They are divided into five clans, one living in the neighborhood of the Soyot, another near the Kamasin, and another near the Buryat. Although now speaking a Turkish language of which we have an excellent grammar by Castrén, and closely resembling their Turkish neighbors in costume and manners, their methods of hunting and reindeer-keeping, as well as their winter tents made of reindeer-skins, are identical with those of the Samoyed. Also their physical habitus, several of their family names, and the survival of many Samoyed words in their speech, clearly bespeak their origin. The Soyot or Soyon, styling themselves Tuba and designated by the Mongols Urangkhai (see above, p. 109), inhabit northwestern Mongolia and a small strip of country along the Russian frontier from the sources of the river Kobdo as far as lake Koso. A great number of them who live farther south on the slopes of the Tangnu mountains are completely converted into Mongols. According to Castrén, many Soyot clan-names agree with those of the Samoyed; and the Soyot clan Mattar, according to traditions, originated from the Mator, who decidedly were Samoyed; he argued also that several Yenisei-Ostyak clans had become Soyot. Radloff[2] regards them as a medley of Kirgiz, Samoyed, and Yenisei-Ostyak; Katanov,[3] as consisting of Mongol, Turkish, and Samoyed elements. At present their language is Turkish, but among many tribes Buddhism and Mongol speech have spread so widely, that the Turkish element is threatened with extinction.

G. Radde,[4] in 1862, outlined the following sketch of the distri-

[1] *Ethnographische Uebersicht der Türkstämme*, p. 6. Regarding the Sayan tribes compare also the interesting article of N. F. Katanov, "Predaniya Prisayanskix plemen o prežnix d'älax i l'ud'ax," in *Sbornik v čest' semides'atil'ätiya G. N. Potanina*, pp. 265–288.

[2] *L. c.*, p. 17.

[3] In *Sbornik Potanina*, p. 286.

[4] *Reisen im Süden von Ost-Sibirien*, vol. I, *Säugetierfauna*, p. 287.

bution of the reindeer in the Baikal region. South of Ilchir lake the tame reindeer, together with the horse and frequently also with cattle, is found among the mountain tribes. During the summer a division of these herds becomes necessary, the reindeer being driven into the high mountains of an altitude of seven or eight thousand feet, the horses and cattle grazing in the deeper valleys of four or five thousand feet. In the Baikal regions the reindeer is ubiquitous; in the southwestern parts, however, it is sparse now. In the mountains where the river Jida takes its source, south of Turansk, it is met among the Uryankhai, who inhabit there the space between the Russian and Chinese frontiers. It is excluded from the Selenga valley, the upper part of which, on the Russian side, is inhabited by Buryat engaged in the rearing of sheep, cattle, and horses. In the northeastern corner of lake Baikal it increases in frequency, but even there the Tungusians become impoverished in consequence of the decrease of the stock. In regard to the Soyot and Jot, he observes that they rear reindeer in large numbers (up to three hundred). The wild species still occurs farther to the south as an inhabitant of the upper zones of the forest boundary, and beyond as far as the snow-line. Hahn[1] has made the correct observation that in the Sayan mountains, the source of the Amur, the reindeer reaches the southernmost point of its diffusion, and comes there in contact with the camel and tame yak; but he draws from this fact no conclusion whatever as to the home of the domestication, but offers solely the commonplace remark that any of the migratory tribes of northeastern Asia may have been pushed back into an inhospitable country, and, losing its stock of cattle and pack-animals owing to the unfavorable climate, tamed the reindeer as a substitute.

The Soyot were visited and studied in the summer of 1914 by Ørjan Olsen, who published interesting information on the tribe.[2] According to this author, the breeding of reindeer constitutes a secondary industry among the Soyot, who also keep horses and dogs, in opposition to the Lapp and Samoyed. Their herds are

[1] *Haustiere*, p. 266.

[2] *Et primitivt folk. De mongolske Rennomader* (Kristiania, 1915). Compare the analysis of Ch. Rabot in *La Géographie*, vol. XXXI (1916–17), pp. 42–46.

not very numerous. The most fortunate among the inhabitants of the Sayan mountains (and there are few) own no more than four hundred animals; in general, the herds count from ten to fifteen heads, at least on the banks of the Sesti-Kem. The people, accordingly, cannot live exclusively on the flesh of their herds; and those on the upper Yenisei loathe to slaughter their animals, unless compelled to do so by famine. The only alimentary product is the milk, consumed either fresh or in the shape of butter or cheese. One or two large cupfuls are obtained from each operation, which is performed twice a day in an enclosure formed by wooden palisades. The reindeer is used by the Soyot as a pack and riding animal.[1] It is not attached to a sledge. The animal belongs to a very sturdy breed, the largest being able to carry loads from eighty to one hundred and ten kilo; with such a load, they make five to six kilometers an hour.

Among the Soyot, the domestication of the reindeer has progressed further than among any North-Asiatic tribe. Although they capture wild reindeer and cross these with their domesticated individuals, this offspring is remarkably little savage. Whereas other reindeer must be lassoed in order to be caught for duty, the Soyot reindeer allow themselves to be caught by hand, and follow their master like dogs, licking his hand with the expectation of a bit of salt. When pasturing in the woodland, a call from their owner is sufficient to make them return immediately. It is a notable feature also that the domestic reindeer of the Soyot territory is capable of standing the extreme summer heat. At that time the wild reindeer, which likewise occurs in the region of the sources of the Yenisei, take refuge in the snow zone of the high mountains. The domesticated herds constantly remain in the forest, in the proximity of human habitations, without suffering

[1] Compare the illustrations in Olsen, pp. 52, 73. According to I. Pesterev *Magasin asiatique*, by J. Klaproth (Paris, 1825), vol. 1, p. 126, who was commanded to the Russian-Chinese frontier in the districts Udinsk and Abakansk from 1772 to 1781, the nomadic tribes near the fort of Udinsk (then belonging to the government of Tobolsk), divided into four sections, Silpigursk, Udinsk, Karagansk, and Kamgatsk, kept domestic reindeer from oldest times, the richest possessing a hundred animals; seven years before his time they lost the greatest part. He states also that the stags were used for the hunt and mounted by the hunters.

from the heat. During the hot hours they rest under thickly foliated trees. In order to protect the fawns from the blaze of the sun, the Soyot erect hedges around large cedars.

The culture of the Soyot, like that of any other people in northern and central Asia, is in a state of complete disintegration, and original conditions can no longer be expected. What we find at present is merely the weak echo of a former glory which still eloquently speaks to us from the brief accounts of Marco Polo, Rashid-eddīn, and the Chinese annals. It is difficult, if not impossible, to credit any domestication to a certain people, or even to a certain stock of peoples. In the majority of cases we must be content to trace the beginnings of a domestication to a more or less securely defined geographical area. In the present case it can be positively stated only that the primeval domestication of the reindeer took place in the Baikal region; but, if the original domestication of the reindeer is to be attached to the name of a tribal group, I should venture to say it was the southern Samoyed, or the Samoyed in the early period of their history, before migrating into their present northern habitats. I do not say, of course, that the present Soyot were the domesticators: our knowledge of the history of this tribe is altogether too vague to admit of such an interpretation. The Soyot are simply remnants and epigones of that once extended and powerful family in the midst of which this fact was accomplished.

The history of the domestication can now be clearly conceived. From the Samoyed it spread eastward to the Tungusians; from the latter to the Yakut, Chukchi, and Koryak; westward to the Ugrian tribes of the Ural and the Lapp.[1] Applied to the reindeer, this result means that the woodland reindeer was domesticated in times prior to the tundra reindeer. When the Samoyed moved northward, they naturally took along their woodland reindeer, and gradually replenished and improved their old stock by capturing wild tundra reindeer (by the methods described in the following

[1] The peculiar boat-shaped sledges of the Lapp, to which G. Hatt, "Lappiske slædeformer," *Geografisk Tidskrift*, vol. XXII (1913), pp. 139–145, has devoted a special study, in my opinion are derived from the Samoyed; for A. Olearius, *Reise-Beschreibungen* (Hamburg, 1696), p. 81, already mentions the reindeer-sledges of the Samoyed, which are shaped like half canoes or boats.

chapter), until a point was reached when the latter breed preponderated or prevailed exclusively.

The Ainu of Saghalin do not keep reindeer, but only know the animal (styled by them *tonakai*) in the possession of the Tungusian Oročon. It would hardly be necessary to emphasize this fact, were it not that A. E. von Nordenskiöld[1] has published the sketch of an Ainu standing on large snowshoes, and pulled along by a reindeer the bridle of which is tied to his belt. This illustration is said to be derived from a Japanese book published in 1804. In regard to such an employment of the reindeer on the part of the Ainu I learned nothing on Saghalin, nor can I find any reference to it in the literature on the Ainu. Even the Japanese traveler Mamia Rinsō, who visited Saghalin in 1808, and whose valuable account has been made accessible by Ph. von Siebold, gives no information on this point; on the contrary, he mentions the reindeer only in the possession of the Orotsuko (Orokko, Oroki, Oročon).[2] The sketch in question, accordingly, is either based on an incidental and isolated occurrence, or, which is more probable, represents a purely imaginative artistic production in which two features foreign to the Japanese—snowshoes and reindeer—were arbitrarily combined.

Process of Domestication

We have no contemporaneous records showing how the initial domestication of the reindeer was brought into effect. In order to obtain some idea as to how this was done, or might have been done, we must rely upon a reconstructive method. One means to this end is furnished by present-day observations of the training of individual animals. The schooling of the individual is typical of the entire breed, and the course of lessons through which each animal has to run at present must have been valid, with some variations perhaps, also ages ago.

In regard to the training of the animals, S. Jackson [3] has the following observation:

[1] *Umsegelung Asiens und Europas auf der Vega*, vol. II, p. 101.

[2] Ph. von Siebold, *Nippon*, new ed., vol. II, p. 229.

[3] *Fourteenth Annual Report on Introduction of Domestic Reindeer into Alaska* (Washington, 1905), p. 126.

The training begins when the deer is three years old. Generally the stoutest males and geldings are selected. Females are also trained, but they are smaller and less enduring. The training begins by lassoing the selected animals, thus separating them from the herd. The poor beasts are much scared, and jump about in frantic efforts to escape. The trainer advances hand over hand on the rawhide lasso till the head is reached. They are then sometimes given a little salt, of which they are fond; they are then led about for some time or tied to a post to accustom them to confinement and, the lesson over, again released. This is repeated day by day, and when sufficiently tamed they are harnessed and in the same manner gradually accustomed to draw light loads. This takes a long time and persistent work. They should not be worked before they are three years old. At six or seven they reach their prime and then gradually decline.

The Eskimo selected by the government as apprentices to learn the art of breeding reindeer from expert Lapp reindeer-men enter into an agreement to remain from two to five years, or until sufficient skill to handle a herd is acquired.[1] This affords some idea as to the time required for a man to develop into a herder.

Although the reindeer is the only species of the deer family that has been brought into the state of domestication, there are many examples known of other members of the family *Cervidae* which develop a great adaptability to domestication and have been tamed to a high degree. Yet domestication has succeeded only in the case of the reindeer. The efforts to raise other kinds of deer are interesting to the student of reindeer-domestication as affording an object-lesson and showing us the possibilities in the initial stages preceding the state of true domestication.

All of the deer family are easily tamed. The moose has often been reared and tamed in this country; but I know of no systematic attempt to domesticate them, nor have I ever heard of their breeding in domestication. They have been sometimes broken to the harness and proved themselves able to draw good loads; and yet I know of no regular effort that has been made to reduce them to servitude. When tamed, they are reasonably docile, except the males during the rutting season, when, as might be suspected, they become ferocious, and should be kept in close quarters where they can do no harm. If castrated young, and early taught obedience to man, we may not doubt that they would readily submit to his dominion, and their great strength would give promise of useful

[1] *Ibid.*, p. 128.

beasts of draught, especially in countries where deep snows prevail, through which they pass with facility where ordinary cattle could make no progress.[1]

A highly interesting notice on deer-farming has been written by D. E. Lantz.[2] In the United States, the wapiti or Rocky Mountain elk (*Cervus canadensis*) and the Virginia deer (*Odocoileus virginianus*) are managed and reared in enclosures, chiefly for profit in the sale of venison; but also the desire to preserve our vanishing game has caused the confinement of small herds under private ownership in many places. The elk readily adapts itself to any environment. It proves especially useful in clearing out under-brush from thickets, in which they are more useful than goats, since they browse higher. The increase of elk, while kept in preserves with surroundings as nearly natural as possible, is equal to that of cattle: fully ninety per cent. of the females produce healthy young. The male elk is ordinarily docile, but in the rutting season the older ones often become ill-tempered and dangerous. The remedy for viciousness is castration, the effects of which are that the animal is made docile, and its value for venison is greatly enhanced. The stocking of parks and preserves with deer merely for sport or aesthetic purposes appeals much more to a sensitive mind. The idea of raising beautiful animals like deer merely for slaughtering purposes is revolting and unsportsmanlike, and for this reason has no future. A vigorous propaganda in favor of the destruction of some of our finest game-animals, which we have every reason to wish to see preserved, should be combated in all ways possible.

Examples of tame deer can be gathered from all parts of the world and from all times. In ancient Italy herdsmen reared does (*caprea*) on sheep's milk, and the wealthy Romans were fond of keeping them in their parks together with chamois and gazelles.[3]

[1] J. D. Caton, *The Antelope and Deer in America*, p. 277. This author, further, has interesting notes on efforts to tame caribou, elk, and other deer.

[2] "Deer Farming in the United States," published by the U. S. Department of Agriculture, *Farmer's Bulletin 330* (Washington, 1908), p. 20. Compare also the same author's "Raising Deer and Other Large Game Animals in the United States," published by the U. S. Department of Agriculture, *Biological Survey, Bull. No. 36* (Washington, 1910), p. 62.

[3] O. Keller, *Tiere des classischen Altertums*, p. 103.

Tame stags are frequently mentioned by Greeks and Romans. Sertorius owned in Spain a white deer, which, he made the people believe, communicated prophesies to him. Vergil[1] tells how a stately stag was bathed, combed, and adorned with flowers by Silvia, the daughter of the head pastor Tyrrhus, and how the animal became accustomed to the master's hand and table. In art, neck-collars and girth are repeatedly represented on stags. Apollo, Artemis, and Amor drive in chariots drawn by stags or deer. Heliogabalus possessed a chariot pulled by four powerful stags; and Aurelian, in his triumph over Zenobia, drove with a team of four tame stags which had once belonged to a king of the Goth.[2] Columella[3] says that wild animals, like roes, antelopes, stags, and boars, are kept either for one's pleasure or for sale and profit. In the former case, any hedged place near one's homestead is sufficient, and the animals receive food and drink from one's hand; a plot of woodland with running water, walled around or fenced with pallisades, must be set aside for the game.

The genus *Dama*, which originally appears to have been restricted to the Mediterranean countries and Persia, has been introduced into western and central Europe, where it exists in a semi-domesticated condition as far north as the British Islands and the south of Sweden.

> Owing to long domestication [read "taming"], the fallow deer of the British parks frequently display great variation from the original type of coloration, and a uniformly dark brown breed has been long established, while white or whitish varieties are far from uncommon.[4]

Tamed deer were kept and fed by the hermits of ancient India. The deer-park near Rājagṛiha in which Buddha used to dwell is familiar to all readers of Buddhist literature. The kings of India built special stables for deer on the west side of their palaces.[5]

West of Tokmak the Turkish Khans of the seventh century maintained a summer residence with a park of tame harts provided

[1] *Aeneis*, VII, 483.

[2] Keller, *l. c.*, p. 90; and *Antike Tierwelt*, vol. I, p. 278.

[3] *De re rustica*, IX, I.

[4] R. Lydekker, *Catalogue of the Ungulate Mammals in the British Museum* (London, 1915), vol. IV, p. 229.

[5] B. K. Sarkar, *The Sukranīti* (Allahabad, 1914), p. 30.

with bells and rings,—in the words of the Buddhist pilgrim Hüan Tsang, " familiar with men and not fleeing at their sight." The Khan, being very fond of them, forbade his subjects to kill them on pain of death without remission.[1]

The Island Mijo, or Aki-no Mijo (so called from the neighborhood of the province Aki), is famous for a particular breed of deer, which they say are very tame and familiar with the inhabitants. It is contrary to the laws of the country to chase and to kill them.[2]

In several places of the Altai, the maral (*Cervus elaphus*) is reared in captivity in consequence of the large demand for its antlers on the part of the Chinese, who are said to pay as much as 150 rubles for a pair, and employ it for medicinal purposes. Taming and feeding the animals are said to be easy; the antlers are cut off in their third year, the operation being without harm for the animals.[3] The Chinese have many stories in regard to tame deer, which were even used for drawing carriages. In mythology, gods and fairies ride on deers' backs.[4] Some tribes of Formosa practised the capturing of harts alive, and dexterity in this feat was regarded as a manly virtue highly extolled by folk-songs.[5]

It is not necessary to multiply these examples. Those given illustrate sufficiently the fact that many species of deer exhibit a high degree of adaptability, and that in diverse parts of the world and at different times efforts have been made to tame them and to keep them as pets in parks mainly for aesthetic reasons. In the case of every domestication, the animal deserves as much credit as man; an animal unqualified for the status, and without sympathetic instincts for man, cannot be domesticated.

[1] S. Julien, *Mémoires sur les contrées occidentales*, vol. I, p. 14; S. Beal, *Records of Western Countries*, vol. I, p. 28; Chavannes, *Documents sur les Tou-kiue (Turcs) occidentaux*, p. 120.

[2] E. Kaempfer, *History of Japan* (Glasgow edition), vol. I, p. 200.

[3] A. Printz, *Erman's Archiv für wissenschaftl. Kunde von Russland*, vol. XXV, 1867, p. 294; A. Jarilow, *Beitrag zur Landwirtschaft in Sibirien*, p. 319.

[4] An interesting article on Chinese notions of cervines is by M. Cibot, " Notice sur le cerf," in *Mémoires concernant les Chinois* (Paris, 1788), vol. XIII, pp. 402–408.

[5] K. Florenz, " Formosanische Volkslieder," *Mitt. D. Ges. Ostasiens*, vol. VII, (1898–99), p. 122.

The following case presents a good example as to how primitive man may have managed to get possession of wild reindeer alive. The ancient Kitan and Jurči (Niüči) of Manchuria had a peculiar method of hunting deer by imitating its belling, and killing with arrow-shots the animal thus allured.[1]

A lively description of this manner of hunting was given in the eighteenth century by C. Visdelou[2] as follows:

The Niüči were always celebrated for a sort of hunting peculiar to their nation. The same method is still appropriate solely to the Manchu. These tell the following story as a well-substantiated fact. Briefly before the rutting-season each stag will establish a seraglio of does and occupy a stretch of forest or mountain. After this division there are stags left who either did not receive their share or were robbed of their spoils. Each is intent on acquiring a territory by right of conquest. He invades the district of one of his neighbors. On entering it he utters a cry as a challenge for combat. A courageous owner does not await another call, but will pounce on the intruder instantaneously. Meanwhile the does will line themselves up in two rows to watch the duel. The adversary being put to flight or thrown to the ground, his does will pass over to the victor. The Manchu take a stag's head with the antlers, hollow it out, and place it over their own head. With a hidden decoy whistle they imitate the call of a stag so perfectly that the animal is deceived. They crouch in the thicket, and at the sound of the whistle the stag comes out in the open for an attack, sometimes so precipitately and furiously that the hunter has no time to make use of his weapons. He who is thus surprised is usually lost and torn to pieces. During his youth the Emperor K'ang-hi once risked his life on such a hunt, which takes place annually. The Manchu affirm that the best, largest, and strongest

[1] H. C. v. d. Gabelentz, *Geschichte der grossen Liao*, pp. 98, 154; Chavannes, "Voyageurs chinois chez les Khitan," *Journal asiatique* (mai-juin, 1897), p. 404; also Klaproth, *Tableaux historiques de l'Asie*, p. 90. In the latter's translation appears a zoölogical puzzle by which no one as yet seems to have been struck. According to Klaproth, the Jurči subsisted on the flesh of the stags, and prepared an intoxicating beverage from the milk of the does. The question as to how it was possible to milk a wild animal did not alarm the learned sinologue. In fact, the Chinese author, the traveler Hu Kiao, who lived among the barbarians of the north from 947 to 953, did not write this nonsense. The text of the *Wu tai shi* (ch. 73, p. 3 b), in which his account is embodied, simply contains a misprint (*mi* 麋, *Cervus davidianus*, instead of *mi*, "millet"); and the passage means, as rendered by Chavannes, "They make a fermented beverage from a decoction of millet." Schlegel (*T'oung Pao*, vol. III (1894), p. 141), citing the same passage after Ma Tuan-lin, arbitrarily takes the term *mi* in the sense of "reindeer," and thinks that the Jurči distilled an alcoholic beverage from reindeer's milk. As to the other animals mentioned by Hu Kiao in this region, the "wild dogs" (*ye kou*), I believe, represent *Canis procyonoides*.

[2] In d'Herbelot, *Bibliothèque orientale* (La Haye, 1789), vol. IV, p. 292.

stags are brought in from these hunting-expeditions, and that there is no finer sight than the majesty, pride, and intrepidity of these animals when coming forward to fight,—a quality less conspicuous at other times.

G. Radde[1] reports that during the rutting season the hunters of the Sayan, Baikal, Yabloko, and Chingan mountains avail themselves of slightly curved horns made from fir or larch wood;[2] on the left bank of the Amur they use also the thick, hollow stems of the Kongola-Umbelle (*Calisace daurica*). At this time the stag is not timid, and approaches the hidden sportsman at a short distance. Old stags, however, do not easily accept this challenge to battle, and are said to discriminate well between the call of the hunters and their own kind. The Mongols avail themselves of a whistle (called *urum* or *urum-dal*) to attract the hart, or also imitate his cry.

According to the reminiscences of the Lapp, they received their domesticated reindeer from the wild animal. Johan Turi narrates in his autobiography (p. 64),

In ancient times there were many wild reindeer, and there was no one who cared to guard reindeer. And the Lapp learned how to make the wild reindeer feel safe, so that they remained in his herd. When a wild reindeer has joined the herd, it is necessary to go cautiously around the herd and to allow it to walk ahead quite a distance, that the wild reindeer does not know that men are near. When the wild animal has visited the herd, it is familiar with it, and does not move away even when seeing men. Not all wild reindeer, however, are equally bold; some never become confiding, however long they may remain with the herd, but some it takes only little time to become accustomed to reindeer and man; neither does it run away unless it should drift into a troup of wild reindeer; in this case it follows the wild ones. The timid ones can never be tamed. The wild ones are much larger than the domesticated stock, and more glossy, as though having silver hair. A few of those which cannot be rendered tame were obtained in this manner, that a wild reindeer bull visited the herd in the rutting season. And when a wild reindeer is in the herd, the latter need not be guarded.

Johan Turi continues (p. 65),

A Lapp sojourned in the vicinity of Koutokäino, and he would annually

[1] *Reisen im Süden von Ost-Sibirien*, vol. I, p. 284. Radde has transcribed the call in notes. See also A. von Middendorff, *Sibirische Reise*, vol. IV, p. 1390.

[2] According to the *Ta Kin kuo chi* (ch. 39, p. 1; written in 1234 by Yü-wen Mou-chao), the Jurči made horns from birchbark, on which they produced sounds like *yu-yu*, in order to allure the harts (*mi-lu*), and then to shoot them with bow and arrow. *Yu-yu* is a Chinese term of endearment for a tame deer.

allow the reindeer to mate on a strip of land in the Elf. A wild reindeer always appeared during the mating-season for several years, and he did not kill him. Somebody, however, killed him at last. And the Lapp regarded this as a much more deplorable loss than if it had been one of his own bulls. Yet he received offspring from the wild reindeer. His deer became as glossy and slender as wild reindeer; it was quite extraordinary reindeer, and every one envied him for his reindeer, since they were much finer than others.

While this account proves nothing for the origin of the domestication, it shows clearly that the old stock was renewed and recruited from wild material, and that a great number of wild animals were gradually absorbed by the Lapp. In this respect also Ohthere's account given above (p. 95) is of fundamental value.

Aside from battues, the Samoyed have conceived a peculiar method of capturing wild reindeer. They train four or five tame, usually female, reindeer in such a manner that they walk together around the hunter in a certain order. One walks ahead, being held by a rope many fathoms long, the others going at the side of the hunter, who fastens to his girdle the ropes of all animals. The hunter, clad in reindeer-skins and bending low, steals along as near as he can to the wild herd, and picks out the best specimen for his shot. During the rutting season the Samoyed select a strong, ungelded buck, and look for a wild herd. When such is sighted, slings are laid around the antlers of the buck and attached by means of loose bast. Thus he is set on the wild herd. The wild stag, being aware of the alien rival, challenges him to a duel. During the brawl, his antlers become entangled in the slings of the tame pseudo-opponent, who will press his antlers toward the ground, and thus hold the adversary till the hunter arrives.[1]

The Ostyak have developed a similar method, or rather adopted it from the Samoyed. They fasten to their tame deer a strap between the upper tips of the antlers, and allow them to disperse near a herd of wild ones. These rush on the strangers, and, during the struggle, entangle their antlers in the straps prepared, being held till the arrival of their captors.[2] A similar method prevails

[1] P. S. Pallas, *Reise durch verschiedene Provinzen des russischen Reichs* (1776), vol. III, p. 91.

[2] A. Erman, *Reise um die Welt*, vol. I, pt. I, p. 653.

among the Amur tribes. In the autumn and the spring the native hunters indulge in the chase of wild reindeer by means of tame ones. The latter are let loose, but held with a long strap by the hunter, who cautiously follows behind in their trail. According to his will, the reindeer is made to pasture, to lie down, to stand up, and to turn round in this or that direction. The skilful hunter can thus slay many wild deer before his presence is suspected by the herd.[1] V. Jochelson[2] has described the same procedure for the inhabitants of the Kolyma district, where the decoy animal is known under the name *man'ščik* (probably from Russian *manit'*, "to lure"), in the language of the Lamut *ondadá*. Thus the practice is universal throughout Siberia. This method may illustrate how the decoys of Ohthere were used (p. 95), and how primitive man at all times understood how to add a fresh supply to his stock. What method he employed in detail for breaking his deer certainly escapes our knowledge. Some of his methods have been alluded to, as gelding and the imitation of processes gained by experience with other domestications.

An interesting problem is whether reindeer-driving is to be conceived as an imitation of the method of driving on dog-sledges. In regard to the latter we possess unfortunately little historical material. We have seen that dog-sleighs were known among the Kirgiz in the thirteenth century (p. 111) and in northwestern Siberia in 1499 (p. 97), and that they even extended to the west of the Ural in ancient times.[3] Driving with dogs is practised throughout Siberia. As is well known, the dog was originally the sole domestic animal kept by the so-called Palaeo-Asiatic peoples, the Ainu, Gilyak, Kamchadal, Yenisei-Ostyak,[4] Yukagir, Koryak, and Chuk-

[1] Grum-Gržimailo, *Opisanie Amurskoi Oblasti*, pp. 334, 335.

[2] *Očerk zv'äropromyšlennosti i torgovli m'äxami v Kolymskom okrug'ä* (Sketch of the Animal Industry and Fur Trade in the District of Kolyma), p. 44.

[3] These data escaped L. von Schrenck (*Reisen und Forschungen im Amur-Lande*, vol. III, p. 488) in his discussion as to the time when the Russians became acquainted with dog-driving; he does not go beyond the seventeenth century. S. von Herberstein (*Notes upon Russia*, vol. II, p. 46) mentions large dogs used as beasts of burden, "which are very useful for this purpose, with which they convey baggage in carriages, in the same manner as will be hereafter described in speaking of the deer."—Compare above, p. 99.

[4] J. Klaproth, *Asia polyglotta* (Paris, 1823), p. 167, stated that the Yenisei-Ostyak

chi; and dog-sleighs represent the exclusive means of land transportation among these tribes. The same condition is found among the Eskimo, while the tame reindeer is unknown to them. From this wide geographical distribution covering the Old and New Worlds it necessarily follows that the employment of the dog for the sledge is far older in time than that of the reindeer for the same purpose. Although strictly mathematical proof cannot be put forward, the ethnographical facts well warrant the conclusion that the reindeer-sledge is based on the dog-sledge, and that reindeer-driving sprang into existence as a perfectly conscious and volitional imitation of driving with dogs. This being the case, it is clear that the reindeer people must have profited from the experiences of the dog-drivers, and reproduced many of their methods.[1]

subsist on fishing, hunting, and to a small extent reindeer-breeding. Recent authors say nothing about this point, but mention only fishing and hunting, with the dog as the exclusive domestic animal (S. Patkanov, *Essai d'une statistique et d'une géographie des peuples palae-asiatiques* (St.-Pétersbourg, 1903), p. 9). The peculiar language of this group has been studied by M. A. Castrén, *Versuch einer Yenisej-ostjakischen und kottischen Sprachlehre* (St.-Petersburg, 1858). G. I. Ramstedt "Ueber den Ursprung der sog. Jenisej-Ostjaken," *Journal de la Société finno-ougrienne*, vol. XXIV, 1907, pp. 1–6, has made the singular attempt to compare the Yenisei-Ostyak numerals from two to ten with those of Tibetan and Chinese, and to proclaim on the basis of this result the Yenisei-Ostyak as a branch of the Indo-Chinese family. The alleged coincidences are by no means convincing, and either do not exist at all, or are mere resemblances on paper; not phonetical, however. It would hardly be worth while to call attention to this fantasy if the author were not a good philologist, whose contributions to Mongol phonology and dialects command respect.

[1] Two extraordinary statements in respect to reindeer-driving are made by the Jesuit Philippe Avril, *Travels into Divers Parts of Europe and Asia* (London, 1693), p. 172, English translation of his *Voyage en divers états d'Europe et d'Asie* (Utrecht, 1673, and Paris, 1692). "To make the reine-deer go more swift, they tie a great dog behind, that scaring the poor beast with his barking, sets her a running with that speed, as to draw her burthen no less then forty leagues a day." "But that which is more wonderful as to these sort of sledds, they are also driven along by the wind sometimes over the land cover'd with snow, sometimes over the ice of frozen rivers, as our vessels, that sail upon the sea. For in regard the country beyond Siberia is open and extreamly level as far as Mount Caucasus, the people who inhabit it making use of this advantage to spare their beasts, have so order'd their sledds, as either to be drawn along by the reine-deer, or else to carry sails, when the wind favours 'em." I cannot find any confirmation of this dog contrivance and of sail sledges in any other source. Avril was commissioned by the then King of France to discover a new way by land into China, left Marseilles in 1664, reached Moscow, where he

The advantages of reindeer over dog keeping are obvious. The reindeer feeds itself, the dog must be fed. In traveling, food must be carried for the dogs. The maintenance of dogs develops into a burdensome task. In case of emergency the reindeer will furnish food to his master.

As soon as the wind blows a little, the dog cannot travel; especially is this so if the wind happens to be in the face. The deer does not mind the wind in the least, from whatever direction it comes; it rather enjoys travelling against the wind. It costs nothing for feed; it faces all weather, and makes its way where the driver can hardly walk without snowshoes. It goes uphill and downhill alike. Trail or no trail, it will haul its two hundred pounds or more day after day, even week after week.[1]

It is not to the point that, as asserted by G. Mortillet[2] after K. Vogt, reindeer-breeding is impossible without the use of the watch-dog. In fact, only the western group of reindeer-tribes—Lapp, Wogul, Ostyak, and Samoyed—have their herds managed by dogs; while neither the Tungus nor the Koryak and Chukchi have their reindeer chaperoned by dogs; on the contrary, they keep these away from the herds.[3] With me it is not an open question, as stated by Bogoras, whether reindeer-breeding was begun with dogs or without them. The dog, in my estimation at least, had nothing to do with the incipient process. He is merely an incidental accessory, being transferred from his office previously held in other herds to the guarding of reindeer long after the latter's domestication was completed.

As regards the employment of the reindeer for riding purposes, there can be no doubt that it existed at least as early as the thirteenth century in the Baikal region (p. 107). The only moot point

was compelled to return, and traveled by way of Warsaw to Constantinople, reaching Toulon in 1670. The information supplied by him on Siberia was gathered in Russia, for the most part from oral accounts. In his biography (*Biographie universelle, Supplément*, vol. LVI, p. 605) it is said, " Ce qu'il dit sur l'histoire naturelle montre que ses connaissances en ce genre n'étaient pas très étendues." Nevertheless his book is full of interest and teems with curious information (see, for instance, *T'oung Pao* (1916), p. 363).

[1] *Fourteenth Annual Report on Introduction of Domestic Reindeer into Alaska* (Washington, 1905), p. 105.

[2] *La Préhistorique*, p. 439.

[3] Bogoras, *Jesup North Pacific Expedition*, vol. VII, p. 71.

is whether this practice is primarily due to the southern Samoyed or to Tungusians.[1] Since the northern Samoyed do not ride the reindeer, it would seem that the claim of the Tungusians merits preference; but this conclusion would be fallacious. The northern Samoyed are mentioned as early as 1096 in the Russian chronicle of Nestor, and it is therefore conceivable that the northward migration of this stock was an accomplished fact at a time when the reindeer was not yet trained to the saddle in their southern home; or, in other words, that the riding of the reindeer in the Baikal region came into existence after the separation of the Samoyed tribes, and for this reason never reached the northern group. Thus the question as to the particular people which first mounted the reindeer must remain undecided; assuredly it was a tribe that had gained some experience with horses. It is said that it takes the reindeer only a very short time to become accustomed to the saddle.[2]

Although truly in a state of domestication, it can by no means be asserted that the reindeer has been brought fully under the control of man. On the contrary, the reindeer controls man to a much higher degree than man has sway over the animal, and in fact determines his whole manner of life. In this respect reindeer-keeping differs radically from cattle or horse breeding. Cattle and horse have been subordinated to human will so completely that they cannot subsist without being provided by man with fodder and shelter. They share man's habitation, and stable-feeding has made them the close associates and friends of his home. To the reindeer man does not furnish lodging and board. It remains independent, and pursues its natural instincts along the question of nutrition; it is not sheltered from the inclemencies of the weather by house or tent, but spends the night like its wild congener. In short, it makes and lives its own life, only to answer its master's call when occasion for labor arises. It performs its duties willingly

[1] N. V. Latkin, *Yeniseiskaya Guberniya*, p. 169, includes also the Dolgan and Yakut among the reindeer-riders. If this is the case, it is certainly due to Tungusian influence. An example of reindeer-riding Yakut is found in the autobiography of the Yakut Uvarovski (O. Böhtlingk, *Sprache der Jakuten*, pp. 26, 49).

[2] Latkin, *l. c.*

and submissively; but as soon as the short working-hours are past, it demands its freedom, and must be released for grazing and browsing: it cannot be held in socage indefinitely. The reindeer's life is bound to a well-defined geographical area with specific floristic characteristics, and it cannot be removed to other quarters without its existence becoming endangered. Individuals taken into our zoölogical parks, even if provided with moss, do not thrive long, and are usually doomed after a few years; while transplantations of herds into Switzerland, for instance, have proved failures. The reindeer cannot live in captivity, it cannot be acclimated to uncongenial zones, and will never approach that state of true domesticity attained in cattle and horse. If domestication be taken in the true sense of the word, "habituation to home-life," the reindeer has certainly not reached it, quite in conformity with its master.

In view of the reindeer's economic independence, the interesting question arises: What forces bind the animal to man? If it receives from him neither food nor shelter, by what factors is it induced to maintain such a seemingly unprofitable association? Indeed, the reindeer's position is singular. Examining other domestic breeds, we plainly recognize the foundation of their social contract with man, which is based on an unwritten law of reciprocity, that on both sides has developed into the quality of faithfulness. Dog, cat, and swine have reserved to themselves a certain degree of independence in the choice of their diet, and if forsaken by man, or even while under his care, may hunt for a meal on their own initiative; nevertheless they will always appreciate more what is offered them by man. Reindeer are fond of salt and sugar, and a bit of these articles may accelerate their run; but they are so rarely given to them, that this could hardly be thought of as an inducement for them to keep up companionship with man.[1] It may be, then, that it believes in man as a superior being, that it trusts in his power and strength, and looks up to him as his guardian from perils threatening from wild animals, chiefly its arch-enemy

[1] Hahn, *Haustiere*, pp. 558–559, regards the animal's craving for salt, satisfied by human urine, as the strongest bond that binds it to the service of man,—doubtless an exaggeration.

the wolf. But even this argument, weighty as it may be, does not seem to me sufficient to explain the whole scale of the reindeer's relation to man. It seems to me that psychic qualities both in the animal and in man must be made responsible for the final result. There is man's aesthetic pleasure in animals, and the entire deer family is attractive to every human soul. This sympathy is doubtless reciprocated by the reindeer. Above all, there is the social instinct developed both in deer and man, and in the loneliness of the arctic regions these social bonds are doubtless intensified. The deer is a highly social creature, impressing its friendship on man. It is of gentle disposition, and is loved by children. Those of the Tungus are fond of decorating their riding-deer with ribbons to which are sewed glass beads or buttons.[1]

Not much positive information is available in regard to feral reindeer. The Lapp, Johan Turi, in his fascinating autobiography (p. 40), speaks of the savage character of the bulls during the rutting season, when they even pounce on men, and observes that the " bulls of the wilderness " (that is, animals which have segregated from the herd and lived long in the wilderness without man's care) particularly are prone to attack people.

In accordance with the history of the domestication, the tending of the herd, and the care of everything connected with it, are everywhere the business of man. Among the Chukchi, labor is divided between man and wife in this manner: that all domestic affairs, inclusive of preparation of hides, yarn, and clothing, fall to the lot of woman; while man looks after the herd, harnesses or unharnesses the deer, and, if necessary, slaughters it. This is man's sole business, but his time is fully occupied with it.[2]

To dilate on the effects of reindeer-breeding is beyond the scope of this article. This would mean to set forth in detail the economic features of the culture of the tribes in question, which has been done in a number of excellent monographs. I should like to emphasize merely a single point; and that is, that in my estimation the reindeer-breeders have developed higher psychic qualities

[1] Pekarski and Tsv'ätkov, *Očerki byta Priayanskix Tungusov*, p. 39.

[2] G. Maydell, *Reisen und Forschungen im Jakutskischen Gebiet Ostribiriens*, pt. 1, pp. 186–187.

than the Palaeo-Asiatic dog-breeders, owing to the fact that the latter, as agreed upon by all observers, have no inward relations to their dogs, and their savage dogs lack all superior traits of the civilized dog, while there is mutual affection between man and reindeer. I do not believe in generalizations nor in comparisons, still less in dogmas of racial superiority and inferiority or of good and evil, and I am very far from extolling the reindeer tribes at the expense of the dog-breeders. Of these the Gilyak and Ainu are known to me from personal experience, also the Olča and Golde on the Amur, whose culture is partially based on the maintenance of dogs. I gained a deep respect and sympathy for these people, for their manliness and good nature, their hospitality, and their intellect. I felt more at home, however, with the reindeer-breeding Tungusians, who are more alert, open-minded, straightforward, and psychically more developed, and I found that A. von Middendorff was perfectly right in styling them the aristocracy of Siberia. There can be no doubt that constant intercourse with an animal as noble, civil, and civilized as the reindeer has a psychical value, and exerts a beneficial and ennobling influence on the hearts of the people. Let me quote the experience of a Finnish author. Among the Lapp, songs are particularly cultivated by the reindeer-breeders; and in the opinion of Armas Launis,[1] who has published a comprehensive collection of such songs with their musical notations, they may be regarded as the originators of the songs which receive their natural explanation from the life of the herder. At home he is reserved and taciturn, and he scarcely sings otherwise than during his sojourn on the tundra, where he tends his herd. Confronted with the wide panorama of lakes and the blue mountains bordering the horizon, he will remember a good friend or brood evil against an enemy. The reminiscence assumes shape in words and tones, and a tune thus arises on the subject of his thought. While he looks over his herd with a feeling of content, he gives vent to his sentiments, and, muttering the words "*čăbba ællo čăbba ællo*" (handsome herd, handsome herd), he will finally compose a melody in praise of his flock.

[1] "Lappische Juoigos-Melodien," *Mémoires de la Société finno-ougrienne* (Helsingfors, 1908), vol. XXVI.

Hahn[1] says that the economic value of the reindeer has been overstated, as would follow from the fact that it is restricted everywhere to the aborigines; while Europeans did not take to its breeding, even there, where the animal would be important. Again he thinks that "the reindeer is not sufficient to man of European descent and culture, or that the latter has not the patience required for it; in this point he is surpassed by the 'savage.'" But Ohthere, the Norseman, was a European (p. 94). The Russians, when advancing and settling in northern regions, where horses do not thrive, easily took to reindeer-breeding. P. S. Pallas[2] reported in 1772, in regard to the district of Obdorsk, that horses imported there did not live a year, and that the reindeer-herds,—which, despite numerous diseases and wild animals, increase rapidly—form a not unimportant wealth both of the Russian and Pagan inhabitants of those northern countries.[3]

Erroneous also is Hahn's statement that

> the reindeer has never followed the European, as particularly shown by the introduction in 1770 into Iceland of reindeer which were supposed to give new domestic animals to that poor country.

What was introduced into Iceland in 1771 and 1777 (not in 1770) were not domestic, but wild reindeer from Norway, which were gradually shot, and are now almost exterminated.[4]

The reindeer introduced into Alaska at the end of the last century are as useful to the whites as to the Eskimo. Says Dr.

[1] *Haustiere*, pp. 264, 267.

[2] *Reise durch verschiedene Provinzen des russischen Reichs*, vol. III, p. 23.

[3] The Russian nomenclature relating to the reindeer, chiefly in the dialects of northern Russia, is borrowed from Finno-Ugrian languages: thus *pyžik* (young reindeer, fawn), that already occurs in Old Russian, from Syryän *pež*, Wotyak *pužey*, Wogul *pežka*, Ostyak *peži; vážatka* or *váženka* (doe) from Syryän *važenka*, Lapp *važ, važa;* to the same root belong in the dialect of Archangel *vačegal'* (to tend a reindeer-herd), *vačuga* (reindeer relais), *vačužn'a* (reindeer-herd); *hora* (reindeer-bull) from Samoyed *hora*, Syryän *kora; girvas* (male reindeer) from Finnish *hirvas; gigna, higna* (leash in the reindeer-harness) from Finnish *hihna; loima* (a herd of reindeer) from Finnish *lauma*, etc. (compare R. Meckelein, *Die finnisch-ugrischen Elemente im Russischen*, p. 20).

[4] See the interesting account of A. Gebhardt (after Th. Thoroddsen), "Die Rentiere auf Island," *Globus*, vol. 86 (1904), pp. 261–263.

Jackson, the father of this new economic movement, on this point:[1]

The industrial pursuit which nature has mapped out for the native population of arctic and subarctic Alaska is the breeding and herding of reindeer and the use of the deer as a means of transportation and intercommunication. During the past season the influx of miners into the Yukon region has made a very urgent call for reindeer for freighting-purposes. In the original plan for the purchase and distribution of reindeer, reference was mainly had to securing a new food-supply for the famishing Eskimo; but it is now found that the reindeer are as essential to the white men as to the Eskimo. The wonderful placer mines of the Yukon region are situated from 25 to 100 miles from the great Yukon River. The provisions brought from the south and landed upon the banks of the river are with great difficulty transported to the mines. So great was the extremity last winter, that mongrel Indian dogs cost $100 to $200 each for transportation purposes, and the freight charges from the river to the mines, 30 miles, ranged from 15 to 20 cents per pound. The difficulty experienced in providing the miners with the necessaries of life has demonstrated the necessity of reindeer-transportation, and that the development of the large mining interests of that region will be dependent upon the more rapid introduction of reindeer for freighting. There are no roads in Alaska, and off of the rivers no transportation facilities to any great extent. In the limited traveling of the past, dogs have been used for that purpose; but dog-teams are slow, and must be burdened with the food for their own maintenance. On the other hand, trained reindeer make in a day two or three times the distance covered by a dog-team, and at the end of the day can be turned loose to gather their support from the moss, which is always accessible to them.

On the other hand, it is stated,

The ordinary white man is unwilling to undergo the drudgery of herding in that rigorous climate, and unwilling to work for the small compensation that is paid for such services. He can do better. . . . With the increase of domestic reindeer in Alaska, it will become possible for white men to own large herds; but the men that will do the herding and teaming will always be Eskimo and Lapp.[2]

Hahn's gloomy prophesy of the ultimate extinction of the reindeer jointly with the "miserable" tribes of the Ostyak, Wogul, and Samoyed, has happily not been fulfilled. He who is but superficially posted on the subject knows that the Samoyed are not a dying people, but vigorously spread and thrive.[3] So does

[1] *Report of the Commissioner of Education for the year 1895–96* (Washington, 1897), vol. II, pt. 2, p. 1454.

[2] *Report of the Commissioner of Education for the Year 1903*, vol. II, p. 2375.

[3] See, for instance, W. Crahmer in *Zeitschrift für Ethnologie* (1913), p. 543.

the reindeer. It is protected by the Russian government, and the study of the improvement of the economy has been entrusted to special commissions. The reindeer is gaining ground, and will claim more importance and attention in the future. It has conquered Alaska and parts of Canada. The successful introduction of domestic reindeer into Alaska has led to their introduction into Newfoundland. Dr. Grenfell, who in 1892 organized a medical mission among the fishermen off the shores of Newfoundland and Labrador, on reading the U. S. reindeer reports, became convinced that he had to have reindeer for his winter trips, and January 7, 1908, landed safely three hundred head at the village of Cremeliere, two miles from St. Anthony, on the northern coast of Newfoundland.[1]

The ethnologist will watch with interest the gradual transformation of the Alaskan Eskimo into reindeer-breeders.[2] History repeats itself: it is the same process that reshaped the life of the Chukchi and Koryak. The introduction in 1890 of reindeer into Alaska was inspired by a desire to provide a new and more permanent food-supply for the half-famishing Eskimo. Up to 1902 there were sixty individual holders of domestic reindeer in Alaska, forty-four of these being Eskimo, the majority of whom had served a five-years' apprenticeship and gained a competent knowledge of the management and care of reindeer. In 1903 sixty-eight Eskimo and one Indian owned 2,841 deer. From the 1,280 Siberian reindeer imported between 1892 and 1903, and from their natural increase, 7,983 fawns have been born in Alaska.

The Eskimo has always been skilful in driving dogs, and now, under instruction, is proving equally skilful in driving reindeer, and upon various occasions, when the opportunity has offered, has invariably demonstrated his ability to successfully transport with reindeer mails, freight, and passengers between mining-camps.[3]

[1] *Sixteenth Annual Report on Introduction of Domestic Reindeer into Alaska* (Washington, 1908), p. 42.

[2] Compare E. W. Hawkes, "Transforming the Eskimo into a Herder," *Anthropos*, vol. VIII (1913), pp. 359–362.

[3] From Dr. S. Jackson's report, in *Report of the Commissioner of Education for the Year 1903* (Washington, 1905), vol. II, p. 2374.

In view of the opportunities and facilities granted in Alaska, it is a matter of surprise that biologists have not yet seen fit to take up the study of breeding problems in connection with the reindeer, either for theoretical purposes or with a view to improving the races. We are anxious to know, for instance, why the Tungus reindeer is larger and sturdier than that of Lapland, and why most of the wild deer are larger than the domesticated. As to the question of color variation in the domestic stocks we have merely vague descriptions of laymen, and the differentiations of the various stocks have not yet been determined scientifically. Likewise the following observation would offer a problem to the biologist.

No deterioration in the herds on account of inbreeding has been noted. On the contrary, the chief of the Alaska division maintains that the reindeer now in Alaska are larger animals than those which comprised the original stock imported from Siberia, that Alaska affords a better range than Siberia, and that the climate is better adapted to the reindeer industry. The herds in Alaska average more than seven hundred reindeer each, so that the danger of inbreeding cannot be serious. The introduction of wild caribou into some of the herds has increased the size of the reindeer in those herds.[1]

[1] "Report on the Work of the Bureau of Education for the Natives of Alaska, 1913–14," p. 10 (1915), *Bulletin*, No. 48.

139

虾夷语中数字的二十进制和十进制——附论虾夷语音韵学

JOURNAL

OF THE

AMERICAN ORIENTAL SOCIETY

EDITED BY

JAMES A. MONTGOMERY
Professor in the
University of Pennsylvania

GEORGE C. O. HAAS
Sometime Instructor in the
College of the City of New York

VOLUME 37

PUBLISHED FOR THE AMERICAN ORIENTAL SOCIETY
YALE UNIVERSITY PRESS
NEW HAVEN, CONNECTICUT, U. S. A.
1917

THE VIGESIMAL AND DECIMAL SYSTEMS IN THE AINU NUMERALS

WITH SOME REMARKS ON AINU PHONOLOGY

BERTHOLD LAUFER

FIELD MUSEUM OF NATURAL HISTORY, CHICAGO

The vigesimal character of the numeral system of the Ainu was first recognized clearly by the great philologist A. F. Pott,[1] although he had at his disposal only the scanty and deficient vocabularies of A. J. v. Krusenstern[2] and Klaproth (*Asia polyglotta*). On the basis of a Japanese collection of Ainu words, the *Moshiogusa,* A. Pfizmaier[3] arrived at the same conclusion a few years later. J. Batchelor,[4] the patient and meritorious investigator of the Yezo Ainu, has refrained from giving an analysis of the numerals, being content to observe that 'twenty, more literally a "score," is the highest unit ever present to the Ainu mind when counting. Thus, forty is "two score," sixty is "three score," eighty is "four score," and a hundred is "five score."' An interesting analysis of the numerals from the pen of B. H. Chamberlain, however, is inserted in his Grammar.

The cardinal numerals from one to five are *šine, tu, re, ine,* and *ašikne;* or properly, *-ne* being a suffix, as will presently be recognized,[5] they are *ši, tu, re, i, ašik.* The word for the number 5, *ašik,* is doubtless associated with the nouns, Yezo *aške,* from **ašike* 'hand', *aškororo* 'a handful', *ašikipet,* Saghalin *askipit*

[1] *Die quinare und vigesimale Zählmethode bei Völkern aller Welttheile,* p. 85 (Halle, 1847).

[2] *Wörtersammlungen aus den Sprachen einiger Völker des östlichen Asiens* (St. Petersburg, 1813).

[3] 'Untersuchungen über den Bau der Aino-Sprache,' p. 26 (*Sb. Wiener Akad.* 1851). In 1883 Pfizmaier adopted the only correct spelling 'Ainu' (see his 'Untersuchungen über Ainu-Gegenstände,' p. 1).

[4] *A Grammar of the Ainu Language,* p. 47 ff. (Yokohama, 1903); reprinted also at the end of his *Ainu-English-Japanese Dictionary* (2d ed., Tokyo, 1905).

[5] The same suffix is employed also in adjectives: *kuras-ne* 'black', *on-ne* 'old', *tan-ne* 'long', *tak-ne* 'short'.

'finger'.[6] The designation for 'foot' (*kema*) is not met with in the numeral system.

The numbers six to nine are formed by subtraction from 10, *wan*,[7] as follows:—

iwan, i-wan (*i* 4, *wan* 10), 10 — 4 = 6.
arawan, a-ra-wan[8] (*a* prefix, *ra* = *re* 3, *wan* 10), 10 — 3 = 7.
tupesan, tu-pe-san (*tu* 2, *pe* 'thing,' *san* 10), 10 — 2 = 8.
šinepesan, ši-ne-pe-san (*ši* 1, *ne* suffix, *pe* 'thing,' *san* 10), 10 — 1 = 9.[9]

---

[6] This seems to me the only rational explanation in opposition to B. H. Chamberlain (*The Language, Mythology, . . . of Japan Viewed in the Light of Ainu Studies*, p. 9), who interprets *ašik-ne* as 'possibly "new four" (*aširi ine*).' This is artificial and runs counter to phonetic requirements. Pott (*l.c.*) had already remarked that the relationship of the numeral 5 to 4 in the sense of 'a beyond it' is merely deceptive.

[7] Batchelor writes *wa*(*n*). On Saghalin I heard only *wan* or in composition with *pe* 'thing': *wam-pe*. I. Radliński ('Sl'ownik narzecza Ainów,' p. 67, Kraków, 1891) gives for the Kuril dialect *wam-pi-y* or *vam-pi-kasma*. The materials of this Polish author have not been utilized by Batchelor.

[8] On Saghalin only *a-ru-wan*. Batchelor (*Dictionary*, p. 44) gives for Yezo both *arawan* and *aruwan* on equal footing; the *Moshiogusa*, according to Pfizmaier, only *aruwan*. Kuril Ainu (Radliński) *arwa* (from **aruwa*).

[9] Chamberlain (*l.c.*) analyzes *tupesan* as 'two (*tu*) things (*pe*) come down (*san*) [from ten],' and similarly *šinepesan*. True it is, Batchelor has on record a word *san* with the meaning 'to descend, to flow along as a river, to go down'; but there is nothing to indicate that it conveys the notion of subtraction. I prefer to assume that *san* in the numbers 8 and 9 appears in lieu of *wan*, and signifies 'ten.' The question, however, is not of a phonetic change, an alternation of *s* and *w* being otherwise unknown in Ainu, but we are bound to suppose that *san* is an independent stem or base with the meaning 'ten' on a par with *wan*. Also the languages of primitive tribes are no longer extant in their original forms, and especially in the numerals far-going modifications and re-adjustments of various systems have doubtless taken place. In Friedrich Müller's *Grundriss der Sprachwissenschaft* (2. 1. 145), where a rather poor and in many respects incorrect sketch of Ainu is given, we read literally as follows: '8 *tu-be-šan* (2 + 5); 9 *šne-be-šan* (1 + 8).' The element *san* cannot be compared with the numeral 5 *ašik, ašik-ne*, for, as is evidenced by the word for 'finger,' from which the numeral is derived, the final *k* is part of the stem. Moreover, if we are not mistaken, even in Müller's time (1882) 2 + 5 was 7, and not 8, as he makes out. To be consistent, Müller should have explained *šne-be-šan* 9 as 1 + 5, but it will not do to conceive the element *be-šan* as 8. Pott had already recognized the true condition of affairs, saying that the numbers from 6 to 9 raise the suspicion of having originated retrospectively through deduction from 10, and that there is no doubt of this in 8 and 9. Even Ph. von Siebold (*Nippon*, new ed., 2. 255) gave a correct explanation of the Ainu number 8. The first edition of his work, incomplete, appeared in seven parts in Leiden, 1832-52.

The numbers eleven to nineteen are formed on the scheme $1 + 10$, $2 + 10$, *šine ikašima wan;* on Saghalin simply *šinä ikašima* $= 1+$. The unit of all higher counting is represented by the figure 20: Yezo *hot-ne* (the same suffix *-ne* as in the numbers 1, 4, 5, and possibly the mobile *-n* of *wan* 10), Kuril *ot,* Saghalin *ox, otsi.* The number 30 is expressed by $10 - 2 \times 20$[10] (*wan-e-tu-hot-ne*), 31 by $1 + (10 - 2 \times 20)$, $40 = 2 \times 20$, $50 = 10 - 3 \times 20$, $60 = 3 \times 20$, $70 = 10 - 4 \times 20$, $80 = 4 \times 20$, $90 = 10 - 5 \times 20$, $100 = 5 \times 20$ (*ašikne hotne,* Kuril *askinot*), $110 = 10 - 6 \times 20$, $120 = 6 \times 20$, $200 = 10 \times 20$, $1000 = 5 \times 10 \times 20$, etc.

In its origin, this numeral system accordingly was quaternary, the numbers one to four being indivisible and undefinable roots. The number five was derived from the designation of the hand. It plays no role in the formation of higher number-conceptions. The words for ten and twenty are simple and unanalyzable stems. From eleven to nineteen the numbers follow the decimal principle, while from twenty onward a vigesimal system is carried through with clear consistency. Similar conditions are found in American languages.[11]

This method of reckoning is remarkable for its complexity, and bespeaks no small degree of mental effort for such simple folk as the Ainu. We are quite ready to believe Batchelor that in actual practice the higher numbers are rarely, if ever, met with, nor is it surprising to learn from the same authority that at the present time the simpler Japanese method (that is, a purely decimal system) is rapidly supplanting the cumbrous native system. Such transformations are always interesting to note and worth keeping in mind, especially in view of the conventional opinion that the life of primitives should be unchangeable.

It has not yet become known, however, that the Ainu of Saghalin, at least part of them, have advanced toward a purely decimal system of counting, but, while the impetus to this progressive movement was doubtless received from an outside quar-

---

[10] To be understood, of course, as $(2 \times 20) - 10$.

[11] See chiefly the interesting study by R. B. Dixon and A. L. Kroeber, 'Numeral Systems of the Languages of California,' *American Anthropologist,* 9 (1907), p. 663-690; and J. A. Mason, 'Ethnology of the Salinan Indians,' *Univ. of Cal. Publ. in Am. Arch.* 10 (1912), p. 134-136.

ter, they have recruited elements of their own language to this end. Among the Ainu on the southeast coast of Saghalin Island, I recorded the numerals in January 1899 as follows:—

1	*ši-nä'*	10	*wam-pe*	100	*ši-nä-tańku*
2	*tū*	20	*tū-kúńkŭtu*	200	*tū-tańku*
		21	*tū-kúńkŭtu šinä ikašima*		
3	*rē*	30	*rē-kúńkŭtu*	300	*rē-tańku*
4	*ī-nẹ*	40	*ī-nẹ-kúńkŭtu*	400	*ī-ne-tańku*
5	*aši'k, ašiṣ-ne*	50	*ašis-ne-kúńkŭtu*	500	*ašiṣ-ne-tańku*
6	*i-wan, i-wam-pe*	60	*i-wan-kúńkŭtu*	600	*i-wan-tańku*
7	*a-ru-wam-pe*	70	*a-ru-wan-kúńkŭtu*	700	*a-ru-wan-tańku*
8	*tu-pẹ-sam-pe*	80	*tu-pẹ-san-kúńkŭtu*	800	*tu-pẹ-san-tańku*
9	*ši-nä-pẹ-sam-pe*	90	*ši-nä-pẹ-san-kúńkŭtu*	900	*ši-nä-pẹ-san-tańku*
				1000	*wan-tańku*

It is clear that this system, based on the multiplication of 10, is logically decimal pure and simple. How far it is propagated among the Ainu of Saghalin I am unprepared to say, as my sojourn among them was limited to a few days, but it was given me by my Ainu informant as the mode of counting then generally in vogue. There is no doubt that also the ancient vigesimal system still holds sway on Saghalin, as stated by M. M. Dobrotvorski and B. Pilsulski. Dobrotvorski was stationed on Saghalin as Russian military surgeon from 1867-71, and his *Ainu-Russian Dictionary*[12] was published on his death by one of his brothers in Kazan, 1875. In the appendix of this work (p. 15), which contains a criticism of Pfizmaier's treatise cited above, the author speaks exclusively of the vigesimal character of the numerals. In the body of the dictionary, however (p. 153), he remarks that *kunkutu* (thus spelled instead of *kuńkutu*) is a counting-word for sables with the meaning 'ten sables,' also *sne* (= *ši-nä*) *kunkutu* being used in this sense; *tu-kunkutu*, 'twenty sables', etc. It is quite possible and, as will be noted, plausible that this method was originally inaugurated in connection with the calculation of sable-skins; but it is certain that *kuńkutu* does not mean 'sable,' either in Ainu or in any language of the peoples surrounding them. The sable is called by the Ainu both on Yezo and Saghalin only *hoinu* or *hoino*. Under

[12] By the way, a rather mediocre and from a phonetic viewpoint unsatisfactory work.

tańku Dobrotvorski (p. 317) notes that this signifies 'a hundred snares in catching sables.' *Tańku,* however, means simply 'hundred.'[13]

The word *tańku* for hundred occurs in an Ainu story recorded by B. Piłsudski[14] and describing an incident of Tungus life. It is avowedly the reproduction of an Orok tradition. With reference to *tańku* Piłsudski remarks that this is not a word of the Ainu, who denotes hundred by *ašišne hot* 'five score'; 'it is taken,' he continues, 'from the Oltchy [read Olča] tribes, from whom they learned to set snares for pine-martens, and counted the number of snares by hundreds in that language.' Yet *tańku* is not peculiar to the Olča, but the common word for hundred in Manchu (*tańgô*), in the ancient language of the Jučen (*tańgu*), and among all Tungusian and Amur tribes. The Ainu were for two centuries under the rule of the Manchu, and my impression in the matter has always been that they adopted this numeral from their Manchu rulers. This conclusion is amply confirmed by the fact that the annual tribute to be paid to them by the Ainu, as was the case with all the tribes of the Amur region, consisted in sable-skins and other peltry. The Chinese classified all these peoples under the category 'those with an annual tribute of sable-skins' (*sui tsin tiao p'i*). The Ainu ranked in this class,

---

[13] Dobrotvorski (p. 228) notes also a word *opispe* with the meaning 'ten snares in the catch of sables,' used in the same manner as *kuńkutu.* But *opispe* is very far from having in its origin this narrow significance. Eliminating the element *-pe* 'thing,' we have *opis* which was recorded by Steller in the eighteenth century with the spelling *ūpӯhs* as the numeral 10 among the Ainu at the southern end of Kamchatka (see his vocabulary published by J. Klaproth, *Asia polyglotta,* p. 302, or *Aperçu général des trois royaumes,* p. 254, Paris, 1832). Further, Batchelor has noted on Yezo a word *upiš* meaning 'number.' It is therefore probable that *upiš, opis,* or *opiš,* assumed the significance of a high number, and was finally utilized to convey the notion 'ten.' What Dobrotvorski noted is merely a specific case or an applied example. For this reason I am inclined to infer also that the expression *kuńkutu* at the outset had no relation to the business of sable-catching, but, whatever its primeval meaning may have been, is a genuine Ainu word denoting the numeral 10. On Yezo there is a similar word for ten, used only in the counting of animals, *atuita;* for example, *tu atuita* 'twenty animals'—sufficient evidence that the Ainu language does not lack expressions for ten.

[14] *Materials for the Study of the Ainu Language and Folklore,* p. 139 (Publication of the Imperial Academy of Sciences, Cracow, 1912).

as stated in chapter 3 of the *Huang ts'ing chi kung t'u*, 'The Tribute-bearing Nations of the Manchu Dynasty,' an official work published under the reign of K'ien-lung in 1773. Here the Ainu are illustrated and described under the name *K'u-ye* (*Hou-ye* of the Jesuits of the eighteenth century), which is a reproduction of Tungus *Kūgi*, the Tungus and Gilyak designation of the Ainu.

As to Manchu-Ainu relations we are well informed also by Japanese authors. One of these, who wrote in 1786, mentions tobacco-pipes provided with inscriptions in Manchu characters and traded to Karafuto (Saghalin), also Chinese stuffs obtained by the Manchu in Peking and shipped thither.[15] Above all, we have an excellent source of information on Saghalin and the Amur region in the account of Mamia Rinsō, translated by Ph. von Siebold.[16] Rinsō traveled in those regions in 1808, and left a vivid description of Manchu administration in Saghalin and the taxes paid by the Ainu in furs. He also saw on the east coast near Taraika a boundary-stake inscribed with Manchu characters. A Manchu document is still preserved by an Ainu chieftain of Naiero.[17]

The reminiscence of their former dependence on the Manchu is still preserved even in the Ainu traditions of Yezo, in which are allusions to journeys of the people to the governor of Manchuria to pay their respects. Batchelor,[18] who has recorded such a story, comments on this occasion that the ancient Ainu used to go yearly to Manchuria to render homage to the governor of that country, and on their way used to pass through Saghalin; that they used also to do business with the Manchu particularly when at war with the Japanese; and that possibly the Ainu were subject to Manchuria in very ancient times. This chronological definition is somewhat exaggerated. Saghalin became known to the Manchu only as late as during the reign of the Emperor K'ang-hi (1662-1722).[19] It follows therefrom that the

[15] Klaproth, *Aperçu général des trois royaumes*, p. 190.

[16] *Nippon*, new ed., 2, p. 207-235; see chiefly p. 219-221.

[17] Laufer, *Keleti Szemle*, 1908, p. 5.

[18] *Transactions of the Asiatic Society of Japan*, 18 (1890), p. 42.

[19] See Du Halde, *Description of the Empire of China*, 2. 247, or the original French edition, 4. 15 (this report relates to the year 1709); C. Ritter, *Asien*, 3. 450.

Ainu decimal system cannot be older than about the middle of the eighteenth century, when Manchu sovereignty over them was more firmly established. It hardly requires special mention that the numeral system of the Manchu is strictly decimal.

Piłsudski, in his interesting work previously quoted (p. 1-11), is the first author to offer some remarks on the phonetics of the Ainu language. Batchelor has almost neglected this fundamental part of the language, and his transcription of Ainu is no more than an attempt at adapting the English alphabet to the writing of Ainu. And then it is possible to compare with Hebrew and Indo-European, and even to stamp as Indo-European, a language the sounds of which are not yet accurately ascertained. Piłsudski says that Abbé Rousselot studied the phonology of Ainu with some individuals from Yezo at the Anglo-Japanese Exhibition in London, 1910, and communicates some of his results. I have been waiting for their publication on the part of Rousselot, but have not yet seen it. In 1900, shortly after my return from Siberia, I prepared a small Ainu grammar which for some reason or other was never published. In the interest of the progress of Ainu studies I deem it useful to check off my data and conclusions with those obtained by Piłsudski and to state the points in which we agree and those in which we differ.

One of the most interesting experiences in the study of Ainu phonology was to me the fact that all sonants in the series both of the explosives and spirants are lacking. As I was familiar with this phenomenon in many other languages, I naturally paid especial attention to it in examining the Ainu consonantal system. I was able to hear the guttural, palatal, dental, and labial *k, č, t,* and *p* only as pure surds, and summarized the result of these observations in my Ainu grammar literally as follows: 'To the ear the surds may sometimes sound like sonants, but even in this case no laryngeal intonation takes place. Indeed an Ainu is not able to articulate the sonants of the Russian and Japanese languages, and will invariably transform these into the corresponding surds. Russian *dal'še* "farther," for instance, is pronounced by them *tarše;* Russian *gul'ai* "to walk"[20] like

[20] In the Pidgin-Russian as spoken by the aboriginal tribes and the Chinese and Koreans of eastern Siberia, the Russian verb is usually employed in the imperative, regardless of the real form required.

kurai; Japanese *baka* "fool" becomes *paka; ōgi* "fan," *aunki; azuki* "a kind of bean," *antuki.'* Ainu *kumaška* 'ruble' is the reproduction of Russian *bumažka* 'banknote, paper bill.' All close observers are indeed agreed on the one point that the sounds in question, both as initials and finals, are downright surds; this is the opinion, although not expressed by this strict formula, of Dobrotvorski, Batchelor, and also Piłsudski.[21] Batchelor remarks that 'no sonant letter begins a sentence, but in composition surds are sometimes changed into sonants, *k* turning into *g*, *p* into *b*, *t* into *d*.' This would be a sort of sandhi which occurs in exactly the same manner in Japanese, and which, owing to the long and familiar intercourse of both peoples, may conclusively be attributed to the influence exerted by the Japanese upon the Ainu language. Japanese likewise, as is well known, lacks the sonant explosives, and has developed them but secondarily in composition (the so-called *nigori*). It is thus not impossible, I concluded in 1900, that in a further stage of development Ainu will also develop such secondary sonants. On the southeast coast of Saghalin I had little occasion to note this change; on the contrary I recorded many examples with surds in composition, where a sonant is offered by Batchelor; for instance, *inumbe* 'wooden framework round a fireplace'—Saghalin *inumpä; humbe* 'whale'—Saghalin *humpe; rai-ge* 'to kill'—Saghalin *rai-ke* (*-ke* is a suffix forming causative verbs; *rai* 'to die').

Piłsudski formulates his observations as follows: 'The explosives are *k, t, p; g, d, b*. These two groups are not unrelated. In Ainu there is really only one group; if the sounds occur at the beginning of a word, their normal sound is *k, t, p*. In the middle of a word, the sound wavers between the former, the voiceless group, and the voiced group *g, d, b*. Strictly speaking, these are not identical with their Indo-European corresponding consonants. They are, I should say, neither *fortes* or *lenes;* they are *between*. And then, which is yet more important, their conditions of combination are to be noticed. For instance, after *m*, these consonants readily acquire a certain sonorousness of tone, which probably does not last during the whole time of their

---

[21] F. Müller (*l.c.* p. 143) has added *g, d, b* to the consonantal system of the Ainu, for which there was no occasion even at his time; he had accordingly not read Dobrotvorski.

articulation. The outcome of this was that in very many cases I was unable to determine the nature of the consonant, as I heard a sound that could not be identified either with the former group or with the latter. At all events, among the Ainu of Saghalin, the normal and primary group is *k, t, p* (voiceless), possibly less strongly articulated in certain connections. Their corresponding sounds (*g, d, b*) more or less voiced appear only as secondary variations. On the western shore of Saghalin the latter group is more often to be met with than on the eastern shore.'

The last observation accounts for the fact that on the east coast I heard so few *g, d,* and *b;* I had no occasion to visit the southwestern shore of the island. Although Piłsudski expresses himself somewhat differently, I believe that I am perfectly in accord with him as to the facts in the case, save that I am not yet convinced that the Saghalin dialect possesses genuine sonants. In my estimation, these sonants are also voiceless.[22] With respect to the Yezo dialect I do not hazard an opinion, not having had an opportunity of hearing it.

I concur with Piłsudski in the observation that the explosives are capable of palatalization, except that I do not believe with him in the existence of *b′* and *g′*, and have to add *t′* to his *k′* and *p′*. Palatalized *t′* alternates with palatal *č* (see below, p. 204-5).

Piłsudski asserts that the palatal sonant *j* also occurs, but only in very few words after a nasal, as in *unji* 'fire', *tunji* 'interpreter'—cases already cited by Dobrotvorski. The latter example proves little, as it is a loan-word; Batchelor writes it *tunči,* and in my own collectanea I have *tunčinē ainu* 'interpreter': it is Sinico-Japanese *tsūji,* Chinese *t'uṅ(t'ung)-ši.* This word has been carried by the Chinese all over Eastern and Central Asia; it is heard in Tibet as well as in Mongolia and Manchuria (Manchu *tuṅse,* Golde *tuṅsiko,* Oročon *tuṅksa*). It is curious that the first element of the Ainu loan-word agrees with the Manchu form, the second element with Japanese. At any rate this example is not conclusive as to the existence of an original *j* in Ainu. In regard to *unji,* I myself heard only *unči,* and Batchelor gives both *unči* and *unji,* so that this *j* represents

[22] Compare Sievers, *Phonetik,* § 348.

č, and is again inspired by an imitation of the Japanese *nigori.* A Japanese initial *j* is transformed by the Ainu into the palatal surd; for instance, *jo* 'lock' becomes *čo.*

As final consonants occur the three explosives, the four nasals *ṅ, n, n'*, and *m;* and *s, r.* In regard to the final explosives I made the curious observation on Saghalin that they were about to disappear, that they were dropped altogether by most individuals, while a few in some cases pronounced them with a rather obscure articulation, the preceding vowel being greatly shortened and uttered harshly and abruptly. Thus:—

YEZO	SAGHALIN
yuk 'stag'	*yŭ'*
tek 'hand'	*tĕ'*
šiuk 'bear'	*išŏ'*
marek 'spear for salmon'	*marĕ'*
upok 'to wrestle'	*upŏ'*
čup 'sun, moon'	*čŭ'*
ikaiop 'quiver'	*ikaiŏ'*
onnep 'a large seal'	*onnĕ'*
čep 'fish'	*če'*
at 'flying squirrel'	*a'*

Piłsudski states that certain final consonants are not completely articulated and only very faintly heard, but his description of the process is not quite clear. In all probability the history of this event was such that the final explosives were first changed into the spirant *x* (see below, p. 202-3), which is now gradually giving way. We have, for instance, Yezo *etok* 'source, origin, limit', Saghalin *etox* and *eto; mat* or *max* 'woman' becomes *ma* in composition: *kaṣi-ma* 'old woman,' *koṣ-ma* 'daughter-in-law.'

In the combinations *pk* and *pt,* when occurring as medial sounds, the labial explosive is eliminated in the dialect of Saghalin:—

YEZO	SAGHALIN
ataye-yupke 'expensive'	*ataiyuki*
aptoran 'it rains'	*atoran*
irangarapte 'a greeting'	*irańkaratä*

Medial double *k* of the Yezo dialect corresponds to *sk* on Saghalin: Yezo *ikka* 'to steal,' Saghalin and Kuril *iska.*

Pk and *kk* interchange: Yezo *kupka* and *kukka* 'mattock.'

Of nasals, Ainu possesses at present four—the guttural *ṅ*, palatal *n'* or *ñ*, the dental *n*, and the labial *m*. Only the two last-named may be considered as original constituents of the language. The guttural nasal *ṅ* (*ng*) has originated from dental *n* before the guttural explosive:—

Kusuṅ-kotan, the town Korsakovsk
ahun + *kani* = *ahuṅkani* 'to enter'
ahun + *ke* = *ahuṅke* 'to let enter'
ehan + *ke* = *ehaṅke* 'near'
itaṅki 'teacup'; Batchelor spells *itangi*
kuṅkani 'gold,' Japanese *kogane*
aṅ-kutihi 'metal girdle'
toṅkori 'a musical instrument'

As equivalent of Yezo *šinnam* 'cold, frost' I noted on Saghalin *šiṅnamai*.

As a final, *ṅ* occurs very seldom; for instance, *kakuṅ* 'pouch,' *kamiyuṅ* 'thunder.'

The palatal *n'* occurs only before *e* or as a final, and the palatalization is weak and almost imperceptible. *N* changes into *m* before labials: *tan* + *pe* = *tampe* 'this thing.' Yezo final *m* sometimes becomes *n* or *ṅ* in Saghalin: Yezo *haram* 'lizard,' Saghalin *harian;* Yezo *hum* 'voice,' Saghalin *huṅ*. Final *n* and *s* after *o* and *u* may be dissolved into *i*, thus forming a diphthong, or being lengthened:—

pon čika(*p*) 'small bird' becomes *poi* and *pō čika*(*p*)
wen ainu 'a bad man'—*wei ainu*
išo rui 'bear-skin', for *rus*
tonči and *toiči* 'pit, dilapidated habitation'
Yezo *setan-ni* and *setai-ni* '*Pyrus toringo*'

The spirant *x* has been observed by me in the same manner as by Piłsudski and in the Kuril dialect by Radliński. It occurs as initial, medial, and final, corresponding not only to *p*, *t*, or *k* of Yezo, but also to *h*, *č*, and *ra*.

SAGHALIN	YEZO
oax 'one of a pair'	*oara*
max 'wife'	*mat, mači*
kux 'belt, girdle'	*kučihi* (Kuril *kut*)

axto 'rain'	*apto*
oyaxta 'abroad, away'	*oyakta*
čux, čup 'sun'	*čup*
suroxte 'they sit' from *rok* 'to sit'	
ṣinox and *ṣinot* 'to play'	

There is no doubt that Yezo has preserved the original condition, and that *x*, which is absent in Yezo, presents a secondary development on Saghalin. Sometimes *x* appears as a euphonic insertion, as in *repoxpe* 'a sea-animal' from *rep* ('sea') + *ox* (instead of *o*, 'in, inside') + *pe* ('thing, creature'), or in *pinoxponne* 'stealthily' from *pi* ('secret') + *no* + *ponne* (two adverbial suffixes).

Piłsudski explains that *f* occurs but rarely, and as a secondary sound, produced by the influence of the neighboring vowel (*kuf, kux, kuči* 'girdle,' original form *kut* or *kut'*) ; *p*, when weakened, sometimes becomes *f*, but is always accompanied by *u* (*čup* or *čuf* 'luminous body, sun, moon'). I heard *f* in *čufčikin* 'east,' but *čupahun* 'west.' In *utufta* 'between,' from *uturu* 'interval' and *oxta* 'in,' *f* seems to be evolved from *x;* compare also *ekoxpe* and *yokofpe* 'a single rock in the sea.' Batchelor says that 'the letter *f* resembles the true labial in sound, it being softer than the English labiodental *f;* it is always slightly aspirated as though indeed it were *h*.' On Saghalin I heard *f* and *w* as bilabials, seldom as dentolabials, and only in the combination *fu*. All examples of initial *f* given in Batchelor's *Dictionary* and occupying but two pages are indeed of the type *fu*, and several cases show an alternation of *fu* and *hu: fuči—huči* 'fire,' *fura—hura* 'scent,' *furu—huru* 'hill,' *fuško-toita—huško-toita* 'anciently,' *futtat—huttat* 'bamboo grass.' Dobrotvorski enumerates after doubtful older sources a few words beginning with *fa, fe, fi, fo*, but all these can be easily traced to initial *h, p*, or *w;* for instance, Dobrotvorski's *faibo* 'mother' in fact is *habo, faigar* 'spring' is *paikara, fambe* 'ten' is *wambe, fets* 'river' is *pet*. None of these examples speaks in favor of an original *f*. It is plain that the use of this fricative is very restricted, and, as justly emphasized by Piłsudski, is secondary. When Batchelor adds that it is often found in words which appear to be of Japanese origin (this observation was made also by Dobrotvorski), I believe that this points to the real source of the consonant in Ainu, which in my estimation was adopted by them from the Japanese in comparatively late historical times. This assumption would harmonize with the fact that in the dialect of Kamchatka and the Kuriles *f* is absent; Radliński at least does not give any word with initial *f*. In Japanese also, *f* occurs only before the vowel *u*, *h* being substituted for it with the other four vowels; or rather the rule should be formulated that *h* before *u* becomes *f*. In Japanese likewise, *f* and *h* (probably developed from *p*) are interrelated.

In regard to *h* and *w* I have nothing to add to the remarks of Piłsudski, except that I am not inclined to accept his view that *w* (or, as he writes it, *v*) is always voiced.

Of sibilants I distinguished in Saghalin Ainu three—the dental sibilant *s*, the palatal sibilant *š*, and an intermediary sound transcribed *ṣ*, in the formation of which the tip of the tongue moves farther down than in the two former. This *ṣ* I regarded not as an independent sound, but as secondarily developed from *s* before certain vowels within a word, and as perhaps representing merely an individual variation, as some persons pronounced a plain *s* in the place of *ṣ*. In all probability it is developed from a palatalized *s* (*s'*). Piłsudski denies and rejects *š* entirely, and replaces it by *ś*, equalizing the latter with Polish *ś*,[23] and defining it as between *s* and *š* with a distinct palatalization, or an approach to the position in which *i* is articulated. Abbé Rousselot remarks that *s* is formed by the tip of the tongue held somewhat downward, and its upper surface (dorsum) raised toward the palate. I have no doubt that Piłsudski's *ś* (not heard or noted by Batchelor) coincides exactly with my *ṣ*, especially as his examples of *ś* agree with my records of *ṣ* (for instance, *sam* 'to marry,' *i-śam* 'to marry me'—where Piłsudski justly attributes the origin of this *ś* to the influence of the preceding vowel *i*),[24] but I am convinced also that a genuine palatal *š*, as recorded by Batchelor for the Yezo dialect, likewise exists on Saghalin. This observation is confirmed by the fact of a phonetic alternation of *t*, *t'*, *č*, and *š*.

Compare the following examples:—

---

[23] The same observation was already made by A. Pfizmaier, 'Erörterungen und Aufklärungen über Ainu,' p. 30, *Sb. Wiener Akad.* 1882).

[24] Some examples noted by me are *seta* 'dog,' but *pō-ṣeta* 'small dog' (*po* originated from *pon, poi*); Yezo *sesek, šešek* 'warm'—Saghalin *sēṣĕ'*; *rus* and *ruṣ* 'skin'; *ṣiṣ* 'eye'; *čiṣ* 'to weep, to grieve.'

Kuril *t'eonatarp* 'green' *t'eonatorpa* 'yellow' (Klaproth: *t'euninua*)	Saghalin and Yezo *šiunin* 'green, yellow'
Saghalin *t'iše, t'ise, t'iṣä* 'house'	Yezo *čisei*
Yezo *inuye, šinuye* 'to tattoo, carve'	Saghalin *inuye, činuye, šinuye*
Kuril *kut* 'girdle'	Yezo *kučihi* (*kutšihi*)
Kuril *po-mat, e-po-moč* 'daughter'	Yezo *mat, mači* (*matši*) 'wife' *matne* 'female' of animals
	Yezo *etu, eči, čietu* 'spout, handle'

Of liquidae Ainu possesses only *r*. *L* is absent, as in Japanese. In Russian loan-words *n* is substituted for initial *l*, while Russian medial *l* becomes *r* or is dropped entirely. The Russians are called by the Ainu *Nuča* instead of *Luča,* the general name for the Russians among the Amur tribes. Russian *gul'ai* 'to walk' is pronounced by the Ainu *kurai.* Ainu *čaṅki* 'chief, commander, superior' (address to all Russian gentlemen) is derived from Russian *načal'nik.* Piłsudski affirms that he heard clearly *l* instead of *r* pronounced by many persons on Yezo in the village Piratori and still more frequently in Shiravoi. This observation is confirmed by Abbé Rousselot, who says that *l* exists only as a modification of *r*. Accordingly it is a mere local variation, and cannot be credited to the fundamental phonetic system of Ainu. *R* has its normal articulation; only as an initial it is, according to Rousselot, semi-occlusive, yielding such variations as *r, tr, kr, tl.* Piłsudski heard *tr* or *dr* only after *n,* and noted a frequent interchange of *t* and *r,* particularly among the Ainu of the north. I heard *tåṣoku* 'candle' for Japanese *rosoku,* and *tetara* 'white' for Yezo *retara.*

The consonantal system of modern Saghalin Ainu is accordingly composed as follows:—

	Explosives	Palatalized	Nasals	Spirants	
Gutturals	*k*	*k'*	*ṅ*	*x*	*h*
Palatals	*c*		*n'*	*š*	
Dentals	*t*	*t'*	*n*	*s*	*ṣ*
Labials	*p*	*p'*	*m*	*f*	*w*
Liquids	*r*				

Eliminating the secondary, more or less modern, developments, we obtain the following:—

	Explosives	Nasals	Spirants
Gutturals	*k*	..	*h*
Palatals	*č*	..	*š*
Dentals	*t*	*n*	*s*
Labials	*p*	*m*	*w*
Liquids	*r*		

That this limited inventory of eleven sounds bears no relation to Altaic, Indo-European, Semitic, or Bask, with all of which Ainu has thoughtlessly been compared, must be patent to every one. Ainu is an isolated language at present, its congeners, if they ever existed, being extinct long ago.

Of all sounds the vowels have been most unsatisfactorily fixed in the Ainu texts hitherto placed on record. Batchelor and Piłsudski note merely *a, e, i, o, u,* while Abbé Rousselot points out that *a, e,* and *o* may have the three different qualities of timbre found in French. In the speech of Saghalin I discerned eight vowels—*a, ä, e, ẹ* (*e* in gardener), *i, o, å* (English *aw*), *u,* and the semi-vowels *y* and *u̯*. In the articulation of *å* the larynx is lowered, the tip of the tongue is pressed downward, and the orifice is rounded. This vowel is important, as it sometimes occurs in the same word beside ordinary *o,* and as there are homonyms distinguished only by these two timbres of *o;* for example, *porå* 'seal,' *på* 'to boil' (intr.), but *po* 'child.' The diphthongs are *ai, ao, au, eo, eu, ou, oi, åi, ui, oa, ua, u̯a, ea.* As the language has no accentual stress, but only a musical accent (as in Japanese or French), the distinction between short and long vowels is very slight. There are no naturally long vowels, but all vowels may be artificially lengthened under the force of the chromatic accent. In conversation, the word *pirika* 'good, well,' for instance, may be heard according to circumstances in three different ways—*pírika, piríka,* and *piriká.* Monosyllables terminating in a vowel as a rule evince a tendency to being somewhat lengthened; for instance, *kū* 'bow,' *tū* 'two.' Lengthened vowels, moreover, arise from contraction of two vowels into one or from elision of consonants: *či* + *okai* yields *čōkai;* Yezo *ataye-hauke* 'cheap' becomes *atǟhauki* on Saghalin; *pon seta* 'small dog' develops into *poi seta* and *pōṣeta; pūrai* 'window' co-exists with *puyara.* Many vowels between consonants show a tendency to evaporate and to be almost eliminated: *seta* 'dog'—*s^eta,* Kuril *sta; šiken* 'sledge'—*š^iken,* Kuril *skini;* Yezo *čikap* 'bird' becomes on Saghalin *čika, čka, čkapu,* and *čkap.* This fact accounts for the many consonantal combinations in the Kuril dialect, like *st, sk, kr,* and others, which are otherwise foreign to the language.

Piłsudski observes: 'It seems that the Ainu make no fixed distinction between short and long vowels; that is, they know nothing of quantity properly so-called. We can only say that an accented syllable is longer, and may be simply termed long; but this length is in strict connection with the accent. However, we do meet with fixed differences in quantity in certain words the sound of which would otherwise be the same; their only distinctive quality is the length of articulation.' As examples he cites *ē* 'to eat' and *ĕ* 'to come,' *rū* 'way' and *rŭ* 'ice in the river' or 'a flock of birds.' Piłsudski has further made a new and interesting observation, namely that a few homonyms change their accents to bring out a change of meaning; thus, *átai* 'chair'—*atái* 'payment'; *án-koro,* possessive pronoun—*an-koró* 'I have'; *ṣíri* 'earth'—*ṣirí* 'payment'; *úma* 'horse'—*umá* 'also.' The same phenomenon is encountered also in Japanese: *áme* 'rain'—*amé* 'a kind of sweetmeat'; *háši* 'chopsticks'—*haší* 'bridge,' etc.[25] It would not be surprising that the Ainu, as in so many other cases, should have imitated the Japanese model.

Some vocalic changes in the various dialects are noteworthy. Final *a,* for example, is eliminated in the Kuril dialect:—

KURIL	YEZO
rip 'high'	*ripa* (Saghalin *ripa*)
rer 'wind'	*rera*
rar 'eyebrows'	*rara*
čar 'mouth'	*čara, čaro*
mukar 'ax'	*mukara*

The Saghalin and Kuril dialects have sometimes preserved a final *u* which is dropped in that of Yezo:—

Saghalin and Kuril *erumu* 'rat'	Yezo *erum*
Saghalin *ihoku* 'to buy'	Yezo *ihok*

[25] For other examples see B. H. Chamberlain, *Handbook of Colloquial Japanese,* 3d ed., p. 20. The accent is so extremely slight that it will be hardly noticed by an untrained ear, but it really exists, as I had many times occasion to convince myself. It cannot be compared in strength with the energetic tonic accents of Russian in such pairs of words as *zámok* 'castle'—*zamók* 'lock'; *múka* 'grief'—*muká* 'flour'; *óbraz* 'manner'—*obráz* 'pattern'; *pólnoči* 'midnight'—*polnóči* 'half a night.'

When more exact records of the various dialects are placed at our disposal (and there are none thus far of the Kuril dialects), it will be possible to establish a greater number of phonetic laws and to trace the history of Ainu speech, possibly leading also to a clue as to tribal migrations. The fact that the Yezo and Saghalin dialects are closely related was, of course, known long ago; but the theory that the idiom of Saghalin is purer or more archaic must be disputed. Despite the possibly larger variety of vowels, diphthongs, and spirants (*x* and *ṣ*, both of secondary origin), the phonetic system of this dialect shows decided evidence of a far more advanced state of disintegration and even deterioration. The dialectic differentiations are largely phonetic and lexicographical; accidence and structure appear to be the same everywhere. According to statements made to me by natives of Saghalin, their language is not divided into dialects, but is spoken with a high degree of uniformity. Local variations of words are frequent, particularly in the names of animals: an eagle is designated in Naiero *furä,* in Naibuči *pisetteri;* Naiero *samakka* (explained as 'a black sea-eagle with a red-tipped beak') answers to *onnim* of Naibuči; a strap of sea-lion skin used for carrying loads is styled *ečikä* in Ottašam, but *tara* in Naiero and Taraika. There are likewise identical words with different meanings on Yezo and Saghalin; for instance, *hoinu* on Yezo means 'marten,' on Saghalin 'sable.' We need a complete dictionary of the Saghalin dialect for further comparative study; we need a good grammar of the language, not after the fashion of the Latin grammar, but one interpreting the spirit and laws of the language from within. We have had enough theories and fancies about the Ainu; it is time to get at the facts.

140

月氏语或为印度-西徐亚语考

The Language of the Yüe-chi or Indo-Scythians

by

BERTHOLD LAUFER, Ph.D.
Curator, Department of Anthropology,
Field Museum of Natural History

CHICAGO
R. R. DONNELLEY & SONS COMPANY
1917

FIFTY COPIES PRINTED
FOR THE AUTHOR

The Language of the Yüe-chi or Indo-Scythians

By Berthold Laufer

The question of the nationality of the ancient Yüe-či is still unsettled. It is known that KLAPROTH first classified them with Tibetans, but subsequently became converted to the theory of their Indo-European origin, identifying them with the Goths.[1] The Pan-Turks who have done so much mischief to the history of Central Asia did not fail to claim the Yüe-či as their property.[2] This speculation is exploded not only by the very remains of the Yüe-či language itself, but also by the formal statement of the Chinese annalists to the effect that the Yüe-či were different from the Hiuṅ-nu; they belonged to the group of Hu, that is, Iranians.[3] Most writers on the Yüe-či (and there is a goodly number of them) did not commit themselves to any opinion as to the ethnical position of the tribe.[4] Nationality is based on language: I propose to examine the few remains of the ancient Yüe-či language (that is, in times prior to the foundation of the Indo-Scythian empire) preserved to us in the records of the Chinese and to offer some conclusions with regard to the position of their language.

When in A.D. 87 the king of the Yüe-či asked for a Chinese princess in marriage, he sent as gift to the Emperor Čaṅ of the Han dynasty precious jewels and two kinds of animals hitherto unknown to the Chinese, *ši* ("lion") and *fu-pa*.[5] It is a common experience that the

---

[1] Tableaux historiques de l'Asie, pp. 132, 287–289. It is regrettable that F. JUSTI in his history of Iran (Grundriss, Vol. II, p. 489) still speaks of "the Tibetan Yüe-či or Tochar," and that even to E. H. MINNS (Scythians and Greeks, p. 110) they "appear rather to have been nomad Tibetans." Polyandry is not ascribed to the Yüe-či in any document, as asserted by Minns. The Tibetan hypothesis has been well refuted by O. FRANKE (Zur Kenntnis der Türkvölker und Skythen, pp. 25–27).

[2] F. HIRTH, Nachworte, p. 48; H. G. RAWLINSON, Bactria, p. 128; A. STEIN, Khotan, p. 50 ("the Yüe-či probably spoke a language of the Turkī-Mongolian family").

[3] *Hou Han šu*, Ch. 117, p. 27b.

[4] In regard to the older theories, which are all defective and inacceptable, see E. SPECHT, *Journal asiatique*, 1883, nov.-déc., p. 320; it is superfluous to discuss these anew in the present state of science.

[5] *Hou Han šu*, Ch. 127, translated by E. CHAVANNES, *T'oung Pao*, 1906, p. 232.

Chinese, whenever foreign products were brought to them for the first time, adopted together with these their foreign designations. Thus it is in the present case: *ši* and *fu-pa* are actual words received from the language of the Yüe-či.

1. 師 and subsequently 獅, *ši*, *š'i, lion. On a former occasion I remarked that this word originally hailed from some East-Iranian language and was transmitted to China through the medium of the Yüe-či.[1] This opinion should now be modified by the formula that the word *š'i, šı or šē, actually represents a Yüe-či word with the meaning "lion," and that this Yüe-či word is closely related to its Iranian congeners.

2. 符拔 *fu-pa*, *fu-bwaδ, fu-bwal, fubal. As is known, this word has been identified by A. v. GUTSCHMID[2] with Greek βουβαλίς or βούβαλις,[3] but he has merely added the Greek word in parenthesis to *fu-pa* by way of explanation without discussing the philological basis of the case. First of all, it must be stated that *fubal is the Yüe-či designation of an animal, and that this word may be related to βουβαλίς, in the same manner as other words in Indo-European languages. Certainly *fubal is not a Greek loan-word in Yüe-či. Moreover, the animal of the Yüe-či did not represent the same species as the bubalis of the ancients, as plainly follows from a close comparison of the classical and Chinese traditions. The bubalis of the ancients has been identified with *Bubalis mauretanica* of northern Africa with long tail and short, lyre-shaped antlers, as well as with other kinds of antelope.[4] Aeschylus[5] is the first author to speak of "the young bubalis serving as food to the lion" (λεοντοχόρταν βούβαλιν νεαίτερον). Herodotus (IV, 192) places bubalis among the animals occurring in the Libyan desert, and Polybius (XII, 3, § 5) praises their beauty.[6] The point of interest is that in the opinion of the ancients the lion and the bubalis were arch-enemies; they were often represented jointly on engraved gems.[7] This notion of a contest between the two creatures may have been prevalent also in the minds of the Yüe-či, and their gift to the Chinese Court

---

[1] *T'oung Pao*, 1916, p. 81.

[2] Geschichte Irans, p. 140.

[3] Wrongly written by him βούβαλος (that is, *Bos sylvestris*, urochs).

[4] O. KELLER, Antike Tierwelt, Vol. I, p. 294.

[5] Fragm. 322 Nauck.

[6] See further ARISTOTLE, Hist. an. (ed. of Aubert and Wimmer, Vol. I, p. 64); AELIAN, Hist. an., XIV, 14; PLINY, VIII, 15.

[7] IMHOOF-BLUMER and KELLER, Tier- und Pflanzenbilder auf Münzen und Gemmen, Plate XVII, 43.

savors strongly of a political allegory (the weak swallowed by the powerful). Yet the *fubal of the Yüe-či was an animal different from the bubalis. Certainly the Yüe-či had not exported it from northern Africa, but it was an antelopine species indigenous in the steppes of Central Asia. According to the Han Annals, the *fu-pa* occurred in the country Wu-yi-šan-li 烏弋山離,[1] and together with lions (or a lion) in A.D. 87 was also sent as tribute from Parthia 安 息 (*An-sik, Ar-sik), on which occasion it is described as having the shape of a *lin* 麟, but without antlers.[2] In a late dictionary, the *Er ya i* of the twelfth century, the *fu-pa* is defined as "resembling a stag, and being provided with a long tail and a single horn." The word *fubal, accordingly, was not only Yüe-či, but also belonged to the speech of the Parthians. Again, the affinity of the Yüe-či language points to Iran.

3. Another Yüe-či animal-name is handed down, not in the official annals, but in the *Hüan čuṅ ki* 玄 中 記 written by Kuo 郭 (his personal name is unknown) in the fifth century or earlier. This is the Yüe-či term for the "ox," transcribed in Chinese by means of the character 及 *ki*, anciently *g'iep. The text of the *Hüan čuṅ ki*, as far as I know, is not preserved, and we have to rely on extracts from it given by later writers. The fact that the name of the animal was 及 is guaranteed by the very careful work *T'ai p'iṅ huan yü ki*[3] published by Yo Ši 樂 史 in the latter part of the tenth century. The Chinese ascribe a miracle to this peculiar cattle-breed of the Yüe-či: three or four pounds of its flesh may be sliced off, yet the wound will heal in the course of a day (in other texts, on the following day), and the animal is then restored to its normal size.[4] In the most detailed version of the story that I have been able to trace[5] it is added under the name of the *Hüan čuṅ ki*, "Chinese who

[1] Regarded by CHAVANNES (*T'oung Pao*, 1904, p. 555) as a transcription of Alexandria (*U-yir-šan-ri) and identified with Strabo's Alexandria in Aria.

[2] *Hou Han šu*, Ch. 118, p. 4 (see CHAVANNES, *T'oung Pao*, 1907, p. 177); repeated in *T'ai p'iṅ huan yü ki*, Ch. 184, p. 6b.

[3] Ch. 80, p. 7.

[4] Cf., for instance, *T'ai p'iṅ yü lan*, Ch. 900, p. 2b, with the misprint 反 for 及. As will be seen from the *T'ai p'iṅ huan yü ki* (*l. c.*), the spurious work *Po wu či* erroneously ascribes this ox to the district Yüe-sui 越 巂 in Se-č'uan. The *Pen ts'ao kaṅ mu* (Ch. 51 A, p. 7) cites the "ox of the Yüe-či" as a special variety, and simply quotes the text of the *Hüan čuṅ ki*, but without giving the native name. Some texts say that the animal occurs in the country of the Ta Yüe-či (*T'ai p'iṅ yü lan* writes 月 支) and the western Hu 西 胡 (that is, Iranians), others omit the latter term. Suṅ Yiṅ 宋 膺, in his *I wu či* 異 物 志, adopted a humane attitude toward the story, and had the tail ten pounds in weight of the Yüe-či ox cut off and restored.

[5] *T'u šu tsi čeṅ*, sub 牛. Here the name of the animal is given as 日 及, "the ox *ki* of the day," with reference to its recuperative powers gained in a day.

entered this country [Ta Yüe-či] saw the ox without being aware of the fact that it is there regarded as a precious rarity. The Chinese said that in their country there were silkworms the size of a finger, feeding on the leaves of mulberry-trees and producing silk for the benefit of man, but those foreigners would not believe in the existence of silkworms."

It does not require much sagacity to recognize in the transcription *g'iep, g'iev an Indo-European word and in particular one of Iranian characteristics,— Avestan *gav-*, Middle Persian *gav*, *gō*, New Persian *gāv*, Armenian *kov*, Sanskrit *gáv-*. Above all, however, the Yüe-či form agrees closely, also in its vocalism, with Yazgulami *γēw* ("taureau") from *γāwa, recently disclosed by R. GAUTHIOT;[1] Ossetian *gäwd*, and Scythian *godi* (from *gowdi). It is a Scytho-Iranian type of word.

There can be no doubt, either, that the notion of the decreasing and increasing bull of the Yüe-či answers to a mythical conception of specifically Iranian type, which the sober and prosaic Chinese were of course unable to grasp: it is the waxing and waning of the moon that is symbolized in the image of the bull.[2] Compare Avestan *aēvōdāta* and *gaočiθra*.

4. 翖 or 翕侯 *hi-hou*, *h'iep (or hep)-gou, hiev-gou. Title of the five satraps of the Yüe-či, wrongly read *yap-hau* by HIRTH[3] and identified by him with the Turkish title 葉護 *ye-hu*, *yab (džab, šab)-gu.[4] The two titles, however, have nothing to do with each other. The title *hi-hou* is

---

[1] Notes sur le Yazgoulami, dialecte iranien des confins du Pamir (*Journal asiatique*, 1916, mars-avril, p. 264).

[2] Cf. for instance, DARMESTETER, Etudes iraniennes, Vol. II, p. 292; L. H. Gray, Spiegel Memorial Volume, pp. 160–168; G. HÜSING, Iranische Überlieferung, pp. 23–54.— The ox appears to have been an important domestic animal among the Yüe-či. According to the *T'uṅ tien* 通典, written by Tu Yu 杜佑 from A.D. 766 to 801, the Great Yüe-či availed themselves of four-wheeled carts (unknown to the Chinese), which in proportion to their size were drawn by four, six, or eight oxen.

[3] Nachworte, p. 47. The foundation of this reading is the modern Cantonese dialect, but it is erroneous to identify the latter with ancient Chinese (see my Sino-Iranica, No. 11). There is no reason to assume that 翕 ever had the reading *yap in ancient times; *yap* is merely a development peculiar to Cantonese. The *fan-ts'ie* of the character in question is indicated in K'aṅ-hi by 許及 and 迄及, the sound being 吸, that is, *h'iep* or *hep*. Moreover, Hirth's identification of the title *hi-hou* with *ye-hu* is entirely arbitrary, not being supported by any Chinese text. If the two transcriptions, which phonetically are different, were intended to render the same foreign word, the ancient commentators would certainly not have failed to call attention to it.

[4] Regarding the phonology of *ye* cf. PELLIOT, *Bull. de l'Ecole française*, Vol. IV, pp. 267–269.

applied by the Chinese also to Hiuṅ-nu, Wu-sun, and Sogdians. There is no reason to assume with the Turkomaniacs that it should be of Turkish origin: the Chinese themselves say nothing to this effect; but if the term is equally found among three Scythian or Turanian groups, compared with a single Turkish tribe, the greater probability is that the title is of Turanian origin and a Turanian loan-word in Hiuṅ-nu. In my opinion the word itself is of Scythian origin, the first element being connected with Armenian *šahap*, from Iranian *šarhap, šahrap (Old Persian *xsaθrapāvan*, *σατράπης*).[1] Again, we observe the peculiar vocalism of Yüe-či: the vocalization *hiep* or *hiev*, compared with Iranian *-hap*, corresponds exactly to *giev* – Iranian *gav* (No. 3).

In regard to the second element *gou*, I have not yet arrived at a positive conclusion, but will offer merely a suggestion. It is well known that Young-Avestan *gava* is used as a synonyme of Sogdiana, and that the Pahlavi translation explains this word by *dašt* ("plain"). DARMESTETER[2] has therefore conceived *gava* as a noun with the meaning "plain," and compared it with Gothic *gawi* ("county, country"), Old High German *gewi*, *gouwi*, Middle High German *göu*, *gou*. The Yüe-či word *gou* may be related to this Germanic word, and the term *hap-gou* may signify as much as "county-prefect."

5. From the royal names Kaniṣka, Huṣka, Huviṣka, Vāsuṣka, we may well infer that *-ṣka* was an ending peculiar to the language of the Yüe-či. S. LÉVI[3] has joined to these forms the tribal name Turuṣka, which in fact is based on the name "Turk," but also serves for the designation of the Kuṣana or Indo-Scythians. In 1896 I indicated from the Mahāvyutpatti the Sanskrit-Tibetan term *turuṣka* or *turuka* for the designation of frankincense (*Gummi olibanum* or *Thus orientale*). The *Pen ts'ao kaṅ mu* gives Sanskrit *turuṣkam* as a synonyme of *su-ho* ("storax"), but evidently a confusion with frankincense has here arisen.[4] Turuṣka, however, does not mean, as believed by RHYS DAVIDS, "Turkish incense,"[5] but "incense of the Indo-Scythians." In the chapter *Žui yiṅ t'u* 瑞應圖 of the *Suṅ šu* it is on record that in A.D. 458 the country Yüe-či 月支 sent as tribute divine incense 神香, which

[1] HÜBSCHMANN, Armenische Grammatik, Vol. I, p. 208; cf. also the note of ANDREAS in A. Christensen, L'Empire des Sassanides, p. 113.

[2] Le Zend-Avesta, Vol. II, p. 7.

[3] *Journal asiatique*, 1897, janv.-févr., p. 11.

[4] PELLIOT, *T'oung Pao*, 1912, p. 478.

[5] See PELLIOT, *Journal asiatique*, 1914, sept.-oct., p. 418. As to *tarukkha*, the Pāli equivalent for *turuṣka*, adopted by Rhys Davids, it should be remarked that A. WEBER (*Abh. B. Ak.*, 1871, p. 85) had already explained *turuṣka* from *turukhka.

was examined by the Emperor Hiao Wu 孝武. It had the appearance of swallow-eggs, and there were three lumps altogether, in size resembling a jujube. The emperor refused to burn it, and had it transferred to the treasury. Subsequently an epidemic broke out in the capital Č'aṅ-ṅan. The officials were infected and requested the emperor to burn a lump of the precious incense in order to ward off the pestilence. The emperor then burned it, whereby those sick in the palace were relieved. At a distance of a hundred *li* around Č'aṅ-ṅan, the odor of this incense was perceptible, and even after nine months, had not yet gone. Hence Chinese writers on incense have established the term "Yüe-či incense" 月支香.[1] It follows from this story that only real frankincense can be involved. The ending *-ṣka* certainly is not Turkish, but Scytho-Iranian.

6. The tribal name Yüe-či has been much discussed, but a phonetically correct restoration has not yet been secured. The name in the writing 月氏 appears in the first part of the second century B.C., and probably was first committed to writing in the memoranda and documents of General Čaṅ K'ien himself. We are confronted, accordingly, with the transcription of a foreign name attempted in the early Han period; and, as is well known, we are practically ignorant of what the phonetic condition of the Chinese language was in that era. The philological science of the Chinese permits us to restore the structure of words to the speech of the T'aṅ period, but beyond this we tread upon unsafe ground. Yet there is hope that the progress of comparative Indo-Chinese philology will also reveal to us some day the sounds of the language of the Han. In view of this state of affairs it behooves us the more to proceed cautiously and to heed all available data in the attempt to reconstruct the name by which the Yüe-či designated their nation. Especially Chinese comment bearing on it must not be taken lightly, as was done by A. v. STAËL-HOLSTEIN.[2] Unfortunately the

---

[1] *Hiaṅ p'u* 香譜 by Huṅ Č'u 洪芻 of the Suṅ period, p. 11b (ed. of *T'aṅ Suṅ ts'uṅ šu*); *Miṅ hiaṅ p'u* 名香譜, cited in *Pien tse lei pien*, Ch. 7, p. 6b.

[2] The speculations of A. v. STAËL-HOLSTEIN (SPAW, 1914, pp. 643–650 and repeated in JRAS, 1914, p. 754) are entirely inadmissible. He has a rather comfortable method of discarding any evidence that is opposed to his preconceived theory of the identity of 月氏 with his artificial *Kuṣi, alleged to be the nominative singular of Kuṣa. In order to suit the purpose of this fantasy, the reading *či* 支 for *ši* demanded by the Chinese philologists must be senselessly sacrificed: it is branded as "unauthoritative," while Wylie is heralded as an "authority," for he consistently transcribes Yuĕ-she. In one of the old dialects, according to Staël-Holstein, 月氏 was pronounced Gur-ṣi or Kur-ṣi (again on p. 650: Kuṣi). The authority of Yen Ši-ku in matters of Chinese philology is still to be regarded at least as high as that of Wylie and Staël-Holstein. In favor of his theory the data of Chinese history must

ancient commentators, while they give us positive information as to the phonetic value of the second element, fail to enlighten us on the first part. There is, however, a full interpretation of the name in a work of mediæval date, which has been overlooked by previous writers. Yo Ši 樂史, author of the *T'ai p'iṅ huan yü ki*[1] in the latter part of the tenth century, explains the pronunciation of 月氏 by means of the characters 肉支; that is, *žuk-či or *n'iuk-či (d'i). This, of course, is striking, since the character 月 was anciently possessed of a final dental, but never of a final guttural. Yo Ši adduces no source or authority for his comment, but as he proves himself well informed and appears to have utilized original documents of the T'aṅ period, he may have derived this suggestion from a T'aṅ source. The idea underlying his explanation is that the character is not taken by him as the classifier 73 ('moon'), but as the classifier 130 肉, which is written also with a variant 月 (somewhat different in shape from 月 'moon') that appears in combinations with this classifier. It is difficult to believe that, if right from the beginning the character 月 of Yüe-či should have conveyed the phonetic *žuk or *n'iuk, the ancient commentators should not have drawn attention to this anomaly. The opinion of Yo Ši leads us nowhere, but it merits to be kept in mind.

The direction of the commentators is that the second element 氏 *ši* should be read 支 *či*,[2] and later works have indeed substituted this character for the former.[3] The fact that the verdict of these old philologists is not arbitrary, as arbitrarily asserted by A. v. Staël-Holstein, is plainly to be seen from other names, for instance, the transcription 閼氏 *yen-či*, designating the queen of the Hiuṅ-nu, where

---

be discredited and turned upside down. It is perfectly obvious, however, that the two names Yüe-či and Kuṣana, both of which were known to the Chinese, are by no means etymologically interrelated, but thoroughly independent. Kuṣana was known to the Chinese in the form 貴霜 *Kwi-saṅ, and they were aware of the fact that *Kwi-saṅ was one of five satrapies or principalities, and that 丘就却 K'iu-tsiu-k'io (Kuzulakadphises; regarding the Chinese transcription see PELLIOT, *Journal asiatique*, 1914, sept.-oct., p. 401), after the subjection of the four other satrapies, established himself as king of *Kwi-saṅ. Kuṣana, as pointed out by F. W. THOMAS (JRAS, 1906, p. 203) is not a tribal name, but a family or dynastic title; otherwise we should not have an Indian inscription describing Kaniṣka as "propagator of the Kushan stock." See also J. MARQUART, Chronologie der alttürkischen Inschriften, p. 59.

[1] Ch. 184, p. 8.

[2] *Ts'ien Han šu*, Ch. 61, p. 1; *Hou Han šu*, Ch. 118, p. 5.

[3] The writing 月支 appears in a document from Niya (CHAVANNES, in Stein, Khotan, p. 540).

Yen Ši-ku clearly formulates the rule that anciently the character 氏 had the sound *či* (昔氏音支).[1]

This verdict cannot be overruled in favor of any hypothesis to be attached to the name Yüe-či. Likewise in the name of the district 烏氏, *ši* is to be pronounced *či*. The question arises, however, as to how the character 支 was articulated in early times. The opinion of KLAPROTH, who adopted the reading Yüe-ti, that *t* may often be replaced by *č*, cannot be set aside so completely, as has been done by O. FRANKE;[2] save that Klaproth did not express himself very clearly; he doubtless meant to say that palatal *č* or *tš* may develop from dental *t;* and this, in fact, is a common phenomenon in Indo-Chinese. Moreover, it is justly emphasized by PELLIOT[3] that the small dash differentiating at present the symbol 氐 *ti* from 氏 is a comparatively recent affair, so that formerly the latter character might have been read *ši* as well as *ti*. Pelliot is further right in concluding that under the Former Han the character 氏 was sounded with a dental initial more or less palatalized (mouillé), but not with a palatal and still less with a fricative. As Wań Č'uń offers the variant 焉提 Yen-t'i in lieu of 閼氏 Yen-či, and as *t'i* answers to an ancient *di, there is good reason to assume that also 氏 and likewise 支 were at that period articulated *d'i*, *di*, or *ti*.[4] Thus there is also reason to believe that this element *-di* or *-ti* was assimilated to the final dental

[1] The restoration of the transcription *yen-či* presents a complex problem, as the commentators offer various means of reconstructing the prototype. The *Ši ki so yin* states that *yen-či* should be read 曷氏 *had-di or *hat-ti, which would indeed lead to Turkish *xatun* (*qatun*, *khatun*). In ancient times this word had several phonetic variants: we have in T'u-küe *kahatun 可賀敦 (*k'o-ho-tun*) and *katun 可敦 (*k'o-tun*), in T'u-yü-hun *katsun 恪尊 (*k'o-tsun*), and in T'o-pa *kasun 可孫 (*k'o-sun*). It seems to me that the Hiuń-nu word *haddi (= *haddun) represents the primeval form, and that *katsun and *kasun are subsequent developments. It is difficult to see, however, why the Chinese wrote *yen*, if the sound phenomenon *had* was intended; but whatever the basis of this identification may be, there can be no doubt of the existence of the Hiuń-nu word *had-di itself. It is obvious that Yen Ši-ku visualized a different term of the Hiuń-nu language when (*Ts'ien Han šu*, Ch. 94 A, p. 5) he defines the *fan-ts'ie* of *yen* by means of 於連 (*yien*). It is singular that K. SHIRATORI (Sprache des Hiung-nu Stammes, p. 4), in dealing with this word, does not heed the Chinese indications, although he quotes them, and identifies the word with Uigur *abeči* or *evči*, which in my opinion is impossible.

[2] Zur Kenntnis der Türkvölker, pp. 22–23. The variant 氐 does not occur quite so rarely, as assumed by Franke. It is employed in K'ań-hi's Dictionary, and may be seen in the *Šu kien* 蜀鑑 edited by Kuo Yün-t'ao 郭允蹈 in 1236 (Ch. 9, p. 3; ed. of *Šou šan ko ts'uń šu*). At any rate 氐 is merely a graphic, not a phonetic variant.

[3] *Journal asiatique*, 1912, juillet-août, p. 169.

[4] See also PELLIOT, *Bull. de l'Ecole française*, Vol. V, p. 428.

of the first part, 月, and that this final may have been not a surd, but a sonant. In the same manner as in Tibetan, it will be shown that also in Chinese the final explosives were originally all sonants (partially also liquids).

The ancient phonetic formation of 月 is somewhat complex and very far from being such a simple affair as *get*, as confidently asserted by former sinologues, in order to fall into the trap of the Getae and Massagetae. The original initial was not a guttural sonant, but the guttural nasal *ṅ*, which is plainly indicated by the *fan-ts'ie* 魚 厥 (*ṅi k'iud) and the sound equivalent 軏, which, in the same manner as 月, still has in Sino-Annamese the pronunciation *ṅüet*. Further, this initial *ṅ* was palatalized (mouillé) and labialized (provided with so-called *ho-k'ou* 合 口); that is, phonetically written, *ṅ'wiet or ṅ'wied, ṅ'wieδ.[1]

The initial *ṅ* seems to have had a tendency to develop into *g* during the T'aṅ period. In an Uigur Sutra the name Yüe-či is said to be transcribed Kitsi or Ketsi.[2] Of course, this transcription made after a Chinese mode of articulating the term in the T'aṅ period (provided the identification were correct, which is doubtful) would have no absolute value for the restoration of the ancient form of the name which makes its début some eight hundred years earlier. The fact remains that initial *ṅ* (now *y*) generally corresponds to *g* in the transcriptions made in the age of the T'aṅ;[3] and if the transcription Yüe-či had originated in that period, we should be perfectly justified in restoring it to a form with initial *g*. But we must not lose sight of the fact that the transcription was made early in the second century B.C. under the Han; and that the same rules then prevailed as under the T'aṅ, no one can affirm. The greater probability is that phonetic conditions were then somewhat different and perhaps more complicated than in mediæval times. In Iranian and Scythian names we have always to reckon with double

---

[1] In this correct form it is transcribed, for instance, by H. MASPERO, Etudes sur la phonétique historique de la langue annamite, p. 94. See also PELLIOT, *Bull. de l'Ecole française*, Vol. V, p. 443. The *ho-k'ou* is still preserved in Fu-kien *ṅwok* and Japanese *gwatsu*. It is only through this *ho-k'ou* that the utilization of 月 in the transcriptions of Sanskrit for *vit* and *vut* becomes intelligible (see, for instance, examples in VOLPICELLI, Prononciation ancienne du chinois, p. 179; and SCHLEGEL, Secret of the Chinese Method, p. 98). Concerning the use of 月 in the transcription of an Iranian word see the writer's Sino-Iranica, No. 18.

[2] According to F. W. K. MÜLLER, Uigurica, p. 15; but according to PELLIOT (Traité manichéen, p. 29), the Chinese equivalent corresponding to Kitsi is the name 義 淨 *Ṅi-tsiṅ (Yi-tsiṅ).

[3] Examples may be seen in CHAVANNES and PELLIOT, Traité manichéen, pp. 29, 42.

consonants which it was difficult for Chinese to reproduce. From my Sino-Iranica it will be seen that Chinese initial *s* and *š* may correspond to Iranian *xs* and *xš*.[1] A somewhat vacillating initial as that of 月 indicates very well that it should answer to a combination of foreign sounds unfamiliar to a Chinese ear. This assumption being made, there are two hypothetical reconstructions possible: *ṅ′wied-di would lead either to *aṅ′wied-di or to *sgwied-di. The latter is the more probable one, and bears all the characteristics of a Scytho-Iranian name. It comes very near to the Suguda of the Old Persian inscriptions (Avestan Suγδa), the name of the Sogdoi or Sogdians. I do not mean to say that the two names are physically identical, but only that there is a linguistic relationship between the two.

As regards the element *di* (*d′i*, or eventually even *ti*), I hold that it should be conceived as the ending of the plural, and that the plural suffix *-di* is on a par with the plural-suffixes, *-tä* of Ossetian, *-ται* of Scythian, and *-t* or *-y-t* of Sogdian and Yagnōbi.[2] The identification of the name Yüe-či with that of the Getae and Massagetae, in my opinion, is out of the question. Not only phonetic, but also geographical and historical reasons run counter to this assumption.[3] Also the identification with Ἰάτιοι[4] must be rejected, likewise any alleged relation of the name to that of the Ye-t'a (*Yep-dal, Ebdal, Abdal) or Ephtalites. The latter, however, are not Huns, as wrongly asserted by SPECHT, but in the same manner as the Yüe-či, are Indo-Europeans, that is, Scythic Iranians. Likewise so were the ancient Wu-sun, as I hope to demonstrate in a subsequent article. Turkistan, before being settled by Turki, was a country of Iranians.

O. FRANKE[5] has justly called attention to the fact that very close relations and intermarriage existed between the Yüe-či and the Sogdians (K'aṅ-kü 康居); and the kings of Sogdiana are said to have descended from the Yüe-či and to have gloried in this extraction. It seems to me

---

[1] A similar phenomenon obtains in the Sanskrit transcriptions: for instance, Ki-pin, based on *Ki-spir = Ptolemy's Kasparia. In the same manner, the Sino-Iranian word 鑌 *pin* ("a fine steel imported from Persia") is based on Iranian *spin (Sariqolī *spin*, Afgan *ōspīnah* or *ōspanah*, Ossetian *afseinäg*, "iron").

[2] Cf. W. MILLER, Sprache der Osseten, p. 42; J. MARQUART, Untersuchungen zur Geschichte von Eran, II, pp. 77–79; R. GAUTHIOT, Du pluriel persan en *-hā* (*Mém. Soc. de Linguistique de Paris*, Vol. XX, 1916, pp. 74–75).

[3] On this point I concur with the opinion of MARQUART (Eranšahr, p. 206).

[4] TOMASCHEK, Sogdiana, SWA, 1877, p. 159.

[5] *L. c.*, p. 67.

that the cause of this mutual good feeling was given in the linguistic relationship of the two peoples.[1]

When Sieg and Siegling published their memorable study of one of the Indo-European languages rediscovered from ancient manuscripts of Turkistan, they styled this language Tokharian and further defined it in the very title of their publication as "the language of the Indo-Scythians."[2] It has been recognized long ago that both these designations are hazardous.[3] "Indo-Scythian" is out of the question as a

---

[1] Regarding Indo-Scythian proper names see F. W. THOMAS, JRAS, 1906, pp. 204–216. I do not believe that 謝 Sie (the vice-roy of the Yüe-či, vanquished by Pan Č'ao in A.D. 90) represents the title *sāhi* (S. LÉVI, *Journal asiatique*, 1915, janv.-févr., p. 86), as *sie* answers to an ancient *zie. I take the liberty of calling attention to some contradictions in the history of the Kushan dynasty of India, as conceived by our scholars, and Chinese accounts of the Yüe-či. According to V. A. SMITH (Early History of India, 3d ed., 1914, p. 272), the reign of the last Kushan ruler, Vāsudeva, terminated according to the chronology now tentatively adopted, in A.D. 178, and the year 226 denotes the collapse of the Kushan power in India (p. 278). The *Wei lio*, however, informs us that during the period of the Three Kingdoms (San kuo, A.D. 221–277) Kashmir (Ki-pin), Bactria (Ta-hia), Kabul (Kao-fu) and India (T'ien-ču) were all subject to the Great Yüe-či (*San kuo či, Wei či*, Ch. 30, p. 12b; and CHAVANNES' translation, *T'oung Pao*, 1905, pp. 538, 539; CHAVANNES remarks, "Thus, in the middle of the third century, the power of the Kushan kings, was at its climax." See also J. KENNEDY, JRAS, 1913, p. 1057, who called attention to the text of the *Wei lio*). Moreover, we have in the Annals of the Wei dynasty (*Wei či*, Ch. 3, p. 3) the record of an embassy sent in the twelfth month of the winter of the fourth year of the period T'ai-ho 太和 (A.D. 230) by the king of the Great Yüe-či called 波調 Po-tiao, *Pwa-div; that is, Vasudeva. Of course, this Vasudeva may be different from the one of V. A. Smith, but the Chinese text shows us that as late as A.D. 230 at least the Kushan dynasty was still in power. According to Smith, historical material for the third century is completely lacking in India, and nothing definite is recorded concerning the dynasties of northern India, excluding the Panjab, during that period. See also CHAVANNES, *T'oung Pao*, 1904, p. 489.

[2] E. SIEG and W. SIEGLING, Tocharisch, die Sprache der Indoskythen, SPA, 1908, pp. 915–934. Neither the determination Tokharian nor Indo-Scythian is due to these authors, but to F. W. K. MÜLLER (SPA, 1907, p. 960). S. LÉVI always shifts the responsibility on Sieg and Siegling (*Journal asiatique*, 1911, mai-juin, p. 432; and JRAS, 1914, p. 959).

[3] Whereas A. MEILLET has determined the historical position of "Tokharian" with as much acumen as circumspect scholarship, without committing himself to any nomenclature and any theory, German scholars hastened to make the "Tokharians," whose very name in this connection is not yet assured, subservient to their wild speculations regarding the alleged primeval home of the Indo-Europeans. E. MEYER (Geschichte des Altertums, 3d ed., Vol. I, pt. 2, pp. 892–893) popularizes Müller's nomenclature, which he accepts without restraint, and proclaims that the old hypothesis of the origin of the Indo-Europeans from Asia has gained considerably from this discovery. The question of the "Tokharian" language, which is one of

label for so-called Tokharian; the two are entirely different things. This can now be actually demonstrated by referring to the word for "ox." This is *okso* in so-called Tokharian, while, as we have seen, it is *g'iev* or *gev* in Yüe-či or Indo-Scythian. The Tokharian word, accordingly, bears a strictly European character; the Yüe-či word, a Scytho-Iranian character. We further note that Yüe-či possesses initial and final sonants, which are lacking in Tokharian. The two languages, in consequence, belong to two sharply distinct groups of Indo-European types of speech. Yüe-či is a member of the same group as Scythian, Sogdian, Ossetian, and Yagnōbi.

---

mediæval form, has nothing to do with this problem. Also the alleged identity of the "Tokharian" suffix *-aššäl* with Hittite *-aššil* (MEYER, *l.c.;* S. FEIST, Kultur der Indogermanen, p. 431) must be rejected.

141

古代中国的宗教和艺术思想

$3.00 THE YEAR 50 CENTS THE COPY

ART AND ARCHAEOLOGY

An Illustrated Monthly Magazine

PUBLISHED AT WASHINGTON, D.C., BY

THE ARCHAEOLOGICAL INSTITUTE OF AMERICA

(TITLE REGISTERED U.S. PATENT OFFICE)

VOLUME VI DECEMBER, 1917 NUMBER 6

CONTENTS

Remittances should be addressed to ART AND ARCHAEOLOGY, The Octagon, Washington, D. C. Manuscripts, books for review, material for notes and news, and exchanges, should be sent to David M. Robinson, Johns Hopkins University, Baltimore, Md.

Entered at the Postoffice at Washington, D. C., as second-class mail matter

RELIGIOUS AND ARTISTIC THOUGHT IN ANCIENT CHINA

BERTHOLD LAUFER

THE culture of China represents a development of four or five millenniums accompanied by an uninterrupted flow of tradition which partially is still in full operation at the present day. The artistic achievements of the Chinese are closely associated with their religious notions and inspired by religious sentiments nourished from the same fountainhead throughout the long course of their history,—nature-worship, deep reverence for the dead, worship of the ancestors, and an insatiable craving for salvation and immortality. The peculiar conception and happy blending of these elements, combined with a pantheistic philosophy of nature, gave them the best conceivable preparation for artistic accomplishments, and resulted in a unity and harmony of thought and life unattained by any other human society. Whereas our mental culture is based on disconnected ideas,—Semitic, Greek, and Roman,—which have no direct inward organic relation to our national consciousness, Chinese civilization is a unit cast of one mould in which religion, philosophy, poetry, and art are one and the same, emanating from a sound conception of man in his relation to life and nature. New forms and expressions of art were created at all times, but, despite all changing influences, the fundamental ideas underlying their significance persisted with conservative force throughout the ages.

About 3000 B.C. China was a comparatively small country, hardly comprising one-fifth of her present area, chiefly located in the north-western portion of her present home. The climatic and physical conditions of the country were, to some extent, very different from what they are now: the mountain-ranges were still crowned by dense forests haunted by numerous wild beasts like tiger, rhinoceros, tapir, buffalo, hunted by man only with bow and arrow; and the gradually advancing farmer made slow headway in clearing the jungle. Yangtse River was populated by huge alligators, the terror of the rice-growing villages, and, like the crocodile in the valley of the Nile, the alligator soon became the object of religious worship. In carvings of bone the formidable reptile was well portrayed. Every one is familiar with the conventional design of the Chinese dragon; the origin of this fabulous creature has been the object of much discussion. These realistic carvings give us a clue in pointing to the alligator as the prototype from which the mythical figure of the dragon seems to have developed. These glyptic works present the earliest attempt of the Chinese in the line of sculpture, being utilized for purposes of divination. Divining was practised by scorching animal bones, the shoulderblade of cattle and deer, or the carapace of a tortoise, and from the designs formed by the cracks in the bone the future was prophesied. When the process of fortune-telling was completed, a record of the oracle and the reply were engraved upon the bone-carved figure

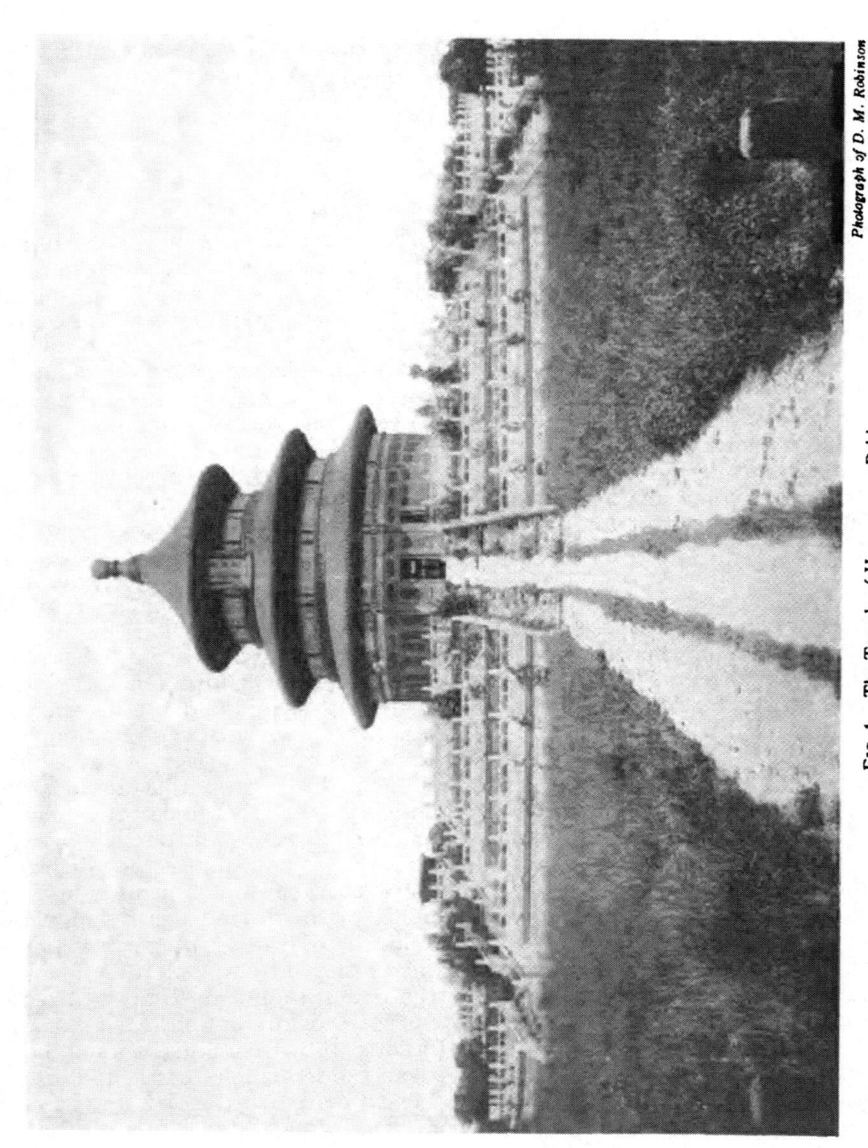

Photograph of D. M. Robinson

FIG. 1.—The Temple of Heaven near Peking

Photograph of D. M Robinson

FIG. 2.—The Round Pavilion of the Temple of Heaven, near Peking

of an animal, and large deposits of such bone archives were buried under ground. One such deposit was excavated in Honan Province about two decades ago, and the Field Museum secured over a hundred pieces, including most of the best carvings from the collection of the late F. H. Chalfant. This material embodies the oldest forms of Chinese writing, but most of the inscriptions are still undeciphered. One of the finest carvings is a charm worn by a princess and stained a turquois-blue through chemical agencies underground. It represents a snake in the centre and a bird in full figure at each end.

Like the nations of western Asia and the prehistoric peoples of Europe, the Chinese passed through a so-called bronze age, during which all implements, weapons, and vessels were made of copper or bronze, to the exclusion of iron. Iron gradually came into use from about 500 B.C. The great stimulus to the development of early art was the unceasing care for the departed ancestors, who were constantly alive and awake in the minds of the people, culminating in a minutely ritualistic cult that created an epoch of highly artistic vases utilized during the ceremonies.

An enormous bronze colander from about 2000 B.C. presents a magnificent example of the high technical skill reached by the early bronze-founders, the enormous vessel being cast from a single mould. It represents the combination of a stove with a cooking-vessel used for steaming grain and herbs in the ancestral cult. A fire was built by means of charcoal in the hollow tripod base which is separated by a hinged grate from the upper receptacle holding the articles to be steamed. The ornamentation of the surface is formed by chased meander bands; and conventionalized figures of monsters stand out in undercut relief. The meander or fret is symbolic of thunder and lightning, while the monster suggests the representation of a storm-god. This composition is intended to illustrate an atmospheric phenomenon, emblematic of fertile rain. The ancient Chinese were a nation of agriculturists, and being deeply interested in weather and wind, their attention was turned toward the observation of the sky and the stars; and this at an early date yielded a surprising advance in the knowledge of astronomy.

A bronze goblet with a marvelous lustrous patina comes down from the same archaic period, the Shang dynasty. Wine was poured out of it in the rituals relating to the worship of the great nature-deities Heaven and Earth. Subsequently goblets of this type were used in the marital ceremony when bride and groom partook of wine from the same cup.

Any trace of realism is absent in that archaic epoch. The human figure does

FIG. 3

not yet appear in decorative art, all designs being purely geometric and receiving a symbolic interpretation evolved from the minds of farmers. No names of individual artists are on record from those remote days; art was subconscious and strictly national.

The process of casting was that *à cire perdue*, and vessels (except the lid) were usually made in a single cast, inclusive of the handle or handles and bottom. The bronze kettle in Fig. 3 is gilt all over, and is the only ancient bronze piece known with such a complete gilding. The handle terminates at both ends in figures of sheep, the favorite domestic animal of the ancient Chinese and a symbol of beauty and gentleness. Such kettles filled with wine were bestowed as a mark of distinction by the sovereign on deserving vassal princes or meritorious statesmen.

The religion of this primitive period was simple and mainly consisted in nature-worship. There was no officially recognized priesthood; the father was the priest of the family, the prince the priest of his kingdom, and the emperor the *pontifex maximus* of the nation. The great cosmic powers, Heaven and Earth, were the chief deities, and their interaction was believed to have prompted the creation of man and nature. The highly intellectual mind of the Chinese was always keen along the lines of mathematics and astronomy, and everything in their rituals was reduced to fixed numbers and categories reflected in celestial phenomena. This peculiar trend of mind called into existence a geometric construction of the supreme deities. Heaven was conceived of as circular, and the image of this all-powerful deity was represented in the shape of a perforated disk carved from jade, which was valued as the most precious material. At the same time, it served as an emblem to the Emperor who was believed to receive his mandate from Heaven and ruled by his command as the Son of Heaven. A most striking feature is the imposing simplicity of this nature-worship.

The notion of round heaven is still expressed by the building of the Temple of Heaven in Peking, the only known example of the circular principle in Chinese architecture, where till the end of the monarchy the Emperor annually used to offer prayers and sacrifices to Heaven. See Figures 1 and 2, pages 296, 297.

The deity Earth, next in importance to Heaven, was revered under the image of a jade tube rounded in the interior, but square outside, because the earth was conceived to be square. This object referred to female power, and accordingly was the sovereign emblem of the empress. In the grave, it was placed on the chest of the corpse, and likewise symbolized the deity Earth, while the disk representing Heaven was placed under the back of the body. Man was the creation of the combined forces of Heaven and Earth, so he should not be separated from the two, and should also rest between them in his subterranean slumber.

Other jade implements are ceremonial emblems connected with solar worship. Types like that of a large jade knife, the perforations forming a constellation, were originally images of the solar deity, and shared in the quality of sun-light to dispel darkness and demons. For this reason, they were interred in the graves as efficient weapons in warding off from the dead all evil influences. A similar implement was the imperial emblem of sovereignty which the emperor held in his hands, while he worshipped the sun early in the morning.

Simultaneously we meet in that archaic epoch shamanistic practices of exorcisms and healing performed by a class of male and female sorcerers or medicine-men. The shaman was conspicuous in the funeral ceremonies when, clad with a scale armor and helmet and equipped with a spear and shield, he marched with pompous steps in front of the hearse, and four times wielded his spear over the open grave in the four directions of the compass, before the coffin was lowered. He opened the way to the departed soul on its peregrination into the beyond, safeguarded it from dangers, and insured the living ones against the return of the soul. Thus, the custom developed that a clay statuette of the shaman as the most efficient exorciser of demons was buried with the dead. In case of disease, the shamans performed a dance, quite in the manner of their present Siberian colleagues, and the making of crude clay images of males and females with movable arms, into which the spirit of the disease was banished, and which were subsequently interred, was an essential part of the ceremony.

About a thousand years later, we come down to the Han period covering the time around the Christian era. In this memorable epoch marking the transition from antiquity to the middle ages, the graves of the people were laid out in large sepulchral chambers composed of flat stone slabs. These formed a vault, sheltering the coffin, and were covered with fine sculptures in bas-relief depicting favorite incidents of ancient history or mythological subjects and forming one of our fundamental sources for the study of ancient culture-life.

The "Battle of the Fishes" is perhaps the most unique subject represented in the plastic art of the Han, reminding one of the Homeric "Battle of the Frogs and Mice". In this Chinese fish-epic we see a whole army of fishes going to war, fish riding on fish, or warriors astride fishes, armed with bucklers, swords, and spears, apparently making ready for a fierce aquatic battle. The king or god of the watery element is driving in a chariot drawn by three large sea-fishes. The significance of this curious representation may be interpreted from ancient Chinese lore. At the time of the first Emperor Tsin Shi (third century B.C.) a belief prevailed in the existence of three Isles of the Blest supposed to be far off in the eastern ocean; there grew a drug, capable of preventing death and securing immortality. The desire of the Emperor to possess this drug, prompted him to send out an expedition, under the leadership of a magician, in search of the Fortunate Islands. But the drug of immortality was carefully guarded by the god of the ocean (here represented in the chariot) and his militant army of fishes. The magician had an interview with him (he is seen kneeling in front of the fish-chariot), to negotiate for the precious remedy; the marine god sent his army toward the coast of Shantung to caution the Emperor against his high-minded ambitions. The Emperor was lying in ambush on the shore, killed one of the fishes with a repeating crossbow, and died a few days afterwards.

A century later, the belief in these Fortunate Islands in the eastern ocean received a fresh stimulus under the influence of alchemy. As later on among the Hindu and Arabs and in mediæval Europe, the notion was entertained that cinnabar could be transmuted into gold in the furnace, and that immortality could be attained by him who should eat and drink out of such vessels made of such gold. The Han Emperor Wu, instigated by the adepts of the black art, again despatched expeditions over the sea to discover the Fortunate Isles, the conception of which held the imagination of the people deeply enthralled, and was a prominent feature in the religious beliefs of those days. To these feelings of their contemporaries, the artists of the time lent expression, by moulding mortuary jars with covers, shaped like the hilly Islands of the Blest emerging from the sea. Such jars were interred in the graves and implied the mourner's wish that his beloved deceased might reach the land of bliss and attain immortality on the Fortunate Isles.

The mortuary pottery of the Han period is a microcosm of the life and culture of that age. All the property dear to the living ones was reproduced in clay of miniature size and confided to the grave, as houses, towers, farm-sheds, barn-yards, mills, grain-crushers, sheepfolds, stoves, and the favorite domestic animals like pigs and dogs. The likeness of an object suggested a living reality, and the occupant of the tomb was believed to enjoy the durable clay offerings as if they were the real thing.

Models of pottery show us the common farm house of the Han period. The gabled roof is covered with solid tiles; some houses show agricultural implements like rice-crushers in the interior. Other houses are provided with double roofs, and people are looking out of the window.

In the west, ancient China bordered on the nomadic Turkish tribes of central Asia which harassed and overflowed the country for many centuries. One of the means of checking the inroads of these restless hordes was the building of the Great Wall and numerous watch-towers along the frontier. These were occupied by soldiers, en-

FIG. 4.—Model in Clay of a Watch-tower

gaged in spying the movements of the enemy from a distance, and repelling him, if necessary. Renowned officers who had deserved well of the country in the frontier-wars were buried with a model in clay of such a military watch-tower indicating their former profession. On the two parapets and roofs the sentinels are engaged in showering from their crossbows a volley of darts

FIG. 5.—A Pottery Jar Representing a Draw-well

on an advancing column of Turkish scouts. This unique bit of pottery (Fig. 4) is in the collection of Mr. Charles L. Freer of Detroit.

The meaning of death was to the Chinese a continuation of this life in another form. The spirits of the departed were therefore obliged to continue their cooking during the other life, and thousands of models of kitchen ranges have been discovered in the ancient graves. A mortuary stove in the Field Museum is a unique specimen and presents one of the earliest attempts of the Chinese to cast iron. On their first acquaintance with iron, they did not begin with forging it, but treated it in the same manner as bronze by casting it in sand-moulds, at a time when this process was unknown in the west. Drink was as necessary to the inmates of the grave as food. Clay jars (Fig. 5) bearing out the idea of a draw-well were lowered into the grave to furnish a constant supply of fresh water. The jar itself represents the well-curb; the well-frame is erected over its edge. The frame above contains the pulley, a small wheel with a deep groove, over which the rope passes, and from the ends of this rope the water-buckets are suspended. The pulley is protected by a tiled roof, and in some jars a water-bucket is placed on the edge of the well. It will be recognized that the art of the Han is different in principle from that of the early archaic period. The rigid formalism and geometric symbolism of the latter is felicitously replaced by an idealism of sentiment expressing ideas in a straightforward way with a personal human touch. It was in fact the great epoch of Chinese idealism.

The notions of immortality find their most curious expression in the utilization of jade for the benefit of the dead. Jade as the most highly prized substance of the Chinese was endowed with the property of preserving the body from decay, and prompting its resurrection. The last service rendered to a departed friend was to send a piece of jade which was placed on his tongue. These protecting amulets had either the shape of this organ or were carved in the form of a cicada. The cicada plays a prominent rôle in the folk-lore of the Chinese who were deeply impressed by the long and complicated life-history of this interesting insect. As the larva creeps into the ground and rises again in the state of the pupa till finally the cicada emerges, so the soul of the dead is to fly out of the old body and to awaken to a new life. The cicada, accordingly, is a symbol of resurrection.

FIG. 6.—Jade Girdle-Ornament Worn as a Love-token

This longing for resurrection is still more forcibly brought out in the jade girdle-ornaments worn by women. These were love-tokens bestowed on them by their husbands and carved from the finest qualities of jade in magnificent colors, in which three modes of technique, engraving, carving in relief and *à jour*, are happily combined, with a great variety of design. The underlying principle is that two animals, birds, fishes, or dragons, are represented as engaged in the love-play of nature. The Chinese cherished the belief that marital relations would be resumed in the other life, and such ornaments were buried with women as symbols of a future resurrection and re-union. See Fig. 6.

China is the land of unlimited possibilities for the archæologist, and we never experienced a greater surprise than some seven or eight years ago when, during the construction of railroads, graves of the middle ages were opened for the first time and yielded an unexpected harvest of clay figurines of a bewildering variety of forms. The most notable feature cropping out of these finds is the rich personal element speaking to us with eloquent language. Under the T'ang dynasty (A.D. 618-906), from which most of these figurines come down, life had assumed new forms, and was enriched by a noble refinement in social customs as well as in poetry, painting, and sculpture. The feminine ideal of that epoch is illustrated in numerous graceful statuettes, some with snail-like chignon and dressed with brown jacket and green shawl,—tranquil, a bit dreamy, reserved, modest, as Chinese women are. They made faithful companions of their masters in the grave, and the great variety of style in their costumes and hair-dressings, varying according to local usage, render them a live source for the study of ancient fashions. A certain style of hair-dressing is similar to that of Japanese

FIG. 7.—Attendants of a High Official on Horseback

women of the present time. The culture of Japan is chiefly based on the Chinese customs of the T'ang period, which in some cases are well preserved in Japan, while they are lost or modified in China.

When a high official is going out on business or a call, his carriage is accompanied by outriders in front and in the rear. In the same manner, we find the coffin in the grave flanked at both ends by male or female attendants on horseback. See Fig. 7.

Powerful knights armed to the teeth, clad with iron armor, protected the lord from demons or malignant intruders whose avarice might have disturbed the peace of his burial-place. And he derived personal solace from edifying conversation with the priests of the Tao doctrine (Fig. 9). In viewing the Chinese entirely under the influence of their rigid moral system, we are prone to make them out as a very serious and even a pedantic people. It should not be forgotten, however, that there always was a merry old China fond of good shows and entertaining games. Acrobats, jugglers, dancers, and musicians are carved on the walls of the grave chambers, and statuettes (Fig. 8). of quaint mimes and actors, providing amusement for the dead, have arisen from the tombs. Some are represented in the midst of reciting a monologue, others are modelled in highly dramatic poses, gesticulating with lively motions as if acting on the stage; others are portrayed with such impressive realism and individual expression that we feel

FIG. 8.—Clay Statuettes of Mimes and Actors

almost tempted to name them after favorite casts familiar to us. The features of these figures show a decidedly Aryan cast. In my opinion, these actors hailed from Kucha in Turkestan. From manuscripts discovered in the sands of Turkestan we know that music and dramatic art were eagerly cultivated by the people of Kucha, who spoke an Indo-European language of the European type known as Tokharian B. Actors from Kucha frequently visited China and were favorite guests at the Imperial Court.

In ancient times, dwarfs were noted in China for their wit and sagacity, and

FIG. 9.—Priests of the Tao Doctrine

FIG. 10.—Yama, God of Death

were frequently employed as jesters and court-fools. We possess records in the contemporaneous Chinese Annals to the effect that Negritos from Malakka, Java, and Sumatra were traded to China as slaves and well-preserved specimens of this class are represented in several T'ang clay figures.

Yama, the ancient Hindu God of Death, was a favorite conception of the people of the T'ang period. He stands either over a demon or a reclining bull which is his emblem, and appears as a mighty warrior with heavy armor and plumed helmet,—an efficient guardian of the grave. His image (Fig. 10) has a large number of different forms; besides the human form, there is a zoomorphic one of Çivaitic origin, with flamed bull-head and eagle's claws, such as still occurs in Tibetan Lamaism. Bull-carts were employed to carry the coffin and paraphernalia at the funeral to the burial-place. See Fig. 11.

All domestic animals, with the exception of the cat, were represented in clay.

FIG. 11.—Bull-cart represented carrying the coffin to the grave

The powerful Tibetan mastiff (Fig.12), a remote ancestor of the English bull-dog pricks up his ears in the attitude of watching; he is provided with neck-collar and belly-band as he is still guided by his master in Tibet. The lover of horses had his noble steeds immortalized in clay, which, complete with their harness, were ready for his immediate use. Many of them (Fig. 13) are remarkable for the spirit and character with which they are portrayed, as though mourning for their deceased master. The big wholesale-dealer who carried on a large trade in the goods of Central Asia and Persia had his grave furnished with the emblem of commerce, a camel (Fig. 14), loaded with the goods of distant lands. Chinese enterprise at that period encompassed the entire world of Asia, their navigation extended over the Indian Ocean to Java, India, the Persian Gulf, and the

FIG. 12.—A Powerful Tibetan Mastiff

Fig. 13.—A Harnessed Horse in Clay

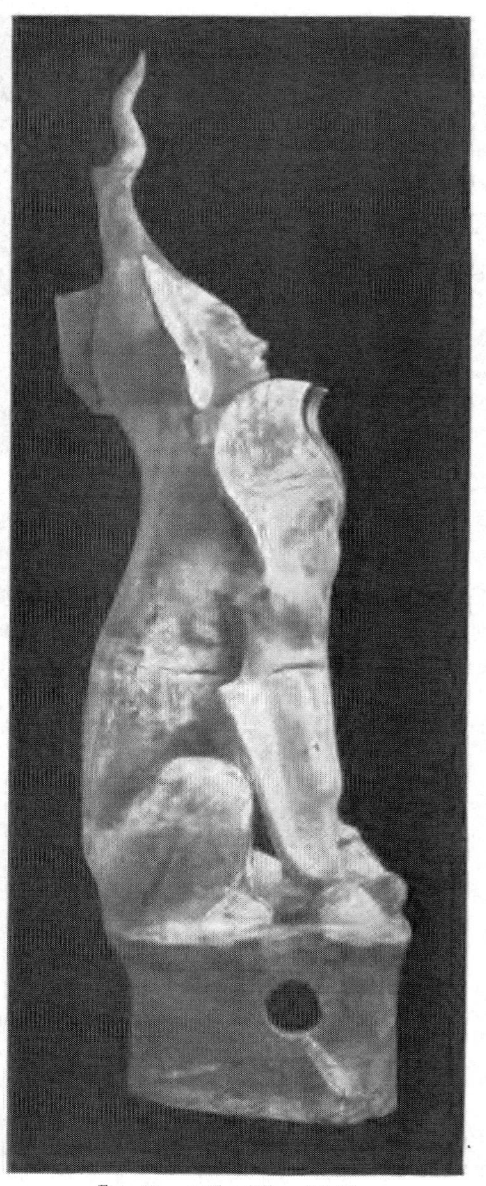

Fig. 15.—A Clay Winged Sphinx

coasts of Egypt; and their caravans penetrated the deserts of Siberia and Turkestan as far as the shores of the Caspian Sea. This little art-work (Fig. 14) is conspicuous for its clever modeling. The guide, a Turk, astride the animal, is giving the signal for the start. Taking a last deep breath and showing the straining of its muscles, the camel endeavors to rise under its heavy burden.

Fig. 14.—A Clay Camel loaded with Merchandise

Other clay statuettes even more intensely evoke memories of China's lively intercourse with Persia, where the great dynasty of the Sassanians had brought to life peculiar and highly developed art-industries. The charming floral designs on Persian rugs and silk textiles, and their engarved gems, awoke a responsive echo in China and the priests of the ancient religion of Zoroaster, the Magi, or Mazdaznians, met with a warm reception in the fatherland of Confucius. They were permitted to found in the capital Si-ngan a temple for the worship of the Sun and the Fire. These facts may account for figures of lions and sphinxes (Fig. 15) with bull-feet, lion-bodies, wings and pointed goat's horn, such as the imagination of the Egyptians and Assyrians had fostered, or fantastic winged unicorn lions, such as greet us from the palaces of Assurbanipal or the Persian kings of Susa and Persepolis.

The religious history of China presents a unique spectacle in that all the great religions of the world have found a hospitable shelter within her domain, and even flourished there at one time or another. The attitude of the Chinese toward foreign religions has always been dictated by a liberal and broad-minded policy of tolerance, and their Government has never persecuted, or encouraged any mental tyranny or stifled any free opinion that keeps clear of State policy, scandal, or libel. In the third century A.D. a new religion arose in Persia, founded by the sage Māni, with ideas borrowed from Christianity of the Gnostic type and overlaid with a dualistic system based on Babylonian and Persian ideas. Māni was crucified in A.D. 275 at the instigation of jealous Zoroastrian priests, and his followers soon dispersed over the Roman Empire, Turkestan, and China. By the end of the fourth century, the religion of Māni had grown into a world-system com-

FIG. 16.—The Effigy of a Manichean Priest. Notice the descending dove on the head-dress

peting with Christianity for religious and intellectual leadership. But the opposition of the ruling Church made the adherents of Māni suffer bitter persecution,—and while the Christian Emperor Justinian condemned Manicheans to death and had their sacred writings committed to the flames, their co-religionists in China enjoyed perfect peace and liberty under the enlightened Emperor T'ai-tsung of the T'ang dynasty and translated their religious books into Chinese. Some of these have recently been rediscovered, and it is to the glory of the Chinese that they preserved to us in their language the books of a Christian sect which had been mercilessly destroyed by the barbarism of Europe. It appears from one of these Chinese Manichean treatises that the adherents of Māni conceived the Holy Ghost as a white dove. On the high head-dress of one statue (Fig. 16) we see in relief the design of a descending dove, and for this reason I believe to be justified in recognizing in it the effigy of a Manichean priest.

Field Museum of Natural History, Chicago.

142

书评四则

AMERICAN ANTHROPOLOGIST

NEW SERIES

ORGAN OF THE AMERICAN ANTHROPOLOGICAL ASSOCIATION,
THE ANTHROPOLOGICAL SOCIETY OF WASHINGTON,
AND THE AMERICAN ETHNOLOGICAL
SOCIETY OF NEW YORK

VOLUME 19

LANCASTER, PA., U. S. A.
PUBLISHED FOR
THE AMERICAN ANTHROPOLOGICAL ASSOCIATION

1917

Antike Technik. Sechs Vorträge. H. DIELS. Teubner: Leipzig and Berlin, 1914. 140 pp., 50 figs., 9 pls.

H. Diels belongs to that sane and refreshing type of classical philologist that likes to cross the Chinese wall of the special field and by his broad outlook comes in contact with representatives of many other disciplines. In 1908 the edition of a new Greek work, ascribed to Melampus, and treating of divination from the throbbing or pricking of certain bodily organs (known to every one, for instance, from Shakespeare's Macbeth: "By the pricking of my thumbs, Something wicked this way comes") gave him occasion to study the same subject with the assistance of a staff of able collaborators, and to edit related books from Russian, Serbian, Bulgarian, Rumanian, Arabic, Hebrew, Turkish, Indian, German, English, and French,—a contribution to folklore of the first order.[1] His fruitful interest in the history of natural sciences and medicine is well known, and the attempt of such a high-standing authority to acquaint a larger public with some of the chief results of his science in concise and elegant form must be received with great satisfaction. He was mainly actuated by the desire to demonstrate by way of selected examples that classical antiquity, even in its technical achievements, is much more intimately associated with the modern world than the intervening middle ages, and that innumerable threads visible and unseen link these two worlds together. The struggle of engineering and natural science against humanism in the last century was based on a deplorable mutual ignorance and semi-culture of the two conflicting parties: the humanists, in the grip of a nebulous idealism, were unfamiliar with the material civilization of the ancients and unable to grasp its connection with our modern life, while the adversaries flatly rejected any reminiscence of our indebtedness to the ancients. This state of affairs has changed, for the classical philologists, though they may still belong to the best-hated species of mankind, a veritable *odium generis humani*, have now deeply penetrated into the realities of antique culture and pay as much attention to them as to its canon of beautiful forms and its ideal thoughts. The author's six lectures deal with science and technics among the Greeks; antique doors and locks; steam-engine, automaton, and taxameter; ancient telegraphy; ancient artillery; and ancient chemistry. Diels has been the first to reconstruct correctly the Homeric system of closing a door by means of lock and key, and his study of this subject is of great ethnological interest, as, for

[1] Beiträge zur Zuckungsliteratur des Okzidents und Orients, *Abhandlungen der preussischen Akademie*, two parts, 1908–09.

instance, analogies to the so-called *balanos* lock, which is found in Egypt, as early as the time of Ramses II (1292–25), still exist in China and Tibet. It might be worth the trouble to pursue this object in its distribution all over Asia. Our taxameter is modeled after the principle of Heron's hodometer, and certain modern automatic devices may well be traceable to ideas first conceived by Greek mechanicians. Carrier-pigeons were known in Hellas during the fifth century B.C., and were utilized by the Romans in beleaguered cities and for the transmission of sport news after races. As justly remarked by Diels, this practice was derived from the Orient. A Chinese author of the ninth century, it may incidentally be added here, reports that the Persians kept pigeons on board their ships and sent them back with letters to their homes over distances of several hundred miles (*Yu yang tsa tsu*, ch. 16, p. 6). Noah's dove, however, was not a carrier-pigeon, as asserted by Diels, but represents a distinct affair: it belongs to the category of land-spying birds dispatched by navigators in search of land, frequently mentioned in India and Babylonia, also by Pliny (VI, 22; see my notes in *T'oung Pao*, 1914, p. 11).

The chapter on the telegraphy of the ancients is full of interest. Long-distance communication was well developed by fire signals, a water telegraph, and torch signals expressive of the letters of the alphabet. In Aeschylus' *Agamemnon* we have the earliest example of a rapid transmission of a message from Asia to Greece: the conquest of Troy is signaled by huge bonfires from Mount Ida in Asia Minor by way of the island of Lemnos to Mount Athos, then flashed southward over Euboea into Boeotia and Mount Kithaerum, thence across the Isthmus to Epidaurus and ultimately to the castle of Mycenae. From Herodotus' statement that Mardonius after the battle of Salamis expected to send to Xerxes the news of the capture of Athens by fire-signals along the Aegaean islands to Asia, Diels rightly infers that such like institutions existed in Asia; in fact, they are alluded to in Babylonian texts. The most ingenious instrument of the ancients was the signal telegraph described and improved by the historian Polybius. Each station was formed by two battlemented turrets, each wall having five crenelles or embrasures, two feet distant from one another, in which torches could be placed. The system was based on a chiffre according to which the twenty-four letters of the Greek alphabet were distributed over five tables; in each table the letters were numbered in succession. Two operations were thus required for telegraphing a single letter, one turret serving to indicate the number of the table, the other, the number of the letter in question within this table. The letter K, for instance, was listed in the second table; accordingly, two torches were shown in the embrasures of the first bastion. K being the fifth letter of the second table, five torches were placed in the second bastion. It has been calculated that a dispatch of only three words, like "hundred Kretans deserted" would require a hundred and seventy-three torch-signals, and that the transmission could be accomplished within a half hour. It is obvious that this system contains the germ idea of electric telegraphy. It had no practical success, however, which Diels attributes to various causes, as the scanty range in the radiation of the torches, the great number of stations required, the complexity and costliness of the enterprise. It seems to me that the main cause for its failure must rather be sought for in the republican institutions of the Greeks, which lacked a centralized will-power and the strong hand of a ruler interested in the maintenance of such a service; for in all Oriental empires postal and intelligence organization was far superior to that of Greece. In ancient China, optical telegraphy by fire-signals was at a high stage of development as early as the third century B.C. Many of these ancient watch-towers have survived, particularly in Turkistan. They were erected in defence of the empire, the news of an invasion of nomadic tribes being at once signaled to the capital, two, three, or four firebrands in succession indicating the strength of the hostile forces. At the same time, and similarly as in the Persian empire, a staff of rapid couriers changing horses every fifth kilimeter or so was maintained. Marco Polo (chap. XXVI) and other mediaeval travelers have left us vivid descriptions of this magnificent postal system.

Artillery (by an irony of fate this word is derived from *artes*, "arts") was well known to the ancients, and from the descriptions of the historians and the writers on military matters the various types of antique cannon have been reconstructed and examined as to their range and effect. Remains of them have not survived, as they were chiefly constructed from wood; but missiles from sandstone, three to ten pounds in weight, and arrow-points have been unearthed in Spain on the site of the ancient Numantia, and are supposed to have been hurled into the city on the occasion of its gallant defense against Scipio the Younger in 133 B.C. For the first time we hear of the employment of artillery around 400 B.C. in Syracuse under Dionysus the Elder (I hardly think we can say with Diels that it was "invented" in that year; dates for inventions are usually fallacious). According to a theory of Diels, the ancient cannon was developed from the bow by way of the crossbow. With all their ingenuity the ancients have nothing to oppose to the Chinese repeating crossbow with a magazine holding up to ten darts which may be shot consecutively and rapidly, the darts sinking through their own weight,—mentioned in the Annals as early as 211 B.C.,—the prototype of our magazine rifle. Diels frankly admits that the modern development of weapons is not directly connected with Hellenic tradition; but "an imperceptible transition is visible in the so-called Greek fire derived from Greek tradition in its relation to modern fire-weapons." Indeed, he heralds the Greek fire as the precursor of gunpowder, without any regard to the well-founded claim of the Chinese to this invention and the admission of the Arabs that it was received by them from the Chinese. It is unproved that the Greek fire, first mentioned by Marcus Graecus in his *Liber ignium* toward 1250, has any connection with ancient Greece; moreover, it was a combustible or incendiary, but not an explosive like gunpowder. In China we can trace a rational development of gunpowder from the humble firecracker (known in the sixth century of our era), which was originally employed in religious ceremonies, to the launching of fiery projectiles in warfare as early as the twelfth century and the full development of fire-weapons under the Mongols in the thirteenth and fourteenth centuries. In Java, Sumatra, and the Philippines fire-arms were known and used prior to the arrival of Europeans. It is therefore out of place to say that as early as the beginning of the fifteenth century cannon advanced from Europe to China (p. 101). An Arabic treatise on gunpowder written in 1249 and preserved in the Library of the Escurial, styles saltpetre "Chinese snow" (*thelg-as-Sīn*) and the rocket "Chinese arrow" (*sahm khatāī*); saltpetre was first exported by the Arabs from China, where it occurs as a natural efflorescence of the soil in Szechuan, Shansi, and Shantung. Arabic writers mention fire-arms as far back as 1312, and in 1323 they were used at the siege of Baza. It is certainly due to Moorish influence when they make their first appearance in Spain. In the opinion of Diels, "this progress (?) is due neither to the Arabs whose claims are certainly unjustifiable, nor to the Chinese, but to the Germans," because the Germans were regarded by Byzantinians and Italians as the inventors of this new "barbaric" technique, and because in the fourteenth and fifteenth centuries the Germans exclusively possessed a literature on artillery, and German gun-smiths were everywhere in demand. All this is imaginary or inconclusive. The Germans may have had their share in the further development of fire-arms, but crediting them with their "invention" and even with the invention of gunpowder is a misstatement of facts.

In his discourse on chemistry Diels rejects the current etymology of this term from the Egyptian. He adopts the forms *chymes* and *chymeia* as a basis and derives them from Greek χύμα (metal-cast), metal-casting being in the center of antique chemical technology. A very interesting extract is given from the *Papyrus Holmiensis* in Upsala concerning Egypto-Hellenistic alchemy. At the end the author comes back to his former theory that alcohol should have been known to the Greeks, and was to a larger extent manufactured toward the close of the thirteenth century after a recipe of probably Greek origin. The article of J. Ruska in which this opinion is convincingly refuted should at least have been mentioned; also it would be interesting to know what Diels' attitude toward his opponent's arguments is.[1] The reviewer sides with Ruska and E. v. Lippmann in the opinion that the isolation of alcohol is not due to the Greeks or the Arabs, but a discovery made in Europe toward the close of the middle ages. It is justly emphasized by Ruska that with all respect for Alexandrine and Arabic science we are compelled to admit that the age of discoveries begins in the West at an earlier date than is usually assumed, and that we are not justified in contesting to the closing middle ages the discoveries then mentioned for the first time, although frequently under a false flag. The middle ages were not quite so dark and backward as believed by Diels and many others. Though we respectfully dissent from the author as to his result in the alcohol problem, we gladly concede that his study contains a great amount of new and useful information, and in the same manner as the present volume, is an important contribution to the history of science.

B. LAUFER

ASIA

Indian Shipping. A History of the Sea-Borne Trade and Maritime Activity of the Indians from the Earliest Times. RADHAKUMUD MOOKERJI. Longmans, Green and Co: London and New York, 1912. XXVIII, 283 pp.

Books move slowly these days. A copy of Professor Mookerji's work reached me only in August, 1916. The history of navigation and shipping is a subject of primary importance for the study of early international relations: especially in the sphere of the Indian Ocean, tribal

[1] Compare H. Diels, "Entdeckung des Alkohols," *Abhandlungen der preussischen Akademie*, 1913, no. 3; J. Ruska, "Ein neuer Beitrag zur Geschichte des Alkohols," *Der Islam*, vol. IV, 1913, pp. 320–324, and "Alkohol and Al-kohl, zur Geschichte der Entdeckung und des Namens," *Aus der Natur*, vol. X, pp. 97–111.

migrations, ethnic movements, commercial conditions, intellectual exchanges, can be fully comprehended only by close attention to the rôle which the high seas have played in connecting India with Africa, the Archipelago, and the Far East. Mookerji is the first to give us in a handsomely printed volume a coherent survey of Indian navigation from earliest times down to the end of the Moghul period, and his account is interesting and readable from beginning to end. While it is chiefly based on well known English translations of Indian sources, the author has succeeded in finding a new Sanskrit text on ship-building by Bhoja, in which some information on various types of boats is given. A task which the ethnologist would like to see performed in this connection—a detailed study of the present-day types of Indian vessels—is unfortunately not attempted; but ships figured in ancient Indian sculpture, painting, and coins are studied to some extent and accompanied by good illustrations.

Bhoja recommends that no iron should be used in the holding or joining together of the planks of sea-going vessels, for the iron will inevitably expose them to the influence of magnetic rocks in the sea, or bring them within a magnetic field and so lead them to risks. Mr. Mookerji thinks that this warning is worth carefully noting, and that this rather quaint direction was perhaps necessary in an age when Indian ships plied in deep waters on the main. There is, of course, no reality involved here, but we face the well known legend of the Loadstone Mountain which is believed to attract toward it ships wrought with iron and to smash them to pieces. René Basset has treated this subject in an interesting manner under the title "La Montagne d'Aimant" (*Revue des traditions populaires*, 1894, pp. 377–380). Jordanus, Montecorvino, Friar Odoric, and Marco Polo have described these vessels that have no iron in their frame and are only stitched together with twine (see Yule, *Cathay*, new ed., vol. II, p. 113). In all likelihood, the legend was carried to India by the Mohammedans.

From the number of passengers given for ships in the Buddhist birth-stories (*jātaka*) Mookerji is inclined to draw inferences as to the size of the ships. This conclusion is not admissible, as such numbers as 500 or 800 are quite typical in Buddhist stories and mean nothing more than any indefinite large number. If the boat carrying the Buddha to the Sea of the Seven Gems was loaded with 700 merchants besides himself, it is evident that the figure 700 is symbolically chosen to correspond to the Seven Gems.

The discourse on the traditions of the gold-digging ants (pp. 96, 97) lags somewhat behind the times. This legend is not Indian in its origin, but, as shown by the reviewer on a previous occasion, arose in Central Asia, where it was anciently known, and was diffused from there to India. At any rate, the fantastic explanations of Ball and Schiern, which Mookerji accepts in good faith, are far from the point.

The contrivance on the boat reproduced from the sculptures of Borobudur (plate opposite p. 46, no. 1) is not, as explained, a compass, which was not yet in existence at that time, but is simply a reel or pulley, over which the ropes for towing the vessel are run, as is plainly demonstrated in figure 5. The same device is placed both on the prow and stern of the boat (fig. 3).

The author throughout speaks of India as a unit, and ardent Indian patriot as he is, he tells us a great deal of Hindu imperialism in ancient times and India's command of the sea for ages. Here again the ethnologist will pause to raise the question what the share of the manifold tribes of India was in the navigation of the ocean, what the Aryans owe to the Dravidas, and what foreign influences on Indian boatgear and methods of shipping and navigating may have been in operation. The ancient Egyptians, Persians, Malayans, Cambodjans, Chinese, and Arabs were all sailing the Indian Ocean, and it is difficult to believe that these various peoples should not have learned from one another. This subject remains to be studied at close range. One lesson is worthy of being retained, and this is that the idea of an exclusive mastery of the sea and the destruction of rivals never entered into the minds of any of those nations, and that they plodded along one beside the other in peaceful competition till—the first Europeans arrived. The Mediterranean spectacle of Carthage and Rome was not repeated along the expanse of the Indian Ocean.

In his conclusion, Mookerji laments the want of a fully developed Indian shipping at the present time, and makes an appeal to the Government, and all who are interested in the material progress of India, "to be fully alive to the importance and necessity of reviving and restoring on modern lines a lost industry that rendered such a brilliant service in the past, and with which are so vitally bound up the prospects of Indian economic advancement."

The author should by all means be encouraged to continue his meritorious studies, but they should be founded on a broader basis. His knowledge, for instance, of the history of Java, Farther India, and China is not up-to-date, chiefly owing to his failure to read French literature. As to the Arabic and Persian sources, G. Ferrand's *Relations de voyages et textes geographiques relatifs à l'Extrême-Orient* must be consulted. The failure to utilize Chao Ju-kua in the translation of Hirth and Rockhill, where numerous references to shipping and sailing could have been gathered, is a serious drawback. Chao Ju-kua is referred to but once (p. 170), and wrongly designated as a Chinese traveler of the thirteenth century. He did not travel at all, however, but collected his notes on foreign trade and peoples, while stationed as Inspector of Maritime Trade at the port of Ts'üan-chou in Fu-kien Province.

B. LAUFER

Guide au Musée de l'École française d'Extrême-Orient. H. PARMENTIER. Hanoi: Imprimerie d'Extrême-Orient, 1915. 136 pp. 32 pls.

Few modern seats of learning have had a more glorious history than the École française d'Extrême-Orient, founded in 1898 at Hanoi, French Indo-China, by Governor General Doumer at the instigation of the Institut de France. In the first line the École is a research institute entrusted with the task of exploring the archaeology and ethnology of Indo-China, which, owing to the peculiar cultural position of this region, necessitates close study of both India and China and the interrelations of the two countries. The publications of the School are hence devoted to indology as well as sinology, the center of gravitation being directed toward Farther India. The annual Bulletin, of ambitious size, fifteen volumes of which have been issued, has earned the well-deserved reputation of being the leading and most solid periodical in all scientific matters relating to the East: every contribution contained in it is of importance and signals a decided advance in our knowledge of the subject. Besides this Journal, the School has published twenty special volumes concerned with philology, bibliography, numismatics, and archaeology,—all fundamental works of permanent value. The practical duty imposed upon it is to preserve the historical monuments of Indo-China, to propose the necessary measures for their protection, and to watch the execution of the orders insuring their safeguarding. Two precious instruments of work are at the disposal of the institution,—a library particularly rich in manuscripts and rubbings of inscriptions, and an ethnographical and archaeological museum. Not many among us may know that the city of Hanoi is even favored with two museums, the other being the Musée Agricole et Commercial, which harbors selected examples of the modern industrial art of Indo-China. The museum of the École is under the able guidance of H. Parmentier, an architect and art student of note, who directs the archaeological work of the institution. We owe to him an *Inventaire descriptif des monuments čams de l'Annam* (with a separate volume of plates), a monograph on the Temple of Vat Phu, and two ingenious studies bearing on the interpretation of architecture in the bas reliefs of ancient Java and Cambodja. A clear-minded scholar and an artist with a vision, he has stamped his personality on the arrangement and interpretation of the collections in the Museum of Hanoi, to the treasures of which we are introduced by him in a well written little guidebook. The prehistoric archaeology of Indo-China is well represented. Any palaeolithic types of stone implements (*pierre taillée par éclats*) have not yet been discovered in its soil; the prehistory opens with an epoch of polished stone, in the same manner as in China. Also in Annam the folklore interpretation of stone implements as thunderbolts is prevalent, and the same belief attaches to prehistoric bronze hatchets; in southern Annam a tradition is current that the latter should be produced only in places where a living being has been struck, the metal being silver or gold in the case of a man. The stone artifacts of Tonking, in general, are carelessly wrought and of mediocre material, save numerous small pieces executed in phtanite. The interpretation of the spade-shaped utensils from Tortoise Island as mattocks for hoeing the rice fields is doubtless correct (compare the reviewer's *Jade*, pp. 47, 78). No information, unfortunately, is given on the potsherds of the stone age of Annam. The few observations made on the prehistoric pottery of Laos and Cambodja whet our appetite for more and make us wish for a full publication of this important material. With a certain degree of chagrin we note that the present inhabitants exploit the shell heaps of Samrong-sen on their own hook for the purpose of extracting an excellent lime, and find there quantities of stone implements and shell ornaments. This reckless unprofessional digging should be stopped; these shell heaps ought to be systematically explored and to receive the same governmental protection as the historic monuments. The Museum owns a very large and interesting series of bronze hatchets from Tonking, abounding in unusual forms and technically of great perfection. It appears plausible that metal in this area was largely reserved for weapons, while stone continued to be utilized for household utensils; and this conclusion is supported by the observation that no stone arrowhead or spear has ever been found in this region. The engraving on a bronze fragment of two stags with large antlers under a dugout with two rowers is emphasized as a unique specimen of the prehistoric art of Indo-China.

Coming to historical times, the arts of Cambodja, Laos, Siam, Burma, Annam, China, and Tibet; with minor sections bearing on Corea and

Japan, are represented by well selected material. A good sketch of the prominent figures of the Lamaic pantheon is inserted. The unexplained female statuette (plate XXIX and p. 91) appears to have formed part of a triad, and to represent one of the consorts of Padmasambhava, probably Mandārava; at any rate, it is the type of an Indian, not a Tibetan woman. A piece of evidence of Japanese relations with Indo-China is presented by a Japanese sword guard found at Angkor Vat. Some terra cotta statuettes discovered in the marshes or rice fields of villages in the proximity of Hanoi are said to be relatively modern and of Canton manufacture; but one representing a lion with human head is regarded as older than the others. This type is well known among the T'ang clay figurines from Shen-si Province, being doubtless modeled after an Iranian prototype.

From plate XIX we glean with some consternation that a fine set of sacrificial bronze vessels of the K'ien-lung period is not cased. Chinese bronze, it is true, in general possesses a much greater power of resistance to atmospherical influences than Egyptian, Greek, or Roman bronze; but even if sheltered in a fairly air-tight case, it demands the constant, watchful care of a museum curator as to possible formation of malignant patinas. In the humidity of the Hanoi climate any open exhibition of whatever character would seem unsafe. We entertain the best wishes for the future growth and prosperous development of the Hanoi Museum, to which this attractive guidebook will assuredly win many new friends.

B. LAUFER

Chinese Clay Figures. Part I. *Prolegomena on the History of Defensive Armor.* BERTHOLD LAUFER. (Field Museum of Natural History, Publication 177, Anthropological Series, vol. XIII, no. 2.) Chicago, 1914. Pp. 69–315, 64 pls., 55 figs.

The basis of this series of studies, which the author modestly terms prolegomena, is afforded by certain ancient clay figures from the provinces of Honan and Shen-si—the region, in other words, where the old Chinese culture first took on its historical form. As is the case with all of Dr. Laufer's writings, the subject is treated with the utmost thoroughness, and light is thrown on it from almost every conceivable angle. In doing this the author has availed himself of his very wide knowledge of Chinese records and customs, and has based his conclusions upon evidence which seems incontrovertible. It is fortunate that Dr. Laufer is an ethnologist of thorough training and wide experience, for his consistent adherence to the modern anthropological point of view gives his work a quality which is lacking in much that has been done in the same field.

of the Strait of Belle Isle, and facing the Labrador Coast, which is distinctly visible from here; being only about nine miles distant. This path is called "Chemin de Sauvage." There is also a place on the same shore called Savage Cove, which is probably the supposed place of their departure. This would seem to bear out the statement of the Micmacs. Again, in the English Coast Pilot for 1755, there is a place near Hawkes' Bay or Point Riche, called "Passage de Savages."

It is only fair to the future ethnologist to point out that while we have no definite reason to suppose that descendants of the Beothuk do actually exist anywhere now, there is, nevertheless, a strong force of opinion to render this an affirmative possibility. There is still a hope that a source of information on culture may be found. The ethnology of the interior of Labrador, especially that of the eastern section, is practically unknown and it would be very unsafe to say at this time what sub-types of northeastern culture and dialects might not be found there.

Whether or not the hope of learning more of the culture of the supposedly extinct tribe is ever realized there is a grave doubt if Mr. Howley's monograph will ever be superseded.

FRANK G. SPECK

ASIA

Grammaire de la langue khmère (cambodgien). GEORGES MASPERO. Imprimerie nationale: Paris, 1915. Ouvrage publié sous le patronage de l'Ecole française d'Extrême-Orient, VIII, 490 pp.

This excellent grammar of Khmer, the fruit of twenty years of labor among the people, by an official in the Civil Service of Indo-China, marks an important event not only for the philologist, but also commands the serious attention of the ethnologist. The Khmer, whom we style Cambodjians (from Sanskrit Kamboja, the official designation of the country in native documents), number at present about 1,700,000 individuals distributed over Cambodja and the adjoining territories of Cochin-China, Siam, and Laos. The tribal name is of ancient date, being preserved in the transcription Ki-mao in the Chinese annals of the T'ang dynasty (618–906), and being recorded as Comar by the Arabic travelers from the ninth century onward. In former epochs the area covered by the Khmer was far more extended, and comprised the lower basins of the Mekong (from Luang-Prabang) and the Menam (from the wall of Kamphëng-Phet). Khmer domination in the latter region up to the eleventh or twelfth century is attested by ancient inscriptions found in Siam. The entire country now designated Cochin-

China was exclusively Cambodjian up to 1658 when it was overrun by the armies of the Annamese. The theory has been proposed that the Khmer should not be indigenous to their present habitats, but should be regarded as immigrants hailing from western regions, probably from India, while their present home, inclusive of Cochin-China, Annam, Champa, and the littoral of southern Indo-China, should have been the cradle of the ancestors of the Malayo-Polynesian stock, the Cham being its direct offspring, and the Malayans their emigrant descendants. Maspero points out that this thesis is based on purely linguistic data, and that no historical fact has as yet come to the fore to corroborate it. Profound anthropological studies will be required before the discussion of this problem may assume a definite shape. For the time being we can set the Khmer in relation only with a certain number of tribes surrounding them, on the basis of linguistic evidence. These are the Pear and Cong on the right bank of the Mekong; the Kuoy and Suoy or Sue on both sides of this river; the Stieng, Pnong, Krol, Tiom-Pueun, Rmang, Brao, Bahnar, Sedang, Boloven, Kaseng, Alak, Ve, Kon-Tu, Ta Hoi, and Leung on the left bank of the Mekong. The ethnology of these primitive tribes, aside from some notes on the Pnong by Dr. Harmand, has not yet been studied. This group is enclosed in the north and west by peoples of the T'ai family (Siamese and Laos); in the south, north, and east by Annamese; and in the east, first, by peoples of Malayan speech, the Cham and some savage mountain tribes like the Raglai, Churu, Pih, Rade, and Jarai, second, by the important tribe of the Cho-Ma whose language seems to be intermediary between the Malayan and Khmer types of speech.

Further, Khmer and its dialectic variations form with Mon and a number of allied savage dialects a linguistic group which has received the name Mon-Khmer. The Mon on whom we bestow the designation Peguans, and whom the Burmese style Talaing were the inhabitants of the kingdom of Pegu that occupied the delta region of the Irawaddy, the north of the peninsula of Malakka, and probably part of the lower basin of the Menam, where it was contiguous to the Khmer kingdom. Conquered and subjected by the Burmese, the Mon since 1286 have ceased to form a distinct nation. The close relationship of their language to that of the Khmer is incontestably proved;[1] likewise to that of the Palaung, Wa, and Riang, primitive tribes in the basin of the Salwen.[2]

[1] P. W. Schmidt, "Grundzüge einer Lautlehre der Mon-Khmer-Sprachen," *Denkschriften Wiener Akademie*, vol. LI (1905), pp. 1–233.

[2] Compare the appendix concerning these languages in P. W. Schmidt, "Grundzüge einer Lautlehre der Khasi-Sprache," *Abhandlungen der bayerischen Akademie*, vol. XXII (1904), pp. 778–805.

An attempt has been made also to link Annamese and Cham into the same group with Mon-Khmer, and philologists like Fr. Müller, E. Kuhn, and P. W. Schmidt have heralded the existence of a Mon-Khmer-Annam family. The students who attended the general linguistic courses at Columbia University in 1906–07 will recall the interest then evoked by a discussion of P. W. Schmidt's theories.[1] It has been apparent for some time that these, partially at least, are not built on a sound foundation, and to some extent are in need of a revision. Among the propositions submitted by the writer to the Research Committee of the American Association for the Advancement of Science at its last meeting was therefore included a recommendation that the program developed by P. W. Schmidt be made the guiding motive of a research worker in the field in question, chiefly as Schmidt's conclusions are partially based on limited and unsatisfactory material. In regard to Annamese, Henri Maspero,[2] a brother of Georges Maspero, has conclusively proved in a remarkable study that Annamese cannot be attached to the Mon-Khmer family. The alleged relationship was merely suggested by the coincidence of the numerals from six to nine.[3] Phonetically and grammatically, Annamese exhibits radical differences from Mon-Khmer. It does not share in the prefixes and infixes of this group, and moreover has developed a system of tones which in minute details agrees with that of the T'ai languages, whereas there is no trace of similar or any tones in Mon-Khmer. Whether Annamese for this reason should be classified in the T'ai family it is difficult to decide in our present state of knowledge. Modern Annamese presents an extremely complex structure, being the final product of a mixture of quite a number of different languages. Tied up between Khmer in the south, T'ai in the west, and Chinese in the north, it has been subject to an overwhelming influence from each of these groups. Primeval Annamese, with a foundation of its own, appears to have been evolved from the fusion of a Mon-Khmer

[1] "Die Mon-Khmer-Völker, ein Bindeglied zwischen Völkern Zentralasiens und Austronesiens," *Archiv für Anthropologie*, vol. V (1906), pp. 59–109.

[2] "Etudes sur la phonétique historique de la langue annamite," *Bulletin de l'Ecole française d'Extrême-Orient*, vol. XII, no. 1.

[3] Hence E. Kuhn in his fundamental study, "Beiträge zur Sprachenkunde Hinterindiens" (*Sitzungsberichte der bayerischen Akademie*, 1889, p. 219), rightly inferred, "We are therefore justified in concluding that originally Annamese was foreign to the cycle of the other languages, and that its coincidences with these are traceable to subsequent influences,—a supposition which is well in keeping with historical conditions, as the Annamese appear to have started from the extreme northeast of the territory now inhabited by them."

and a T'ai dialect, even with the addition perhaps of a form of speech as yet unknown, and subsequently with the layer of an enormous mass of Chinese words. The dominating influence to which Annamese owes its modern characteristics is to be attributed to a T'ai language, and certain it is that Annamese is not Mon-Khmer.

In opposition to Schmidt, Cham and its congeners must likewise be detached from Mon-Khmer and assigned to the Malayan group; the Khmer elements in Cham are simply loans due to historical contact. I concur with G. Maspero in his scepticism as to the alleged relationships of Mon-Khmer on the one hand and Khasi, the Kolarian languages of Central India, Nicobar, Senoi and Semang on the other. The material at our disposal is still so incomplete and deficient that Schmidt's conclusions must still be regarded as somewhat premature, or at least be accepted with caution. I regret that G. Maspero has not pronounced an opinion on Schmidt's further theory of a connection of Mon-Khmer with Malayan (his so-called Austronesian group).

The earliest document of the Khmer language, known at present, is a twelve-line inscription yielding the date A.D. 629. Three hundred and fifty inscriptions and a large number of manuscripts written on palm leaves allow us to pursue the development of the language from the seventh century till the present. During this interval it has undergone no fundamental change, not even in its vocabulary; the noticeable modifications are all phonetic and of a normal character. In consequence of its monosyllabic formation, the language follows the tendency to reduce polysyllabic foreign words by contraction to a single syllable; for instance, *suor*, "heaven," from Sanskrit *svarga; pros*, "male," from Sanskrit *purusha; sas*, "religion," from Sanskrit *çāsana*.

The most interesting feature in the structure of Khmer is the formation of words by means of prefixes and infixes. In a former stage also suffixes seem to have been employed to a limited extent, but only few traces of these can be recognized (for instance, *leu*, "above," *leuṅ*, "to mount"); the language has lost consciousness of them. Some examples of prefix formation are:

leṅ, to play	— *p'-leṅ*, orchestra.
em, sweet	— *pa-em*, sugared.
cap, to adhere	— *k'-cap*, adhering.
san, peace	— *k'-san*, peaceful.
cuor, furrow	— *p'-cuor*, to till the ground.
ṅap, death	— *ro-ṅap*, to be extinguished.
sa, unconstant	— *saṅ-sa*, light-minded.

tiet, other — *kan-tiet*, stranger.
seum, humid — *an-seum*, dew.
bauk, hump of oxen — *dam-bauk*, conical hill.
heṅ, dry — *kra-heṅ*, cleft in dry earth.
mat, mouth — *pra-mat*, to insult.
mul, together — *sra-mul*, to join.
he, to follow in procession — *daṅ-he*, to accompany the king.

Infixes are derived in the following manner:

krup, enough — *k-om-rup*, to be sufficient.
kan, to hold — *k'-m-an*, one who holds.
cam, to guard — *c-m-an*, guardian.
deu, to march — *t'-m-eu*, traveller.
ar, to saw — *a-na-r*, a saw.
kal, to support — *k'-n-al*, a support.
khe, month — *k-aṅ-he*, season.
kem, varied — *k'-l-em*, mixed.
tiṅ, balance — *t'-l-iṅ*, to weigh — *t-aml-iṅ*, weight.
pēk, division — *p-ṅ-ēk*, portion, *p-amp-ēk*, to divide, *p-amn-ēk*, piece, *p-rap-ēk*, division.
k'vak, blindness — *k-aṅ-vak*, blind.
k'laṅ, strong — *k-am-laṅ*, strength, force.
k'cil, lazy — *k-am-cil*, laziness.
krol, to roll from above downward — *ke-d-rol*, cascade.
braṅ, black — *be-d-raṅ*, soot.
pran, to bait — *pe-d-ran*, bait.
kaṅ, circle — *k'-v-aṅ*, curved.
dal, to husk rice — *t-b-al*, mortar for husking rice.
cik, to dig — *c-ran-ik*, hoe.

In respect to social customs and conventions, the excessive use of appellative nouns in the place of personal pronouns is of great interest. Their judicious employment is a veritable crux of the language and must be observed with rigorous attention. Maspero enumerates no less than forty-four forms of address varying according to sex, age, title, or function of the speaker or the person accosted. In addressing children different words are used for boys and girls, for babies, infants, small boys and girls, boys and girls in the stage of puberty, while the children have different pronominal expressions in speaking to father and mother. Functionaries, dignitaries, bonzes, kings, slaves, women have all distinct forms of address.

From a consideration of the numerals it may be inferred that Khmer has run through a threefold stage of development,—a quaternary, quinary, and vigesimal. Primarily the Khmer counted only up to four, for solely the numbers from one to four are common to all the members of the family. After their separation the Khmer proceeded to form a word for five, a monosyllable incapable of analysis. By means of the quinary system thus attained they succeeded in counting up to nine by the process of addition: five one, five two, five three, five four. In creating an unanalyzable root-word for ten they completed a decimal system which is logically carried out in the numbers from 11 to 19, framed after the scheme 1 + 10, 2 + 10, etc. Finally a monosyllabic expression for twenty, neither related to five nor to ten, was called into existence, and this represents their highest unit. Hence it should be supposed that all higher decimal units were based on twenty, but whether this was the case or not escapes our knowledge, as from the end of the thirteenth century the Khmer adopted the Siamese numerals from thirty upward. All other languages of the group (Mon, Stieng, Bahnar, Sedang, and Boloven) have common forms for the numerals from six to nine (but different from those of Khmer), and have evolved a decimal system. Like many other languages, Khmer, too, abounds in numeratives styled by Maspero *déterminatifs spécifiques*.

The verb is little developed, and does not offer much of general interest. A useful list of principal roots and their most common derivatives concludes the volume. The typographical work is perfect, and the new font of Khmer type designed, engraved, and cast by the Imprimerie Nationale, is very creditable. G. Maspero is also the author of a history of Champa (*Le royaume de Champa*, Leiden, 1914) and of an excellent history of the Khmer (*L'Empire khmèr, histoire et documents*, Phnom-Penh, 1904).

B. LAUFER

AFRICA

Anthropological Report on Sierra Leone. Part i. Law and Custom of the Timne and Other Tribes. Part ii. Timne-English Dictionary. Part iii. Timne Grammar and Stories. NORTHCOTE W. THOMAS. Harrison and Sons: London, 1916. III vols., 8°, pp. 196, viii, 139, xxx, 86.

Specimens of Languages from Sierra Leone. NORTHCOTE W. THOMAS. Harrison and Sons: London, 1916. Large 8°, pp. 62.

These reports by Northcote W. Thomas, government anthropologist, are in form and character like his earlier reports on Nigerian peoples. Sierra Leone, with an area of about thirty-one thousand square miles has a native population of about one and a quarter million. The tribes

143

印度支那的图腾踪迹

VOL. XXX. OCTOBER—DECEMBER, 1917. No. CXVIII.

THE

JOURNAL OF

AMERICAN FOLK-LORE.

EDITED BY

FRANZ BOAS,

ASSISTED BY GEORGE LYMAN KITTREDGE, AURELIO M. ESPINOSA, C.-MARIUS BARBEAU, AND ELSIE CLEWS PARSONS.

CONTENTS.

LANCASTER, PA., AND NEW YORK:

PUBLISHED BY THE AMERICAN FOLK-LORE SOCIETY.

G. E. STECHERT & CO., NEW YORK, AGENTS.

THE JOURNAL OF

AMERICAN FOLK-LORE.

TOTEMIC TRACES AMONG THE INDO-CHINESE.

BY BERTHOLD LAUFER.

A RECENT article by Henri Maspero [1] affords me a welcome occasion to acquaint American students of anthropology with some data relative to certain totemic traces to be found among the Indo-Chinese stock of tribes. As these facts are not generally known or have hardly transgressed the boundaries of the sinological domain, it is hoped that they may prove of some utility to American anthropologists interested in the much-ventilated subject of totemism.

Maspero's new information is based on the communications of an individual from the so-called "Black Tai." The whole Tai population of Tonking and southern China is familiar with the usage of family names. This practice is not borrowed by them from the Chinese or Annamese, but, on the contrary, represents an indigenous custom which is reflected in the religious life of the Tai. The Black Tai believe that after death the soul of the individual is divided, — one remaining in the house, another resting in the grave, and another residing in heaven. Every family inhabits a special place in the villages of the souls, where it resides under the rule of its particular god of the dead, a Celestial Father (*Pú-then*), who bears the family name of the family under his sway. Thus there is a *Pú-then* styled Kwàng for the family Kwàng. Every Black-Tai family adheres to particular alimentary restrictions. Some of these, according to Maspero, are connected with the name by means of simple homonymy; others appear to bear no relation to the name. The family Lò-kằm, for instance, which forms the aristocratic family that supplies the village and district chiefs, is forbidden to eat the flesh of the bird *tang-lo*, fruits of the tree *tang* (this word agrees in tone with the element *tang* of *tang-lo*), and fungi growing on the same tree or at its foot. The family Kà must abstain from eating the flesh of the "cock of the pagodas" (*nôk kôt-ka*), as well as the buds of a certain flower termed

[1] De quelques interdits en relation avec les noms de famille chez les Tai-Noirs (Bull. de l'Ecole française d'Extrême-Orient, 16 [1916] : 29–34).

kā. Young bamboo-sprouts (*nó lău*) are interdicted to the family Lầu. The bird *me* and the fish *me* (*nôk me, pa me*) form the taboo of the family Mè. The members of the family Tòng must avoid eating the turtle-dove (*nôk său-tong*), and must not wear on their caps a copper point (*tong*). The family Ma is not allowed to feed on the flesh of the horse (*tô ma*). The use of a fan (*vi*) is forbidden in the family Vi when rice is served during meals. While the linguistic relation of these interdictions is easily grasped, it is not apparent, however, or obscured in the following cases. The family Lèo is not allowed to eat the blackbird *nôk iêng* and the water-fowl *nôk hăk*. The family Lüong abstains from fungi growing on the trunk of a branchless tree. Or the family Kwàng does not partake of the flesh of cat, tiger, and panther. If one of these families eats any things tabooed, even unknowingly, he will lose his teeth. There is no expiatory ceremony known, and no rite is practised to raise the taboo.

In regard to the family Kwàng, to which his informant belonged, Maspero gives more particulars. This family owes its superior rank to the concept that its ancestor was the first to emanate from the primeval gourd which produced mankind; immediately after him appeared the ancestor of the Lüong. The Kwàng belong to the family of the tiger, which they name by a term of respect, "grandfather" (*pu*). The degree of relationship is not ascertained: they do not descend from a tiger, nor are the tigers descendants of a transformed ancestor of their own; but it is certain that there is some sort of affinity. For this reason cat and tiger flesh are prohibited; the cat represents a highly prized dish of the Black Tai. The members of the family are immune from attacks of the tiger, and are not allowed to attack him or to take part in a tiger-hunt. Solely as an act of self-defence may they kill him. When they note a dead tiger on their road, or when the villagers carrying a slain tiger pass their habitation, they must without delay perform a minor ceremony. Taking a small piece of white cloth and throwing it over the corpse, they signify by this act that they have entered into mourning in his honor, and that the term of mourning is over. The prayer said on this occasion is of great interest, for it reveals the inner relations of the family to the tiger and the latter's influence on their welfare and that of their progeny. It runs as follows: —

"The grandfather is dead, leaving his children and grandchildren behind. The children and grandchildren ought to wear mourning in conformity with the rites, but the children and grandchildren were not able to go into mourning; the children and grandchildren terminate their mourning for the grandfather. There you are! [The piece of white cloth is then thrown over the tiger's corpse.] Protect your children, protect your grandchildren!

Those of you who survive, make them grow, let them prosper! In their work let them succeed, in their affairs let them do well! In their journeyings may they be without accident, wherever they may be, bless them! May they never see what is wrong, and never know bad omens! Let your children and grandchildren live long, ten thousand years, a hundred thousand harvests, eternally!"

On the other hand, the affiliation with the tiger also has its drawbacks. It causes the spirits to detest the members of this family. They have to keep aloof from sacred places. The field where the district festival (*lông tông*) is held to commemorate the commencement of agricultural pursuits, and the spot consecrated to the spirit of the district (Fi müòng), are interdicted to them at all times. During the festivals they take part in the offerings; but they are not permitted to enter, and may attend only outside. At their village ceremonies they have to keep behind the other families, and the functions of master of ceremony occupied by the old men are closed to them. Finally the priest of the district, whose office is hereditary in the Lüòng family for all the Black-Tai regions, must not marry a woman of the Kwàng family; even his brothers fall within this rule. However, the affinity with the tiger is not transmitted by the mother, but solely by the father. Whether similar beliefs and ceremonies with reference to the taboos prevail among the other families, says Maspero, is not known to him; in the case of the family Vi it appears to him difficult to admit that the fan might play there the same rôle as the Kwàng assign to the tiger. He thinks that among all peoples of southern China and northern Indo-China the tiger, from a religious viewpoint, is an animal so different from others, that it would be unwise to conclude the existence of similar rites in other families. This caution is praiseworthy, as is also the author's reserve in drawing any conclusions from his notes. He even avoids the terms "totem" and "totemism" and any theoretical discussion. His data, needless to add, are of intense interest to anthropology, and, if occasion offers, should by all means be completed. A complete list of all these Black-Tai families should be drawn up, and their ancestral traditions should be placed on record. Meanwhile it may be useful to render accessible the available data on real or apparent totemic phenomena within the Indo-Chinese group.

Aside from the Black Tai, actual observations of totemic phenomena, as far as the Indo-Chinese are concerned, were only made among the Lolo, first by A. Henry.[1] According to this author, "Lolo surnames [2] always signify the name of a tree or animal, or both tree and animal;

[1] Journal Anthropological Institute, 33 (1903) : 105.

[2] It is not correct to speak of Lolo surnames. The Lolo, like the Tibetans, did not have family names before contact with Chinese. The Sinicized Lolo adopted Chinese surnames.

and these are considered as the ancestors of the family bearing the name. This name is often archaic. Thus the surname Bu-luh-beh is explained as follows: *Bu-luh* is said to be an ancient name for the citron, which is now known as *sa-lu*.[1] The common way of asking a person what his surname is, is to inquire, 'What is it you do not touch?' and a person of the surname just mentioned would reply, 'We do not touch the *sa-lu*, or citron.' People cannot eat or touch in any way the plant or animal, or both, which enters into their surname. The plant or animal is not, however, worshipped in any way." The Lolo are a widely extended group of tribes, and those studied by Henry are those of Se-mao and Meng-tse in Yün-nan.

The term "totemism" with reference to the Lolo was then actually employed by Bonifacy,[2] who believed that certain animal legends, traces of exogamy, and certain taboos, might be considered as survivals of a very ancient totemic organization, but that the proofs are lacking. In my opinion, the data offered by the author reveal no survivals allowing of any conclusion as to former totemism. If, for instance, the newly-weds among the Lolo are not allowed to cut bamboo or to eat the young bamboo-sprouts, this is easily explained from the legend of the first couple who performed their marriage under a bamboo that made speech to them. Bonifacy's material on the Lolo, especially as to social and religious life, belongs to the best we have.

In the "Notes ethnographiques sur les tribus de Kouy-tcheou" (Kuei-chou), by A. Schotter,[3] which must be taken with great reserve, we meet a heading "Totémisme chez les He-miao" (Hei Miao), but the notes appearing under this catch-word are disappointing. The author learned that a certain family of the tribe, Pan, abstains from beef, and received as explanation thereof the following story. One of the ancestors of the Pan was much taken by the charms of a young girl of the family Tien of the same tribe, whose hand was refused him nine times. Finally the condition was imposed on him that he should sacrifice an ox, but not partake of its flesh. The Pan family went beyond this request, and all its descendants avoid the meat of any sacrificed ox. Another piece of evidence: the Tien do not eat dog-flesh. A young mother died, leaving a small girl about to die for lack of milk. She was suckled by a bitch, and, out of gratitude to her nurse, never touched canine flesh, cursing those of her descendants who would not imitate her example. It is obvious that these two cases are simple taboos, the legends being invented in order to explain

[1] This word is related to Nyi Lolo *č'u-se-ma* and Tibetan *ts'a-lum-pa* (see T'oung Pao, 17 [1916] : 45).

[2] Bull. de l'Ecole française, 8 (1908) : 550.

[3] Anthropos, 6 (1911) : 321.

them, and bear no relation to totemism. Finally also N. Matsokin,[1] with reference to Schotter and some other sources, has spoken about totemism among the Lolo and Miao.[2] It is notable that the two men who were best familiar with the life of the Lolo — Vial and Liétard, two Catholic missionaries — have nothing to report that might be interpreted as totemism. At all events, if totemism ever existed among the Lolo, only scant survivals of it have remained. The independent Lolo, who are not yet explored, may offer better guaranties in contributing to this problem.

I now proceed to place before the reader in literal translation some ancient Chinese records that speak for themselves, and that have the advantage of not being biased by any modern totemic theory. The numerous aboriginal tribes inhabiting the territory of southern and southwestern China are designated by the Chinese by the generic term "Man" or "Nan Man" ("southern Man"). The following legend is told in the Han Annals concerning the origin of the Man.[3]

"In times of old, Kao-sin Shi [4] suffered from the robberies of the K'üan Jung.[5] The Emperor, being grieved at their raids and outrages, attempted to smite them by open attack, but failed to destroy them. Thereupon he issued a proclamation throughout the empire: 'Whoever shall be able to capture the head of General Wu, the commander of the K'üan Jung, will be offered a reward of twenty thousand ounces of gold, a township comprising ten thousand families, and my youngest daughter as wife.' At that time the Emperor had raised a dog whose hair was of five colors [that is, manicolored], and whose name was P'an-hu.[6] After the issue of

[1] Materinskaya filiatsiya v vostočnoi i tsentralnoi Asii (The Matriarchate in Eastern and Central Asia), pt. 2 : 94–96 (Vladivostok, 1911).

[2] Several conclusions of this author are inadmissible, owing to his blind faith in Schotter's uncritical data. He accepts from him the statement that "the antique form of the Chinese character for Miao represented a cat's head and signified a cat." Hence in Matsokin's mind the cat becomes a totem of the Miao. This is a sad illusion. The tribal name Miao is a native Miao word, and its significance cannot be interpreted from any arbitrary manner in which the Chinese please to convey this word to their writing. In fact, neither the word nor the Chinese character with which it is written has anything to do with the cat, which is *mao*, but not *miao*, in Chinese; and, even if the Chinese should etymologize the name in the sense of "cat," the conclusion as to a cat-totem among the Miao would be an utter failure. Nor is it correct, as asserted by Matsokin, that the eagle is a totem of the Miao.

[3] Hou Han shu, Ch. 116, p. 1.

[4] One of the early legendary emperors of China, alleged to have reigned about 2436 B.C.

[5] That is, "Dog Jung." "Jung" was a generic term for barbarous tribes in the west of China.

[6] The characters representing this name have the meaning "tray" or "plate" and "gourd." In explanation of this name, the Wei lio, written by Yü Huan in the third century A.D., has this anecdote: "At the time of Kao-sin Shi there was an old woman living in a house belonging to the Emperor. She contracted a disease of the ear, and, when the object causing the complaint was removed, it turned out to be as large as a silkworm-

this order, P'an-hu appeared at the gate of the palace, holding a man's head in his jaws. The officials were surprised, and examined the case. In fact, it was the head of General Wu. The Emperor was greatly pleased, but considered that P'an-hu could not be married to a woman or be invested with a dignity. He deliberated, as he was anxious to show his gratitude, but did not know what was fitting to do. The Emperor's daughter heard thereof, and held that the pledge which the Emperor had made by the proclamation of his order should not be broken. She urged him to keep his word; and the Emperor, seeing no other expedient, united the woman with P'an-hu. P'an-hu took her, set her on his back, and ran away into the southern mountains, where he stopped in a stone house situated over a precipice inaccessible to the footsteps of man.[1] Thereupon the woman cast off her royal dress, tied her hair into a *p'u-kien* knot, and put on *tu-li* clothes.[2] The Emperor was grieved, and longed for her. He sent messengers out to make a search for his daughter. Suddenly arose wind, rain, thunder, and darkness, so that the messengers were unable to proceed. After the lapse of three years she gave birth to twelve children, — six boys and six girls. After P'an-hu had died, the six boys married the six girls. They used the bark of trees for weaving, and dyed this stuff by means of plant-seeds. They were fond of manicolored clothes, and cut them out in the form of a tail. Their mother subsequently returned home and told the story to her father. The Emperor thereupon sent messengers to bring all the children. Their clothes were striped like orchids, and their speech sounded like *chu-li*.[3] As they were fond of roaming over hills and ravines, but did not care for level country, the Emperor, in conformity with this trend of mind, assigned to them renowned mountains and extensive marshes. Subsequently they increased and ramified, and were called Man Barbarians. Outwardly they appeared like simple folk, but inwardly they were clever."

cocoon. The woman placed it in a gourd, which she covered with a tray. In a moment it was transformed into a manicolored dog, which hence received the name P'an-hu." Compare also Chavannes, T'oung Pao, 6 (1905) : 521. This is etymological play made after the event, and is without relation to the original form of the legend. In all probability, P'an-hu is a word derived from a language of the Man, with a quite different meaning. The ancient pronunciation of the word was *Ban-ku, and *ku* is a Man word meaning "dog." The term will be treated in detail in a forthcoming study of the writer on the languages of the Man.

[1] The Commentary adds the following. This place is identical with what at present is called Mount Wu in the district of Lu-k'i in Ch'en chou (in Hu-nan Province). According to the Wu ling ki, by Huang Min, this mountain is about ten thousand *li* high [Chinese determine the height of mountains by measuring the length of the road leading from the foot to the summit]. Half way on the mountain there is the stone house of P'an-hu, which can hold ten thousand people. Within there is his lair, where his footprints are still left. At present, in front of the caves of Mount Ngan, are to be found ancient remains of stone sheep and other stone animals, which are indeed very curious. Also many rock caves as spacious as a three-roomed house may be seen there. The Yao hold that these stones resemble the shape of a dog. According to The Traditions of the Customs of the Man (Man su hiang chuan), they represent the image of P'an-hu.

[2] This means that she adopted the hair-dressing and costume of the indigenous Man tribes. The commentary admits that the two terms *p'u-kien* and *tu-li* are unexplained; they doubtless represent words derived from a language of the Man.

[3] The commentator remarks that *chu-li* is the sound of the speech of the Man barbarians. The meaning is that their speech was crude and uncultivated.

This tradition makes a dog the ancestor of the Man; and his descendants cut their clothes out in the form of a dog's tail, their coat-of-arms. The relationship of the Man to the Chinese is emphasized; their languages, in fact, are closely allied. They are characterized as hunters in the mountains and marshes, where they have fields cultivated by very primitive methods, while the plains are reserved for the agriculture of the colonizing Chinese. The modern Man have preserved this tradition with some variants. Some tribes still abstain from the flesh of the dog. Among the Man Tien, who style themselves "Kim Mien" (Mien = Chinese Man, that is, "man"), they have images representing the creator Pien-Kan seated on a throne and holding a flower in his hand; beneath him is shown a dog being carried on a palanquin by two men. A man-dog appears in their decorative art. The Man Kao-lan still profess to have descended from the ancestor-dog P'an-hu. They state that the lozenges embroidered on the shoulders of their women's dress indicate the spot where the paws of the ancestor rested when he cohabited with the princess.[1] The chiefs of the Yao retained P'an as their name: thus there was a Yao chief P'an Kuei in the beginning of the fifteenth century.[2] They also sacrificed to P'an-hu at New-Year offerings of meat, rice, and wine. There is a peculiar tribe of several hundred families living fifteen miles east of Fu-chou, in Fu-kien, called Sia. They are said to be descendants of a dog-headed ancestor, styled Go Sing Da, whose image is worshipped in the ancestral hall on the fifteenth of the eighth month and on New-Year's Day. After this it is kept locked up, as they are ashamed to let others see it.[3]

One of the powerful kingdoms of the Southwestern Man at the time of the Later Han dynasty (A.D. 25–220) was called Ye-lang, bordering in the east on Kiao-chi (Tonking). The Chinese have preserved to us the following ancient tradition with reference to the origin of royal power among this people.

"In the beginning, a woman was bathing in the T'un River, when a large bamboo consisting of three joints came floating along and entered between the woman's legs. She pushed it, but it did not move. She heard an infant's voice inside, took the bamboo up, and, returning home, split it. She found in it a male child, and reared him till he had grown up. He developed warlike abilities and established himself as Marquis of Ye-lang, assuming the family name Chu [that is, Bamboo]." [4]

The foundation of the kingdom of Nan-chao in Yün-nan, the

[1] E. Lunet de Lajonquière, Ethnographie du Tonkin septentrional, pp. 210, 252, 253, 272, 280.

[2] G. Devéria, La Frontière sino-annamite, p. 90.

[3] F. Ohlinger, Chinese Recorder, 17 (1886) : 265, 266.

[4] Hou Han shu (Annals of the Later Han Dynasty), Ch. 116, p. 6 b; Hua yang kuo chi, Ch. 4, p. 1 b.

populace of which belonged to the T'ai family, is thus narrated in the Han Annals:[1] —

"The ancestor of the Ngai (or Ai-)Lao barbarians was a woman, Sha-yi by name, who dwelt on the Lao mountain.[2] Once when she was engaged in catching fish, she came in contact with a drifting piece of wood, which caused her a feeling as if she had conceived. Accordingly she became pregnant, and, after the lapse of ten months, gave birth to ten sons. Subsequently the drifting log was transformed into a dragon, who appeared on the surface of the water. All of a sudden Sha-yi heard the dragon speak thus: 'Those sons begotten by me, where are they now?' Nine of the sons became frightened at sight of the dragon and fled. Solely the youngest child, who was unable to run away, set himself on the back of the dragon, so that the dragon could lick him. In the mother's native [literally, 'bird'] language, 'back' is termed *kiu*, 'to sit' is called *lung*:[3] hence the name 'Kiu Lung' was conferred on the child. When he had grown up, his elder brothers inferred from Kiu Lung's strength that he had been licked by his father, and, on account of his cleverness, proceeded to elect him king. Afterwards there was a couple living at the foot of Mount Lao. Ten daughters were born to them. These were taken as wives by Kiu Lung and his brothers. At a later time, when they had gradually increased in number, all the tribesmen cut and painted [that is, tattooed] their bodies with designs representing a dragon, and wore coats with tails. After Kiu Lung's death, several generations succeeded to him. Eventually the tribe was divided under the rule of petty kings, and habitually dwelt in places scattered in the ravines and valleys far beyond the boundaries of China. While, intercepted by mountains and rivers, the populace strongly increased, it had never held any intercourse with China."

The term "Dragon-Tails" (*lung wei*) was still applied to the later dynasty Nan-chao. The dragon-tail is an analogon to the dog-tail of the Nan Man.

In 1635 a Chinese, Kuang Lu, who had been in the service of a female chieftain of the Miao, published a small book under the title "Ch'i ya," which belongs to the most interesting and instructive documents that we have on the Miao. This author (Ch. 1, p. 17 b) mentions a tribe under the name "Tan," who lived on river-boats, subsisting on fish, without engaging in agriculture and intermarrying with other people. They called themselves "dragon-tribe" (*lung chung*) or "men of the dragon-god" (*lung shen jen*). They painted a

[1] Hou Han shu, Ch. 116, p. 7 b.

[2] A native tradition is more explicit on the origin of Sha-yi. She was the wife of Mong Kia Tu, who was the fifth son of Ti Mong Tsü, son of Piao Tsü Ti, who is identified with King Açoka of Magadha. One day when Mong Kia Tu was fishing in Lake Yi-lo, south of the city of Yung-ch'ang, he was drowned, whereupon Sha-yi came to this place to weep (see E. Rocher, T'oung Pao, 10 [1899] : 12; Devéria, La Frontière sino-annamite, p. 118; C. Sainson, Histoire du Nan-tchao, p. 25).

[3] Modern Chinese *kiu* was in Old Chinese **gu*, and *gu* is a typical Indo-Chinese word for "back" (see T'oung Pao, 17 [1916] : 52). *Lang* or *lung* in Siamese means "to sit." The compound signifies "sitting on the back" (namely, of the dragon).

snake on their temples for purposes of worship, and the records of population they styled "dragon-doors" (*lung hu*).[1] The remains of the Tan are still to be found in the floating river-population of Canton.[2]

The Western K'iang (Si K'iang) were a large group of nomadic tribes, the present province of Kan-su forming the centre of their habitat, who must be regarded as the forefathers of the Tibetans. A brief notice on their social organization is preserved in the Han Annals.[3]

"There was no fixed distinction of families and clans: the designations of tribes were derived from the personal name of the father or from the family name of the mother. After the twelfth generation, marriages were permitted in the same clan. On the father's death, the son married his step-mother.[4] When an elder brother died, a new marriage was arranged for his widow; so that there were no widows in their country. Their tribal divisions were numerous, but they did not have any institution like princes and officials. They did not take regard of elders, but it was the strongest man who was elected chief by the tribes. When he weakened, he was relegated to the common people; and then they vied with one another in a contest of strength to find out who was the bravest."

A division of the K'iang bore the name Wu-yi Yüan-kien. The designation Wu-yi (*Mu-yit) is explained by a gloss to mean "slaves," as they were held in serfdom by the Duke Li of Ts'in in the fifth century B.C. Subsequently they were split into several tribes, each with a special appellation. One of these was called the "Yak (*li-niu*) Tribe;" these were the K'iang of Yüe-si. The K'iang of Kuang-han styled themselves "White-Horse (*pai ma*) Tribe;" those of Wu-tu had the name "Wolf (*ts'an lang*) Tribe."[5] The annalist then continues, —

"Jen and his younger brother Wu alone remained in Huang-chung [in the present prefecture of Si-ning in Kan-su], and took many wives. Jen had nine sons, who formed nine tribes. Wu had seventeen sons, who formed seventeen tribes. The rising power of the K'iang began from this time."

Yü Huan, in his "Wei lio," written in the third century A.D., enumerates the following three clans of the K'iang, — the Ts'ung-ts'e[6]

[1] The Chinese count the number of families by doors.

[2] See Notes and Queries on China and Japan, 1 : 15, 28, 107.

[3] Hou Han shu, Ch. 117, p. 1.

[4] The same custom is related by the Chinese in regard to the ancient Hiung-nu (Huns) and T'u-küe (Turks). It means, of course, that it was bound up by the law of inheritance of these peoples, and that the son fell heir to his father's entire property, inclusive of his women, slaves, etc. See also G. Soulié (Bull. de l'Ecole française, 8 [1908] : 362, note 2).

[5] Hou Han shu, Ch. 117, p. 3. The term *ts'an* seems to refer to a particular species of wolf, but its meaning is not explained. This account relates to the fourth century B.C.

[6] The compound consists of two plant-names, — *ts'ung* referring to garlic (see T'oung Pao, 17 [1916] : 96), and *ts'e*, to a plant yielding a red dye (*Lithospermum officinale*). It is

(Garlic) K'iang, the Pai-ma (White Horse) K'iang, and the Huang-niu (Yellow Ox) K'iang, — adding that each of these tribes has its chiefs, and that among the last-named the women give birth to a child after six months.[1] The same author speaks of another group of tribes, called "Ti," the descendants of the Si Jung, and related to the K'iang in language and customs. Some divisions of this people were termed by the Chinese "Green and White Ti," from the color of their costume; but another clan styled itself "Ti Jan," the latter word designating a reptile under which it was classed.[2]

From a passage in the Annals of the Sui Dynasty,[3] we note that a clan of the K'iang, scattered in the country Fu (2000 *li* northwest of Se-ch'uan), was named "Pai Kou" (White Dog).

In the age of the Sui dynasty (A.D. 590–617) a tribal group of the K'iang became known to the Chinese under the name "Tang-hiang," the element Tang appearing as Tangud or Tangut (*-ud* being a Mongol termination of the plural), the Turkish and Mongol designation of the Tibetans. To the Tang-hiang belonged the Tang-ch'ang and Pai-lang (White Wolves), who conferred on themselves the name "Monkey Tribe" (Mi-hou Chung).[4] In fact, the monkey belonged to the sacred animals of the ancient Tibetans, and was sacrificed with sheep and dogs once a year, when the officers assembled for the ceremony of the minor oath of fealty.[5] In their own traditions the Tibetans have preserved at great length the story of how they descended from the alliance of a monkey with a female giant (Rākshasī).[6] But there is no evidence that the monkey ever was the totem of a Tibetan clan, or that a Tibetan clan named itself for the monkey; the latter, however, as shown by the Chinese account of the Tang-hiang, may have been the case in ancient times.

In regard to the Chinese, the existence of totemism is denied by some authors, while others are inclined to uphold it.[7] Neither the one nor the other can be asserted in our present state of knowledge. We must not forget, of course, that Confucius, who made the Chinese what a

more probable, however, that *ts'ung-ts'e* relates solely to a single species, presumably to a wild *Allium*.

[1] Chavannes, T'oung Pao, 6 (1905) : 528.

[2] *Ibid.*, pp. 521–522.

[3] Sui shu, Ch. 83, p. 8 b.

[4] *Ibid.*, p. 2 b.

[5] Kiu T'ang shu, Ch. 196 A, p. 1.

[6] See, for instance, Rockhill, Land of the Lamas, pp. 355–361. For a complete bibliography of the subject, see Laufer (T'oung Pao, 2 [1901] : 27–28).

[7] A. Conrady, "China" (in Pflugk-Harttung's Weltgeschichte, p. 491). The evidence merely rests on the interpretation of names. Conrady's popular history of China is modelled on Lamprechtian ideas of evolution, which are interpreted, and partially in a very forced way, into the given material. This method is not to be taken seriously; the critical anthropologist will understand without comment.

French writer aptly styled *affreusement bourgeois*, has spoiled China completely for the ethnologist. Certainly the Chinese never were those angels of virtue that we are prone to make them out in reading the tenets of their moral creed. Morals look well on paper always and everywhere. There was a prehistoric age when also the Chinese, like their congeners the T'ai, Miao, and Tibetans, did not pose as the champions of morality, but behaved like real and natural men. This has been very clearly shown in a most interesting study by M. Granet.[1] While no positive data are as yet available, from which conclusions as to a former totemic organization could be drawn, there are some indications which may be suggestive. Unfortunately the development of social organization in China has never been investigated by modern scientific methods.

The number of family names derived from words designating plants and animals is comparatively large. Following is an alphabetical list of the more common ones: —

FAMILY NAMES BASED ON PLANTS.

CH'I, white jasmine (*Jasminum sambac*).
CHU, bamboo.
CH'U, hay, straw.
HING, apricot.
HU, gourd, calabash.
HUA, flower.
HUAI, *Sophora japonica*.
HUAN, *Sapindus mukorossi*.
JANG, stalk of grain.
JUI, small budding plants.
JUNG-KÜAN, family of the *Hibiscus*.
KI, thistles.
KI, several species of *Rhamnus* and *Zizyphus*.
KEN, root.
KU, cereals.
KUA-T'IEN, gourd-field.
K'UAI, a rush (*Scirpus cyperinus*).
KUEI, cinnamon-tree (*Cinnamomum cassia*).
K'UEI, *Amarantus*.
KÜ, chrysanthemum.
KUO, fruit.
LAI, goosefoot (*Chenopodium album*).
LI, plum (*Prunus triflora*).
LI, lichee (*Nephelium litchi*).
LI, chestnut (*Castanea vulgaris*).
LIAO, *Polygonum*.
LIU, willow (*Salix babylonica*).
LU, a reed (*Phragmites*).
MA, hemp.
MAI, wheat.
MANG, a grass (*Erianthus japonicus*).
MAO, reeds, rush.
MEI, plum (*Prunus mume*).
MI, hulled rice.
MOU, barley.
MU, tree.
NGAI, *Artemisia vulgaris*.
PO, thickly growing vegetation.
PO, arbor-vitæ (*Thuja orientalis*).
SANG, mulberry-tree.
SING, a marshy plant.
SU, grain.
T'AN, *Dalbergia hupeana*.
T'ANG, *Pyrus*.
T'AO, peach.
T'AO, rice.
TI, *Prunus japonica*.
TOU, beans.
TSAO, various aquatic plants.
TSAO, jujube (*Zizyphus vulgaris*).
TSE, *Cudrania triloba*.
TSI, panicled millet.

[1] "Coutumes matrimoniales de la Chine antique" (T'oung Pao, 13 [1912]: 517–558).

Ts'UNG, a conifer.
T'UNG, *Paulownia imperialis.*
Ts'E, calthrop (*Tribulus terrestris*).
WEI, grass.
YANG, poplar.
YÜ, elm (*Ulmus campestris*).

FAMILY NAMES BASED ON ANIMALS.

CHI, leech.
CHI, ringed pheasant.
CH'I, worm.
CHUI, piebald horse.
CH'UNG, general term for reptiles and insects.
FANG, bream.
FU, wild duck.
FUNG, male phœnix.
HIAO, owl.
HIUNG, bear.
HO, crane.
HU, fox.
HU, tiger.
HUI, venomous snake.
JAN, boa.
KI, fowl, chicken.
K'I, piebald horse.
K'IN, birds in general.
KOU, dog.
KU, heron.
KÜ, colt of a horse.
LANG, wolf.
LIN, fish-scales.
LO, white horse with black mane.
LU, stag.
LUNG, dragon.
MA, horse.
MONG, tree-frog.
NGO, moth.
NIU, ox.
PAI-MA, white horse.
PAO, dried fish.
PAO, panther.
PAO-P'I, panther's skin.
PEI, cowrie-shell.
PIAO, tiger-cat; stripes of a tiger.
PIE, fresh-water turtle (*Trionyx sinensis*).
SE, team of four horses.
SHE, serpent.
SIA, chrysalis of a mantis.
TIAO, sable.
TS'ING-NIU, dark ox.
TS'ING-WU, dark raven.
TSOU, small fishes, minnows.
TSOU, a fabulous beast.
TSÜ, fish-hawk, osprey.
TS'UI, bird-down.
TS'UI, kingfisher.
T'UN, sucking-pig.
WU, raven.
YANG, sheep.
YANG-SHE, sheep-tongue.
YEN, swallow.
YEN, wild goose.
YU, polecat.
YÜ, fish.

It should be understood, of course, that it is by no means implied that the foregoing names had a totemic origin. This remains to be investigated by tracing in detail the history of these families bearing such names. In some cases it is certain that such names are not connected with a totem, but have a quite different origin. For instance, a man in the sixth century B.C. bore the family name Chuan, a word designating a large fish found in the Tung-t'ing lake. He killed Wang Liao, prince of Wu, with a poisoned dagger which was concealed in the belly of this fish served to him at dinner. This story plainly accounts for the origin of the family name. The list of these plant and animal family names, however, is interesting in itself, and, it is hoped, may prove a stimulus to serious investigation.

FIELD MUSEUM OF NATURAL HISTORY,
CHICAGO, ILL.

144

藏文的来源

JOURNAL

OF THE

AMERICAN ORIENTAL SOCIETY

EDITED BY

JAMES A. MONTGOMERY
Professor in the
University of Pennsylvania

GEORGE C. O. HAAS
Sometime Instructor in the
College of the City of New York

FRANKLIN EDGERTON
Professor in the
University of Pennsylvania

VOLUME 38

PUBLISHED FOR THE AMERICAN ORIENTAL SOCIETY

YALE UNIVERSITY PRESS

NEW HAVEN, CONNECTICUT, U. S. A.

1918

ORIGIN OF TIBETAN WRITING

BERTHOLD LAUFER

FIELD MUSEUM OF NATURAL HISTORY, CHICAGO

THE CHINESE ANNALS of the T'ang Dynasty (A. D. 618-906) report that the ancient Tibetans (*T'u-fan*) possessed no writing, but that they availed themselves of notched tallies and knotted strings (quippus) in concluding treaties.[1] This account evidently refers to the people at large, but not to the government of Lhasa; for continuing our reading of the annals we notice sufficient evidence for the existence of some form of actual writing as a means of official communication. We are informed that in A. D. 634 the king (*btsan-p'o*) K'i-tsuṅ luṅ-tsan or K'i-su-nuṅ (corresponding to Tibetan Sroṅ-btsan sgam-po) sent envoys with tribute to the Chinese emperor, and subsequently despatched to him a respectful letter petitioning for a matrimonial alliance. In A. D. 641 he received in marriage the Chinese princess (*kung chu*) Wen-ch'eng, and gradually adopted Chinese customs and manners. He invited scholars from China to compose his official reports to the emperor. After his successful participation in Wang Hüan-ts'e's campaign in central India (A. D. 648)[2] he applied to the emperor for work-

[1] *Kiu T'ang shu,* ch. 196 A, p. 1; *Sin T'ang shu,* ch. 216 A, p. 1b; *T'ang hui yao,* ch. 97, p. 2b. The correctness of this tradition was called into doubt by Abel-Rémusat (*Recherches sur les langues tartares,* p. 67-68), who gathered his information from the compiler Ma Tuan-lin of the thirteenth century, and was led to the belief that this one referred the quippu tradition of the *Yi king* to a people little known to him. Ma Tuan-lin, of course, excerpted the T'ang Annals, and the latter were based on contemporaneous state documents of the T'ang dynasty. Tallies and mnemotechnic knots were universally known in ancient times, and still survive to a great extent. There is no reason to doubt their occurrence in ancient Tibet. Tallies and quippus are ascribed also to another Tibetan tribe, the Ta-yang-t'ung (*T'ang hui yao,* ch. 99, p. 13b). The Annals of the Sui Dynasty (*Sui shu,* ch. 81, p. 10b) state in regard to the ancient Japanese that 'they have no script, but only carve notches in wood and tie knots in cords.'

[2] Regarding the missions of Wang Hüan-ts'e see S. Levi, *JA* 1900, 1. 297-341, 401-468; *T'oung Pao,* 1912. 307-309; Pelliot, *T'oung Pao,* 1912. 351-380.

men to manufacture paper and ink,[3] and the request was granted—a sure symptom of the fact that writing then existed and was practised. Under the successors of Luṅ-tsan, who died in A. D. 650, the official correspondence between Tibet and China increased in volume, and a chancery for the transaction of such business was established in the capital Lhasa. Several Sino-Tibetan documents, notably the celebrated treaty solemnized in A. D. 822, are still preserved on stone tablets in Lhasa.

While there is thus no doubt of the existence of writing under the first powerful king, the Chinese annals are reticent as to the character and origin of this writing. This is by no means striking, since the Chinese historians were chiefly interested in the political relations of the country to their own, and not in its inner cultural development; they do not tell us either of that great religious movement which swept Tibet in those days—the introduction of Buddhism from India.[4]

According to the tradition of the Tibetans, King Sroṅ-btsan sgam-po in A. D. 632[5] sent T'on-mi or T'ou-mi, the son of ,A-nu, subsequently honored by the cognomen Saṁbhoṭa, to India to study Sanskrit and Buddhist literature and to gather materials for the formation of an alphabet adapted to the Tibetan language. On his return to Lhasa he formed two Tibetan alphabets, one 'with heads' (*bdu-čan*) out of the Lāñcā script, and another 'headless' (*dbu-med*) out of the Wartula characters. The details of this tradition, to which there will be occasion to revert, vary to some extent in different accounts, but the principal elements of it are identical both in historical and grammatical works. It somewhat lacks in precision and detail, and we must not forget that it comes down to us from a comparatively late period, and that the contemporaneous, original form of the tradition is lost.

As regards the time of the introduction of writing, it follows from the Chinese annals that it indeed existed under the reign

---

[3] According to the *T'ang hui yao:* paper and writing-brushes.

[4] Only the New History of the T'ang Dynasty says that the Tibetans are fond of the doctrine of Buddha, and that the Buddhist clergy was consulted on all important state affairs.

[5] According to the chronology adopted by the Mongol prince and annalist Sanaṅ Setsen. The History of the Tibetan Kings sets no exact date for the event, except that it is recorded in the beginning of the king's reign.

of King Sroṅ-btsan sgam-po. It is clear from both the Chinese and Tibetan annals (the latter stating the fact implicitly) that prior to his era there was no writing. The Chinese annals do not impart the date of his accession to the throne: they give us the year 634 as that of his first mission sent to China and 650 as the year of his death. Sanaṅ Setsen states that he was born in 617 and assumed the reign in 629 in his thirteenth year[6]; this would agree with the Chinese statement that he was a minor at the time of his succession. The foundation of the national system of writing, accordingly, must have taken place between the years 630 and 648; for the latter date must be regarded as the *terminus ad quem,* since in that year the request for paper and ink manufacturers was submitted to China. As this event followed immediately the punitive expedition of Wang Hüan-ts'e against Magadha, who was then assisted by a Tibetan army, suspicion is ripe that this enterprise may have had a causal connection with the inauguration of writing in Tibet. At any rate, the case illustrates the fact that the road from Lhasa to Magadha was known to the Tibetans, and that there is nothing surprising or incredible in regard to T'on-mi's mission.

The time spent by T'on-mi in India is variously given. According to Chandra Das[7] he should have resided in Magadha from A. D. 630 to 650—doubtless an exaggeration and contradictory to Chinese chronology, according to which King Sroṅ-btsan died in A. D. 650; and according to the Tibetan accounts he profited from his emissary's instructions and himself composed several books.

The substance of the Tibetan tradition was clearly known as early as the eighteenth century: it was recorded by the Augustinian Pater A. Georgi,[8] who gave the name of the founder of writing in the corrupted form Samtan-Pontra, and who styles his Indian instructor the Brahman Lecin (that is Le-čin, according to the Tibetan pronunciation Li-j'in). P. S. Pallas already set forth rather sensible views on the Tibetan alphabet, recog-

[6] According to the chronological table published by Csoma (*Grammar of the Tibetan Language,* p. 183) he should have been born in that year (the European dates of Csoma are wrongly calculated and have to be increased by two); this is evidently an inadvertence of the Tibetan author.

[7] 'The Sacred and Ornamental Characters of Tibet,' *JASB* 57 (1888). 41.

[8] *Alphabetum Tibetanum,* p. 290 (Rome, 1762).

nizing its similarity with the Devanāgarī, and opposing Georgi's speculation that it should have sprung from the Syrian Nestorians.[9] With respect to the Tibetan tradition, Abel-Rémusat remarked:[10] 'Cette tradition n'a rien d'invraisemblable en elle-même.' He emphasized the connection of Tibetan script with the Devanāgarī and other Indian alphabets in Farther India and the Archipelago. Klaproth, an orientalist and historian of great critical acumen, likewise accepted the Tibetan tradition, and so did Koeppen and Lassen.[11]

In 1829 I. J. Schmidt devoted a thorough investigation to the origin of Tibetan writing.[12] This was in the same year when Schmidt published his edition and translation of the Mongol chronicle of Sanaṅ Setsen, which for the first time disclosed the native tradition relative to the introduction of writing into Tibet.[13] Schmidt compared the Tibetan alphabet with that utilized in an Indian inscription found in a rock-cave of Gayā and on a pillar of Allahabad.[14] The combination of these alphabets reproduced by him on a plate is in all ways convincing. Schmidt further held that Tibetan writing was not modeled after the Lāñcā, but owed its origin to an older and obsolete form of script.

The best summary of the problem is given by T. de Lacouperie.[15] He treats the Tibetan tradition with sound and sensible criticism and arrives at this conclusion: 'As to the Tibetan

---

[9] *Samlungen historischer Nachrichten über die mongolischen Völkerschaften,* 2. 359 (St. Petersburg, 1801).

[10] *Recherches sur les langues tartares,* p. 343 (Paris, 1820).

[11] J. Klaproth, *Tableaux historiques de l'Asie,* p. 158 (Paris, 1826), cf. also some observations on the Tibetan alphabet in *JA* 10 (1827). 132; C. F. Koeppen, *Lamaische Hierarchie und Kirche,* p. 56 (Berlin, 1859); C. Lassen, *Indische Alterthumskunde,* 4. 714.

[12] 'Über den Ursprung der tibetischen Schrift,' *Mémoires de l'Acad. Imp. de St.-Pétersbourg,* 6th series, 1 (1829). 41-52. This treatise has not been consulted by the recent theorists on Tibetan writing, A. H. Francke and A. F. R. Hoernle.

[13] *Geschichte der Ost-Mongolen und ihres Fürstenhauses,* p. 29-31, 325-328 (St. Petersburg, 1829).

[14] A similar observation is made by Csoma, *Grammar of the Tibetan Language,* p. 204 (Calcutta, 1834).

[15] *Beginnings of Writing in Central and Western Asia,* p. 56-67 (London, 1894).

expedition, there is no apparent reason to doubt it, with the exception of the additions and embellishments which have been added by the historians. Let us remember that we have no contemporary records nor annals of the time, and that all the knowledge we have from the Tibetan history is derived from native compilation, if not of a late date, at least made many centuries after the events they purpose to record.'

The discoveries made in Turkistan have also enriched Tibetan philology; and ancient Tibetan inscriptions, manuscripts, and business documents will contribute a large quota to our knowledge of Tibetan palaeography, language, and literature. Under the influence of these finds the theory has been advanced by A. H. Francke that the Tibetan tradition relative to the introduction of writing from India is unfounded, and that writing was introduced into Tibet from Turkistan, more particularly from Khotan. A. H. Francke is somewhat handicapped by lack of scientific training and unfortunately more endowed with imagination than with sound and cautious scholarship. My opinion on his theory I have briefly set forth in the *T'oung Pao* (1914, p. 67), where I declared myself wholly in accord with Lieut.-Colonel Waddell, who vigorously and successfully opposed this alleged discovery.[16] Even now I would not deem it worth while to submit Francke's hypothesis to a detailed discussion, were it not that recently it has been officially indorsed by a serious scholar of the type of A. F. R. Hoernle.[17] In his last work[18] Hoernle even elaborates a complex theory based on the fancies and figments of A. H. Francke. It is deplorable that a scholar to whom we all look with respect, and to whom we owe so many great things could be led astray by such vague and unfounded speculations, and that the pages of a work which is essentially devoted to the presentation of new and important documentary material are thus marred.

The notions of A. H. Francke center around two points, a new etymology of the name Li-byin and real or alleged coincidences between the Tibetan and Khotan alphabets. According to the Tibetan tradition the Brāhmaṇa consulted by T'on-mi

---

[16] *JRAS* 1909. 945-947.

[17] *JRAS* 1915. 493.

[18] *Manuscript Remains of Buddhist Literature found in Eastern Turkestan,* p. XVI-XXXII (Oxford, 1916).

in India was styled Li-byin.[19] E. Schlagintweit[20] observed that this name seems to allude to the art of writing and to be a Tibetanized form of Sanskrit *lipi* 'writing.' W. W. Rockhill[21] conceived the name as a corruption of *lipikara* 'scribe'; this explanation was accepted by de Lacouperie, Waddell, and Huth.[22] I hold the same opinion save that I do not accept the restitution *Lipikāra* or *Livikāra*,[23] but take Tibetan *Libyin* (properly *Lib-yin*) as the transcription of a Prākrit or vernacular form Lipyin or Livyin. As shown in my forthcoming study 'Loan-Words in Tibetan,' a large number of these is derived, not from Sanskrit, but from the Prākrits, more particularly from the Apabhraṃśas.

Now A. H. Francke, without taking account of this reasonable interpretation, dogmatically proclaims: 'This name (*Li-byin*) has always been wrongly translated. It has to be translated "Glory" (or blessing) of the land "Li." Li-byin had apparently received his name, because the land Li had reason to be proud of him. The land Li is either a country near Nepal or Turkistan. I am convinced that it here signifies Turkistan; for there is some probability that it was in the Turkistan monasteries that Tibetan was first reduced to writing, and T'on-mi simply reaped the fruit of such learning.'[24] All very simple indeed: a magic word of Francke is sufficient to upset any tradition and all history. Historical conclusions cannot be based on any subjective etymologies, however ingenious they

---

[19] According to I. J. Schmidt (*Forschungen*, p. 221) also the form *Lha-byin* occurs. This, if correct, would render Sanskrit *Devadatta*.

[20] *Könige von Tibet*, p. 839, note 4.

[21] *Life of the Buddha*, p. 212.

[22] T. de Lacouperie, *Beginnings of Writing*, p. 63; L. A. Waddell, *Buddhism of Tibet*, p. 22; G. Huth, *Geschichte des Buddhismus in der Mongolei*, 2. 8. Moreover, in the grammatical work *Si-tui sum-rtags* (p. 3, ed. of Chandra Das) the name of the Brahman appears in Tibetan transcription as *Li-bi-ka-ra*, i. e. Skt. *Lipikara*. Cf. also E. J. Thomas, *JRAS* 1916. 857.

[23] Tibetan *byin* in Sanskrit words is always the perfect of the verb *sbyin-pa* 'to give' and corresponds to Sanskrit *datta*, not, however, to *kara* or *kāra*. A restoration *Lipidatta* would, of course, be impossible. In fact, the element *byin* does not represent a Tibetan word, but forms part of the transcription.

[24] *JASB* 6 (1910). 97; repeated in *Epigraphia Indica*, 11 (1912). 269, and adopted by Hoernle.

may be. Francke's explanation of Li-byin is solely invented to suit his case and his own conveniences; it is not borne out or upheld by any Tibetan tradition, it is even impossible in the spirit of the Tibetan language. The word *li,* it is true, designates 'Khotan,' but it has other meanings also: it signifies 'bell-metal' and 'apple'; with the suffix *-ka* it denotes a certain tree and with the suffix *-ba* it means 'squinting'; it appears in a number of compounds, and further transcribes several Chinese characters reading *li.* The word *byin* never has the meaning 'glory'; it means 'blessing' only in certain fixed combinations, as *byin-gyis rlob-pa, byin brlabs,* etc. ('to bless'). It is never used, however, in the absolute or purely abstract sense of 'blessing,' as Francke would have us believe. As previously stated, the element *byin* in proper names either represents a translation of Sanskrit *datta,* as, for instance, *gSaṅ-ba byin* = *Guhyadatta* (Tāranātha, 147), *Ye-šes byin* = *Jñānadatta* (*ibid.* 212), *Ts'aṅs byin* = *Brahmadatta, Mya-ṅan med-kyis byin-pa* = *Açokadatta, gSer byin* = *Hemadatta,* etc.; but it is never the noun *byin* visualized by Francke. A name of such a type as 'Blessing of Khotan' has no analogy in Tibetan literature, and is a plain absurdity on the very face of it. It is merely a personal fancy, but Francke and Hoernle are so enraptured with it that they accept as a well substantiated fact what at the best might be regarded as a bold hypothesis. Says Dr. Hoernle literally: 'He (T'on-mi) had come into contact with a Brāhman from Khotan, whom the Tibetan tradition calls Li-byin or 'Blessing of Khotan,' and that Brahman taught him the alphabet of his own country. This, in effect, means that the alphabet, as introduced into Tibet, is the alphabet of Khotan, Li being the well-known Tibetan name of Khotan. It is not the alphabet of India . . . To judge by the Tibetan tradition he (T'on-mi) was saved the completion of his journey through the lucky accident of meeting, on his way in Kashmir, with a learned Brāhman from Khotan, who could supply him with the information he was in search of.' Again, he speaks of the Khotanese Brahman Li-byin from whom the Tibetan scholar T'on-mi is said to have learned his alphabet. Further he hazards the assertion: 'It has been stated already that Tibetan tradition distinctly refers to Li-yul, the land of Li, i. e. Khotan, as the country of origin of its alphabet.' This statement is downright fiction: Tibetan tradition has nothing whatever about Li-yul in the history of writing. This manner of argumentation is baffling and beyond my comprehension: Dr. Hoernle fearlessly advances as historical facts what is merely inferred from the imaginary and arbitrary dissection of a name—a singular instance of history-making!

The only documentary evidence on which Francke's conclusions are based is presented by the Tibetan chronicle of the Kings of Ladākh in the edition of E. Schlagintweit. This work is widely different from the older and more complete *rGyal rabs gsal-bai me-loṅ*[25] of Central Tibet, and as far as the history of the Central-Tibetan kings is concerned, gives merely a much abridged and corrupted version of the older standard book, written in A. D. 1328.[26] Now we have known for a long time

[25] This title does not mean, as translated by Francke and Hoernle 'Bright mirror of the line of kings,' but 'Mirror clearly setting forth the genealogy of kings.'

[26] Francke, for the benefit of his speculations, argues that the West-Tibetan record strikes him as being the more original of the two. He pleads also that 'the West-Tibetan account makes mention of the Indian Nāgarī alphabet, it is true, but this passage looks like a later interpolation' (*Epigraphia Indica,* 11. 267). This argumentation is inadmissible: it is a sound principle of historical criticism that the older source is the purer source, and that the original merits preference over the later work copied after it. It is a comfortable method to brand as interpolation what does not suit one's preconceived idea.—A strange assertion occurs on p. 269 of the same article. Here Francke states that 'we have a single testimony of history for the early use of Indian characters in Western Tibet, in the Chinese *Sui shu,* where it is stated that such characters were used in the empire [*sic*] of the Eastern Women (Guge), etc.' The source is not quoted; the *Sui shu* contains nothing of the kind, and in fact maintains silence as to any writing in the Women's Kingdom, as every one may convince himself from reading this chapter in Rockhill's translation (*Land of the Lamas,* p. 339). In the *T'ang shu* it is said that the written characters of the Women's Kingdom are the same as those of India (see, for instance, Bushell, *Early History of Tibet,* p. 98); but this is merely due to the well-known confusion of the two Women's Kingdoms and the information of Hüan Tsang misplaced and smuggled into the New History of the T'ang, as has been shown particularly by Pelliot (*T'oung Pao,* 1912. 358). This reference to writing in fact has nothing to do with the Eastern Women's Kingdom. Moreover Francke is wrong in placing it in Western Tibet; on the contrary, it embraced parts of Eastern Tibet, bordering in the east on Mao-chou in Se-ch'uan and the Taṅ-hiaṅ, in the south-east on Ya-chou in Se-ch'uan.

how the matter about Schlagintweit's text stands. K. Marx, a Moravian missionary than whom no one was more intimately familiar with the history of Ladākh, has shown with able criticism that this copy was specially prepared for his brother H. Schlagintweit by three Lamas, and that from folio 30 on 'the text is merely a meaningless jumble of words, culled at random from the original and put together in such a way that only a careful examination of the text by one who knows the language could reveal the fraud.'[27] Not only in that portion pointed out by Marx, but also in the preceding portions, the Schlagintweit text is so hopelessly faulty, mutilated, and corrupt that it forfeits any claim to historical value. It must be positively denied that any such far-reaching conclusions to which Francke and his champion Hoernle are inclined can be deduced from it. Without being aware of the criticism of Marx, Francke even thought it a useful task to publish a new translation of Schlagintweit's text, for which no other editions were consulted.[28] Such lack of critical faculty can only lead to error and disaster. It is solely Schlagintweit's text in which it is stated that T'on-mi on his mission betook himself to Kashmir (*K'a-č'e*), while all texts of the large and real edition of the *rGyal-rabs,* inclusive of its Mongol and Kalmuk translations, agree on the reading that he traveled to India (*rGya-gar*). If the Schlagintweit text be correct, this is merely the local Ladākh, not the general Tibetan, tradition. Marx justly observed: 'Any MS, specially prepared by a native of Ladākh for a foreigner, is apt to be less reliable than others of independent origin, for the reason, which would especially be true regarding historical documents, that the copyist will have a tendency to slightly alter the text, in the interest of his master, religion, or country, suppressing such facts as may seem derogatory to their fame, and substituting for phrases liable to be misunderstood others of a less equivocal character.' It is not difficult to see how the Ladākh tradition may have arisen. Sum-pa mk'an-po, in his remarkable work *dPag bsam ljon bzaṅ,*[29] has T'on-mi go to India, and says that on his return to Tibet he prepared the alphabet *dbu-čan* in the

---

[27] *JASB* 60, pt. 1 (1891). 97-98.
[28] *JASB* 6 (1910). 393.
[29] Ed. by Sarat Chandra Das, p. 167.

royal castle Ma-ru of Lhasa by taking as model the forms of the letters of Kashmir, and instituted the *dbu-med* writing in harmony with the Wartu script. It is plausible to a high degree that T'on-mi concluded his work in Lhasa, after submitting his scheme to the approval of his royal master. Certainly it was not necessary for him to make a trip to Kashmir in order to get hold of Kashmir writing; that was procurable as well in Magadha.

The sentence from the Schlagintweit text to which Hoernle[30] attributes so much importance meets with no exact parallel in the large *rGyal-rabs:* it is simply corrupt, and the word *riṅs* is meaningless; probably we have to read *raṅ* (*drug raṅ bčos-nas* 'he himself made six new letters,' for this is required in accordance with the text of the large *rGyal-rabs*).[31] Francke's translation 'they formed 24 *gsal-byed* [consonants] and 6 *riṅs*'[32] demonstrates that he is ignorant of the elementary rules of Tibetan grammar: for the numeral is always placed behind the noun (as we have in this very sentence *gsal-byed ñi šu rtsa bži*), or, if the numeral precedes the noun, which rarely occurs, it must be followed by the suffix of the genitive.[33] What Hoernle distils from this sentence is purely fantastic.

In 1905 A. H. Francke pointed out certain similarities between the Tibetan alphabet and the Brāhmī of Kashgar.[34] A sensible French critic[35] remarked with reference to these surface comparisons: 'This proves nothing for the origin of one or the other; the resemblance disclosed by Dr. P. Cordier between the Tibetan alphabet and that of the Gupta of the seventh century A. D. are interesting otherwise.' In the same manner Dr. L. A. Waddell[36] justly remarks that the forms of the Tibetan letters themselves declare their origin from the developing Indian Devanāgarī characters at the stage to which they had attained

[30] *Manuscript Remains of Buddhist Lit.* p. XXXII.

[31] The chapter concerning the introduction of writing is reprinted in Si-tui Sum-rtags' Tibetan Grammar, 139 *et seq.* (Bengal Secretariat Press, 1895). See also I. J. Schmidt, *Geschichte der Ost-Mongolen,* 327.

[32] *Epigraphia Indica,* 11 (1912). 267.

[33] Foucaux, *Grammaire de la langue tibétaine,* § 49.

[34] *Memoirs As. Soc. of Bengal,* 1. 43-45.

[35] *Bull. de l'École française,* 6. 446.

[36] *JRAS* 1909. 946.

in mid-India in the seventh century A. D., and, it would appear, not any earlier, as a reference to the fine photographic illustrations of Indian inscriptions of that period in Fleet's *Corpus Inscriptionum Indicarum*, vol. 3, will show. These help to make it clear that the so-called Tibetan letters bear a strong family resemblance to those of the somewhat florid style which Fleet has called 'the Kuṭila variety of the Magadha alphabet of the seventh century A. D.' Many of the letters are identical in shape. Sten Konow[37] rightly observes, in a note to Francke's article: 'The correspondence between Central Asian Gupta and Tibetan is not so great that it is necessary to assume that they have been developed in the same locality. They have both been developed from the same source, and that explains the similarity.' And Dr. Vogel, after careful study, presents the conclusion that the Tibetan alphabet is derived from the Northern Indian script which was used in the seventh century. This evidence has not been discussed or even antagonized by Dr. Hoernle.[38] On the other hand, his juxtaposition of the Khotanese and Tibetan alphabets is by no means convincing in proving a close relationship between the two. A glance at plate IV of Bühler's *Indische Palaeographie* and the work cited by Dr. Waddell is sufficient to show that the Tibetan alphabet stands much closer to those of mid-India than of Khotan, and that the Tibetan tradition in its general features is perfectly correct. In all his theoretical speculations and his eagerness to prove his unfortunate theory, Dr. Hoernle entirely loses sight of the fact that the Khotanese alphabet itself hails from India. His investigation, moreover, is vitiated by a methodological error. The writing of Khotan is throughout compared with the Tibetan alphabet in its modern printed form instead of with the oldest accessible forms of the inscriptions and the manuscripts of the ninth century. No regard, for instance, is taken of the fact that in the beginning the plain consonant did not imply the letter *a*, but that *a* was written alongside it,[39] and that there were two graphic forms of the vowel *i*. Further, we have to be mindful of the

---

[37] *Epigraphia Indica*, 11. 269.

[38] *Manuscript Remains of Buddhist Lit.* p. XIX.

[39] Csoma, *Grammar of the Tibetan Language*, p. 5, who says that this was the case also with the other vowels; Laufer, *T'oung Pao*, 1914. 52.

fact that we do not yet possess a single specimen of Tibetan writing of the seventh century, so that it is premature to render a positive verdict on what this writing was.

The historical proof on which the Khotanese theory is founded is likewise a failure. Hoernle asserts that according to Tibetan tradition Khotan fell under the domination of Tibet in the seventh century under Sroṅ-btsan sgam-po, invoking as his authority Rockhill's *Life of the Buddha* (p. 211). True it is, Rockhill writes in this passage: 'Sroṅ-btsan ascended the throne of Tibet in his thirteenth year, and the neighboring states recognized him as their sovereign, so that his rule extended over the whole of Tibet, to the north as far as Khotan, which during his reign became subject to China, and to the east to China.'[40] This statement, however, is at the best merely an illogical combination of Chinese accounts with the erroneous Tibetan chronology, which makes Sroṅ-btsan live up to A. D. 698, while in fact, according to the Chinese annals, he died in A. D. 650. Neither Sanaṅ Setsen nor the Bodhimör, the Kalmuk translation of the Tibetan *rGyal rabs,* the only native sources which, in the translation of I. J. Schmidt, Rockhill utilized for his sketch of Tibetan history, make any mention of Khotan with reference to Sroṅ-btsan's reign, nor does the Tibetan *rGyal-rabs.* The Chinese annals likewise are perfectly silent as to Khotan in the report of the life and deeds of Luṅ-tsan (= Sroṅ-btsan). In reality, the relations of Tibet with Khotan begin only from A. D. 670 when the Tibetans conquered the Four Garrisons (Kucha, Khotan, Tokmak, and Kashgar), which they lost again to the Chinese in A. D. 692.[41] Thus Khotan was entirely beyond the reach of the Tibetans during the lifetime of King Sroṅ-btsan, and Hoernle's theory is a fallacy.

Finally we may raise the question: if the theory of Francke and Hoernle is true, why does a tradition to this effect not crop out in the literature of the Tibetans? Or, in other words, why should such a tradition, if it ever existed, have been suppressed? As is well known, there are Tibetan works on Khotan embodied

---

[40] For the rest Rockhill follows an utterly impossible chronology as to the life of the king, placing his birth in A. D. 600 and T'on-mi's missions to India in A. D. 616.

[41] Chavannes, *Documents sur les Tou-kiue (Turcs) occidentaux,* p. 114, 281.

in the Kanjur and Tanjur[42]; the *Gośṛṅgavyākaraṇa* in the Kanjur was translated from the language of Khotan, and Śīladharma, a Bhikshu from Khotan, collaborated in the translation of the Kanjur work no. 242.[43] Architects were summoned from Khotan by King K'ri-lde sroṅ-btsan for the building of a monastery.[44] The Tibetans do not shy at admitting their debt to Khotan whenever occasion arises; but they are persistent in pointing to India as the cradle of their writing and literature. It was from India that Sanskrit Buddhist literature was transmitted to Tibet, it was from India and Kashmir that Buddhist missionaries entered Tibet to preach the gospel of Buddha. The role of Khotan in this respect was reduced to a minimum. Surely, Turkistanitis is a new form of learned disease.

---

[42] Rockhill, *Life of the Buddha*, p. 231.

[43] Regarding Śīladharma see Pelliot, *Journal asiatique*, 1914, 2. 135.

[44] Laufer, *T'oung Pao*, 1908. 5.

145

沙畹讣告

ÉDOUARD CHAVANNES

B. Laufer

Field Museum of Natural History, Chicago

Édouard Chavannes, professor of Chinese literature at the Collège de France, died in Paris on January 29, 1918, at the age of fifty-two years. Born at Lyons on October 5, 1865, he was sent on a scientific mission to China in 1889, being attached to the French Legation at Peking till 1893. In 1893 he was appointed professor at the Collège de France, where he opened his courses with a lecture entitled 'Du Rôle social de la littérature chinoise' (published in the *Revue bleue,* 1893). In 1903 he became a member of the Institut de France. He was also directeur d'études honoraire à l'École des Hautes Études, corresponding member of the Russian Academy of Sciences, and an honorary member of the Société Franco-Japonaise of Paris, the Société Finno-Ougrienne, the Royal Asiatic Society, and our own Society (elected last year). His premature death is an irreparable loss to the scientific world, and will be regretted by the entire community of orientalists, for the magnitude of Chavannes' work rests on the fact that he was not merely a sinologue in the narrow, old-fashioned sense of this misused word, but an orientalist and historian of eminent learning and insight, with a broad-minded vision and unusual intellectual powers coupled with almost superhuman activity and unbounded capacity for research. Of all great sinologues whom France has produced, he was doubtless the most vigorous, the most intelligent, and the most successful. There is no branch of sinology to which he has not made profound contributions of permanent value. His memory will live, and his immense labor will bear fruit, as long as there is an oriental science in this world.

Chavannes' first literary production was 'Le Traité sur les sacrifices Fong et Chan de Se-ma Ts'ien, traduit en français,' published in the third volume of the now defunct *Journal of the Peking Oriental Society* (1890). This work already displays the characteristics of the mature scholar: the tendency to open new and original resources, mastery of Chinese style, accuracy and elegance of translation, and critical, philological treatment of the subject. This first essay matured in him the magnificent plan of elaborating a complete translation of Se-ma Ts'ien's *Shi ki,* the oldest of the twenty-four Chinese Annals. The first volume of this work, *Les Mémoires historiques de Se-ma Ts'ien traduits et annotés,* appeared in Paris, 1895, with an introduction of 249 pages, which is a masterpiece of historical and critical analysis and is not surpassed by anything of this character written before or after him. Five volumes of this monumental work, consisting altogether of 3051 pages, were brought out, the last being published in 1905. The translation comprises the first 47 of Se-ma Ts'ien's 130 chapters, and is accompanied by a full commentary and indices. It is a fundamental source-book for the ancient history of China and a marvelous storehouse of erudition. There are many appendices dealing with special problems or subjects of general interest, like the essay 'Des Rapports de la musique grecque avec la musique chinoise' (3. 630).

Chavannes not only placed historical studies on a new and solid basis, but also inaugurated sound archaeological research by his volume *La Sculpture sur pierre en Chine au temps des deux dynasties Han* (1893). In 1907 he paid his second visit to China, chiefly for the study of ancient monuments and inscriptions. The important results of this mission were published in a sumptuous album (*Mission archéologique dans la Chine septentrionale,* 1909), consisting of 488 plates. Of the descriptive portion two volumes have thus far appeared, *La Sculpture à l'époque des Han* (1913) and *La Sculpture bouddhique* (1915). It is hoped that more of this material will be published from his posthumous papers. One of his greatest achievements is presented by the decipherment and translation of the business documents written on wood and found in Turkistan (*Les Documents chinois découverts par A. Stein dans les sables du Turkestan oriental,* Oxford, 1913). In connection with R. Petrucci he studied the Chinese paintings of the Musée Cernuschi (*La Peinture chinoise au Musée Cernuschi,* 1914); another briefer study is again devoted to Buddhist art (*Six Monuments de la sculpture chinoise,* 1914). In his *Documents sur les Tou-kiue (Turcs) occidentaux* (1903) he gave a complete collection of all Chinese sources concerning the history of the

Western Turks and a correlation of the Chinese with all available occidental documents.

Chavannes was interested also in the great religions, Buddhism, Confucianism, Taoism, Nestorianism, and Manicheism. In 1894 he published his *Mémoire composé à l'époque de la grande dynastie T'ang sur les religieux éminents qui allèrent chercher la loi dans les pays d'occident par I-Tsing,* which contains the biographies and travels of sixty (mostly Chinese) monks who went to India in the second half of the seventh century in search of Sanskrit books. In co-operation with S. Lévi he translated the itinerary of Wu K'ung (*Journal asiatique,* 1895). His 'Voyage de Song Yun dans l'Udyāna et le Gandhāra' appeared in the *Bulletin de l'École française* (1903). The best fruit of his labors in this field is represented by the monumental work *Cinq cents contes et apologues extraits du Tripiṭaka chinois,* published in three volumes (Paris, 1910-11); a fourth volume containing notes and indices has been promised and, I believe, prepared for the press. This fine collection of Indian stories has given many a stimulus to the comparative study of folk-lore. Also his translations of the life of Guṇavarman, Jinagupta, and Seng-Hui (*T'oung Pao,* 1904, 1905, 1909), his 'Quelques titres énigmatiques dans la hiérarchie ecclésiastique du bouddhisme indien' and 'Les Seize Arhat protecteurs de la loi' (*JA* 1915, 1916, the two last-named in collaboration with S. Lévi) should be mentioned in this connection.

His book *Le T'ai Chan, essai de monographie d'un culte chinois* (Paris, 1910, 591 p.) is devoted to the indigenous religion of China and represents a wonderfully complete and fundamental study of an ancient mountain-cult, based on personal investigation and on all available documents both literary and epigraphical. In 1897 he contributed to the *Journal asiatique* a remarkable study on 'Le Nestorianisme et l'inscription de Kara-Balgassoun.' In collaboration with P. Pelliot he edited and translated in 1912 a Manichean treatise, written in Chinese and discovered by Pelliot in the caves of Tun-huang, Kan-su. This is perhaps the most brilliant achievement of modern sinology.

As an epigraphist, Chavannes deserves the highest praise: in this branch of research he was truly a pioneer and reformer, the first European scholar who approached this difficult subject with sound and critical methods and undisputed success. In 1893 he published in the *Journal asiatique* a study on 'Les Inscriptions des Ts'in' (re-edited in his *Mémoires de Se-ma Ts'ien,* 2. 544). The Chinese inscriptions of Bodh-Gayā engaged his attention in two articles (*Revue de l'histoire des religions,* 34, 36). When Prince Roland Bonaparte edited his luxurious work *Documents de l'époque mongole,* Chavannes undertook the translation of the Chinese portion of the inscription of Kiü-yung-kuan. In 1902 the Académie des Inscriptions issued his *Dix inscriptions chinoises de l'Asie centrale.* His 'Inscriptions et pièces de chancellerie chinoises de l'époque mongole' (*T'oung Pao,* 1904, 1905, 1908) contain seventy-six documents in text and translation; he was the first to penetrate successfully into the peculiar official style of the Mongol epoch. The inscriptions of Yün-nan occupied him in 'Une Inscription du royaume de Nan-tchao' (*JA* 1900), 'Quatre inscriptions du Yun-nan' (*JA* 1909), and 'Trois inscriptions relevées par M. Sylvain Charria' (*T'oung Pao,* 1906). His 'Les Deux plus anciens spécimens de la cartographie chinoise' (*Bull. de l'École française,* 3) is a most important contribution to the history of cartography.

The number of articles written by Chavannes is legion. Special mention may be made of his 'Voyageurs chinois chez les Khitan et les Joutchen' (*JA* 1897-98), 'Les Prix de vertu en Chine' (published by the Institut de France, 1904), 'Les Livres chinois avant l'invention du papier' (*JA* 1905), 'Les Pays d'occident d'après le Wei lio' (*T'oung Pao,* 1905), 'Le Cycle turc des douze animaux,' and 'Trois généraux chinois' (*ib.* 1906), 'Les Pays d'occident d'après le Heou Han chou' (*ib.* 1907), 'Le Royaume de Wou et de Yüe' (*ib.* 1916), and 'L'Instruction d'un futur empereur de Chine en 1193' (*Mémoires concernant l'Asie orientale,* 1. 1913). For the series *La Science française* published by L. Poincaré for the Panama-Pacific Exposition he wrote a brief sketch on the development of sinology in France. In conjunction with H. Cordier, who founded the *T'oung Pao* in 1890, he edited that Journal from 1904 onward till his departure.

146

罗伯特·罗维《文化与民族学》书评

AMERICAN ANTHROPOLOGIST

NEW SERIES

ORGAN OF THE AMERICAN ANTHROPOLOGICAL ASSOCIATION,
THE ANTHROPOLOGICAL SOCIETY OF WASHINGTON,
AND THE AMERICAN ETHNOLOGICAL
SOCIETY OF NEW YORK

VOLUME 20

LANCASTER, PA., U. S. A.
PUBLISHED FOR
THE AMERICAN ANTHROPOLOGICAL ASSOCIATION

1918

BOOK REVIEWS

METHODS AND PRINCIPLES

Culture and Ethnology. ROBERT H. LOWIE. Douglas C. McMurtrie: New York, 1917. 190 pp.

Although the methods and principles of modern ethnology, as conceived and cultivated in America, are well known to the small band of workers, a coherent and concise summary of the ideas for which the science of ethnology stands, and which could serve for the information of a wider public, has hitherto been lacking. This long-felt want is felicitously supplied by Dr. Lowie's booklet which has grown out of a series of lectures delivered in 1917 in the American Museum of Natural History. The cause of ethnology is here pleaded with eloquence and well-balanced judgment, and a difficult subject is presented in an attractive and lucid exposition. This work will form an excellent introduction and guidance to the raw recruit along the thorny path of ethnological research, and it will be read by the veterans with as much pleasure and profit. It should be the permanent possession of every one, together with Boas's *Mind of Primitive Man* and Wissler's new book on *The American Indian.* So many confusions, misconceptions, and prejudices in regard to what ethnology constitutes are still prevelant, even in the minds of scientists in general, especially among historians, orientalists, art-students, etc., that a book like this one will be warmly welcomed and will help to clarify the atmosphere. Dr. Lowie's universal reading, his keen analytic sense that goes straight at the root of things, and his gift of clear representation, insure his pages a piquant flavor and the constantly growing sympathy of the reader.

In his first lecture, the author endeavors to determine the relation of culture to psychology and to defend the rights of ethnology as an independent science. Psychology, as he aptly expresses it, is as impotent to reduce to really interpretative psychological principles the subjective aspect of cultural phenomena as it is to explain the historical sequence of events, and the principles of psychology are as incapable of accounting for the phenomena of culture as is gravitation to account for architectural styles. Over and above the interpretations given by psychology, there is an irreducible residuum of huge magnitude that calls for special treatment and by its very existence vindicates the *raison d'être* of ethnology; but we need not eschew any help given by scientific psychology for the comprehension of specifically psychological components of cultural phenomena. At this point I wish the author might have made some reference to the growing importance of sexual psychology without the aid of which numerous factors in the life of all peoples cannot adequately be grasped. No sinologue, for instance, has ever been able to advance a correct explanation of the reasons why the Chinese bind (or used to bind) the feet of their women: this is a sex problem to which no historian or ethnologist (not to speak of the Chinese themselves) can make an answer. Moreover, it seems to me that for practical working purposes we are compelled to postulate the existence of a social or national psyche, whether this be correct in a strictly scientific sense or not, this psyche to be conceived, of course, as being composed of ideas and habits historically acquired and developed, and accordingly as a changeable quantity. It is a commonplace observation that the Japanese adopted and absorbed the fundamental principles of Chinese civilization; it is still more interesting to note what they did not adopt from their superior neighbors: they did not take over their costume, their domestic architecture, their furniture and kitchen, sitting on chairs and at tables (in fact retained all their ancient material possessions relative to domestic life), foot-binding, consumption of opium, not to mention numerous other customs and practises. An answer to this why has not yet been given by any ethnologist or historian of Japan. Why did Christianity conquer the peoples of the Indo-European stock, while it did not appeal to the Semites in the midst of whom it was born? Why could it never obtain any permanent influence on any nation of Asia? Why is Buddhism, the product of an Aryan society, extinct in India, the land of its birth, while it flourishes among so many non-Aryan peoples of central and eastern Asia? Or why do the tribes of Africa so easily become Mohammedanized? For each of these questions, hundreds of historical facts can be marshalled to state the merits of the particular case; but these do not bring us nearer to a satisfactory explanation. Here, and in similar problems, it seems to me the application of psychological methods or viewpoints will remain unavoidable. On the other hand, each cultural fact reacts on the tribal or national psyche, and these effects have as yet been little studied. To illustrate, among the European and many Asiatic tribes family life has always centered around the fire of the kitchen or tent; the kitchen is still the bond that unites the members of the family in our civilization (as far as they are not corrupted by the institution of the family hotel). The formation of the family in any other way seems almost inconceivable to us. Yet the Chinese in this respect are radically different from others. The kitchen never formed the center of their house; on the contrary, it was always far removed from it as a separate building and entrusted to the care of male servants. There are no female cooks, and the mother of the house avoids the kitchen, where the family never assembles. While the organization of the Chinese family outwardly is very much like our own, yet the forms of family life and the psychical relations of the family members are diverse, and this difference seems to be directly traceable to influences of material causes, as given by the arrangement of habitations and the location of the kitchen.

The second lecture, "Culture and Race," forms especially attractive reading, and there is no doubt that every ethnologist will subscribe to the author's clear exposition and his statement of the relation of culture to racial traits: culture cannot be adequately explained by race, and the same race varies extraordinarily in culture even within a very narrow space of time. The Chinese and some of our American Indians, such as the ancient Central Americans and Peruvians, did attain a very high level, which may be equated with that of Europe at a relatively recent period. No term has been more misused and sinned against than the catchword "racial mentality." The anthropologist is inclined to assume that mental powers are fairly uniform through mankind, but work in different directions or are projected on different lines of activities. The limitation of activities in a certain people is not the consequence of a limitation of mentality, but merely the outcome of historical agencies and traditional training based thereon. In judging nations we are just as perverse as in judging individuals: both are judged by the average man from their outward success, while, of course, success is not necessarily caused by a higher state of mentality. The best exposure of this shallowness of judgment is to the credit of a Japanese statesman, who during the Russo-Japanese war, when his country held the attention of the world, remarked bitterly, "When we were a nation of artists you called us barbarians; now that we kill men you call us civilized."

In his third lecture Dr. Lowie discusses the relation of culture to geographical environment. Here he is particularly fortunate in the choice of his illustrations by contrasting the cultures of the Hopi and Navajo, which are thoroughly disparate despite the fact that both tribes have occupied for a long period the same part of northeastern Arizona, or by characterizing the Chukchi and Eskimo, the Bushmen and Hottentot. Several interesting sidelights fall in this connection on the history of domestications. The fact that environment has a rather insignificant share in the formation of culture is patent to every one who can think objectively; nevertheless, in the face of the many pretensions made by that infertile pseudo-science, anthropogeography, it becomes our duty to antagonize its hollow dogmas on every suitable occasion. Dr. Lowie very well sums up his standpoint by saying that environment cannot explain culture because the identical environment is consistent with distinct cultures; because cultural traits persist from inertia in an unfavorable environment; because they do not develop where they would be of distinct advantage to a people; and because they may even disappear where one would least expect it on geographical principles.

Psychology, racial differences, geographical environment, are all inadequate for the interpretation of cultural phenomena. Culture is a thing *sui generis* which can be explained only in terms of itself. With this motto in mind, the author finally attacks in his fourth chapter the determinants of culture. He points out the great importance of cultural diffusion and assimilation, as, for instance, in the propagation of maize, the adoption on the part of the white settlers of the entire complex of aboriginal maize cultivation (tobacco and tea are as striking examples), the early traffic in bronze and amber, and the close connection and interrelations of the ancient Asiatic civilizations. Contact of peoples is thus an extraordinary promoter of cultural development. The theory of cultural evolution, to my mind the most inane, sterile, and pernicious theory ever conceived in the history of science (a cheap toy for the amusement of big children), is duly disparaged. It is worthy of especial mention that Dr. Lowie (p. 89) recedes to some extent from his former standpoint taken on the subject of convergencies. The majority of these now appear to him not genuine, but false analogies due to our throwing together diverse facts from ignorance of their true nature. I hope the author will discuss this point somewhat more fully in the near future for the benefit of his fellow-workers. What he has to say about survivals, rationalistic explanations, and the determinants of culture in general, belongs to the best portion of the book and will hardly provoke any contradiction. Culture cannot be forced into the straitjacket of any theory whatever it may be, nor can it be reduced to chemical or mathematical formulas. As nature has no laws, so culture has none. It is as vast and as free as the ocean, throwing its waves and currents in all directions. It is absurd to seek the origin of civilization in any particular region or to trace it to a single nation. In its present aspects, culture is the common good of mankind, the product of human thought of all ages. It is something above or below the nations, but not a thing of their own. There is no nation (inclusive of primitive peoples) that has a culture entirely evolved from its own resources; there is no living nation that has a ghost of a claim to the generation of any fundamentals of culture. Our present and our future lie in our past. All that the practical investigator can hope for, at least for the present, is to study each cultural phenomenon as exactly as possible in its geographical distribution, its historical development, and its relation or association with other kindred ideas. The more theories will be smashed, the more new facts will be established, the better for the progress of our science.

The last chapter of Dr. Lowie's work is taken up by a thorough discussion of kinship terminologies, added as a concrete illustration of the methods propounded in the four preceding lectures. This subject is particularly his own domain, and as is well known, has elicited from his pen many contributions of permanent value.

B. LAUFER.

Organic Evolution. A Text-book. RICHARD SWANN LULL. The Macmillan Company: New York, 1917. XVIII, 729 pp., XXX plates, 253 figures. Price $3.00 net.

The addition of another book to the extensive literature on evolutionary biology is justified by the fact that the author is a paleontologist. Of late, paleontology has fallen into disfavor with experimental biologists and students of heredity. The reason for this is, that paleontologists, seeing at every hand animals in complete agreement with their environment, have shown a tendency to describe the evolution of different organisms in terms which, though they cannot be called Lamarckian, still savor of the Lamarckian theory. Dr. Lull states his attitude as follows:

> The geologic changes and the pulse of life stand to each other in the relation of cause and effect. This statement does not, however, imply the acceptance of the Lamarckian factor any more than that of natural selection, for whether the influence of a changing environment acts directly upon the creature's body, or indirectly through induced habit, or, whether it merely sets a standard to which animals must conform if they would survive, matters not; the fundamental principle remains that changing environmental conditions stimulate the sluggish evolutionary stream to quickened movement.

As the title implies, the scope of the book is unusually comprehensive and it is necessarily somewhat of a compilation. Several chapters are

147

押不芦

LA MANDRAGORE.

PAR

BERTHOLD LAUFER.

Čou Mi 周密 (1230—1320), écrivain célèbre de la fin des Song, nous a transmis une tradition fort curieuse dans ses ouvrages *Kwei sin tsa ši* 癸辛雜識 (續集 上, p. 38, éd. du *Pai hai*) et *Či ya t'an tsa č'ao* 志雅堂雜鈔 (chap. 上, p. 40 b—41 a, éd. du *Yüe ya t'an ts'un šu*).[1]

Le texte du *Kwei sin tsa ši* est ainsi conçu:

回回國之西數千里地產一物極毒全類人形若人參之狀。其酋名之曰押不蘆。生土中深數丈。人或誤觸之著其毒氣必死。取之法先於四旁開大坎可容人。然後以皮條絡之。皮條之系則繫于犬之足旣而用杖擊逐犬。犬逸而根拔起。犬感毒氣隨斃。然後就埋土坎中。經歲然後取出曝乾。別用他藥制之。每以少許磨酒飲人則通身麻痺而死。雖加以刀斧亦不知也。至三日後別以少藥投之卽活。蓋古華陀能刳腸滌胃以治疾者必用此藥也。今聞御藥院中亦儲之。白廷玉聞之盧松厓。或云。今之貪官汚吏贓過盈溢被人所訟則服百日丹者莫非用此。

---

[1] Sur cet auteur, sa vie et son œuvre, cf. Pelliot, *T'oung Pao*, 1913, p. 367—368.

Voici le texte du *Či ya t'aṅ tsa č'ao*:

回回國之西數千里地産一物極毒全似人形如人參之狀。其名押不盧。生於地中深數丈。或從傷其皮則爣。毒之氣著人卽死。取之之法先開大坑令四旁可容人。然後輕手以皮條結絡之。其皮條之前則繫於大犬之足既而用杖打犬。犬奔逸則此物拔起。大 [pour 犬] 感此氣卽斃。然後别埋他土中。經歲後取出暴乾。别用藥以製治其性。以少許磨酒飲之卽通身痳痺而死。雖刀斧加之不知也。然三日别以少藥投之卽活。蓋古者華陀能刳腸滌臟治疾者或用此藥也。聞今御藥院中有二枚此神藥也。白廷玉聞之盧松厓云。Le texte s'interrompt ici et n'est pas terminé.

Ni l'un ni l'autre texte ne semble être en parfait état, mais celui du *Kwei sin tsa ši* (A) est certainement le meilleur et le plus complet. Il est à la base de la traduction qu'on va lire, tandis que les divergences de la rédaction du *Či ya t'aṅ tsa č'ao* (B) sont ajoutées en crochets.

"Quelques milliers de *li* à l'ouest des pays mahométans le sol produit une chose excessivement vénéneuse et pareille dans son ensemble à la figure d'un homme; en effet, elle a l'apparence du ginseng. On l'appelle *ya-pu-lu* (*ya-pou-lou*). Cette plante croît dans la terre jusqu'à une profondeur de plusieurs toises. Si un homme se heurte contre la plante par erreur, il recevra son exhalaison vénéneuse et doit mourir. [B: Quand on la blesse, son écorce brille; l'exhalaison du poison pénètre dans l'homme qui meurt aussitôt.] Voici la méthode de prendre la plante. D'abord, aux quatre côtés (autour de la racine) on creuse un trou assez grand pour recevoir un homme [B: D'abord on creuse une grande fosse dont les quatre côtés soient assez spacieux pour recevoir un homme]. Ensuite on lie la plante au moyen d'une lanière de cuir dont l'extrémité est attachée aux pieds d'un grand chien [B: Ensuite on lie la plante légèrement au moyen d'une lanière de cuir, dont la partie antérieure est attachée aux pieds d'un grand chien]. Avec un bâton on bat et chasse le chien qui s'enfuit en entraînant avec lui la racine. Accablé de l'exhalaison du poison, le chien périt sur le champ. Alors on ensevelit la racine dans un trou du sol [B: dans un autre sol], et au bout d'un an on l'en sort pour la sécher au soleil. Elle est mélangée avec d'autres ingrédients [B: pour dominer sa nature] et en chaque cas on en râpe un peu dans du vin qu'on donne à boire à un homme; le corps entier de celui-ci en sera paralysé, et il tombera en torpeur comme s'il était mort. Même si on lui applique des couteaux ou des haches, il ne s'en apercevra pas. Au bout de trois jours si une petite dose de médecine lui est administrée, il reviendra à la vie. C'est peut-être là le remède employé par Hwa T'o qui anciennement était capable d'ouvrir les intestins et de purger l'estomac pour guérir des malades. [1] Or j'ai entendu dire qu'une provision de cette médecine [B: deux pièces, c'est une médecine divine] est conservée dans la Pharmacie Impériale. [2] C'est Pai T'iṅ-yü qui l'a appris de Lu Suṅ-yai.

[1] Hwa T'o est le célèbre médecin et chirurgien qui mourut en 220 de notre ère. Cf. surtout la notice de Chavannes, *BEFEO*, III, 1903, p. 409. Comme M. Chavannes fait remarquer d'après le *Hou Han šu*, l'anesthétique employé par Hwa T'o était du chanvre infusé dans du vin qui excitait et étourdissait le patient. Donc la conclusion de Čou Mi n'est pas juste. Cf. aussi C. Pétillon, *Allusions littéraires*, p. 380, et *T'oung Pao*, 1898, p. 237—238; S. Julien, *Chirurgie chinoise. Substance anesthétique employée en Chine, dans le commencement du IIIe siècle de notre ère, pour paralyser momentanément la sensibilité* (*Comptes rendus hebdomadaires des séances de l'Académie des Sciences*, XXVIII, 1849, p. 195—198). L'information est extraite de l'ouvrage médical *Ku kin i tuṅ* du commencement du XVIe siècle. Voir aussi Flückiger et Hanbury, *Pharmacographia*, p. 547. Sur les propriétés narcotiques du chanvre connues dans l'Inde, cf. C. Joret, *Les plantes dans l'antiquité*, II, p. 645.

[2] Cf. Bazin, *Notice historique sur le collège médical de Péking*, p. 24—25 (extrait du *Journal asiatique*, 1856).

Quelques uns disent: les officiers avides et les fonctionnaires oppressifs du temps présent, quand ils ont fait des exactions excessives et qu'ils sont accusés, prennent de la drogue dite drogue de cent jours; ne serait-ce pas cette plante dont ils se servent?"

Il semble que Čou Mi soit resté le seul auteur chinois à parler de la plante *ya-pu-lu*. Du moins, Li Ši-čen, dans son *Pen ts'ao kaṅ mu* (chap. 17下, p. 13 b), ne cite-t-il que le texte du *Kwei sin tsa ši* à propos du *ya-pu-lu*; il le cite d'ailleurs assez inexactement, en supprimant le conte du chien et en ajoutant au préambule les mots 漠北 *mo pei*, "au nord du désert Gobi." [1] La dernière phrase il l'a changée ainsi: 貪官汚吏罪甚者則服百日丹皆用此也. C'est à ce texte que se rapporte la brève note de Stuart, [2] qui fait remarquer qu'il n'y a pas de description de la plante, et que son identification demande de nouvelles recherches. De même, J. L. Soubeiran et Dabry de Thiersant [3] ont déjà noté la plante *ya-pu-lu* d'après le *Pen ts'ao* sous le titre *Atropa* (avec point d'interrogation), en disant: "Décrit par le *Pen ts'ao* comme déterminant une anesthésie suffisante pour permettre de faire des opérations. On dit que l'action s'en fait sentir pendant trois jours; il aurait été employé par le chirurgien Houa-to, pour des opérations intéressant les intestins."

Il est surprenant de voir ce que l'encyclopédie *Ko či kiṅ yüan* 格致鏡原 (chap. 69, p. 5 b) a fait du texte du *Kwei sin tsa ši*. Ici la plante est introduite sous le titre "herbe qui réveille de la

---

[1] Cette addition est donnée aussi par le *Yüan kien lei han* (chap. 411, p. 22) dans un autre texte, de seconde main et mal digéré, concernant le *ya-pu-lu*. Le *P'ei wen čai kwaṅ k'ün faṅ p'u* (chap. 97, p. 25 b; Bretschneider, *Bot. Sin.*, I, p. 70) contient le même texte écourté, sans le conte du chien, mais avec l'introduction correcte 回回地方.

[2] *Chinese Materia Medica*, p. 59; voir déjà F. P. Smith, *Contributions towards the Materia Medica of China*, p. 36. Smith dit sans raison que la plante vient du pays des Huns ou des Ouigours.

[3] *La matière médicale ches les Chinois*, p. 190 (Paris, 1874).

mort et qui rétablit la vie" 起死回生草.[1] Naturellement Čou ne veut pas dire que l'homme qui prend la potion meurt réellement et ressuscite au bout de trois jours, mais seulement qu'il reste sans conscience pendant cet intervalle. S'il mourait, l'expérience de frapper le corps avec un couteau n'aurait aucun sens. Qu'il est insensible aux coups c'est la merveille; par conséquent, la vie n'est pas encore éteinte. Le conte du chien est éliminé, et le document entier est abrégé ainsi: 一名押不盧。出回回國以少許磨酒飲人則通身麻痺而死。雖加以刀斧亦所不知。至三日別以少藥投之即活。御苑中亦儲之. Ce texte corrompu et mutilé fut adopté par G. Schlegel,[2] qui fit venir la plante de l'Arabie (au lieu des pays mahométans), l'attribua au palais impérial et induisit en erreur P. J. Veth.[3]

La plante décrite par Čou Mi peut être identifiée sans difficulté avec la mandragore sur la base de la transcription *ya-pu-lu*, laquelle correspond exactement à l'arabe-persan *abruh* ابروح ou *yabruh* يبروح, désignation pour le fruit de cette plante.[4] Elle-même s'appelle en

[1] Selon le *P'ei wen čai kwań k'ün fań p'u* et le *T'u šu tsi č'eń* cette définition émane du *Tien tsai ki* 滇載記.

[2] *Nederlandsch-chineesch Woordenboek*, IV (supplément), p. 25.

[3] *Archives internat. d'ethnographie*, VII, 1894, p. 82.

[4] Selon d'Herbelot (*Bibliothèque orientale*, I, p. 72) les Persans appellent aussi communément cette plante *esterenk* [*astereng*] et les botaniques arabes ont formé par corruption les noms d'*iabroug* et d'*iabrouh* qu'ils lui donnent, du mot persan *abrou*. L'origine perse du mot est plaidée aussi par Wetzstein (*ZE*, 1891, p. 891) et Veth (*Archives internat. d'ethnographie*, VII, 1894, p. 200) qui pour cette raison ont hasardé l'opinion que, de même, les notions magiques sur la mandragore auraient pris leur origine en Perse. C'est une hypothèse qui ne s'inspire que de considération purement philologique sur les mots; les preuves historiques font défaut: Il n'y a pas de texte iranien de date ancienne à ce sujet. Pour la première fois la mandragore est mentionnée dans la littérature perse par Abu Mansur, qui autour de l'an 975 écrivit son Livre des Principes Pharmacologiques (traduction d'Achundow, p. 148). Baber écrit dans ses Mémoires que la mandragore se trouve dans les montagnes du Fergana (A. S. Beveridge, *Memoirs of Bābur*, p. 11). Il est difficile de se ranger à l'avis de Wetzstein que l'arabe *yabrūh* serait issu du persan *abrewi*. Voir aussi Horn, *Grundr. iran. Phil.*, I, 2, p. 73.

arabe *toffāh-el-jenn* تفاح لجن ("la pomme des esprits") ou *sirāj el-kotrob* سراج القطرب ("la lampe des lutins"), aussi *la'ba* لعبة et *beiḍ el-jinn* ("œufs des esprits"). [1] En Araméen le fruit est nommé *yawruha* יַבְרוּחָא; et la forme *jerābūh* جرابوح est usuelle en Syrie. Ce nom sémitique paraît être d'une date relativement ancienne; du-moins trouvons-nous dans Dioscoride un terme dit égyptien de la forme ἀπεμούμ laquelle, selon moi, semble être apparentée à l'arabe *abruh*: peut-être ce mot est-il à corriger en ἀπερούμ. La mandragore (ou mandegloire par étymologie populaire) [2] forme un genre de la famille des solanées à la racine fusiforme et souvent bifurquée, aux feuilles radicales d'un vert sombre, aux fleurs purpurines et dont les fruits rouges, semblables à une petite pomme, exhalent une odeur agréable. [3] Les propriétés narcotiques de cette

---

[1] L. Leclerc, *Traité des simples*, II, p. 246; III, p. 240; d'Herbelot, *Bibliothèque orientale*, III, p. 524. Les Arabes ont emprunté aux Grecs aussi le mot *mandragoras* dans la forme *mandaghura* (Leclerc, III, p. 341).

[2] Francisque-Michel, *Recherches sur le commerce, la fabrication et l'usage des étoffes de soie*, II, p. 76, Paris, 1854) a fait cette observation: "Au XVe siècle, ils [nos ancêtres] employaient la soie à conserver certaines amulettes, dont un célèbre prédicateur de l'époque brûla un grand nombre, ce qui valait mieux assurément que de brûler les gens qui y avaient foi. On les appelait *madagoires*, par une altération du mot *mandragores*. 'Aujourd'huy, ajoute l'auteur du Journal du roy Charles VII, le vulgaire les appelle mandegloires, que maintes sottes gens gardoient en lieux de repos, et avoient si grande foy en celle ordure, qu'ils croyoient fermement que tant comme ils l'avoient (mais qu'il fust bien nettement en beaux drapeaux de soye ou de lin envelopé), jamais jour de leur vie ne seroient pauvres.' Dans le dialogue de Mathurine et du jeune du Perron, celui-ci lui dit: 'As-tu point aidé a souffler le feu lent sous la coque d'œuf où est le germe, la soye cramoisie, et cela de quoy les magiciens faisoient leur pâque avec la petite mandragore?' (Confession catholique du sieur de Sancy, liv. II, ch. 1er)." Aussi l'expression *main de gorre* était en usage populaire.

[3] C. Joret, *Les plantes dans l'antiquité et au moyen âge*, I, p. 498. — "La *Mandragora officinarum* est connue sous le nom de *Mandragore femelle*. Elle est très commune dans le midi de la France, on la rencontre en abondance sur les rivages de la Calabre, de la Sicile, de l'île de Crète, de la Cilicie, de l'Afrique, de l'Espagne; elle se plaît dans les lieux ombragés, sur les bords des rivières, à l'entrée des cavernes. Elle fleurit en automne, quelquefois aussi au printemps. Sa racine est grosse, noirâtre extérieurement, blanche à l'intérieur, charnue; ses feuilles sont grandes, les plus extérieures obtuses, les plus intérieures aiguës; leur couleur est un vert bleuâtre, luisant en dessus, terne en dessous. Le pétiole est long; les hampes florales sont longues, rougeâtres, et un peu pentagonales.

plante étaient connues anciennement, et elle était douée de vertus magiques, aphrodisiaques et prolifiques. La racine prend souvent des formes singulières, rappelant plus ou moins le corps de l'homme. Le nom est dérivé du grec *μανδραγόρας*, mot dont l'étymologie est encore inconnue. D'après Littré, ce paraît être un nom d'homme appliqué à une plante, et contenir *μάνδρος* ou *μάνδρα*, nom d'une divinité locale de l'Asie Mineure. L'origine orientale de plusieurs croyances attachées à cette plante, comme nous verrons, paraît certaine. Assurément, le nom n'a rien à voir avec le persan *mardum-giyā* مردم گیا, comme supposent Wetzstein [1] et Schrader. [2]

Ce n'est pas le but de cette notice de retracer toutes les croyances touchant les vertus de la mandragore et accumulées pendant beaucoup de siècles. Un tel travail a été maintes fois tenté, mais, à vrai dire, aucun essai de ce genre n'est tout à fait satisfaisant ou complet dans l'emploi des sources. [3] Une œuvre d'ensemble et

---

La fleur se compose d'un calice à cinq divisions aiguës et lanceolées, d'une corolle trois fois plus longue que le calice, de couleur violette et découpée en cinq lobes oblongs, obovés" (A. Milne Edwards, *De la famille solanacées*, p. 56, Paris, 1864).

[1] *L. c.*

[2] *Reallexikon*, p. 36. Une nouvelle hypothèse sur l'affinité du nom grec se trouve à la conclusion de cet article.

[3] Il y a trois monographies que je n'ai pas eu l'occasion de voir, J. Schmidel, *Dissertatio de mandragora* (Lipsiae, 1671); Granier, *Dissertation botanique et historique sur la mandragore* (Paris, 1788); et Bartolomi, *Commentarii de mandragoris* (Bologna, 1835). Ce sont les traités suivants qui me sont connus: F. v. Luschan a illustré six racines de mandragore de l'Orient proche sculptées en figures humaines (*ZE*, XXIII, 1891, p. 726—728); sa brève notice est accompagnée de notes explicatives par Ascherson et Beyer (p. 729—746) et de notes additionnelles par Wetzstein (p. 890—892). — W. Hertz, *Sage vom Giftmädchen*, traite de la mandragore dans un appendice (*ABAW*, 1893, p. 164—166). — P. J. Veth, *De alruin en de heggerank* (*Archives internationales d'ethnographie*, VII, 1894, p. 81—88) et *De mandragora* (*ibid.*, p. 199—205). — C. B. Randolph, *The Mandragora of the Ancients in Folk-Lore and Medicine* (*Proceedings American Academy of Arts and Sciences*, XL, Boston, 1905, p. 487—537). — E. O. v. Lippmann, *Alraun und schwarzer Hund*, dans ses *Abhandlungen*, I, 1906, p. 190—204. — Comme on suppose qu'il s'agit de la mandragore dans l'Ancien Testament (דּוּדָאִים *dūda'īm*, "plante d'amour", dérivé de *dūd*, "aimer"; *Gen.*, XXX, 14—16, et *Cant.*, VII, 14), on trouve des articles à ce sujet dans les nombreux dictionnaires bibliques; le meilleur que j'aie vu est celui de E. Levesque

de critique reste à faire. Les notes suivantes ne doivent être regardées que comme un commentaire du texte de Čou Mi; toutefois rien d'important n'y est omis.

---

au *Dictionnaire de la Bible* par F. Vigouroux (IV, col. 653—655). Il ne faut pas oublier que cette interprétation du terme hébreux repose sur une hypothèse, d'ailleurs fort vraisemblable, suggérée par les traductions *μῆλα μανδραγόρου* des Septante, *mandragora* de la Vulgate, et *yābruḥin* du Targum d'Onkelos et du syriaque; en outre, la plante est répandue en Palestine. — Le savant japonais Kumagusu Minakata (*Nature*, LI, 1895, p. 608; et LIV, 1896, p. 343—344; cf. *T'oung Pao*, 1895, p. 342) a contribué deux brèves notices à ce sujet en se servant de sources chinoises, mais sans méthode et critique. Je ne veux pas entrer dans une critique détaillée de ce travail, mais je voudrais remarquer seulement que ses rapprochements entre la mandragore et la plante *šań-lu* 商陸 (*Phytolacca acinosa*) ne sont que des parallèles psychologiques, mais non historiques (voir *infra*). Minakata a aussi donné une traduction du conte de Čou Mi avec quelques contre-sens sans consulter le texte meilleur du *Kwei sin tsa ši*, et a fait allusion à Josèphe par des sources de seconde main. Je ne dois rien à cette étude; en effet, j'ai trouvé tous les textes indépendamment, et mon travail était achevé quand par hasard l'article de Minakata est tombé dans mes mains. — Niccolò Macchiavelli (1469—1527) est l'auteur d'une comédie, d'abord intitulée *Comedia di Callimaco et di Lucrezia* (1re édition, *s.l.n.d.*), puis *Mandragola* (1524, etc.; éd. sous mes yeux, Roma, 1688), en cinq actes, en prose, précédée d'un prologue; c'est une satire sur la croyance à la vertu de la mandragore pour féconder une femme. Callimaco dit a Mèsser Nicia (p. 63): "Voi havete a intendere questo, che nõ è cosa piu certa a ingravidare d'une potione fatta di Mandragola, questa è una cosa esperimẽtata da me due para di volte, et trovata sempre vera: e se non era questo, la Reina di Francia sarebbe sterile, e infinite altre principesse di quello stato." La comédie de Machiavel a fourni à J. de la Fontaine le sujet d'un conte rimé qui est intitulé "La Mandragore, nouvelle tirée de Machiavel" (Œuvres de J. de la Fontaine par H. Regnier, tome V, 1889, p. 22, avec une introduction intéressante de l'éditeur).

"Cette recette est une médecine
Faite du jus de certaine racine,
Ayant pour nom mandragore; et ce jus
Pris par la femme opère beaucoup plus
Que ne fit onc nulle ombre monacale
D'aucun couvent de jeunes frères plein".

La *Mandragola* a été imitée par J.-B. Rousseau dans sa comédie *la Mandragore*, également en cinq actes, en prose, "tirée, dit le titre, de l'italien de Machiavel". Andrea Calmo écrivit *la Potione, comedia facetissima et dilettevole*, en quatre actes et un prologue, imitation de la *Mandragola*, écrite dans les dialectes vénitien, bergamasque, italo-grec, etc. (Venise, 1552, réimprimée en 1560, 1561, et 1600). Il y a une nouvelle de Charles Nodier, intitulée *la Fée aux miettes* (1832), dont le héros, pour posséder sa maîtresse, doit trouver "la mandragore qui chante". Une nouvelle allemande *Mandragora*, d'ailleurs assez faible, par de la Motte Fouqué, a paru en 1827.

L'historiette du chien déracinant la plante ne se trouve ni dans Pline ni dans Dioscoride qui l'un et l'autre ont écrit sur la mandragore. La version la plus ancienne que nous connaissions est due à Flavius Josèphe (37—93) qui dans son œuvre *De bello judaico* (VII, 6, § 3), écrit entre les années 75 et 79, s'exprime ainsi:[1] "Or dans ce palais croissait une espèce de rue[2] qui mérite notre admiration à cause de ses dimensions, car elle était aussi large qu'un figuier en ce qui concerne la hauteur et l'épaisseur; et, suivant une tradition, elle avait duré depuis le temps d'Hérode, et probablement elle aurait continué beaucoup plus longtemps si elle n'avait pas été tranchée par les Juifs qui occupaient la place plus tard. Et dans la ravine qui environne la cité [Machaerus] au côté du nord, il y a une certaine place nommée Baaras et produisant une racine du même nom. Sa couleur est semblable à celle du feu, et vers le soir, elle émet un rayon comme un éclair. Elle n'est pas prise aisément par ceux qui s'approchent d'elle et désirent l'enlever, mais elle se retire de leurs mains et n'est pas stationnaire jusqu'à ce que l'urine ou le sang menstrual d'une femme soient versés au-dessus d'elle. Même alors ceux qui la touchent rencontreront une mort certaine s'ils ne portent suspendue à la main une racine de la même espèce. Il y a aussi une autre méthode de l'ôter sans risque, et la voici. Les gens creusent le sol autour de la plante jusqu'à ce que la partie cachée de la racine devienne fort petite. Alors ils y lient un chien, et quand le chien suivra la personne qui l'a lié la racine est arrachée sans difficulté; mais le chien expire infailliblement, comme s'il était une victime au lieu de l'homme qui devait prendre la plante. Après cela, personne n'a besoin de craindre de la prendre dans ses mains. Cependant, après tous ces dangers qu'on court à

---

[1] *Flavii Josephi opera* graece et latine ed. G. Dindorfius, II, p. 316 (Parisiis, 1865).

[2] Une herbe de la famille *Rutaceae*, mentionnée par Luc (XI, 42). Plusieurs espèces sauvages croissent en Palestine, tandis qu'une espèce, *Ruta graveolens*, est cultivée.

l'obtenir, elle n'est recherchée qu'en considération d'une seule propriété qu'elle a, à savoir que, apportée à des malades, elle chassera vite les démons (qui ne sont autres que les esprits des méchants) qui entrent dans les hommes vivants et les tuent, s'ils ne peuvent pas obtenir de secours contre eux."[1]

La cité de Baaras était située en Syrie, sur le bord oriental de la mer Morte. Josèphe ne donne pas le nom de la plante, mais il n'y a pas de doute qu'il ait envisagé la mandragore qui existe en Palestine. Le motif de la racine arrachée par un chien paraît être d'origine orientale, et ensuite fut adopté par l'hellénisme lequel a absorbé tant d'idées orientales.[2]

Un conte semblable est raconté par Élien (*Hist. an.* XIV, 27) qui nomme la plante *cynospastus* (*κυνόσπαστος*, "déraciné par un chien")

---

[1] Ἐπεφύκαι δ' ἐν τοῖς βασιλείοις καὶ πήγανον ἄξιον τοῦ μεγέθους θαυμάσαι· συκῆς γὰρ οὐδεμιᾶς ὕψους καὶ πάχους ἀπελείπετο. Λόγος δ' ἦν ἀπὸ τῶν Ἡρώδου χρόνων αὐτὸ διαρκέσαι, κἂν ἐπὶ πλεῖστον ἴσως ἔμεινεν· ἐξεκόπη δ' ὑπὸ τῶν παραλαβόντων τὸν τόπον Ἰουδαίων. Τῆς φάραγγος δὲ τῆς κατὰ τὴν ἄρκτον περιεχούσης τὴν πόλιν Βαάρας ὀνομάζεταί τις τόπος, φύει τε ῥίζαν ὁμωνύμως λεγομένην αὐτῷ. Αὕτη φλογὶ μὲν τὴν χροιὰν ἔοικε, περὶ δὲ τὰς ἑσπέρας σέλας ἀπαστράπτουσα τοῖς ἐπιοῦσι καὶ βουλομένοις λαβεῖν αὐτὴν οὐκ ἔστιν εὐχείρωτος, ἀλλ' ὑποφεύγει, καὶ οὐ πρότερον ἵσταται πρὶν ἄν τις οὖρον γυναικὸς ἢ τὸ ἔμμηνον αἷμα χέῃ κατ' αὐτῆς· οὐ μὴν ἀλλὰ καὶ τότε τοῖς ἁψαμένοις πρόδηλός ἐστι θάνατος, εἰ μὴ τύχοι τις αὐτὴν ἐκείνην ἐπενεγκάμενος τὴν ῥίζαν ἐκ τῆς χειρὸς ἀπηρτημένην. Ἁλίσκεται δὲ καὶ καθ' ἕτερον τρόπον ἀκινδύνως, ὅς ἐστι τοιόσδε. Κύκλῳ πᾶσαν αὐτὴν περιορύσσουσιν, ὡς εἶναι τὸ κρυπτόμενον τῆς ῥίζης βραχύτατον, εἶτ' ἐξ αὐτῆς ἀποδοῦσι κύνα, κἀκείνου τῷ δήσαντι συνακολουθεῖν ὁρμήσαντος, ἡ μὲν ἀνασπᾶται ῥᾳδίως, θνήσκει δ' εὐθὺς ὁ κύων, ὥσπερ ἀντιδοθεὶς τοῦ μέλλοντος τὴν βοτάνην αναιρήσεσθαι. Φόβος γὰρ οὐδεὶς τοῖς μετὰ ταῦτα λαμβάνουσιν. Ἔστι δὲ μετὰ τοσούτων κινδύνων διὰ μίαν ἰσχὺν περισπούδαστος· τὰ γὰρ καλούμενα δαιμόνια (ταῦτα δὲ πονηρῶν ἐστιν ἀνθρώπων πνεύματα) τοῖς ζῶσιν εἰσδυόμενα καὶ κτείνοντα τοὺς βοηθείας μὴ τυγχάνοντας, αὕτη ταχέως ἐξελαύνει, κἂν προσενεχθῇ μόνον τοῖς νοσοῦσι.

[2] Je m'abstiens d'aborder le problème botanique. Dans la plupart des cas il est impossible d'insister sur une identification trop spécifique. *Mandragora officinalis, Atropa mandragora,* ou même *Atropa belladonna* ont été proposées comme les plantes comprises par les anciens à ce titre. Je ne crois pas cependant qu'une seule espèce y corresponde, car les mêmes idées pouvaient passer d'une plante à l'autre. On sait que la mandragore n'a jamais pénétré au-delà des Alpes sauf dans le midi de la France; néanmoins on a réussi à en trouver des substituts dans l'Europe centrale et septentrionale.

ou *aglaophotis* (*ἀγλαόφωτις*,[1] "lumière brillante"). Selon lui, la plante est cachée au-dessous parmi les autres herbes pendant le jour, tandis que de nuit elle devient visible et luisante comme une étoile, car elle rayonne et ressemble à du feu (*Φλογώδης γάρ ἐστι καὶ ἔοικι πυρί* = 熿 ou 晃). Par conséquent les gens attachent un signe distinctif à la racine et s'éloignent. Sans cette précaution, ils ne peuvent pas se souvenir au jour de la couleur ni de la figure de la plante. Mais ils n'ont pas coutume d'extraire ce végétal eux-mêmes, car on dit que celui qui l'a touché par ignorance de sa nature meurt quelque temps après. On conduit donc un chien jeune et robuste qui n'a point reçu de nourriture pendant quelques jours et qui a une faim violente; on le lie à une corde forte aussi loin que possible, et l'on fait un nœud difficile a dénouer, autour du bas de la tige de l'aglaophotis. Un repas opulent de viande rôtie, d'une odeur suave, est présenté au chien qui, poussé par la faim et attiré forcément par la bonne odeur de la viande, arrache la plante avec la racine. Quand le soleil regarde la racine, le chien mourra aussitôt. Les gens l'ensevelissent à la même place, et ayant rempli quelques cérémonies mystérieuses en honorant le cadavre du chien, parce qu'il a laissé sa vie pour eux, ils osent toucher le végétal et le portent chez eux. Ils l'emploient pour beaucoup de choses utiles, et à ce qu'on dit, ceux qui souffrent de l'épilepsie en sont guéris; elle est bonne aussi pour la maladie des yeux.

Le conte d'Élien, sans doute un peu loquace, n'est pas localisé, et est un peu exagéré: il n'y a guère de lieu pour le repas, à moins que ce ne fût un acte de charité. Le texte d'Élien qui vécut à Praeneste en Italie sans jamais quitter ce pays démontre que le conte fit sa migration de l'Orient en Italie.

Pline, afin d'illustrer les mensonges des magiciens anciens, dit que dans sa jeunesse le grammairien Apion lui parla de la plante

[1] Cf. Pline XXIV, 102.

cynocephalia ("tête de chien"), connue en Egypte sous le nom d'*osiritis*, utile pour la divination et préservatif contre tous les mauvais effets de la magie; mais si quelqu'un l'arrache du sol dans sa totalité, il mourra aussitôt.[1] C'est la même superstition que nous avons trouvé dans Josèphe et Élien, et ici même l'Orient (l'Egypte et les magiciens) paraît en prendre la responsabilité. Si le nom *cynocephalia*, qui avant tout se rapporte à la forme de la plante, permet d'établir un rapprochement avec le chien de Josèphe et d'Élien, c'est ce que je n'ose décider.[2]

La légende occidentale reproduite par Čou Mi présuppose évidemment une version d'origine islamique qui doit s'être répandue en Chine à l'époque des Song. En consultant la vaste compilation d'Ibn al-Baiṭār dans l'excellente traduction de L. Leclerc,[3] nous n'en trouvons pas de trace. Malheureusement, Leclerc a cru bon d'éliminer quelque chose de cet article, car il ajoute: "Quelques passages de ce chapitre, qui tranche par son caractère sur le ton général de l'ouvrage d'Ibn al-Baiṭār, nous ont paru devoir être supprimés." J'ai donc recouru à la traduction de Sontheimer, laquelle, comme on sait, est bien inférieure à celle de Leclerc à tous égards, et j'attends, d'ailleurs, la confirmation de ce texte par un arabisant. Selon Sontheimer,[4] Ibn al-Baiṭār mentionnerait le procédé avec le chien et ajouterait que lui-même en a été témoin, mais qu'il a trouvé faux que le chien y perde sa vie.

---

[1] Quaerat aliquis, quae sint mentiti veteres Magi, cum adulescentibus nobis visus Apion grammaticae artis prodiderit cynocephalian herbam, quae in Aegypto vocaretur osiritis, divinam et contra omnia veneficia, sed si tota erueretur, statim eum, qui eruisset, mori (XXX, 6, § 18).

[2] Dans un autre passage de Pline (VIII, 27, § 101) les fruits de la mandragore sont nuisibles aux ours qui lèchent des fourmis comme antidote (Ursi cum mandragorae mala ustavere, formicas lambunt); cf. Solinus (XXVI, 8): Cum gustavere mandragorae mala, moriuntur: sed eunt obviam, ne malum in perniciem convalescat et formicas vorant ad uperandam sanitatem.

[3] *Traité des simples*, II, p. 246—248.

[4] II, p. 14.

Dans la traduction de Leclerc l'auteur arabe fait dire à Hermès à propos de l'acquisition de la plante qu'on prétend que son extraction est difficile par la raison qu'il faut connaître le temps favorable à l'opération.[1] D'autre part, d'Herbelot[2] a révélé une version qui s'approche assez nettement du texte de l'écrivain chinois. "Luthf-Allah dit qu'il y a du danger d'arracher, ou de couper cette plante, et que pour éviter ce danger, quand on veut la tirer de terre, il faut attacher à sa tige un chien que l'on bat ensuite, afin que faisant des efforts pour s'enfuir, il la déracine." Voilà le trait de battre le chien, étranger à Josèphe et Élien, mais admis dans la version chinoise. Cependant un parallèle arabe plus complet et plus exact reste à chercher. D'ailleurs, autant que je sache, il n'y a pas beaucoup d'originalité dans les notices des Arabes sur la mandragore. Par exemple, tout ce qui est rapporté par Qazwīnī à ce sujet, comme l'a reconnu aussi G. Jacob,[3] n'est que l'écho des traditions hellénistiques. Qazwīnī a copié Avicenne (980—1037), et Avicenne a été répété par les historiographes européens des croisades et d'autres écrivains médiévaux. Enfin, les auteurs byzantins comme Théophane et Kedrenos ne font que reproduire les traditions des anciens.

Pour ce qui est des propriétés lumineuses de la plante, nous les avons vues accentuées par Josèphe et Élien. Le chérif el-Edrisy fait remarquer: "On donne à cette plante le nom de *sirāj el-kotrob*, parce que le *kotrob* est cette petite bête qui luit la nuit comme du feu. Cette plante est bien connue en Syrie où elle croît surtout non loin du littoral. La partie interne de l'écorce de sa tige luit la nuit, tant qu'elle reste humide, au point qu'on la croirait embrasée. Une fois desséchée, elle perd cette propriété. Si on la met

---

[1] L. Leclerc, *Traité des simples*, II, p. 247.

[2] *Bibliothèque orientale*, I, p. 72.

[3] *Studien in arabischen Geographen*, p. 165.

dans un linge mouillé, l'humidité lui rend cette lueur qu'elle perd en se desséchant."[1]

La forme anthropomorphique de la plante (plus correctement de la racine) sur laquelle insiste Čou Mi n'est pas relevée par les auteurs classiques. Dioscoride décrit la racine[2] sans mentionner cette qualité. Cependant, nous apprenons par une citation du *Codex neapolitanus* de Dioscoride que la racine de la mandragore était intitulée *ανθρωπόμορφος* dans l'ouvrage perdu du Pseudo-Pythagore sur les effets des plantes. De même, Columella (*De re rustica* X, 19, 20) en parle au terme *planta semihominis.*

Hermès est cité par Ibn al-Baiṭār comme disant: "La racine souterraine de cette plante a la forme d'une idole debout, avec des pieds et des mains et tous les organes de l'homme. Sa tige et ses feuilles, issues de la tête de cette idole, apparaissent à l'extérieur, et les feuilles ressemblent à celles de la ronce. Elle s'attache aussi aux plantes qui l'avoisinent et s'étale par-dessus."[3]

La qualité soporifique de la plante est signalée par Aristote (*De somno et vigilia*), Théophraste (*Hist. plant.* IX, 9, 1) et Xénophon (*Symp.* II, 24). Dioscoride (IV, 76) dit qu'elle fournit un suc endormant, étourdissant ou même mortel, employé par les médecins comme anesthésique sous forme de vin pour les opérations chirurgicales et qu'elle s'atteste comme aphrodisiaque efficace.

Lucien fait deux allusions à cet effet du remède: "tu dors, comme assoupi par de la mandragore"; et Démosthène réveille, malgré eux, ses concitoyens assoupis comme s'ils avaient bu de la mandragore.[4]

---

[1] L. Leclerc, *Traité des simples,* II, p. 247.

[2] Les racines sont très longues, au nombre de deux ou trois, intriquées l'une dans l'autre, noires en dehors, blanches en dedans et recouvertes d'une écorce épaisse (L. Leclerc, *Traité des simples,* III, p. 419); mais Pline et Dioscoride sont d'accord pour rapporter que la plante se présente sous deux sexes, mâle et femelle.

[3] L. Leclerc, *Traité des simples,* II, p. 247.

[4] E. Talbot, *Oeuvres complètes de Lucien de Samosate,* I, p. 31; II, p. 474 (*Tim.* 2, *Dem. Enc.*, 36).

Pline aussi en signale la force soporifique, mais la dose devait être réglée proportionnellement à la vigueur du malade. De plus, on la buvait contre des morsures de serpents et pour assurer l'insensibilité avant des opérations; l'odeur en suffisait à quelqu'uns pour produire le sommeil.[1] Théosphraste[2] dit qu'elle induit en sommeil, mais que donnée en plus grande quantité, elle est mortelle (*οἱ δ'ὑπνωτικοὶ πλείους δὲ διδόμενοι καὶ θανατηφόροι καθάπερ ὁ μανδραγόρας*). D'après Celsus (III, 18), les anciens avaient l'habitude de mettre le fruit de la plante sous leurs oreillers pour hâter le sommeil.

Hermès, cité par Ibn al-Baiṭār, dit que c'est une plante bénie entre toutes et qu'elle est utile contre toutes les maladies qui affligent l'homme par le fait des génies, des démons (cf. Josèphe) et de Satan. Elle est salutaire aussi contre les graves affections internes, telles que la paralysie, le tic nerveux, l'épilepsie, l'éléphantiasis, l'aliénation mentale, les convulsions et la perte de la mémoire.[3]

Le vin mentionné par Čou Mi et Dioscoride, dans lequel on a fait infuser des racines de mandragores s'appelait mandragorite (Littré). En italien c'est *mandragolato*. L'usage de ce terme remonte jusqu'à Dioscoride (V, 81: *ὁ μανδραγορίτης οἶνος*). Théophraste[4] a déjà fait observer que la racine est administrée dans du vin ou du vinaigre (*διδόασι δ'ἐν οἴνῳ ἢ ὄξει*). Le médecin Galène (131—204) fait remarquer que l'extrait de mandragore, aussi bien que le vin qu'il servait à préparer, étaient chaque an apportés de Crète à Rome. Ajoutons le texte de l'évêque Isidore (Isidorus Hispalensis, ca. 570—636), inséré dans ses *Originum sive etymologiarum libri XX* (XVII, 9): "Mandragora dicta, quod habeat mala suaveolentia in

---

[1] Vis somnifica pro viribus bibentium; media potio cyathi unius. Bibitur et contra serpentes et ante sectiones punctionesque, ne sentiantur; ob haec satis est aliquis somnum odore quaesisse (XXV, 94, § 150).

[2] *De causis plantarum*, VI, 5.

[3] L. Leclerc, *Traité des simples*, II, p. 246.

[4] *Historia plantarum*, IX, 9, 1.

magnitudinem mali Martiani; unde et eam Latini malum terrae vocant. Hanc poetae *ἀνθρωπόμορφον* appellant, quod habeat radicem formam hominis simulantem. *Ἄνθρωπος* enim graece, latine dicitur homo. Cuius cortex vino mixtus ad bibendum datur iis quorum corpus propter curam secandum est, ut soporati dolorem non sentiant. Huius species duae: foemina, foliis lactucae similibus, mala generans in similitudinem prunorum; masculus vero folia betae similia habet." [1]

Nous devons tourner maintenant vers une autre idée attachée à la mandragore, qui ne se trouve pas chez Čou Mi, mais qui se manifeste dans un autre groupe de traditions chinoises. Maimonides (1135—1204) dit à propos du livre *L'Agriculture des Nabatéens* [2] que Adam dans son livre fit mention d'un arbre dans l'Inde, les branches duquel rampent comme un serpent, quand on les jette sur terre; et, de même, d'un autre arbre, la racine duquel a la forme d'un homme et une haute voix et prononce des paroles intelligibles. [3]

Nous lisons dans la matière médicale d'Ibn-al-Baiṭār (1197—1248) sur la plante *luf* لوف (*Arum dracunculus*): "Il y en a trois espèces. L'une s'appelle en grec *dracontion*, ce qui veut dire *arum serpentaire*, الحية, à cause que sa tige tachetée ressemble à une peau de serpent. C'est l'arum long, مستطيل, le grand arum, لوف كبير. Nos compatriotes en Espagne lui donnent le nom de *gargantīa* غرغنتية. D'autres l'appellent *sarrākha* صرّاخة, parce qu'ils prétendent qu'elle jette un cri, *sarkha*, que l'on entend le jour du Mihrijān, c'est-à-dire

[1] L'idée que la mandragore hâte la propagation émane pour la première fois du Physiologus (chap. XIX), où la plante est localisée près du paradis, étant cherchée et mangée par les éléphants avant de s'accoupler. Je ne poursuis pas cette piste ici, parce que cette notion ne joue pas de rôle dans la tradition chinoise.

[2] Cf. E. Renan, *An Essay on the Age and Antiquity of the Book of Nabathæan Agriculture* (London, 1862); A. v. Gutschmid, *ZDMG*, XV, p. 1, et Nöldeke, *ibid.*, XXIX, p. 445. On sait que ce livre (*Falāha nabaṭiya*) qui prétend d'être une traduction arabe d'une ancienne source nabatéenne est une forgerie du dixième siècle.

[3] D. Chwolson, *Ssabier*, II, p. 458.

le jour de la Pentecôte, et, de plus, que celui qui l'entend mourra dans l'année."[1] La même observation est aussi faite par Ibn el-'Awwām de Séville, qui écrivit dans la première moitié du VIe siècle de l'hégire le *Kitāb el-falāha* (Livre de l'agriculture).[2] L'analogie de ce cas avec la mandragore est frappante, et il s'agirait de savoir si le trait de la plante qui pousse un cri et cause la mort d'un homme était à l'origine propre à l'arum, c'est-à-dire, appartenait à un autre cycle de traditions, et a passé de là à la mandragore, ou inversement. En tout cas cette notion légendaire paraît bien être d'origine orientale. Autant que je sache, Maimonides ou plutôt l'œuvre apocryphe qu'il cite présente la source la plus ancienne qui contienne la combinaison de cette attribution avec la mandragore. Dès ce temps-là ce motif ne tarda pas d'être vulgarisé: le cri poussé par la racine de la mandragore au moment qu'elle est arrachée au sol devient fatal à l'auditeur. Le plus fameux passage de ce genre se trouve dans Shakespeare, *Romeo and Juliet* (IV. 3, 47):

> And shrieks, like mandrake's torn out of the earth,
> That living mortals, hearing them, run mad.

Dans *King Henry VI* (II. 3, 2) Suffolk dit à la reine:

> Would curses kill: as doth the mandrake's groan.[3]

---

[1] L. Leclerc, *Traité des simples*, III, p. 248.

[2] C. Huart, *Littérature arabe*, p. 313. L'ouvrage d'Ibn el-'Awwam a été traduit en français par J.-J. Clément-Mullet (*Ibn al Awwâm, livre de l'agriculture*, 2 vols., Paris, 1864—1867). Malheureusement je n'ai pas accès à cette traduction; j'ai tiré le fait en question de I. Löw, *Aramäische Pflanzennamen*, p. 239.

[3] Dans plusieurs autres passages, Shakespeare fait allusion à la mandragore.

> Not Poppy, nor Mandragora,
> Nor all the drowsie Syrrups of the world
> Shall ever medicine thee to that sweet sleep
> Which thou owedst yesterday.
>
> *Othello*, III. 3, 330.
>
> Give me to drink mandragora...
> That I might sleep out this great gap of time.
>
> *Anthony and Cleopatra*, I. 5.

Dans *King Henry IV* (II. 1, 2), Falstaff appelle son petit page "whoreson mandrake"; le

2

Mais hâtons-nous d'ajouter que cette tradition est strictement médiévale. C'est par inadvertence que G. E. Post [1] fait observer, "The ancients also believed that this root gave a demoniacal shriek as it was pulled up." Il n'en est rien: rien de pareil dans aucun document de l'antiquité.

Cette idée bizarre, d'où vient-elle? Nous avons vu que Čou Mi compara la mandragore avec le ginseng (*Panax ginseng*), fameuse panacée de sa patrie. D'autre part, le nouveau dictionnaire anglais d'Oxford régistre le terme "Chinese mandragoras" au sens de ginseng, et le dictionnaire persan-anglais de Steingass donne cette définition de l'expression *mardum-giyā* مردم گیا: "a plant, the produce of China, said to resemble a man and woman, and to which many wonderful effects are attributed; mandrake, colocynth." De cette manière, le mot persan désigne la mandragore aussi bien que le ginseng d'origine chinoise. C'était le P. Martini (1655) qui rapprocha le dernier à la mandragore: "Je ne sçaurois mieux representer cette racine qu'en disant qu'elle est presque semblable à nostre Mandragore; hormis que celle-là est un peu plus petite quoyqu'elle soit de quelcune de ses especes. Pour moy je ne doute point du tout, qu'elle n'ayt ces mesmes qualités et une pareille vertu; puisqu'elle luy ressemble si fort et qu'elles ont toutes deux la mesme figure" [suit une assez longue description de la racine et de ses propriétés]. [2] De même

juge Shallow recevait dans sa jeunesse le sobriquet "mandrake" ("when he was naked, he was...like a forked radish with a head fantastically carved upon it with a knife"; *ibid.*, III. 2). Enfin le passage dans *Macbeth* (I. 3, 84)

Or have we eaten of the insane root
That takes the reason prisoner?

paraît contenir une allusion à la mandragore.

[1] Dans le *Dictionary of the Bible* de J. Hastings, III, p. 234.

[2] A. Kircher, *La Chine illustrée*, p. 241 (Amsterdam, 1670). On voit ainsi que le ginseng était connu en Europe au XVIIe siècle. Je ne m'arrête pas à cette matière sur laquelle tant a été écrit. Il suffit de renvoyer le lecteur à Bretschneider, *Botanicon sinicum*, 3e partie, no. 3; Du Halde, *Description de l'empire de la Chine*, II, p. 150 (ce mémoire est dû au P. Jartoux); *Mémoires concernant les Chinois*, II, p. 428; et voir la bibliographie

que la mandragore, le ginseng est anthropomorphisé et doué de langage par les Chinois. L'ouvrage ancien *Pie lu* 別錄 dit que sa racine est comme la figure de l'homme et a des qualités divines (根如人形者有神); et le *Wu pu p'en ts'ao* 吳普本草, écrit au troisième siècle, attribue à la racine des mains, des pieds et des yeux, tout comme chez l'homme, et la range parmi les choses spirituelles (根有手足而目如人者神).[1] Ensuite le ginseng est capable de crier. Le document le plus ancien à cet égard qui me soit connu est contenu dans les Annales de la dynastie Soui, où nous lisons: "Au temps de Kao Tsu (ou Wen Ti, 590—604)

---

dans H. Cordier, *Bibliotheca sinica*, col. 2969, 3085—6. — L'observation du P. Martini fut relevée par J. F. Lafitau (*Mémoire presenté à son altesse royale Monseigneur le Duc d'Orleans, regent du royaume de France; concernant la précieuse plante du ginseng de Tartarie, découverte en Canada*, 88 p., petit 8°, Paris, chez J. Monge', 1718), missionnaire Jésuite parmi les Iroquois, qui, après avoir lu le mémoire de Jartoux sur le ginseng chinois, découvrit une semblable espèce au Canada. Il dit (p. 71): "Quand j'eus découvert le ginseng, il me vint en pensée que ce pouvoit être une espece de mandragore. J'eus le plaisir de voir que je m'étois rencontré sur cela avec le Pere Martini, qui dans l'endroit que j'ai cité, et qui est rapporté par le Pere Kirker [sic], parle en ces termes. Je ne sçaurois mieux representer cette racine, qu'en disant qu'elle est presque semblable à notre mandragore, hormis que celle-là est un peu plus petite, quoi qu'elle soit de quelqu'une de ses especes. Pour moi, ajoute-t-il, je ne doute point du tout qu'elle n'ait les mêmes qualitez et une pareille vertu, puisqu'elle lui ressemble si fort, et qu'elles ont toutes deux la même figure." Lafitau a raison dans sa critique qui suit: "Si le Pere Martini a eu raison de l'appeller une espece de mandragore à cause de sa figure, il a eu tort de l'appeller ainsi à cause de ses proprietez. Nos especes de mandragore sont narcotiques, rafraîchissantes, et stupéfiantes. Ces qualitez ne conviennent point du tout au ginseng." Alors Lafitau s'efforce de démontrer que la mandragore des anciens n'est pas identique à notre mandragore d'aujourd'hui. Une autre curiosité de l'opuscule de Lafitau c'est qu'il rapproche le nom iroquois du ginseng canadien, *garent oguen* (qu'on dit signifier "cuisses, jambes" + "deux choses séparées") au mot chinois traduit par lui "ressemblance de l'homme". Il en conclut que "la même signification n'avoit pû être appliquée au mot Chinois et au mot Iroquois sans une communication d'idées, et par consequent de personnes. Par là je fus confirmé dans l'opinion que j'avois déja, et qui est fondée sur d'autres préjugez que l'Amerique ne faisoit qu'un même continent avec l'Asie, à qui elle s'unit par la Tartarie au nord de la Chine." Tout cela est excusable et intelligible, eu égard à l'état de la science au temps où vivait l'auteur.

[1] Je ne crois pas que la traduction de Bretschneider ("has hands, feet, a face and eyes like a man possessed of a god") soit correcte; le mot 神 ne se rapporte qu'à la racine même.

il y eut un homme à Šaṅ-taṅ [1] derrière la maison duquel on entendait chaque nuit la voix d'un homme. On le cherchait, mais sans le trouver. En s'écartant un *li* de la maison, tout ce qu'on aperçut fut une plante de ginseng avec les branches et les feuilles hautes et bien développées. On la déracina et on trouva que la racine avait plus de cinq pieds de long, et que toute sa forme imitait le corps d'un homme. Depuis ce moment les cris cessèrent." [2] A en croire le *P'ei wen čai kwaṅ k'ün faṅ p'u* 佩文齋廣羣芳譜 (chap. 93, p. 5 b) il y a encore un texte plus ancien à relever ce trait, le *I yüan* 異苑, attribué à Liu King-šu 劉敬叔 du cinquième siècle; mais n'ayant pas à ma disposition une édition de cet ouvrage, je laisse de côté la question chronologique. Liu Kiṅ-šu dit: "Anciennement il y eut un homme qui, en fouillant le sol, y introduisit sa bêche. Puis il entendit dans la terre des soupirs, et en recherchant le son, obtint de fait un ginseng." [3] Rappelons aussi le fait que les Chinois se servent de ginseng comme aphrodisiaque.

Ces coincidences étant constatées, les ressemblances entre les traditions de la mandragore et du ginseng sont épuisées, et les différences, au contraire, sont plus nombreuses et plus fondamentales. Le ginseng n'est pas une plante vénéneuse, elle rétablit la vie et ne donne jamais la mort comme la mandragore. Il n'est pas dangereux ou fatal de recueillir du ginseng qui n'est point devenu objet de magie. Son cri paraît comme un développement logique

[1] La partie du Šan-si sud-est, toujours fameuse pour son excellent ginseng.

[2] 高祖時上黨有人宅後每夜有人呼聲。求之不得。去宅一里所但見人參一本。枝葉峻茂。因掘去之。其根五尺餘。具體人狀。呼聲遂絶. — *Sui šu*, chap. 32, p. 1. Le *Pen ts'ao kaṅ mu* (chap. 12 A, p. 4 b) a tiré le même texte du *Kwaṅ wu hiṅ ki* 廣五行記, ouvrage du temps des Song.

[3] 昔有人掘之始下鏵便聞土中呻吟聲尋音而取果得人參.

de sa caractéristique anthropomorphe, et qui plus est, n'envoie pas un homme à la tombe. En effet, les Chinois n'ont rien emprunté de cela aux peuples occidentaux; une telle théorie se heurterait sérieusement contre la chronologie. L'anthropomorphisme et la faculté de parler du ginseng sont d'une date plus ancienne en Chine que les notions analogues de la mandragore à l'ouest; et selon toute apparence, la connaissance de la mandragore n'y est pas arrivée avant l'époque des Song. Mais s'il est vrai que le ginseng était un objet de commerce de la Chine à la Perse, la question se pose si le cri de la mandragore qui fait son début au moyen âge n'est pas le résultat direct des contes chinois concernant le ginseng. [1]

Le fait rapporté par Čou Mi que des racines de mandragore étaient importées en Chine aux temps des Song et effectivement employées n'est pas moins intéressant. Cependant il est frappant que ni Čou K'ü-fei ni Čao Žu-kwa ne paraissent connaître ce commerce. [2]

Mais Čou K'ü-fei 周去非 nous a laissé une anecdote sur une autre plante apparentée à la mandragore quant à la composition et à l'effet de son poison et qui pour cela ne manque pas de piquer notre curiosité. Aussi nous donnera-t-elle occasion de formuler

---

[1] Il y a d'autres plantes les racines desquelles sont conçues par les Chinois comme anthropomorphes, par exemple, *Phytolacca acinosa*, *šaṅ lu* 商陸 (cf. Bretschneider, *Chinese Recorder*, III, 1871, p. 219; *Bot. sin.*, II, no. 112, III, no. 131), décrite par le *Pie lu* avec les mêmes expressions que le ginseng (如人形者有神) et appelée aussi *ye hu* 夜呼 ("criant de nuit"). Mais le *Pen ts'ao kaṅ mu* ne contient pas de texte qui fasse allusion à la faculté de crier qu'aurait la racine. Minakata (voir *supra*) ne donne à cet effet qu'un texte écrit en 1610, le *Wu tsa tsu* 五雜俎.

[2] La plante *laṅ-tu* 狼毒 a été identifiée avec une mandragore par Bretschneider (*Bot. sin.*, III, no. 132), qui fonda cette opinion sur un dessein japonais, mais Stuart (*Chinese Materia Medica*, p. 257) regarde cette identification comme douteuse et la description dans les sources chinoises comme insuffisante; à l'avis du même auteur (p. 58) il est douteux aussi que le genre *Atropa* se trouve en Chine. Forbes et Hemsley (*Journal Linnean Society*, XXVI, p. 175) en registrent une *Mandragora caulescens* au Yun-nan d'après Franchet (*Bull. Soc. Bot. de France*, XXXII, p. 26). Quoi qu'il en soit, il est certain qu'aucune mandragore n'est connue à la pharmacopée chinoise.

quelque conclusion à propos du nom *mandragore* lui-même. Dans son *Liṅ wai tai ta* 嶺外代荅, écrit en 1178, l'auteur chinois rapporte ainsi: "La fleur *man-tʻo-lo* de la province de Kwang-si croît partout dans l'état sauvage. Ses feuilles sont larges, les fleurs blanches, et la formation des fruits est comme chez l'aubergine ou la mélongène (*Solanum melongena*). Elle forme partout des petits piquants, et c'est une plante qui sert de remède aux hommes.[1] Des voleurs cueillent la plante, la sèchent et broient. Ils la placent de manière que des hommes la boivent ou mangent; et en ce cas ils en deviennent ivres. Pendant qu'ils sont dans cet état de torpeur, les brigands enlèvent leurs cassettes et prennent la fuite. Les hommes au midi de la Chine se servent de ce remède aussi pour les petits enfants et en amassent de grandes quantités."[2]

Le nom *man-tʻo-lo* 曼陀羅 est contenu dans le *Fan yi miṅ i tsi* (chap. 8, p. 6) et équivaut au sanskrit *mandara*, *mandāra*, *mandāraka*.[3] Il est assez étonnant qu'une plante non-cultivée, qui d'après Li Ši-čen croît aussi au nord de la Chine, soit appelée d'un terme sanskrit. Elle n'apparaît pas dans les documents avant l'époque des Song,[4]

[1] Le mot 藥 a ici la fonction verbale. Cf. 益人草 "une herbe qui fait du bien à l'homme"; 毒人草 "une herbe qui empoisonne l'homme".

[2] 廣西曼陀羅花偏生原野大葉白花結實如茄子而偏生小刺乃藥人草也。盜賊採乾而末之以置人飲食使之。醉悶則挈篋而趍。南人或用爲小兒食藥去積甚峻. — *Liṅ wai tai ta*, chap. 8, p. 14 b; éd. du *Či pu tsu čai tsʻuṅ šu*.

[3] Voir aussi Eitel, *Handbook of Chinese Buddhism*, p. 94.

[4] Du moins pas de texte d'une date plus ancienne m'est-il connu. Le *Tʻu šu tsi čʻeṅ* (section botanique, chap. 124), sous le titre *man-tʻo-lo*, ne fait que citer la notice du *Pen tsʻao kaṅ mu*, puis une brève remarque de Čʻen Yü-i 陳與義 des Song, un conte tiré du *Tʻan Yüan* 談苑 par Yaṅ I 楊億, qui vécut au commencement de l'onzième siècle et collabora au *Tsʻe fu yüan kwei*, et une note très courte du *Lo yaṅ hwa mu ki* 洛陽花木記 ("Mémoires des plantes de Lo-yaṅ"), écrit par Čou Sü 周叙 dans la seconde moitié de l'onzième siècle. Le texte le plus important du *Liṅ wai tai ta*

et pour cela est suspecte d'avoir été importée de l'Inde, quoique le fait d'une telle importation ne soit pas relevé par les textes. La plante se rapporte au genre *Datura*, mais il n'est pas certain si c'est l'espèce *alba* ou *stramonium.*[1] C'est une solanée comme la

---

y est omis. Le *T'u šu tsi č'eṅ* contient aussi un dessin de la plante. Č'en Hao-tse 陳淏子, dans son *Pi fu hwa kiṅ* 秘傳花鏡 (chap. 5, p. 37 b) de l'an 1688, décrit le *man-t'o-lo* comme une fleur du nord de la Chine et dit que le nom est sanskrit.

[1] Stuart, *Chinese Materia Medica*, p. 145—147. "Le Datura stramonium, ou Pomme épineuse [anglais *thorn-apple*], appelé aussi stramoine, endormie, herbe aux sorciers, herbe aux diables, croît communément en France, mais il se rencontre également dans presque toutes les parties du monde, à l'exception de l'Australie; on pense qu'il est originaire de l'Amérique et qu'il s'est propagé de là en Europe. Cependant on le trouve depuis des siècles, en France, en Grèce, dans la région caucasique, dans la Syrie" (A. Milne Edwards, *De la famille des solanacées*, p. 87, Paris, 1864). — Les auteurs chinois de l'*English and Chinese Standard Dictionary* étaient conduits par un sentiment juste, en se servant du mot *man-t'o-lo* pour traduire l'anglais *mandrake*. S. Couvreur (*Dictionnaire français-chinois*, p. 369) donne deux termes *su-hwa* pour *Datura*: *šan ma-tse* 山麻子 ("chanvre sauvage") et *la-pa hwa* 喇叭花 ("fleur à trompette"). Le datura est connu aux Čams sous le nom *salak* et aux Khmers sous le nom *slak* (Aymonier et Cabaton, *Dictionnaire čam-français*, p. 481). Notre mot *datura* est ramené au sanskrit *dhattūra* par Yule (*Hobson-Jobson*, p. 298); Hindī et Hindustānī *dhatūra*, persan *dātūra* داتوره. La mention la plus ancienne du datura qui me soit connue dans la littérature européenne vient de Pierre Belon du Mans, qui dit dans son œuvre *Les Observations de plusieurs singularitez et choses memorables, trouuées en Grece, Asie, Iudée*, etc., fol. 369 (Anvers, 1555) [cf. *T'oung Pao*, 1916, p. 362]: "Les Turcs ont des merueilleuses experiences de plusieurs choses, comme pour faire dormir soudainement. Voudroit on chose plus singulière que de trouurer drogue pour faire incontinent dormir qu'elqu'vn qui ne peut reposer? Ils vont chez vn droguiste (car ils n'ont point d'Apoticaires) auquel demandent pour demie aspre de la semence de Tatoula. Puis la baillent à celuy qui ne peut dormir. Tatoula n'est autre chose que ce que les Arabes appellent Nux metel, et les Grecs Solanum somniferum: de laquelle nous en trouuasmes de sauuage en la plaine de Iericho, prés la fontaine d'Helisee." Le mot *tatoula* est Osmanli *dadula* طاطوله (néo-grec *τάτουλας*), évidemment dérivé du persan. Cette forme du nom n'est pas notée par Littré qui ne donne que datura et le dérive de l'arabe *datora* et du persan *tatula*, en ajoutant "du radical *tal*, piquer, par allusion à l'enveloppe épineuse du fruit." Vu le mot sanskrit, cette étymologie semble être caduque. Christoual Acosta (*Tractado delas drogas y medicinas de las Indias Orientales*, p. 87, Burgos, 1576) s'exprime ainsi: "Lhamase esta planta en el Malabar, Vnmata [Sanskrit *unmatta*] caya: en Canarin, Datyro: los Arabes, Nux Methel, y Marana: los Portugueses, Datura, y la Burladora: los Parsios, y Turcos, Datula: los medicos Indianos graduan esta planta fria, enel grado tercero, y seca enel fin del segundo." Acosta donne une gravure de la plante et contribue des observations intéressantes sur son emploi dans l'Inde et l'Espagne. Le mot *metel* du terme botanique *Datura metel*, originaire de

mandragore, et comme toutes les solanées, contient l'alcaloïde daturine ou atropine, $C_{17} H_{23} O_3$. L'analogie des contes de Čou Mi et de Čou K'ü-fei, bien qu'ils se rapportent à des plantes différentes, est due à la composition chimique analogue et à la même action des deux poisons. Encore de nos jours, les Chinois ont employé cette substance funeste pour des buts artificieux. Crawfurd [1] nous informe que *kučubuṅ* (le mot soundanais pour *Datura ferox*) [2] est donné par les Malais pour produire la plus complète stupeur et "is a powerful engine in the hands of the Chinese for effecting various artifices and tricks in trade." On dit que dans quelques parties de la Chine *Datura alba* s'emploie pour stupéfier et saisir des poissons. [3] La désignation propre de la dernière espèce est *nao-yaṅ* 鬧羊; d'autrui identifient ce terme avec *Datura metel*. Les fleurs, digérées dans le vin, servent d'anesthésique et sont indiquées dans la chorée des enfants; on en fait aussi des lotions contre les éruptions de la face, l'enflure des pieds et la chute du rectum. [4]

---

l'Inde, est dérivé du sanskrit *mātula*, peut-être apparenté à *matta* et *unmatta* ("enivré, insensé", et aussi "datura"). Cette espèce est notée par Loureiro (*Flora Cochinchinensis*, p. 135) pour l'Indochine sous le nom *nao yaṅ hwa* 鬧陽花. *Datura ferox* était connu à Linné comme une plante chinoise; elle est commune dans la Chine septentrionale (Bretschneider, *Early European Researches into the Flora of China*, p. 104). Une autre espèce, *Datura meteloides*, connue en Amérique, est employée par les Indiens Zuñi et Mohave (M. C. Stevenson, *Ethnobotany of the Zuñi Indians, Thirteenth Annual Report Bureau of American Ethnology*, 1908—09, p. 46; W. E. Safford, *Proceedings of the Nineteenth Internat. Congress of Americanists*, p. 28, Washington, 1917).

[1] *History of the Indian Archipelago*, I, p. 466.

[2] Javanais *kačubuṅ*, malais *kečubuṅ*. D'après l'*Encyclopædie van Nederlandsch-Indië* (II, p. 204) ce mot se rapporterait au *Datura alba*.

[3] C. Ford, *Flora of Hainan* (*China Review*, XX, p. 161). Le même auteur fait remarquer que cette plante s'appelle à Hoihow *mui-twa-lo* 門山蘿, évidemment une tentative dialectale de reproduire le mot étranger *man-t'o-lo*. Forbes et Hemsley (*Journal Linnean Society*, XXVI, p. 175) disent que *Datura alba* se trouve dans la Chine méridionale et à Formose, et est cultivée à Peking.

[4] J. L. Soubeiran et Dabry de Thiersant, *La Matière médicale chez les Chinois*, p. 190 (Paris, 1874). Dans *An Epitome of the Reports of the Medical Officers to the Chinese Imperial Maritime Customs Service, from 1871 to 1882*, compilé par C. A. Gordon

E. Perrot et P. Hurrier, [1] deux pharmaciens français qui ajoutent à la nomenclature chinoise le nom japonais *mondaraque*, donnent les renseignements suivants: "Les grains de ce *Datura*, irrégulièrement triangulaires et dont la forme a été comparée à celle de l'oreille humaine, sont d'un brun jaunâtre clair, rugueuses, déprimées au centre. Dans l'Inde, elles servent à préparer un extrait et une teinture très estimés comme narcotiques et sédatifs. Les feuilles s'emploient topiquement comme calmantes. Les fleurs, digérées dans le vin, jouissent d'une grande réputation dans l'épilepsie et l'hydropisie." En effet, plusieurs espèces de *Datura* (*fastuosa*, *metel*, et *stramonium*) croissent dans l'Inde. [2]

C'est dans l'Inde que nous rencontrons aussi le prototype des brigands de Čou K'ü-fei. Nous savons par Garcia da Orta (1563) que les thugs indiens mettaient cette drogue dans la nourriture de leurs victimes, et que l'effet en durait vingt-quatre heures; ceux qui prennent cette médecine perdent leurs sens, rient toujours et sont très généreux, car ils laissent les gens enlever quelconque joaillerie qu'ils choisissent, et ne font que rire ou parlent très peu, et seulement des absurdités. [3] Les cas d'empoisonnement avec le *Datura* sont encore très fréquents dans l'Inde. Mais les fripons

---

(London, 1884), il est dit (p. 231): "The *datura* or *man-t'o-lo* of the Buddhist classics is foreign to China, having, it is said, been introduced from India. When eaten, unconscious laughter is set up, and the person acts as if intoxicated. It may be used as an anæsthetic. It is used in infusion to wash the feet; it is also applied to ulcers of the face, in convulsions of children, and in *prolapsus ani*." Voir aussi G. A. Stuart. *Chinese Materia Medica*, p. 145—147.

[1] *Matière médicale et pharmacopée sino-annamites*, p. 174 (Paris, 1907).

[2] W. Ainslee, *Materia Indica*, I, p. 442—446 (London, 1826); W. Roxburgh, *Flora Indica*, p. 188; G. Watt, *Commercial Products of India*, p. 487—489; Flückiger et Hanbury, *Pharmacographia*, p. 459—463. *Datura alba* est indigène dans l'Inde; il n'est pas certain si ceci est le cas pour *Datura stramonium* (A. de Candolle, *Géographie botanique*, II, p. 731). Toutefois cette espèce se trouve dans l'état sauvage à l'Himalaya de Kachmir à Sikkim.

[3] C. Markham, *Colloquies on the Simples and Drugs of India by Garcia da Orta*, p. 175.

chinois et indiens qui apparaissent si modernes et civilisés dans leurs méthodes et assez congéniaux à nos *chloroform burglars* ne peuvent se vanter d'une grande originalité. La ruse est vieille, hors que les anciens préparaient l'extrait non du *Datura*, mais de la mandragore; c'est toute la même chose. Frontin qui vécut sous les règnes de Vespasien et de ses fils, et mourut dans les premières années du règne de Trajan, raconte dans ses *Stratagèmes* l'anecdote suivante: "Maharbal, envoyé par Carthage contre les Africains révoltés, sachant cette nation très-portée pour le vin, en fit mêler une grande quantité avec de la mandragore, substance qui tient le milieu entre un poison et un soporifique; puis, après une escarmouche, il se retira. Vers le milieu de la nuit il fit semblant de prendre la fuite, laissant quelque bagage et tout le vin empoisonné. L'ennemi se jeta dans le camp; et là, dans la joie de la victoire, ayant bu avec excès de cette mixtion, tandis qu'ils étaient étendus par terre comme des corps morts, Maharbal revint sur ses pas, et en fit un grand massacre". [1] Polyen (Polyainos) de la Macédoine, qui vécut à Rome sous les règnes de Marc Aurel et L. Verus, dit dans son *Strategika* (VIII, chap. XXIII, 1) que le jeune César, en voyage pour l'Orient, tomba dans les mains de pirates ciliciéns pas loin du cap Malea. Il fit venir la rançon demandée de Milet et au même temps un pot rempli d'épées et une quantité de vin empoisonné avec de la mandragore. Il en régala les pirates et ordonna qu'ils fussent massacrés dans leur assoupissement. Dans un autre passage du même ouvrage (V, chap. X, 1) Polyen rapporte un conte sem-

---

[1] Maharbal, missus a Carthaginiensibus adversus Afros rebellantes, quum sciret, gentem avidam esse vini, magnum eius modum mandragora permiscuit, cuius inter venenum ac soporem media vis est. Tunc, proelio levi commisso, ex industria cessit: nocte deinde intempesta, relictis intra castra quibusdam sarcinis, et omni vino infecto, fugam simulavit: quumque barbari occupatis castris, in gaudium effusi, medicatum avide merum hausissent, et in modum defunctorum strati jacerent, reversus aut cepit eos, aut trucidavit (*Strategematicon* II, chap. V, 12). J'ai reproduit la traduction de Th. Baudement dans l'édition d'*Amien Marcellin, Jornandès, Frontin*, etc., par M. Nisard, p. 536 (Paris, 1851).

blable à celui de Frontin à propos du général carthaginien Himilco.

Mais retournons à l'Inde. La connaissance du *dhattūra* y remonte à une époque reculée, car la plante est plusieurs fois mentionnée par Suçruta. Je dois à l'obligeance du Dr. A. F. R. Hoernle d'Oxford les renseignements suivants:

"There are the following references to *dhattūra* in the text-book of Suçruta:—

1. *Cikitsāsthāna*, ch. XVII, verse 35, p. 435 (Jīvīnanda, 5th ed.), where pounded seeds of *dhattūra* and other drugs (*madana*, *kodrava*, etc.) are recommended in the treatment of a sinus (*nāḍī*).

2. *Kalpasthāna*, ch. VI, verse 49, p. 589, roots (or, according to others, seeds) of *dhattūra*, made up, with other drugs, into a paste, is recommended as a *çirovirecana* ('clearing of the head') in the case of hydrophobia.

3. *Ibidem*, verses 51, 52, root of *dhattūra*, made, with other drugs, into a paste, wrapped in leaves of *unmattaka* (synonyme of *dhattūra*), and baked into a cake (*apūpaka*), is recommended in the case of bite by a rabid dog.

"In the text-book of Caraka, *dhattūra* does not occur at all. For the occurrence of *dhattūra* in medical text-books and later authors, such as Cakrapāṇidatta, c. A.D. 1060, and Bhāva Miçra (*Bhāvaprakāça*), 16. cent. A.D., see U. C. Dutt's *Materia Medica*, pp. 207—8.

"The Sanskrit word is spelled variously *dhattūra*, *dhuttūra*, *dhūsstūra*, etc.

"You identify *dhattūra* with *Datura stramonium*; but, as you know, our botanical books mention other varieties also. Sanskrit medical text-books distinguish two varieties, black, or rather dark, flowered, and white flowered (see *Mat. Med.*, p. 202); and they have also two names, *dhattūra* and *unmatta* (or *unmattaka*). The former is expressly called *kṛṣṇa-puṣpa*, 'dark-flowered' (see *Rāja Nighaṇṭu*, 10. varga, p. 135), and appears to be the one usually intended to be used. But *unmatta* seems to be the proper name of the white variety, indicated by a prescription of Cakrapāṇidatta (*Mat. Med.*, p. 207, note 2), where the name *çvetonmatta*, or 'white *dhattūra*' occurs. *D. stramonium* has white flowers, while *D. metel* and *D. fastuosa* have darker flowers. Accordingly the Sanskrit name *dhattūra* would seem to refer to the latter two varieties, while *unmatta* would seem to indicate the variety *D. stramonium*." [1]

---

[1] Le Dr. T. Tanaka au Bureau of Plant Industry, Department of Agriculture, Washington, a eu l'extrême obligeance de traduire pour moi les renseignements suivants sur *Mandara-kwa* (ou *Mandara-ge* selon la prononciation bouddhiste) 曼陀羅花, contenus dans le *Honzō-kōmoku-keimō* 本草綱目啓蒙 by Ono Ranzan 小野蘭山 (revue par Iguči Bōši 井口望之, 1847, chap. 13, p. 28—29):

"*Japanese Nomenclature:* *In Provinces:*

Japanese Nomenclature	In Provinces
Čōsen-asagao (Korean morning glory).	
Yama-nasubi (mountain egg-plant).	
Namban-asagao (morning glory of the Southern Barbarians).	
Hari-nasubi (spiny egg-plant).	Iyo.
Tō-nasubi (Chinese egg-plant).	Iyo.
Gekwa-korosi (sergeon killer).	Sanuki.
Gekwa-dausi (sergeon thrower).	Hōki, Iwami, Iyo.
Tenjiku-nasubi (Indian egg-plant).	Awa.
Iga-nasubi (prickly egg-plant).	Awa.
Giba-sō (meaning uncertain).	Buzen.
Čamera-so (meaning uncertain).	Edo (Tōkyō).
Kičigai-nasubi (insane egg-plant).	Iwami.
Awisu (meaning uncertain).	Bingo.
Iga-nasu (prickly egg-plant).	Nagato.
Ki-asagao (tree [or yellow?] morning glory).	Simoosa.
Čōsen-tabako (Korean tobacco).	Tōtōmi.
Tō-asagao (Chinese morning glory).	
Baramon-sō (herb of the Baramon 波羅門, that is, Brāhmaṇa).	

Chinese synonymes:

		quoted from
佛花	*Fu hwa* ('Buddha's flower').	和幼新書
顛茄	*Tien kie.*	香山縣志
悶陀羅草	*Men t'o lo ts'ao.*	名山勝概
天茄彌陀花	*T'ien k'ie mi t'o hwa.*	花曆百詠

"Spontaneous in the provinces Hōki, Buzen, and Suwō, but not grown in the prefectures

Ainsi l'histoire du genre *Datura* dans l'Inde est assez claire. Quant au mot *mandara*, nous avons noté que les Chinois et les Japonais le rapportent exclusivement au datura. En consultant le dictionnaire sanskrit de Boehtlingk, nous trouvons que *mandara*, *mandāra* ou *mandāraka* signifient en premier lieu *Erythrina indica*, [1] l'arbre de corail, un des cinq arbres du ciel d'Indra, appelé aussi *parijāta*, puis une variété blanche de *Calotropis gigantea*, et enfin la pomme épineuse, c'est-à-dire le genre *Datura*. A l'égard de ces identifications, il est évident que le terme *mandara*, quand il est mentionné dans les textes bouddhistes chinois où la plante tombe des cieux comme une pluie au temps où le Bouddha prêche la loi, est l'*Erythrina* à l'exclusion du *Datura*. [2] D'autre part, l'usage du mot *mandara* chez Čou K'ü-fei à l'époque des Song prouve assez bien que dans l'Inde aussi *mandara* servait d'expression pour le *Datura*.

---

near Kyōto. The seed is planted in the spring. The form of the leaves is like that of the egg-plant (*Solanum melongena*), without spines, green, and alternate. The plant is 2—3 feet high, the way of branching being also similar to that of the egg-plant; it blooms in the summer and autumn. Flower standing in axil of leaves, white, resembling the blossom of the morning glory (*Pharbitis nil*) with elongated tube and united petal. There are five edges on the outer margin of a flower, gradually narrowed into a tube, about 3 *sun* (1 *sun* = 1.193 inch) long. Fruit, about 1 *sun* long, is round and spiny, hence the name *hari-nasubi* is derived; it contains flat, brownish-black seeds. The plant dies out in the autumn, and no part of it thrives until next year.

"If one happens to eat the flower and leaves by mistake, a nervous condition of the nature of insanity will be the consequence, but with the removal of the virus which caused the effect, this condition is gradually overcome, the result being a complete cure without leaving any mental disorder."

[1] W. Roxburgh, *Flora Indica*, p. 541.

[2] Contrairement à ce que Stuart (*Chinese Materia Medica*, p. 145) dit à ce sujet. C'est d'ailleurs Li Ši-čen lui-même qui est responsable pour cette erreur, en introduisant sa notice sur le *man-t'o-lo* avec les mots: "Il est dit dans le *Fa hwa kin* (Saddharma-puṇḍarīka-sūtra) qu'au temps où le Buddha prêcha la loi, le ciel fit pleuvoir des fleurs de *man-t'o-lo*." Il ne savait pas que *mandara* se rapporte dans l'Inde à des plantes différentes.

Voici enfin une question que je me demande et que je voudrais proposer aux étudiants des langues indo-européennes pour leur considération: serait-il possible que le terme sanskrit *mandāraka* et le terme gréco-latin *mandragora(s)* soient anciennement apparentés et descendent d'une racine commune? L'accord est éclatant, et si c'est un accident, l'accident serait extraordinaire.

148

藿叶

JOURNAL ASIATIQUE

RECUEIL DE MÉMOIRES

ET DE NOTICES

RELATIFS AUX ÉTUDES ORIENTALES

PUBLIÉ PAR LA SOCIÉTÉ ASIATIQUE

ONZIÈME SÉRIE

TOME XII

PARIS

IMPRIMERIE NATIONALE

ERNEST LEROUX, ÉDITEUR, RUE BONAPARTE, 28

MDCCCCXVIII

JOURNAL ASIATIQUE.

JUILLET-AOÛT 1918.

MALABATHRON,

PAR

BERTHOLD LAUFER.

Sous le nom de μαλάβαθρον (latin *mālŏbăthron, mālŏbăthrum*) ou « feuille indienne » (φύλλον Ἰνδικόν), les anciens connaissaient la feuille aromatique d'une plante venant de l'Inde et mise sur le marché en rouleaux ou balles. Un onguent précieux était extrait de ces feuilles. Le grand ouvrage de Théophraste sur les plantes n'en fait pas mention. Ce qui montre que cette substance n'est pas venue à la connaissance des Grecs durant l'expédition d'Alexandre dans l'Inde. En fait, les plus anciens témoignages y faisant allusion tombent à l'époque d'Auguste. Pline, Dioscoride et le Périple sont nos principales sources.

Voici le texte de Pline :

« La Syrie nous donne aussi le malobathrum, un arbre à feuillée roulée de la couleur d'une feuille sèche. On en extrait une huile pour onguents. Plus abondant en Égypte; le plus estimé cependant vient de l'Inde. On dit qu'il y pousse dans des marais à la manière de la lentille. Plus odorant que le safran, il est d'apparence noire et grossière, et quelquefois d'un goût salé. La feuille blanche est moins estimée, car elle se moisit promptement en vieillissant. Placé sous la langue, la saveur doit être semblable à celle du nard. Chauffé dans du vin, son odeur surpasse toute autre. Le prix atteint par cette drogue est prodigieux, s'élevant d'un à trois cents deniers la livre, alors que les feuilles mêmes se vendent soixante deniers la livre[1]. »

Dans un autre passage, Pline s'étend sur les propriétés médicales de la feuille ainsi que suit : « Nous avons déjà exposé la nature et les différents genres du malobathrum. Il agit comme diurétique; exprimé dans le vin il est très utilement employé pour les écoulements des yeux. Appliqué sur le front son action est soporifique, plus efficace encore si on en frotte les narines ou s'il est pris dans de l'eau. La feuille placée sous la langue assure à la bouche et à l'haleine une odeur agréable, et mise entre les vêtements les parfume de même[2]. »

Le malabathron était regardé comme le premier des onguents nobles (XII, 1, § 14), et était un des nombreux ingrédients entrant dans la composition de ce qu'on nommait le *regale unguentum*, lequel était préparé pour le roi des Parthes (XIII, 2, § 18); cette substance peut donc ainsi avoir été connue dans l'empire des Arsacides. En outre, mêlé au nard, on en usait dans la préparation des vins aromatiques (XIV, 16, § 108).

Dioscoride (I, 11) sur ce sujet note ce qui suit : « C'est une croyance populaire que le malabathron est la feuille du nard indien, mais c'est une erreur causée par la ressemblance de

[1] Dat et malobathrum Syria, arborem folio convoluto, colore aridi folii, ex quo premitur oleum ad unguenta, fertiliore ejusdem Ægypto. Laudatius tamen ex Indo venit. In paludibus ibi gigni tradunt lentis modo, odoratius croco, nigricans scabrumque, quodam salis gustu. Minus probatur candidum. Celerrime situm in vetustate sentit. Sapor ejus nardo similis esse debet sub lingua. Odor vero in vino subfervefacti antecedit alios. In pretio quidem prodigio simile est, a denaris singulis ad CCC pervenire libras, folium autem ipsum in libras LX (XII, 59, § 129).

[2] Malobathri quoque naturam et genera exposuimus. Urinam ciet, oculorum epiphoris in vino expressum utilissime imponitur, item frontibus dormire volentibus, efficacius, si et nares inlinantur aut si ex aqua bibatur. Oris et halitus suavitatem commendat linguæ subditum folium, sicut et vestium odorem interpositum (XXIII, 48, § 93).

l'odeur. Du reste il y a plusieurs drogues dont l'odeur est pareille à celle du nard, comme la valériane, l'asarum, l'acore et l'iris. C'est pourquoi cette opinion est fausse. Le malabathron est d'un caractère différent. Il pousse dans l'Inde en certaines places couvertes d'eau. C'est une feuille qui apparaît à la surface, à la manière de la lentille d'eau, sans racine Après la récolte, les feuilles sont enfilées sur des fils de lin, séchées et emmagasinées. On dit que dès que l'eau s'est évaporée, durant les chaleurs de l'été, on brûle du bois sur cette partie du sol, car faute d'une telle précaution, la feuille ne pousserait plus. Le meilleur malabathron est celui qui est frais, tourne légèrement au blanc et noir, est complet sans brisure et d'une odeur pénétrante et persistante, rappelant celle du nard indien, dont le parfum n'est pas saumâtre. La feuille qui est faible, brisée, et a la moindre odeur de moisi est sans valeur. »

Suivant le *Périple de la mer Érythrée* (ch. 63), écrit entre 80 et 89 apr. J.-C. (en gros vers 85 apr. J.-C.), le malabathron passait en transit par le grand marché Gange, situé sur le fleuve de ce nom (*ἐμπόριόν ἐστιν ὁμώνυμον τῷ ποταμῷ, ὁ Γάγγης, διὰ οὗ φέρεταί τό τε μαλάβαθρον*). A ce sujet il convient de noter particulièrement que le Périple parle dans ce chapitre de « Chrysê, le dernier pays de l'Est situé près du Gange » (*ὁ Γάγγης καὶ ἡ περὶ αὐτὴν ἐσχάτη τῆς ἀνατολῆς ἤπειρος ἡ Χρυσῆ*); car on verra, d'après les relations écrites par les Chinois citées plus loin, que la patrie du malabathron doit être cherchée dans la péninsule de Malacca. Dans le chapitre 56 du Périple, il est en outre fait allusion aux chargements du malabathron de et pour les ports de l'Inde. Le chapitre 65 contient sur ce sujet le récit très discuté qui reproduit sans doute une tradition mal digerée mise en circulation par les marchands indiens. La voici : « Chaque année une peuplade arrive à la terre frontière de Thinai. D'aspect physique ces gens sont de petite taille; la face est large; comme caractère ils ont les qualités du cœur. Ils sont, dit-on, désignés sous le nom de Besatai (ou Sesatai); leur culture est primitive. Ils viennent avec femmes et enfants, portant de lourds fardeaux et des vanneries assez semblables de couleur à des feuilles de vigne fraîches. Pendant quelque temps ils restent sur la partie frontière qui s'étend entre eux et Thinai. Plusieurs jours durant ils se livrent à des réjouissances, répandant sur le sol les matériaux dont sont faits leurs ouvrages de vannerie. Ils regagnent ensuite leurs demeures situées plus loin dans l'intérieur. Les aborigènes, qui surveillent ces manières d'agir, approchent alors et ramassent ces nattes. Ils extraient les brins des tresses que l'on désigne sous le nom de *petroi*[1], mettant avec soin les feuilles les unes sur les autres, et les roulent en forme de boules qu'ils empilent sur les branches d'osier. Trois sortes de boules de malabathron sont ainsi produites : on appelle grosses celles qui sont faites des plus larges feuilles, moyennes celles pour lesquelles on a pris les feuilles de moyenne dimension et petites celles de plus petites feuilles. Telle est l'origine des trois qualités de malabathron qui de là sont toujours exportées dans l'Inde par ceux qui le préparent[2]. »

Les points les plus remarquables de ce récit consistent en ceci que le trafic du malabathron est lié à deux peuplades primitives vivant hors de l'Inde et en bordure de la Chine du Sud, et que ce produit était importé dans l'Inde. Si l'imporium Gange sur le Gange était son marché de transit, il est probable que des bateaux l'y apportaient de par delà la mer. Le caractère anormal, illogique et quelque peu puéril de ce récit est évident : ainsi les feuilles auraient été arrachées d'objets de van-

(1) Sanscrit *patra* ou *pattra* « feuille ».

(2) Ma traduction est basée sur l'édition critique de B. Fabricius (p. 110); celle de Yule (Cathay, new ed. by H. Cordier, I, p. 182) est faite d'après l'édition maintenant surannée de G. Müller.

nerie jetés par une tribu sauvage et il faudrait supposer que leur fourniture marchande tout entière viendrait de cette source de seconde main, sans qu'aucune tentative n'ait été faite pour les obtenir de la localité où elles poussent. Le mystère qui entoure cette localité témoigne du fait qu'elle était très éloignée de l'Inde, dans un coin reculé, non directement accessible aux trafiquants indiens. Les Besatais ont été identifiés avec les Beseidais (Βησεῖδαι) ou Tiladais, placés par Ptolémée à l'est du Gange. Cette identification est peu convaincante, car la description donnée de cette peuplade par Ptolémée ne concorde pas avec les caractéristiques mises en avant par le Périple. Ptolémée qualifie ce peuple de «difforme, grand, velu, la face large et la peau blanche»; tandis que la peuplade du Périple est une tribu de pygmées. La véritable signification de ces données sera discutée plus tard à la lumière des documents chinois.

Ptolémée (VII, 2, 15) dit que l'on indique comme le plus beau le malabathron produit au pays de Kirrhadia (ὑπὲρ δὲ τὴν Κιῤῥαδίαν, ἐν ᾗ φασι γίγνεσθαι τὸ κάλλιστον μαλάβαθρον[1]), lequel paraît identique au Kirāta de l'Inde[2]. Ce rapport fait sous réserve d'un *on dit* n'a pas grande valeur historique et semble être inspiré par celui des Besatais du Périple.

En dehors des textes cités, on trouve seulement quelques allusions occasionnelles au malabathron qui ne jettent aucune lumière nouvelle sur le sujet, dans Horace[3], Sidoine Apollinaire[4], les médecins Galien (129-199 après J.-C.), Celsus

[1] B. FABRICIUS, *Periplus*, p. 166; G. CŒDÈS, *Textes d'auteurs grecs et latins relatifs à l'Extrême-Orient*, p. 57.

[2] Nom déjà porté à une époque aussi ancienne que celle des Védas par une population vivant dans les creux des montagnes. A une époque ultérieure on les plaçait au Népal oriental (MACDONELL et KEITH, *Vedic Index*, I, p. 157).

[3] Cum quo morantem saepe diem mero Fregi, coronatus nitantes malobathro Syrio capillos (*Od.*, II, 7).

[4] *Carmina*, II, 415. Sidoine vivait aux environs 430-480 apr. J.-C., et était évêque de Clermont depuis 472.

(v, 23), Aetius d'Amida et Paul d'Egine (iv, 48) du viie siècle. Ce dernier, comme le Périple, parle de φύλλον μαλάβαθρον σφαιρία. Il en est encore fait mention dans les *Geoponica* (vi, 6), compilation consacrée à l'agriculture et éditée aux environs de 950 après J.-C., mais basée sur une plus ancienne compilation du vie siècle par Cassianus Bassus.

Contrairement à d'autres drogues et aromates, les traditions des anciens concernant le malabathron ne se perpétuèrent pas durant le moyen âge. Aucun écrivain de cette époque ne mentionne la substance, et son importation de l'Orient paraît s'être arrêtée vers la fin des temps antiques. Garcia da Orta est le premier auteur moderne qui, connaissant la description de la drogue des anciens, en ait recherché la provenance dans l'Inde. Dans ses *Colloquios dos simples e drogas e cousas medicinaes da India*[1], publiés à Goa 1563, il a été le premier à faire remarquer que le terme grec malabathron venait par corruption de l'indien *tamālapatra*[2] et que le terme arabe correspondant est *cadegi Indi*, qui signifie «feuille de l'Inde». Les feuilles de ce nom qu'un droguiste lui a procurées sont décrites comme étant d'apparence identique à celles de l'oranger, sauf qu'elles sont plus effilées, et de couleur vert noir. Elles ont une nervure dans le milieu et deux autres qui se réunissent à la pointe. Le parfum en est très plaisant et n'est pas si fort que celui de l'*espiquenardo* (spicanard), ni de la pomme. Il sent comme le girofle, mais ne dégage pas une odeur aussi forte que la cannelle. D'après Garcia, les renseignements que donnent Dioscoride et Pline sont erronés, car les feuilles en question viennent d'un grand arbre et ne poussent pas dans des marécages comme les

[1] Colloquio xxiii (p. 95 de la réimpression publiée à Lisbonne, 1872).

[2] Cette identification est due à Garcia, non à Yule, comme G. Cœdès (*Textes d'auteurs grecs et latins relatifs à l'Extrême-Orient*, p. xvii) semble le prétendre. Le grec *malabathron* ne peut pas se rapporter au Malabar, comme le suppose H. Estienne (*Thesaurus graecae linguae*, V, p. 539 : «Ferunt apud Indos nasci in ea regione quae Malabar dicitur»).

lentilles d'eau. Cet arbre pousse dans beaucoup d'endroits, entre autres à Cambaya, et la feuille peut être obtenue chez tous les droguistes. Il ajoute encore, au sujet de la feuille, qu'elle n'est pas très différente de celle du cannelier, mais que cette dernière est plus étroite et moins pointue, et qu'elle n'a pas les nervures pareilles à celles de la feuille indienne. Le parfum n'est pas si fort que celui du nard. Les feuilles sont recueillies et mises en paquets qui sont expédiés pour la vente. Il n'est pas nécessaire d'allumer un feu pour les faire pousser. On brûle tout le terrain qui doit être ensemencé, mais non celui qui est laissé aux plantes pour pousser. La couleur est vert clair; elle s'assombrit avec le temps, mais elle tire plutôt sur le noir que sur le vert foncé et il y en a qui n'ont pas l'odeur de *salva*. Il est vrai que la partie intérieure est meilleure, parce que ses vertus y sont mieux gardées; le parfum monte à la tête comme c'est le cas des autres parfums. Elle n'existe pas en Syrie ni en Égypte. La feuille de cinnamome n'en est pas un succédané[1].

Garcia s'en tient là. Malheureusement il ne donne pas de description de l'arbre qu'il ne semble pas avoir vu lui-même. Il connaissait simplement la feuille préparée, qu'il s'était procurée dans une boutique de droguiste. Alors qu'il n'y a aucun doute sur l'identification philologique faite par Garcia du grec *malabathron* avec le sancrit *tamālapattra*, je ne suis pas sûr que la feuille qu'il a décrite soit la même que Dioscoride et Pline avaient en vue. Les anciens parlent d'une huile ou onguent extrait de la feuille; sur ce point Garcia ne dit rien : en fait il garde le silence sur la manière dont la feuille était traitée.

L'ouvrage de Garcia fut continué et dans une certaine mesure amélioré par Christovão ou Christoval da Costa (1540-1599

[1] Cf. C. Markham, *Colloquies on the Simples and Drugs of India by Garcia da Orta*, p. 202-207 (London, 1913). Traduction latine par C. Clusius, p. 94-99 (Antverpiae, 1567).

environ), qui a consacré un chapitre de huit pages au « Folio Indo »[1]. Da Costa montre tout au long que la feuille indienne est absolument différente de celle du bétel, mais ressemble à celle du cannelier et de l'oranger. Il repète alors la description de la feuille donnée par Garcia. Comme Garcia, il a lui aussi recherché si, suivant l'assertion de Pline, on peut la trouver en Syrie; mais marchands et médecins du Caire, Damas et Alep, déclarent qu'il ne s'en trouve pas dans ces endroits et qu'ils ignorent s'il y en a en Syrie ou en Égypte. Il nie expressément que ces feuilles soient identiques à celles du cannelier (las hojas de canela, ellas non son folio indo), mais dit que ces dernières, ainsi que les feuilles de girofle, étaient vendues comme feuilles indiennes. La feuille indienne de Syrie et celle de l'Inde furent montrées à Venise à un certain Antonio Musa, mais il ne les connaissait pas. L'addition la plus intéressante de Da Costa est sa référence à l'ouvrage d'un certain moine intitulé *Modus faciendi*, dans lequel il est dit qu'« au pays du prêtre Jean on trouve la feuille indienne, qu'il a eu entre les mains les feuilles du cannelier et qu'elles ne lui ont pas paru avoir poussé dans l'eau, mais sur un arbre, et qu'à leur défaut, on peut leur substituer le spicanard ou la muscade. Il se peut que ces feuilles fussent des feuilles de cannelier, bien qu'elles en diffèrent, la feuille du cannelier étant un peu plus serrée et

[1] *Tractado de las drogas, y medicinas de las Indias orientales, con sus plantas de buxadas al biuo por Christoual Acosta medico y cirujano que las vio ocularmente*, chap. XIX, p. 139-146 (Burgos, 1578). Une gravure avec son portrait porte comme légende « Christophorus Acosta Africanus ». La traduction italienne a pour titre : *Trattato di Christoforo Acosta Africano Medico et Chirurgo, della Historia, natura, et virtu delle droghe medicinali, et altri semplici rarissimi, che vengono portati dalle Indie orientali in Europa, con le figure delle piante ritratte, et disegnate dal vivo poste a' luoghi proprij*, Venetia, 1585, presso a Francesco Ziletti (Folio indiano, p. 108-113). Il y a aussi une traduction latine par Clusius (Antverpiae, 1605). L'ouvrage de Da Costa ou Acosta est orné de quarante-quatre illustrations représentant trente-neuf espèces, mais la feuille indienne n'y figure pas.

moins aiguë aux pointes que la feuille indienne. Quoi qu'il en soit, il est fort douteux que les feuilles de folio indo ou de cannelier puissent venir du pays du prêtre Jean où personne n'a jamais su, ou entendu, au moins jusqu'à présent, que le cannelier ou folio indo se trouvât. Et il n'y a personne, parmi celles qui y ont été, qui les aient vus, comme cela se sait chaque jour à cause du grand commerce qui s'y fait maintenant plus que jamais auparavant[1] ».

Dans Clusius, édition latine de l'ouvrage de Garcia, nous trouvons une reproduction illustrée de la plante intitulée « Tamalapatra cum suo Ramusculo » (fig. 1)[2]. C'est à l'aide de cette illustration que la plante a été identifiée par Linné avec

[1] Quanto a lo que dize el Frayle (que compuso el modus faciendi) que ay este Folio Indo en las tierras del Preste Ioan, y que a sus manos vinieron estas hojas, intituladas hojas del Arbol de la Canela, y que no le parescian nascidas en el agua, sino en arbol, y que en su defecto se podra poner Espique, o Macis : bien podria ser, que fuessen aquellas hojas de canela, aunque diffieren, en que las hojas de la canela son un poco mas angostas, y menos agudas en las puntas que las del Folio Indo. Tambien se dubda mucho, en como le podriã venir las hojas del Folio Indo, ni de canela de las tierras del Preste Ioan, en las quales no se ha sabido, ni oydo hasta agora, auer canela ni Folio Indo : ni ay persona, que en ellas anduuiesse, que tal vieste, como se sabe cada dia, por el mucho comercio que agora ay mas de lo que nũca ha sido.

[2] C. Markham, dont les illustrations viennent de l'ouvrage de Christoval da Costa (Burgos, 1578), ne donne aucune illustration de cette plante. Ainsi qu'il a été noté antérieurement, elle manque dans le livre de Da Costa. L'édition portugaise originale de Garcia était sans aucune illustration. Acosta dans sa préface adressée au *Christiano y prudente lector* dit à ce sujet : « No falto tãbien otra perfectiõ substancial a la obra, que son las pinturas, y debuxos de las plantas, de que trata : que occupado el Doctor Orta en otras cosas mas graues, y que mas deuian importarle, dexo de inxerirlas en ella. Paresciendome ami, que esta nuestra nacion seria a quel libro de grande prouecho, si se diesse notitia de las cosas buenas, que en el ay, mostrãdose con sus exemplos, y figuras, para mejor conoscerlas, y que esto no lo podria hazer, sino quiẽ ocularmente con sus mismos ojos las huuiese visto, y esperimentado : celoso del bien desta tierra, cõ la charidad que a mis proximos deuo, delibere tomar este trabajo, y debuxar a biuo cada planta, sacada de rayz abueltas de otras muchas cosas, que yo vi, y el Doctor Garcia de Orta no pudo por las causas dichas. »

le genre *Melastoma* (famille *Melastomaceæ*)[1], et l'espèce est dénommée *M. malabathricum*. Watt[2], cependant, observe que cette plante est un «arbuste» (tandis que Garcia insiste sur ce qu'elle est un arbre), poussant avec une grande abondance dans toute l'Inde, et dont le fruit produit une teinture pourpre employée pour les étoffes de coton; mais de la feuille il ne dit pas un mot. Que le malabathron des anciens puisse être un produit de cette plante, reste donc hors de question. Je ne pense pas que l'illustration de Clusius soit d'une importance quelconque ou puisse servir de base à une conclusion sérieuse pour l'identification botanique du *tamālapattra*. Les botanistes peuvent décider jusqu'à quel point elle s'accorde avec la description de Garcia, mais ils doivent se rappeler qu'elle n'a aucun rapport avec l'ouvrage de Garcia ou de Da Costa. Garcia vivait et travaillait dans l'Inde et basait ses observations sur ce qui y était offert à sa vue. Clusius ou Charles de l'Ecluse (1526-1609) vivait en Hollande, et les spécimens qui ont pu parvenir jusqu'à lui sous l'étiquette «tamalapattra» sont sujets à grave suspicion; pour moi, je ne crois pas que ce qui a été reproduit ait à faire en quoi que ce soit avec la feuille en question.

Ch. Lassen[3] a discuté deux fois le problème du malabathron. A juste titre il a combattu l'opinion, émise par Neumann, qu'il s'agirait de la feuille de thé chinoise. Il a identifié le mot grec avec le sanscrit *tamālapatra*, sans référence à Garcia, et sans en avoir apparemment connaissance; et comme *patra* «feuille» est particulièrement employé à propos des feuilles de *Laurus cassia*, qui, dit-il, sont appellés aussi *tamālapatra* «feuilles noires», il en conclut que le malabathron est

[1] Linnæus, *Codex Botanicus*, p. 388, n° dxcvi (*Opera*, ed. H. E. Richter, II, Lipsiae, 1835).

[2] *Dictionary of the Economic Products of India*, V, p. 210.

[3] *Z. Kunde d. Morgenlandes*, II, p. 37; et *Indische Altertumskunde*, I, p. 281-283; III, p. 37-39.

constitué par les feuilles de *Cassia* et autres *Lauraceæ*. Cette déduction quelque peu subjective non seulement est en contradiction avec la description de Garcia et l'illustration de Clusius, qui ne peut en aucune façon représenter une *Lauracea*, mais il est encore raisonnablement impossible de la tirer des données des anciens. Les anciens connaissaient bien la casse et le cannelier, qu'ils tenaient en haute estime [1], et par suite auraient reconnu le véritable caractère du malabathron, s'il avait réellement appartenu à cette famille de plantes. Le fait qu'ils étaient dans l'ignorance de la plante même d'où venaient les feuilles prouve bien qu'il s'agit d'une autre famille. Pourtant l'hypothèse de Lassen, à défaut de quelque chose de mieux, a généralement été adoptée.

D'accord avec la conclusion de Lassen, nos botanistes ont établi une espèce *Cinnamomum tamala* [2] dont les feuilles sont supposées avoir fourni les *folia malabathri*. L'arbre en question est un arbre à feuilles persistantes qui croît sur l'Himalaya, peu nombreux de l'Indus au Sutlej, commun à partir de là dans la direction de l'est entre trois et sept mille pieds jusqu'au Bengale oriental, les monts Kasia et la Birmanie. Les feuilles sont d'usage commun en qualité de condiment, on les emploie aussi dans l'impression des étoffes de coton. Il n'est aucunement prouvé, bien au contraire, il reste excessivement douteux, que cet arbre ait jamais produit le malabathron connu des anciens. Cette identification est en conflit avec les observations de Garcia et de Da Costa qui nient positivement que les feuilles en question soient celles du cannelier. De plus, si l'arbre qui donne le malabathron est si commun dans l'Inde,

(1) PLINE, XII, 42.

(2) WATT, *loc. cit.*, II, p. 321; ou l'article *Cinnamomum tamala* dans l'*Agricultural Ledger*, 1896, n° 38. Il est curieux que les botanistes n'aient jamais reconnu le désaccord entre cette identification et le *Melastoma* de Linné.

cela laisserait inexpliquée la curieuse tradition du Périple, qui marque nettement à la feuille son caractère d'article d'importation introduit dans l'Inde, d'une contrée étrangère. Un autre défaut de cette argumentation est qu'il n'a jamais été démontré que la feuille de l'arbre en question se soit réellement, dans l'Inde, appelée *tamālapattra*. Nous savons simplement que le nom moderne donné dans le pays est *tej-pat* : l'élément *tej* est dérivé du sanscrit *tvac* (sens littéral «peau», «écorce», en particulier «écorce de cassia», et aussi «cannelle» et «cannelier»); l'élément *pat* est le sanscrit *patra* ou *pattra* «feuille». Il n'est pas certain, mais très douteux au contraire, que ce terme ait quelque relation avec le sanscrit *tamālapattra*. Le mot indien moderne pour «cannelle» est *dār-cīnī* ou *dāl-cīnī* («bois chinois», de l'arabe *dār-ṣīnī*); et le terme sanscrit pour la feuille de *Laurus cassia* est *pākarañjana*. Yule[1], dans une note de sa traduction du passage du Périple, remarque que Linschoten décrit avec exactitude le *tamālapattra*, qu'il note son agréable odeur de girofle, et dit sa grande réputation parmi les Hindous comme diurétique, etc., et pour la conservation des vêtements contre les insectes, deux des usages expressément assignés au malabathrum par Dioscoride et Pline[2]. Toutefois ce dernier argument n'offre pas un critérium sûr pour l'identification exacte, car les feuilles de nombreuses autres plantes peuvent être employées pour le même usage. Comme Yule

[1] *Cathay*, nouvelle édition, I, p. 184; voir aussi son *Hobson-Jobson*, p. 543 et son introduction au livre de W. Gill, *River of Golden Sand*, p. 89 (London, 1883).

[2] Le botaniste hollandais de grand renom Mathias de Lobel (*Kruydtboeck*, p. 178, Antverp, 1581), qui représente et décrit la plante d'après Garcia et Clusius sous le nom de «Malabathrum, Tamalapatra Garciae et Clusij», remarque au sujet de son emploi en médecine : «Malabathrum ghestooten ende in wijn warm ghemaekt is zeer goedt gestreken op loopende oogen en inflammatien. Malabathrũ onder de tonge gehouden maeckt eenen goeden eñ soeten aessem. Tselfde gheleydt tusschen de cleederẽ doet de cleeders wel riecken eñ bewaertse vande motten.»

avait connu la feuille en question dans les forêts des montagnes des Kasia, il inclinait à penser que les Kasia, dans leur aspect extérieur, répondraient fort bien aux Besadae ou Sesadae du Périple; mais il observe que ceux-ci ne sont pas des nains, bien que nombre de tribus tibétaines de l'Himalaya soient de très petite taille. Yule était, à juste titre, frappé de ce fait que dans la vie domestique des Anglo-Indiens ce malabathrum, autrefois si prisé que quelques-unes de ses qualités étaient recherchées par les Romains au prix de trois cents deniers la livre, est employé seulement à aromatiser les tartes, crêmes et carys.

F. von Richthofen[1] a avancé l'hypothèse que le peuple de pygmées qui mettait cet article sur le marché était probablement les habitants des montagnes situées entre l'Assam et le Sze-tch'oan. Il parle aussi des Man-tse, qui avaient été repoussés par les Chinois dans les montagnes au delà de leurs frontières, et qui étaient doués d'un degré de culture suffisant pour développer un esprit commercial; grâce à leur supériorité ils étaient capables de maintenir leur droit de passage à travers les territoires des autres tribus montagnardes. A ce propos, conclut F. von Richthofen, il ne faut pas perdre de vue que la casse constitue un produit célèbre de Ning-yuan fou, lieu que Marco Polo appelle Caindu et où il fait mention des boutons de casse giroflée (*garofali*). Ces spéculations hasardeuses ne sont pas, comme on le verra, supportées par les traditions chinoises. O. Schrader[2] pense qu'il est probable que la tradition du Périple fait allusion à une sorte d'échange muet entre Chinois et tribus voisines[3]; mais cela ne peut réellement pas expliquer

[1] *China*, I, p. 507.

[2] *Reallexikon d. indogerm. Altertumskunde*, p. 519.

[3] Telle était aussi l'opinion de C. Muller (voir B. Fabricius, *Periplus*, p. 166). En fait, le Périple ne fait d'ailleurs aucune allusion à un tel échange muet.

avec certitude ce qui doit être entendu par malabathron. La théorie de Lassen lui apparaît comme la plus probable, le malabathron aurait alors désigné une sorte de cannelle et les Chinois auraient été les principaux agents de son commerce même aux temps de l'antiquité. Cela est certainement une erreur, car la cannelle est l'écorce, non la feuille de l'arbre; et si même le malabathron était la feuille d'une espèce de cannelier (ce qu'en fait il n'est pas), cela n'impliquerait aucune allusion à l'écorce du cannelier qui n'est en effet mentionnée dans aucun texte traitant de ce sujet.

Ainsi le résultat jusqu'à présent atteint est négatif devant la philologie classique[1]; à un point de vue botanique, un simple compromis[2]. Les Arabes, comme l'a déjà remarqué L. Leclerc[3], ne nous ont rien enseigné au sujet du malabathron, mais ont simplement répété les traditions grecques. Garcia da Orta remarque avec justesse qu'Avicenne, Sérapion et Rasis n'ont rien connu de plus que les Grecs sur ce médicament : ils ont simplement connu, d'accord avec les Grecs, que malabathron était *folio indo* et ils ont traduit ce que disaient les Grecs, ajoutant seulement quelques petites choses sur ses usages. Tous s'accordent à dire son utilité pour provoquer les urines et contre la mauvaise haleine, et à la fin ajoutent qu'il est bon

[1] Se référant aux passages de Dioscoride et de Pline, Ch. Joret (*Les Plantes dans l'antiquité*, II, *L'Iran et l'Inde*, p. 653) remarque, à juste titre, que la véritable nature et l'origine de la feuille leur étaient inconnues.

[2] Pour Salmasius, Vincent, Heeren et MacCrindle, le malabathron est identique à la feuille de bétel (*Piper betle*). Cet avis a déjà été combattu par Garcia da Orta, Lassen et Yule (*Hobson-Jobson*, p. 83, 543). Dymock (*Pharmacographia Indica*, III, p. 184) va jusqu'à établir, selon le procédé de Vincent, un rapprochement entre malabathron et le sanscrit *tambulapatra* «feuille de bétel», bien que d'autre part il identifie *tamalapatra* avec *Cinnamomum tamala*. Comment les Grecs ont-ils pu extraire un élément *mala-* du sanscrit *tāmbūla* ? il est difficile de le voir. Du reste les anciens ne connaissaient rien du bétel. Le mot «bétel» est dérivé du portugais *betre* qui a pour base le malabarais *veṭṭila*.

[3] *Traité des simples*, II, p. 232.

pour toutes ces choses, comme le spicanard[1]. L'opinion d'Abū Dulaf que le malabathron (arabe *sādedj* ساذج, ou *saliādj*, *sadzādj*) est fourni par les feuilles de rhubarbe[2] est sans fondement.

La littérature sanscrite contient divers renseignements sur le *tamāla*, mais rien de concluant pour le côté botanique du problème. Le *tamāla* est mentionné deux fois par le médecin ancien Suçruta. Dans le *Sūtrasthāna* (XXXIX, 5, 6), différentes drogues sont énumérées en qualité de *çirovirecana*, c'est-à-dire médicaments errhins pour dégager le cerveau; parmi eux le *tamāla*, avec l'explication que l'on doit se servir des feuilles. Dans l'*Uttarasthāna* (XLVII, 61), le *tamāla* est cité avec quelques autres substances aromatiques comme remède en cas d'excès de boisson (*pānātyaya*). Toutefois l'*Uttarasthāna* est une addition postérieure de beaucoup au texte original de Suçruta. Dans l'œuvre médicale de Caraka, qui vivait au IIe siècle après J.-C. et était médecin du roi Kaniṣka, mention est faite du *tamāla* au milieu d'un certain nombre d'arbres poussant en pays marécageux (*anūpadeça*); mais la partie de son ouvrage où ce nom paraît, le *Kalpasthāna* (I, 8), est une addition plus récente due à Dṛiḍhabala, qui vivait entre le VIIe et le IXe siècle après J.-C.[3]. On ne rencontre aucune allusion à la plante dans le manuscrit Bower. Le *Çukranīti* (IV, 115)[4] le compte parmi les grands arbres. En dehors des textes médicaux, il apparaît, par exemple, dans le poème de Kālidāsa, *Raghuvaṃça*

[1] C. Markham, *loc. cit.*, p. 206.

[2] G. Ferrand, *Textes relatifs à l'Extrême-Orient*, p. 226. Le mot araméen est *sadeg*. En arménien il est *satj* ou *sadej* et se trouve dans l'ouvrage sur les fièvres écrit par le médecin arménien Mechithar en 1184 (édition de E. Seidel, p. 207). Voir aussi Achundow, *Abu Mansur*, p. 85.

[3] Pour tout ce qui regarde Suçruta et Caraka, c'est au Dr. A. F. Rudolf Hœrnle de l'Université d'Oxford que j'en suis redevable. Au sujet de la vie et des œuvres des auteurs indiens mentionnés, voir l'introduction de Hœrnle dans son livre *Studies in the Medicine of Ancient India* (Oxford, 1907).

[4] Traduction de B. K. Sarkar, p. 166 (Allahabad, 1914).

(vi, 64). Quant aux œuvres plus récentes, il est mentionné dans le *Rājanighaṇṭu* (vi, 176) par Narahari (xve siècle) et dans le *Bhāvaprakāça* par Bhāvamiçra (xvie siècle)[(1)]. C'est le iie siècle après J.-C. qui nous offre la première date positive pour une mention de la plante dans la littérature indienne.

Le terme *tamālapatra* se trouve deux fois dans le dictionnaire bouddhique *Mahāvyutpatti*, traduit en tibétain dans la première partie du ixe siècle : la première fois dans la liste des fleurs (tibétain *ta-ma-lai lo-ma*, c'est-à-dire « feuille de *tamāla* »); la deuxième dans la liste des aromates (tibétain *ta-ma-lai ḍdab-ma*, c'est-à-dire « pétale ou grande feuille de *tamāla* »). Il apparaît également dans le dictionnaire *Amarakoṣa*, où il est accompagné de la traduction tibétaine *ta-mā-la-ḍdab*[(2)].

En consultant le dictionnaire sanscrit de Boehtlingk, nous trouvons que *tamāla* n'est en aucune façon un terme botanique bien défini, puisque on dit qu'il se rapporte à trois plantes totalement distinctes : 1° *Xanthochymus pictorius*[(3)]; 2° une sorte de *khadira* (c'est-à-dire *Mimosa catechu*); 3° *Crataeva roxburghii*. Pour *tamālapattra*, voici comment il y est expliqué : 1° feuille de *Xanthochymus pictorius*; 2° feuille de *Laurus cassia*; 3° *Xanthochymus pictorius*. Une telle variété d'identifications modernes montre bien que la signification originale du terme a été depuis longtemps oubliée dans l'Inde.

Maintenant que la philologie classique ainsi que l'indienne ont, l'une et l'autre, échoué, il est temps sans doute d'appeler les Chinois et de voir si, par l'étude de leurs documents, nous ne pouvons pas recevoir une aide plus efficace. Dans la section « aromates » du dictionnaire bouddhique chinois *Fan yi ming yi*

(1) Communication du Dr. Hœrnle.

(2) Édition de la Bibl. Ind., p. 170, vers 122.

(3) Cet équivalent pour *tamala* est aussi donné par Wise, *Hindu System of Medicine*, p. 150.

tsi[1], compilé par Fa Yun 法雲 vers le milieu du XIIe siècle, nous rencontrons le terme 多阿摩羅跋陀羅* *ta-a-ma-la pat-do-la* (correspondant au sanscrit *tamālapattra*), suivi de la traduction 性無垢賢 (« sans tache et sage de nature », basée sur une analyse fantastique de *tamāla* mué en *ta-amala* [+ *bhadra*!]) et défini comme étant *ho ye hiang* 藿葉香 « aromate provenant des feuilles de la plante *ho* »; suivant d'autres, la définition serait *tch'e t'ong ye* 赤銅葉 « feuilles de cuivre ». Cette dernière interprétation vient de l'idée fantaisiste de donner au sanscrit *tamāla* le sens de *tāmra* « cuivre », et offre simplement une glose philologique sans valeur botanique. Cependant l'identification avec le *ho ye hiang*[2] est un cas tout à fait différent : nous voilà en face d'un vrai terme botanique et le problème à étudier tourne autour de la question : Quelle est la plante *ho*?

Li Chi-tchen, dans la dernière partie du XVIe siècle, est le premier naturaliste chinois qui ait introduit la terminologie indienne dans la discussion de la nature du *ho hiang*. Son prédécesseur, T'ang Chen-wei, l'auteur du *Tcheng lei pen ts'ao* en 1108, ne s'y prête pas. Li Chi-tchen l'a certainement tirée de la lexicographie sanscrite-chinoise des bouddhistes.

Il donne l'explication suivante des termes : « Les feuilles de fèves sont appelées *ho* 藿, les feuilles de ce dernier ressemblant à celles des fèves, — d'où le nom[3]. Le *Leng yen king* 楞嚴經[4] dit : Devant les autels on use de l'encens *teou-leou-p'o* 兜

[1] Ch. 8, p. 7 (édition imprimée à Nankin).

[2] Cette identification est encore donnée dans le *Hiang p'ou* 香譜 par Hong Tch'ou 洪芻 de la dynastie Song (p. 17 de l'édition du *T'ang Song ts'oung chou*), qui la cite d'après le *Chi chi houi yao* 釋氏會要, ouvrage évidemment bouddhique.

[3] Selon Bretschneider (*Bot. Sin.*, t. II, p. 163), le mot *ho* est expliqué dans le *Chouo wen* par « jeune plante légumineuse »; dans le *I li* par « feuilles de fèves ». Le mot *ho* entre dans la terminologie de plusieurs plantes.

[4] Sūtra bouddhique traduit du sanscrit en chinois en 705 après J.-C. Le

婁婆 pour l'eau chaude employée aux ablutions sacrées [des statues bouddhiques][1]. Le *Fa hwa king* 法華經 (*Saddharma-*

premier chapitre a été traduit par J. Edkins, *Chinese Buddhism*, p. 289-301.

(1) Le terme *teou-leou-p'o* est donné dans le *Fan yi ming yi tsi* (ch. 8, p. 8) comme étant le nom d'un aromate poussant dans le pays des Démons (*Kwei chen kouo* 鬼神國); c'était, suivant la description donnée, une herbe odorante (*hiang ts'ao* 香草), nommée aussi antérieurement *pai mao hiang* 白茅香 (fig. 9 et 11). Pelliot, qui note ce passage (*T'oung Pao*, 1912, p. 478), déclare ne pas connaître l'original sanscrit de la transcription *teou-leou-p'o*, qui n'est pourtant pas difficile à vérifier. En effet nous rencontrons dans le *Fan yi ming yi tsi* (ch. 8, p. 9) une autre manière de transcrire le même mot sous la forme 突婆 *t'ou-p'o* (anciennement **dur-ba*), que l'on regarde comme identique à *mao hiang* 茅香. Cette forme évidemment correspond au sanscrit *dūrvā*, et il en est ainsi pour *teou-leou-p'o* (**du-r-ba*). Dans le dictionnaire sanscrit de Boehtlingk le terme *dūrva* ne correspond qu'à *Panicum dactylon* (herbe utile et commune dans l'Inde, mais non employée comme aromate ou encens). Du reste, dans le cas présent il s'agit du genre *Andropogon*, comme le montrent deux faits évidents. Premièrement, le *Fan yi ming yi tsi* (*loc. cit.* nous donne le mot sanscrit pour la racine de *mao hiang* (茅香根) sous la forme 嗢尸羅 *wen-ši-lo* (**un* ou *u-śi-la*), qui répond au sanscrit *uçīra* : c'est-à-dire la racine aromatique de l'*Andropogon muricatus* («les racines, une fois séchées et ensuite légèrement humectées dégagent un agréable parfum» : W. Roxburgh, *Flora indica*, p. 89; Watt, *loc. cit.*, I, p. 245). En second lieu, l'identité du chinois *pai mao hiang* avec l'*Andropogon schœnanthus* a été correctement établie (Stuart, *Chinese Materia Medica*, p. 40, 207). L'identification de la *Platycaria strobilacea* avec le *teou-leou-p'o* signalée par Pelliot, après Giles, ne se rapporte pas à cette plante, mais seulement au *hwai hiang* 櫰香 (voir Stuart, *loc. cit.*, p. 337). Le *I wou tche* 異物志 est cité dans le *Hiang p'ou* (p. 10) de Hong Tch'ou comme ayant dit que le *teou-leou hiang* vient de pays situés aux bords de la mer (*hai pien kouo*) de la même manière que le *tou-hiang* (voir ci-dessous, p. 27). L'*Andropogon* et le *Panicum*, étant tous deux des herbes, peuvent être aisément confondus; de toutes façons il faut attacher aussi le sens d'*Andropogon* au sanscrit *dūrvā*. L'*Andropogon sch.* est notre jonc odorant, ou barbon, qui a eu autrefois sa place dans la pharmacopée européenne sous le nom de *Juncus odoratus*. C'est le *siri* de Java, le *malatrinaka* du sanscrit. On le cultive beaucoup dans les jardins de la côte de Coromandel; dans le Nord du Bengale de vastes espaces de terrains incultes en sont couverts. Les indigènes des Moluques extraient des feuilles une huile essentielle d'agréable goût, et les Javanais l'estiment comme un aromate doux et stimulant (W. Ainslie, *Materia Medica of the Hindoos*, II, p. 58-59). Cette huile est connue sous le nom d'huile *rūsa*, huile de ginger grass ou géranium (Fluckiger et

puṇḍarīka)[1] le désigne sous le nom 多摩羅跋香 d'aromate *to-mo-lo-pa* (**ta-ma-la-pat*, du sanscrit *tamālapattra*); le *Kin kwang ming king* 金光明經 (*Suvarṇaprabhāsasūtra*) sous celui de 鉢怛羅 aromate *po-tan-lo* (**pat-tan-la*, du sanscrit *pattra*)[2]. Tous ces termes sont des mots sanscrits. Dans la section *Nie-pan* (Nirvaṇa) il est en outre appelé *kia-swan* 迦箅». Le *Fan yi ming yi tsi* (ch. 8, p. 9) donne aussi l'explication du dernier terme qu'il regarde comme identique à *ho hiang*, mais il est écrit 迦箅 *kia-pi*, la prononciation du second caractère étant indiquée par 方爾. Il se peut que ce *kia-pi* corresponde au sanscrit *kapi* que l'on donne comme étant «l'*Emblica officinalis*, une espèce de *karañja* (*Pongamia glabra*), et l'oliban». Je doute qu'aucune de ces plantes puisse avoir quelque relation avec le *tamālapattra*. L'évidence l'emporte en faveur de l'opinion que le *tamālapattra* venait de la même plante ou d'une plante semblable à celle désignée sous le nom de *ho hiang* 藿香 par les Chinois. Cette plante a été reconnue identique au *Lophantus rugosus* (famille *Labiatae*)[3]. Cela est-il vrai pour le *ho hiang*,

Hanbury, *Pharmacographia*, p. 726; voir aussi Yule, *Hobson-Jobson*, p. 514). Il n'est pas croyable que le *dūrvā*, bien qu'il puisse être arrivé jusqu'aux Chinois sous le nom de *ho hiang*, ait un rapport quelconque avec le *tamālapattra*. Cette identification est simplement une opinion personnelle de Li Chi-tchen, dont la citation paraît être empruntée au *Hiang p'ou* de Hong Tch'ou, cité plus haut.

[1] Bunyiu Nanjio, *Catalogue of the Buddhist Tripitaka*, n° 134. Comme cet ouvrage a été traduit du sanscrit en chinois par Kumarajiva sous les Ts'in postérieurs (384-417 après J.-C.), le terme *tamalapattra* doit avoir été connu en Chine vers la fin du IVe ou le commencement du Ve siècle et doit avoir existé dans l'Inde avant cette date.

[2] Cette citation se réfère sans doute au septième chapitre du Sūtra, traduction chinoise de Yi-tsing (mort en 713 après J.-C.), qui contient en transcription et traduction une liste de trente-deux parfums indiens (Pelliot, *T'oung Pao*, 1912, p. 474). Le fait que nous rencontrons ici le nom abrégé *pattra* «feuille» est intéressant en soi, parce qu'il y a là une analogie avec le grec φύλλον «feuille» qui était alors employé pour malabathron. Le *Rajanighaṇṭu* (VI, 176) fait de *patra* ou *pattra* un synonyme de *tamalapatra*.

[3] G. A. Stuart, *Chinese Materia Medica*, p. 247.

tel que l'entendaient les anciens? il reste à l'examiner. Stuart observe que « cette plante ne semble pas être indigène à la Chine, car on la rattache à l'Annam, à l'Inde et à d'autres parties de l'Asie du Sud. Un certain nombre de noms sanscrits et étrangers lui sont donnés dans le *Pen-ts'ao*. La plante est cultivée dans le Ling-nan. Les branches et les feuilles sont employées en médecine, des vertus carminatives et stomachiques leur sont principalement attribuées. On en use aussi dans le choléra et pour dissiper par des lavages les mauvaises odeurs de la bouche[1] ». Cela en général est tout à fait dans la question, mais inexact dans le détail. Stuart ajoute qu'il se peut que la *Betonica officinalis* soit aussi rangée sous le terme *ho hiang*[2]. Je me propose d'analyser tous les textes chinois traitant de cette plante et de présenter les matériaux dans l'ordre historique et géographique, car ce n'est que par cette méthode que l'on peut espérer arriver à un résultat. Dans l'attente où nous sommes de savoir d'où vient ce mystérieux produit appelé malabathron, le point de vue géographique est certainement de première importance.

Le *Wou wai kouo tchouan* 吳外國傳 expose que « Tou-k'oun 都昆 [sur la péninsule malaise] est situé à plus de trois mille *li* au sud de Fou-nan 扶南 (Cambodge) et produit le *ho hiang* (l'aromate *ho*) 藿香 ». Il est probable que le *Wou wai kouo tchouan* est identique aux *Wou chi* 吳時 *wai kouo tchouan* et *Wai*

[1] Matsumura (*Shokubutsu-mei-i*, t. I, p. 211) identifie aussi le *ho hiang* avec le *Lophantus rugosus*, et donne comme synonyme *p'ai ts'ao hiang* 排草香 (voir plus bas, p. 34). Suivant S. T. Dunn (*A Key to the Labiatae of China*, dans *Notes from the R. Botanic Garden* Edinburgh, VI, 1915, p. 165), le *Lophantus rugosus* se rencontrerait dans le Kiang-sou, Tche-li, Fou kien, Chen-si, Tche-kiang, Hou-pei, Kwei-tcheou et Yun-nan. Si cela est correct et si le *Lophantus rugosus* est si commun en Chine, il est clair que le *ho hiang* du Sud oriental de l'Asie ne peut pas être cette plante.

[2] Palladius (*Dictionnaire russe-chinois*) désigne le *ho* comme « une plante aromatique de la famille de la menthe » et le *ho hiang* comme étant la bétoine (en russe *bukvitsa*).

kouo tchouan signalés par Pelliot[1], lequel contient des renseignements obtenus par la mission de K'ang T'ai 康泰 au Founan dans la première moitié du IIIe siècle après J.-C.[2]. Il est plausible en effet que les rapports de cette mémorable ambassade aient porté pour la première fois cette plante à la connaissance des Chinois[3]. Kiang Yen 江淹 (443-505) a écrit une stance de quatre lignes à l'éloge des merveilleux effets du parfum du *ho hiang*.

Des renseignements plus positifs se trouvent dans le *T'ai p'ing hwan yu ki* de Yo Chi[4], où il est dit : « C'est sous les Soui (589-618) que l'on entendit parler pour la première fois des quatre pays Pien-teou 邊斗 (appelé aussi Pan-teou 班斗), Tou-k'oun 都昆 (appelé aussi Tou-kun 都君), Kiu-li 拘利 (appelé aussi Kieou-ya 九雅) et Pi-song 比嵩. On atteint ces quatre

(1) *Bulletin de l'École française d'Extrême-Orient*, IV, p. 270.

(2) Le pays de Tou-k'oun est encore mentionné dans le *Wou chi wai kouo tchouan* comme produisant un aromate dénommé *liou hwang hiang* 流黃香 («aromate sulfureux»). Il est dit dans le *T'ai p'ing yu lan* (ch. 982, p. 2b), où ce texte est cité, qu'il se trouve aussi dans le *Nan tcheou yi wou tche*. Pour les autres textes qui font mention du pays du Tou-k'oun voir Pelliot, *Le Fou-nan* (*Bull. de l'École française*, III, p. 266). Je suis d'accord avec Pelliot pour placer ce pays dans la péninsule malaise. Il y a aussi un arbre nommé *tou-k'oun* 都昆, mentionné seulement dans le *Ts'i min yao chou* (ch. 10, p. 48b, nouv. éd., 1896) par Kia Se-niou du Ve siècle (Bretschneider, *Bot. Sin.*, t. I, p. 77) d'après le *Nan fang ts'ao mou tchoang*, mais non dans une autre œuvre ultérieure. « L'arbre tou-k'oun pousse à l'état sauvage et fleurit dans le second mois. Les fleurs restent unies aux fruits qui mûrissent dans le huitième ou neuvième mois. Ils sont de la dimension des œufs de poule. Les gens du pays les cueillent et les mangent. Le goût de l'écorce et du noyau ressemble au vinaigre. L'habitat de cet arbre est le Kiou-tchen et le Kiao-tchi » (都昆樹野生二月花色仍連著實。八九月熟如雞卵。里民取食之。皮核滋味酢。出九眞交趾). La même phrase 仍連著實 est employée dans le *Nan fang ts'ao mou tchoang* pour la description de l'*Areca catechu* (voir *Ts'i min yao chou*, ch. 10, p. 13b) et de diverses autres plantes.

(3) Pelliot, *loc. cit.*, III, p. 275.

(4) Ch. 177, p. 7 (éd. de Kin-ling chou ku, 1882).

pays en franchissant du Fou-nan, la grande baie de Kin-lin 金隣大灣(1) et en voyageant durant trois mille *li* dans la direction du sud. Pour l'agriculture ces peuples sont identiques à ceux de Kin-lin. Parmi cette population, il y en a beaucoup qui sont blancs de couleur. Tou-k'oun seul produit l'aromate *tsien hiang* 棧香(2). Pour ce qui est de l'arbre *ho-hiang*, il vit un millier d'années. Son tronc et ses racines sont très gros. Une fois coupé, ce bois se pourrit complètement et est détruit dans l'espace de quatre ou cinq ans. Seuls restent durs et sains les nœuds du milieu et il n'y a qu'eux qui conservent une odeur parfumée. On les recueille et les emploie comme aromate. » Ce dernier renseignement est singulier, car tous les écrivains ont insisté sur le fait que seule la feuille est utilisée.

Le terme *ho* est mentionné par Tso Se 左思 du IIIe siècle dans sa poésie sur la capitale de Chou 蜀都賦(3), en association

(1) Cf. PELLIOT, *loc. cit.*, IV, p. 270.

(2) *Tsien hiang* semble n'être qu'une variante de *tchan hiang* 棧香 (écrit aussi [木+箋] dans le *Fan yi ming yi tsi*, ch. 8, p. 10b). Cette identification est donnée dans le *T'ie wei chan ts'oung t'an* 鐵圍山叢談 par Ts'ai T'ao 蔡絛 (ch. 5, p. 21b; éd. de *Tchi pou tsou tchai ts'oung chou*) du XIIe siècle. L'auteur offre trois termes synonymes pour *tch'en hiang* 沈香 («aloès, bois d'aigle», en sanscrit *agaru*, *Aquilaria agallocha*) : *tch'en choui* 沈水 «qui s'enfonce dans l'eau», *p'o ts'ai* 婆萊 (transcription d'un mot barbare), et *tsien hiang* 箋香. Ces trois désignations, remarque-t-il, se rapportent à la même espèce mais à différentes sortes de qualités; on les rencontre dans le Tchan-tch'eng (Tchampa), Tchen-la (Cambodge) et Hai-nan. Le *tsien hiang* est aussi décrit comme un produit de Hai-nan dans le *Ling wai tai ta* (ch. 7, p. 2b), écrit par Tcheou K'u-fei en 1178. Le *Fan yi ming yi tsi* (*loc. cit.*) explique que le *tchan* ou le *tsien hiang* est cette sorte de bois d'aigle dont le cœur est blanc, qui n'est pas très solide, qui ne flotte pas plus sur l'eau qu'il ne s'y enfonce, mais se maintient au même niveau qu'elle. *Tchan hiang* est également donné comme un synonyme de *tch'en hiang* par Stuart (*Chinese Materia Medica*, p. 45). Au sujet du bois d'aigle voir particulièrement les intéressantes notes de A. CABATON, *Nouvelles recherches sur les Chams*, p. 49 et suiv. (Paris, 1901). Voir aussi HIRTH et ROCKHILL, *Chau Ju-kua*, p. 204-206.

(3) Le titre principal de la poésie est *San tou fou* 三都賦 «Poésies sur les trois capitales» (voir PELLIOT, *Bull. de l'École française*, III, p. 280). Il convient de noter que Tso Se fait aussi mention de Kin-lin «la frontière d'Or».

avec une certaine sorte de palmier appelé *na* 𦺨[(1)], et le cardamome *teou k'eou* 豆蔻 (*Alpinia globosum*).

Le *Nan tcheou yi wou tche* 南州異物志[(2)], par Wan Tchen 萬震 du IIIe siècle après J.-C., passe pour déclarer que «le *ho hiang* pousse au pays de K'u-sun 曲遜 et qu'il appartient à la classe d'aromates désignée sous le nom de *fou-fong* 扶風, que la plante a l'apparence du *tou-liang* 梁都 (*Eupatorium*)[(3)], et que l'on peut en user pour la préservation des vêtements». Le *Tcheng lei pen ts'ao* (et ainsi aussi le *Kwang k'un fang p'ou*) cite l'ouvrage en question comme déclarant que le *ho hiang* est produit dans les pays du littoral (海邊國), le reste étant pareil à ce qui précède. Dans le récit de Tchang Yu-si 掌禹錫

(1) *Na* est un synonyme du palmier areca (*Areca catechu*); voir Bretschneider (*Chinese Recorder*, III, 1871, p. 247) et C. Imbault-Huart (*T'oung Pao*, V, 1894, p. 318). Le *Ts'i min yao chou* (ch. 10, p. 20) cite ce qui suit d'après le *Teng Lo-feou chan sou* 登羅浮山疏 («Récit d'une ascension des monts Lo feou», en Kwang-tong) par le moine Tchou Fa Tchen 竺法真: «*Na-tse* 𦺨子 est un synonyme du palmier areca de montagne (*chan-pin lang* 山檳榔). Le tronc de cet arbre ressemble à la canne à sucre; ses feuilles sont pareilles à celles d'un chêne. Dix arbres poussent étroitement groupés et chacun produit dix enveloppes à graines (*fang* 房) au fond desquelles se trouvent plusieurs centaines de grains. On les récolte dans le quatrième mois.»

(2) *T'ai p'ing yu lan*, ch. 982, p. 4; *Kwang k'un fang p'ou*, ch. 95, p. 22; et aussi dans *T'ou chou tsi tch'eng*.

(3) C'est le nom d'une plante dont l'identité n'a pas été reconnue et que ne mentionnent ni Bretschneider, ni Stuart. Le *Hiang p'ou* (*loc. cit.* p. 9), par Hong Tch'ou de l'époque des Song, donne l'information suivante: «Selon le *King tcheou ki* 荊州記 (par Cheng Hong-tche 盛弘之 du ve siècle), il y a dans le district de Tou-liang 都梁縣 une hauteur au sommet de laquelle se trouve un étang (d'après le *T'ai p'ing yu lan*, l'eau de cette montagne est pure et peu profonde). Dans cet étang pousse le *lan ts'ao* 蘭草 [*Eupatorium*; Stuart, *Chinese Materia Medica*, p. 167] qui, du nom de la localité, reçoit celui de parfum *tou-liang* et dont l'apparence ressemble au *ho hiang*.» Ce texte est cité aussi dans le *Yi ts'ie king yin i* (ch. 11, p. 8, éd. de Nanking), compilé par Yuan Ying 元應 vers 649 après J.-C., mais la référence au *ho hiang* manque. Suivant le *Kwang tche* 廣志, le *tou-liang* pousse dans le Hwai-nan 淮南 (le *T'ai p'ing yu lan* écrit Hwai-ngan 淮安) et est aussi appelé *tsien tse ts'ao* 煎澤草.

cette opinion est attribuée au *Kwang tche* 廣志 de Kouo Yi-kong 享義恭 du VIe siècle, où il est dit en outre que les tiges du *ho* ressemblent à celles du *tou-hiang*, tandis que les feuilles sont semblables à celles du *choui sou* 水蘇 (*Stachys aspera*, appelé aussi « *sou* odorant » 香蘇 et « menthe camphrée », *long nao po ho* 龍腦薄荷 [1]); cf. fig. 12.

Le nom du pays mentionné dans le *Nan tcheou yi wou tche* est transmis sous la forme incorrecte K'u-sun. Au lieu de quoi nous devons évidemment lire Tien-soun 典孫 (il est aisé de voir comment peuvent être confondus les caractères 曲 *k'u* et 典 *tien*), nom géographique écrit aussi 頓遜 Toun-sun, et identifié avec Tenasserim sur la presqu'île de Malacca [2]. Nous avons en conséquence deux régions indiquées comme fournissant le *ho hiang*, et toutes deux situées dans la péninsule malaise. Toun-sun présente une signification particulière si nous lisons d'un œil ouvert la notice sur ce pays dans les Annales des Liang [3]. « A trois mille *li* passés, depuis la frontière sud du Fou-nan, se trouve le royaume de Toun-sun qui est situé sur une côte escarpée. Le pays n'a pas plus de mille *li* d'étendue. La ville est à dix *li* de la mer. Il y a cinq rois. Tous sont vassaux du Fou-nan. La partie orientale du territoire de Toun-sun le met en relation avec Kiao-tcheou 交州 (Tonkin). Par son territoire ouest, il est en contact avec l'Inde, la Parthie et les royaumes de l'extrême Thulé. Les marchands y affluent en grand nombre pour échanger les produits. La position géographique de Toun-sun, qui décrit une courbe et s'étend dans la mer sur plus de mille *li*, explique ce fait. L'immense Océan (Tchang-hai 漲海) est sans limite et n'a pas encore été traversé directement. Le marché

(1) Bretschneider, *Bot. Sin.*, t. III, n° 68, Stuart, *Chinese Materia Medica*, p. 422. Le *T'ou chou tsi tch'eng* donne une citation à part tirée du *Kwang tche*, disant que le *ho hiang* est produit dans toutes les régions du Ji-nan 日南 (Tonkin).

(2) Voir Pelliot, *Bull. de l'École française*, III, p. 263; IV, p. 407.

(3) *Liang chou*, ch. 54, p. 2. Cf. Pelliot, *loc. cit.*

est la place où l'Est et l'Ouest se réunissent. Il y a là chaque jour plus de dix mille personnes. Objets rares, marchandises précieuses, on y trouve tout[1]. » Ce récit rappelle à notre esprit la narration du Périple au sujet du marché où l'on se procurait le malabathron. Si le terme *ho hiang* est identique à *tamālapatra* (et il n'y a pas de raison pour rejeter cette identification), si l'habitat du *tamālapatra* se trouvait à Toun-sun ou Ténasserim sur la péninsule de Malacca, et si cette région était le grand marché unissant l'Est et l'Ouest, il est raisonnable de conclure que Toun-sun est le théâtre où s'élabore la trame de ce que raconte le Périple. S'il en est ainsi, le « peuple de petite taille à large face » représente les indigènes Malais et leurs rapports avec le Thinai (Chine) sont indiqués par ceux de Toun-sun avec le Fou-nan d'une part et avec le Kiao-tche d'une autre.

Il y a, en outre, une tradition qui veut que le *ho hiang* ait été cultivé au Tonkin.

On dit que le *Nan fang ts'ao mou tchouang*[2] contient le texte suivant se rapportant à la plante *ho* : « Elle a, paraît-il, une

[1] Au sujet de ces conditions du trafic, la note suivante contenue dans le *Chou i ki* 述異記, et citée dans le *Hiang p'ou* 香譜 de Hong Tch'ou 洪芻 (p. 19), présente de l'intérêt : « Dans les régions du Sud il y a un marché aux aromates où les marchands trafiquent des produits aromatiques. »

[2] Ce texte n'est pas contenu dans l'édition de cet ouvrage tel qu'il a été réimprimé dans le *Han Wei ts'oung chou*, mais est donné dans le *Tcheng lei pen ts'ao* (édit. de 1523, ch. 12, fol. 50b) et le *T'ou chou tsi tch'eng* (section des plantes, ch. 148). Cette dernière version diffère quelque peu du *Tcheng lei*, car aux autres localités sont ajoutés Wou p'ing 武平 et Hing-kou 興古. Le texte est donné par Tchang Yu-si 掌禹錫 de la même manière qu'il est cité dans le *Pen ts'ao kang mou*, sauf que la citation concernant l'habitat y précède la description de la plante. Dans le *T'ai p'ing yu lan* (ch. 982, p. 4) ce texte est donné comme suit : « Le *ho hiang* pousse comme le coudrier (榛). Les gens le plantent et dans le cinquième ou sixième mois font la récolte. Une fois séché au soleil il prend une odeur parfumée. Il pousse en Kiao-tche, Wou-p'ing, Hing-kou et Kiou-tchen. » Le *Tchi wou ming che t'ou k'ao* (ch. 25, p. 57b) relate simplement que le *ho hiang* est mentionné dans le *Nan fang ts'ao mou tchouang*.

saveur âcre et pousse comme un arbre épineux. Elle est cultivée par la population, qui fait la récolte dans le cinquième ou sixième mois et la fait sécher au soleil, après quoi elle prend une odeur parfumée. Son habitat est dans le Kiao-tche 交趾, Kiou-tchen 九眞 et dans tous les pays de cette région. » Comme nous l'avons fait remarquer auparavant, le *Kwang tche* du VIe siècle dit que le *ho hiang* est produit dans toutes les parties du Tonkin.

Sou Song 蘇頌, dans son *T'ou king pen ts'ao* 圖經本草, publié vers la fin du XIe siècle, donne le récit suivant : « Le *ho kiang* est abondant au Ling-nan [1], où il est largement cultivé par la population. Il commence à croître durant le second mois, formant tiges et épines. Les plants sont groupés et épais. Les feuilles ressemblent à celles du mûrier, mais sont un peu plus minces. La récolte se fait dans le sixième et septième mois, mais les feuilles doivent être jaunes pour être recueillies. Le livre *Kin leou tseu* 金樓子 [2] et les *tsien* (« écrits ») 牋 par Yu I-k'i 俞益期 [3] contiennent ce récit : « Les gens du Fou-nan 扶南 ont un dicton qui veut que les cinq sortes d'aromates forment ensemble un seul arbre, ses racines étant en bois de santal, ses nœuds de bois d'aigle 沉香 (Agallochum), de girofle (*ki che* 雞舌, *Caryophyllus aromaticus*) les fleurs [les fruits suivant une autre lecture]. Quant aux feuilles, ce sont celles du *ho hiang* 藿香 et le suc, celui du boswellia (*hiun lou* 薰陸) [4]. De là vient la signification de *t'iao* 條 (« branches »), employé

(1) Les deux provinces Kwang-tong et Kwang-si.

(2) Œuvre de Yuan 元, empereur de la dynastie Liang, qui régna de 552 à 555 après J.-C.

(3) Les *tsien* de cet auteur ainsi que ceux de Han K'ang-po 韓康伯 sont cités dans le *Ts'i min yao chou* (ch. 10, p. 15) du Ve siècle. K'ang-po est le surnom (字) de Han Po 韓伯, dont la biographie est dans le *Tsin chou* (ch. 75, p. 14b-15a)

(4) Ce récit est cité aussi dans le *Pei hou lou* écrit par Twan Kong-lou vers 875 après J.-C. (ch. 3, p. 8; éd. de Lou Sin-yuan), le *Yeou yang tsa tsou* et le *Mong ki pi t'an* (voir *Kwang k'un fang p'ou*, ch. 95, p. 25a), mais la réfé-

dans la langue des Herbiers pour la réunion des cinq sortes d'aromates. Ce qui est appelé à présent *ho hiang* dans le Sud est une sorte d'herbe qui concorde avec ce que dit Ki Han 稽含[1].» Le *Pen ts'ao*, publié durant l'époque Kia-you (1056-1064), fut le premier à admettre officiellement le *ho hiang* dans la pharmacopée[2].

Li Chi-tchen observe ce qui suit : «Le *ho hiang* a une tige carrée et des places creuses dans ses nœuds. Les feuilles ressemblent dans une certaine mesure à celle de l'aubergine (*k'ie* 茄, *Solanum melongena*). Suivant Kie-kou 潔古[3] et Tong-yuan 東垣[4], seules les feuilles sont utilisées, les branches et épines ne sont d'aucun usage. Nos contemporains emploient branches et épines mélangées, car il y a pour les feuilles de nombreux succédanés frauduleux. Le *T'ang che* 唐史[5] expose que le pays de Toun-sun est l'habitat du *ho hiang*, qu'on le propage par boutures et que les feuilles sont pareilles à celles du *tou-liang* 都梁[6]. Liou Hin-k'i 劉欣期,

rence au Fou-nan y est omise. Il a sans doute été apporté en Chine par K'ang T'ai et se trouvait probablement dans ses mémoires, qui sont perdus.

[1] L'auteur du *Nan fang ts'ao mou tchouang* (voir plus haut, p. 29) suit une citation d'un livre intitulé *Ho hiang fang* 合香方 (contenant apparemment des recettes pour la fabrication des aromates), attribué à Fan Ye 范曄 ou Fan Wei-tsong 范蔚宗. Voici comment la donnent le *Tcheng lei* et le *Pen ts'ao kang mou* : 零藿虛燥: le *T'ou chou tsi tch'eng* écrit : 靈藿虛操. Comparer *Wei lio* 緯畧 (ch. 6, p. 1) où l'œuvre est intitulée 和香方 et l'auteur nommé Fan Yu 范昱.

[2] *Tchi wou ming chi t'ou k'ao*, ch. 25, p. 57[b].

[3] Nom littéraire de Tchang Yuan-sou 張元素, médecin de l'époque Kin (1115-1243). Voir Bretschneider, *Bot. Sin.*, t. I, p. 48.

[4] Nom littéraire de Li Kao 李杲, médecin de l'époque Yuan (Bretschneider, *loc. cit.*).

[5] Peut être regardé comme identique au *T'ang chi loun twan* 唐史論斷, étude critique de l'histoire de la dynastie T'ang, écrite par Soun Fou 孫甫 durant le xi[e] siècle (Wylie, *Notes on Chinese Literature*, p. 80).

[6] Dans le *T'ai p'ing hwan yu ki* (ch. 176, p. 10[b]), ce texte est ainsi conçu : 出藿香插枝便生葉似都梁以裹衣.

dans son *Kiao tcheou ki* 交州記, affirme que le *ho hiang* ressemble au storax (*sou ho hiang* 蘇合香); mais cette comparaison ne se rapporte qu'à l'odeur semblable des deux produits et non à l'apparence extérieure des plantes en question.»

Dans le *Wang che t'an lou* 王氏談錄[(1)], écrit par Wang Tchou 王洙 antérieurement à 1056 apr. J.-C., on remarque que le terme *lan houi* 蘭蕙 comprend deux herbes différentes[(2)], que ses contemporains ne faisaient pas cette distinction, et que, dans l'opinion de quelques-uns, le *ho hiang* est un *houi ts'ao* 蕙草 («une espèce odorante d'orchidée des marais»). Je ne crois pas que cette opinion ait en fait le moindre fondement.

Wou K'i-tsun, auteur du *Tchi wou ming che t'ou k'ao*[(3)], conjecture que le *ho hiang* est identique à l'ancien *hiun ts'ao* 薰草 (*Melilotus arvensis*). Il représente et décrit aussi une espèce de *ho hiang* sauvage 野藿香, comme venant en grande quantité dans les monts Nan-ngan 南安山 de la province de Kiang-si. Il ressemble au *ho hiang*, les feuilles sont de couleur vert profond, la nuance des fleurs légèrement pourpre et il est très parfumé d'odeur et de goût (cf. fig. 5 et 6).

Le docteur T. Tanaka du service de l'Agriculture à Washington a été assez bon pour nous fournir une note sur le *ho hiang*, extraite du *Honzō kōmoku keimō* 本草綱目啓蒙, écrit par Ono Ranzan 小野蘭山 (édition revue par Iguchi Bōshi 井口望之 [1847]). La voici :

«*Kwakkō* 藿香. Synonymes chinois : *kwakkyo-byō* (*Jibutsu imyō*); *reirō kwakkyobyō* (*Yaku-fu*).

«Deux différentes sortes de drogues du nom ci-dessus sont

(1) Éd. de *T'ang Song ts'oung chou*, p. 6[b].

(2) Dans le *Neng kai tchai man lou* (ch. 15, p. 17[b]; éd. de *Cheou chan ko ts'oung chou*), par Wou Tseng 吳曾 du milieu du XII[e] siècle, la différence entre *lan* et *houi* est déterminée par la comparaison de *lan* avec l'homme de qualité (君子) et de *houi* avec le vulgaire (小人), parce que dans les bois des montagnes dix *houi* valent un *lan*.

(3) Ch. 25, p. 57[b].

introduites des pays étrangers; l'une d'elles, désignée sous le nom d'*aoba* (ou *aoba no kwakkō, ho hiang* à feuilles vertes), est la véritable. On ne trouve pas cette plante poussant à l'état sauvage au Japon.

«Les feuilles de *kwakkō* sont grandes et épaisses, hirsutes et incisées en cinq places environ et dentelées sur le bord; elles poussent opposées l'une à l'autre sur une ramille et sont odoriférantes.

«Suivant Li Chi-tchen (Ri Jichin), dans les temps anciens seules les feuilles étaient employées, mais aujourd'hui on emploie à la fois branches et ramilles avec les feuilles; parce que cette drogue, contenant les tiges, montrera promptement que les feuilles viennent de la véritable plante; ainsi personne ne sera trompé par de fausses feuilles, comme il arrive si souvent quand les feuilles seules sont employées.

«Maintenant il est difficile de les reconnaître pour de véritables feuilles de *ho hiang* si les vraies branches sont mélangées à de fausses feuilles. Le *Pen ts'ao houi yen* (*Honzō igen*) dit : «Une méthode pour éviter les fausses feuilles est d'employer les branches et tiges mêlées aux feuilles, mais il y a la plus grande difficulté d'échapper à la fraude des branches mêmes. On devrait apprendre à se familiariser avec les caractéristiques des feuilles de coton et d'aubergine qui sont communément mêlées à cette drogue.» Le *Pen ts'ao meng tch'ouan* (*Honzō mōsen*), *Pen ts'ao pi to* (*Honzō hitsu-doku*), et *Pen ts'ao yuan che* (*Honzō genshi*) font aussi mention de ce qui intéresse le mélange de feuilles de coton et d'aubergine, et dans un cas j'ai moi-même découvert qu'il se trouvait dans la drogue quelques fibres de coton, qui sans doute y avaient été mises au moment du mélange des feuilles de coton.

«Une drogue, dénommée *uzumi-gwakkō* (*mai kwakkō, mai ho hiang,* c'est-à-dire «*ho hiang* enterré») ou *tsuchi-gwakkō* (*do-gwakkō, t'ou ho hiang*, c'est-à-dire «*ho hiang* de terre»), est aussi une drogue frauduleuse de pays étrangers : ce sont des feuilles de *haisō* 排草 (*p'ai ts'ao;* nom japonais *kawamidori*, voir fig. 7 et 10), dont j'ai déjà parlé [1]. On prépare aussi cette drogue frauduleuse au Japon.

« Comme le montre le *Pen king fong yuan* 本經逢原 (*Honzō hōyen*), la plante *haisō* devient une bonne drogue d'imitation de *ho hiang* en Chine, et est ordinairement cultivée dans les jardins des habitations. Je suis certain que la plante décrite dans le *Pen ts'ao houi yen* sous le nom de *ho hiang* est sans aucun doute cette plante, et qu'elle est en réalité une jeune plante de *kawamidori*, introduite autrefois de la Chine au Japon sous le nom de *ho hiang*. Quoi qu'il en soit, les feuilles de *p'ai ts'ao*, une fois séchées, n'ont pas d'odeur et la forme de la feuille est différente aussi de la véritable *aoba no kwakkō*. Ainsi nous pouvons dire que le *p'ai ts'ao hiang* (*haisō-kō*) est une drogue falsifiée, nullement égale au véritable *ho hiang*. »

En ce qui concerne l'emploi du *ho hiang*, nous trouvons dans l'encyclopédie bouddhique *Fa yuan tchou lin* [2] que les statues de Bouddha étaient lavées le huitième jour du quatrième mois avec trois sortes de substances aromatiques : *tou-liang*, *ho hiang* et *ngai-na hiang* 艾納香 [3].

[1] Le *p'ai ts'ao* est cité pour la première fois par Fan Tch'eng-ta 范成大 (1126-93) dans son *Kwei hai yu heng tche* 桂海虞衡志 et par Tcheou K'u-fei (*Ling wai tai ta*, ch. 7, p. 3, écrit en 1178) comme étant semblable en apparence au *pai mao hiang* (*Andropogon schœnanthus*, voir *supra*, p. 22) et ayant une odeur pareille au musc. C'est un produit du Ji-nan (Tonkin). Li Chi-tchen observe que « le *p'ai ts'ao* est produit en Kiao-tche (Tonkin). A présent il est dans une certaine mesure planté aussi dans le Ling-nan (Kwangtong). C'est la racine d'une herbe, blanche de couleur, de forme pareille à la racine d'un peuplier élancé. Il est fréquemment adultéré à l'aide d'autres substances ». Stuart (*Chinese Materia Medica*, p. 252) l'identifie avec la *Lysimachia sikokiana*. — [B. L.]

[2] *Pien tseu lei pien*, ch. 177, p. 11.

[3] Suivant le *Kwang tche* par Kouo I-Kong du VIe siècle (*Kwang k'un fang p'ou*, ch. 95, p. 21 b), la plante *ngai na* venait de, ou était produite dans les

En ce qui regarde l'Indo-Chine, comme l'a déjà signalé J. de Loureiro (1715-1794)[1], le *ho hiang* se rapporte à la *Betonica officinalis*. Loureiro dit aussi que la culture de cette plante n'est pas rare en Cochinchine et en Chine. A Pékin, et généralement en Chine, le même nom se rapporte au *Lophantus rugosus*. Les médecins français d'Indo-Chine adoptent les mêmes identifications.

I. L. Soubeiran et Dabry de Thiersant[2] disent au sujet de la Betonica (*ho hiang*) : « Très recommandée pour les ivrognes, dont elle dissipe l'ivresse, la bétoine est aussi employée (sommités et feuilles) comme stomachique, antivomitive et anticholérique. » J. Regnault[3] dit : « Feuilles et tiges contre les coliques et les névralgies. Feuilles pulvérisées : sternutatoire; racine : émétique. Employée contre le choléra, contre les coliques et les vomissements. » E. Perrot et P. Hurrier[4] font sur la bétoine les observations suivantes : « Amère et peu odo-

contrées de l'Ouest; elle ressemble à une jolie artémise (*ngai* 艾). On doit la distinguer d'un lichen vert (*lu i* 綠衣), poussant sur l'écorce des pins et pareillement dénommé *ngai-na*, mais différent du premier. Li Chi-tchen aussi fait cette distinction entre deux différentes sortes de *ngai-na*. Il ne donne pas d'explication du nom, qui apparemment est la transcription d'un terme étranger. Ono Ranzan (*Honzō-kōmoku-keimō*, ch. 10, p. 23[b], 1803) constate que le nom n'est pas encore expliqué. Il donne un synonyme *chi-li* 世黎 comme se trouvant dans le *Suvarṇaprabhāsa-sūtra*. Cela est évidemment le sanscrit *çirī*, une herbe du genre kuça (*Poa cynosuroides*). Dans l'encyclopédie bouddhique *Fa yuan tchou lin*, on signale que l'ouvrage ci-dessus mentionné *Kwang-tche* dit que le *ngai-na* vient du royaume de P'iao 驃 (Birmanie). Voir PELLIOT, *Bull. de l'École française*, IV, p. 175. Ceci se trouve aussi noté dans le *Hai yao pen ts'ao* 海藥本草 du VIIIe siècle (voir *Tchi wou ming che t'ou k'ao*, ch. 22, p. 35). *Ngai-na* peut donc être un mot birman. Il est mentionné dans une poésie de Sou Chi (voir *Neng kai tchai man lou*, ch. 15, p. 4).

[1] *Flora Cochinchinensis* de 1793 (p. 441).

[2] *La Matière médicale chez les Chinois*, p. 181 (Paris, 1874).

[3] *Médecine et pharmacie chez les Chinois et chez les Annamites*, p. 149 (Paris, s. d.).

[4] *Matière médicale et pharmacopée sino-annamites*, p. 180 (Paris, 1907).

3.

rante, cette espèce croît dans les lieux ombragés. On fume et on prise, comme le tabac, ses feuilles douées d'une certaine âcreté et considérées comme sialagogues, sternutatoires; sa racine passait pour évacuante et émétique; mais toute la plante a presque tout perdu de son ancienne renommée. Cependant, elle est encore préconisée comme sudorifique, dans les coliques, les vomissements bilieux, les fièvres et la fétidité d'haleine. » Devant les données des botanistes, il n'y a pas lieu de douter que le *ho hiang* cultivé du Tonkin (au moins pour le présent) est une bétoine. La question se présentera alors de savoir si la bétoine pousse à l'état sauvage à Malacca, particulièrement à Tenasserim. Quoi qu'il en soit, il est hors de question qu'une bétoine puisse avoir constitué le malabathron des anciens, puisque cette plante est commune en Europe et était tout à fait familière à Pline[1]. Un point plus sérieux encore est que les anciennes descriptions chinoises de la plante *ho hiang* ne s'accordent ni avec le *Lophantus* ni avec la *Betonica*. On se rappellera que le *T'ai p'ing houan yu ki* décrit le *ho hiang* comme un arbre de tronc et de racines très grands (ci-dessus, p. 26). Il est encore appelé un arbre épineux dans le *Nan fang ts'ao mou tchouang*. Les caractéristiques de l'arbre sont d'accord avec la description que donne Pline du malabathron et la définition du *tamāla* dans le *Çukranīti*. Il y a certainement aussi quelques contradictions dans les textes chinois, et les diverses comparaisons tirées des feuilles et autres plantes peuvent même soulever la question de savoir si plusieurs espèces différentes ne pourraient pas avoir produit la feuille de *ho hiang* ou si elles n'auraient pas été confondues sous ce terme.

Il est certain que le *ho hiang* n'est pas une des espèces du

[1] xxv, 46. Un peuple ibérien d'Espagne, les Vettones, ont les premiers découvert cette plante, qui était connue en Gaule sous le nom de *vettonica*, sous celui de *serratula* en Italie et, en Grèce, sous ceux de *cestros* ou *psychotrophon*.

Cinnamomum, lequel était bien connu des Chinois sous le nom de *kwei* 桂. Aucun texte chinois ne dit que le *ho hiang* ait une relation quelconque avec le *kwei*. Les feuilles sont comparées avec celles de la *Stachys aspera*, du mûrier et de l'aubergine, non avec celles du cannelier. Il est curieux que F. Porter Smith[1] ait distillé un *Cinnamomum tamala* du *T'ien tchou kwei* 天竺桂, mentionné dans le *Pen ts'ao kang mou*, et prétende hardiment que «les folia tamalapathri [*sic*] (ou malabathri), qui ont une forte senteur aromatique, étaient autrefois exportées de Chine». La source de cette erreur était que Smith acceptait le terme *T'ien-tchou kwei* dans le sens littéral de «cannelier indien». G. A. Stuart[2] signale justement que ce *T'ien-tchou* renvoie à une localité de la préfecture de Tai-tcheou, Tche-kiang, où l'arbre pousse en quantité. Il est mentionné dans le *Pen ts'ao yen i*[3], publié en 1116 par K'eou Tsoung-chi, qui dit que cet arbre est le même que le *mou* 牡 ou *k'un kwei* 菌桂 (c'est-à-dire le *Cinnamomum cassia*)[4]. L'identification du *tamā-*

(1) *Contributions toward the Materia Medica of China*, p. 63.

(2) *Chinese Materia Medica*, p. 109.

(3) Ch. 4, p. 4[b]; éd. de Lou Sin-yuan.

(4) Les renseignements suivants sur la plante sont donnés dans le *T'ien tchou chan tche* 天竺山志 (ch. 11, p. 2), publié par Fen Yuan 芬原 en 1875 : 天竺桂。咸淳志。木樨有黃白紅三色。舊天竺多有之。冷.齋夜話云。天竺桂花中秋特盛非必種出月中地氣使然。張子韶云。天竺桂花六出他州所無。東坡有天竺山送桂花。分贈元素詩云。月闕霜濃細蕋乾此花元屬桂堂仙鷲峯子落驚前夜蟾窟枝空記昔年張九成及郭祥正皆有詩。一天竺桂子。羣芳譜。浙中山桂台州天竺最多子。子如蓮實或二或三離。離下垂。天竺僧稱爲月桂。具花時常不絕枝頭葉底依稀數點亦异種也。本草綱目云月桂落子之說起於武后時。相傳有竺僧云。自天竺靈鷲飛來。故八月常有桂子落於天竺道經謂之不時花. En conséquence, la plus ancienne mention faite de l'arbre est dans le *Leng tchai ye hwa* de la fin du XIe siècle. Reste à prouver s'il se rencon-

lapattra avec un *Cinnamomum*, inspirée seulement par les conjectures de Lassen, doit être positivement rejetée et exclue.

L'auteur n'est pas un botaniste et il est par conséquent obligé de laisser la solution d'un problème botanique aux spécialistes, à ceux particulièrement pour qui la flore de la péninsule malaise et de l'Indo-Chine est familière. Une suggestion cependant, qui pourrait mettre sur la bonne voie, peut être offerte sous toutes réserves. Il me semble que la plante en question, celle qui répond au malabathron des anciens, au *tamālapattra* de l'Inde et au *ho hiang* des Chinois, est une sorte de menthe, et que dans la famille des menthes ce sont les feuilles de patchouli (*Pogostemon*) qui, les premières de toutes sous ce rapport, doivent être prises en considération. Ainsi qu'il est bien connu, c'est une plante odoriférante, originaire de Silhet (en Assam), Penang et de la péninsule malaise, dont les feuilles donnent une huile essentielle, qui est employée à la préparation d'un parfum. Le *Pogostemon heyneanus* se trouve à Malacca et est connu dans l'Inde et à Ceylan comme une plante cultivée ou retournée à l'état sauvage. Un sous-arbrisseau qui s'y ramifie, le *Pogostemon cablin*, haut de deux à cinq pieds, est signalé à Perak et à Penang[1]. Suivant Yule[2], la feuille dénommée *pacapāt* en Bengāli est vendue dans tous les bazars de l'Hindoustan : on l'emploie comme un ingrédient mis dans le tabac à fumer, les femmes s'en servent pour parfumer leurs cheveux et surtout on en garnit les matelas et on le place parmi les vêtements comme nous faisons de la lavande. Le botaniste H. F. Hance[3] a exprimé l'opinion

tra réellement dans l'Inde aux temps anciens, comme il est dit dans le texte ci-dessus. Son nom en tout cas vient du monastère T'ien-tchou-chan. Cf. aussi Bretschneider, *Bot. Sin.*, t. III, p. 453.

(1) G. King, *Materials for a Flora of the Malayan Peninsula* (*Journal As. Soc. Bengal*, LXXIV, 2e partie, 1907, p. 708-709).

(2) *Hobson-Jobson*, p. 684.

(3) *Notes and Queries on China and Japan*, II, 1868, p. 175.

que le *Pogostemon patchouly* Pellet., en dehors de Silhet et de Penang, est aussi originaire de la Chine, bien qu'il ne puisse offrir aucune évidence directe à cet effet, ni aucun nom chinois, mais il signale l'odeur particulière de différentes encres de Chine due, croit-il, à un mélange de cette herbe. Aucun nom chinois pour patchouli n'étant connu, il semblera plausible qu'il soit voilé au moins pour une part sous le terme *ho hiang*. Il y a du reste dans la Chine du Sud et en Indo-Chine différentes plantes produisant le patchouli, qui appartiennent au genre *Microtœna* (*cymosa*, *robusta* et *urticifolia*)[1]. Prain suppose que sa présence dans le pays de Shan et en Assam est due à ce que la culture l'y a propagée de la Chine du Sud-Ouest et du Tonkin. Tous ces faits s'accorderaient bien avec l'histoire du *ho hiang*.

Si le malabathron a été transplanté de la péninsule malaise au Tonkin et finalement y a été cultivé, ce fait peut nous livrer un fil qui nous mènera à l'explication raisonnable de la tradition du Périple. La population de Tou-k'oun et de Toun-sun, comme nous l'avons vu, est identique à cette tribu sauvage de petite taille qui, tous les ans, visite le pays frontière de Thinai pour faire le trafic du malabathron. La coïncidence qu'il y a entre les Bescidais à peau blanche de Ptolémée (p. 9) et la population blanche de Kin-lin, mentionnée dans le *T'ai p'ing hwan yu ki* (p. 26), est vraiment curieuse. Maintenant, ainsi que l'a démontré F. von Richthofen[2], le Kattigara de Ptolémée doit être situé près de la bouche de la rivière Rouge, dans les environs de l'actuelle ville d'Hanoi[3], et le Thinai ou

[1] Forbes et Hemsley, *Journal Linnean Soc.*, XXVI, p. 306-308. *Microtœna cymosa* Prain est reproduite dans le *Kew Bulletin*, 1902, p. 11.

[2] *China*, I, p. 508.

[3] A. Herrmann (*Die alten Verkehrswege zwischen Indien und Süd-China nach Ptolemaus*, dans *Z. Ges. f. Erdkunde*, 1914, p. 11) identifie Kattigara à Ha-tinh.

Sinai correspond ainsi au Tonkin et à la partie sud de la Chine. La part que les habitants du Tonkin tenaient dans le trafic du malabathron est, par suite, parfaitement claire : tout d'abord ils se procurèrent le produit de Malacca et, afin d'augmenter le volume de la production, cultivèrent la plante eux-mêmes, jusqu'à ce qu'ils fussent en position d'exporter une grande quantité de feuilles dans l'Inde et en Chine.

Afin d'aider les botanistes dans l'identification de cette plante, nous reproduisons ici toutes les illustrations qu'en ont données les Chinois, en même temps que celles de diverses autres plantes mentionnées dans cet article.

En terminant cette notice, je désire exprimer mes plus vifs remerciements à M. U. Odin, qui a bien voulu se charger de traduire avec le plus grand soin mon manuscrit de l'anglais en français sous la direction de M. Sylvain Lévi. C'est grâce à l'aimable intérêt de M. Gabriel Ferrand et de M. Sylvain Lévi que ce travail voit le jour, et ma plus sincère reconnaissance à ces Messieurs !

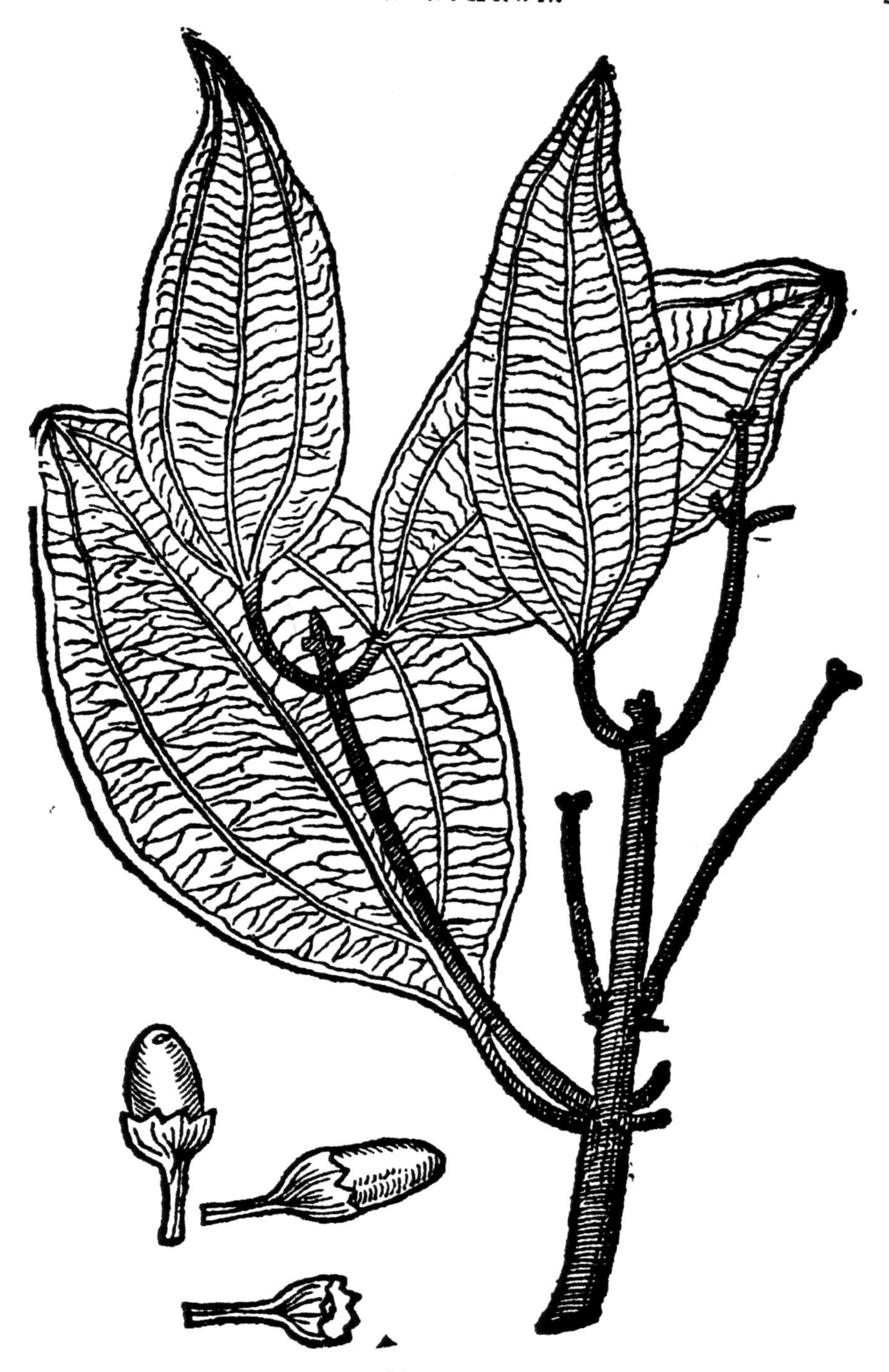

Fig. 1.

«Tamalapatra cum suo Ramusculo» (d'après C. Clusius, 1567).

Fig. 2.

Ho hiang of Mong tcheou.

(Tiré du *Tcheng lei pen ts'ao*, ch. 12, fol. 50; éd, de 1523.)

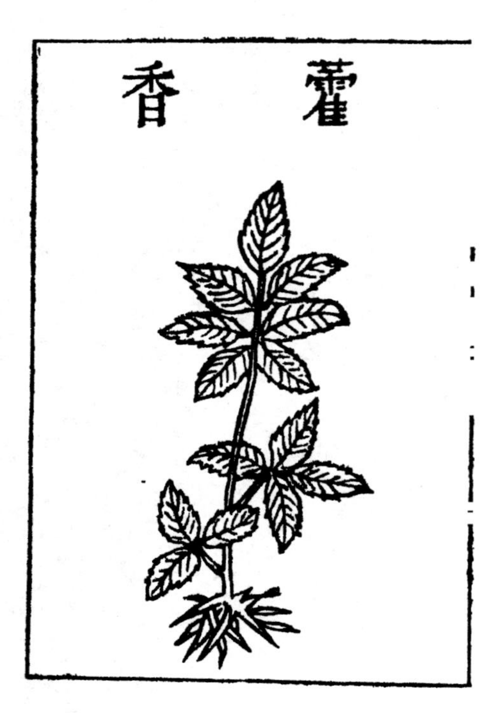

Fig. 3.

Ho hiang.

(Tiré du *Pen ts'ao kang mou.*)

Fig. 4.
Ho hiang. (Tiré du *T'ou chou tsi tch'eng.*)

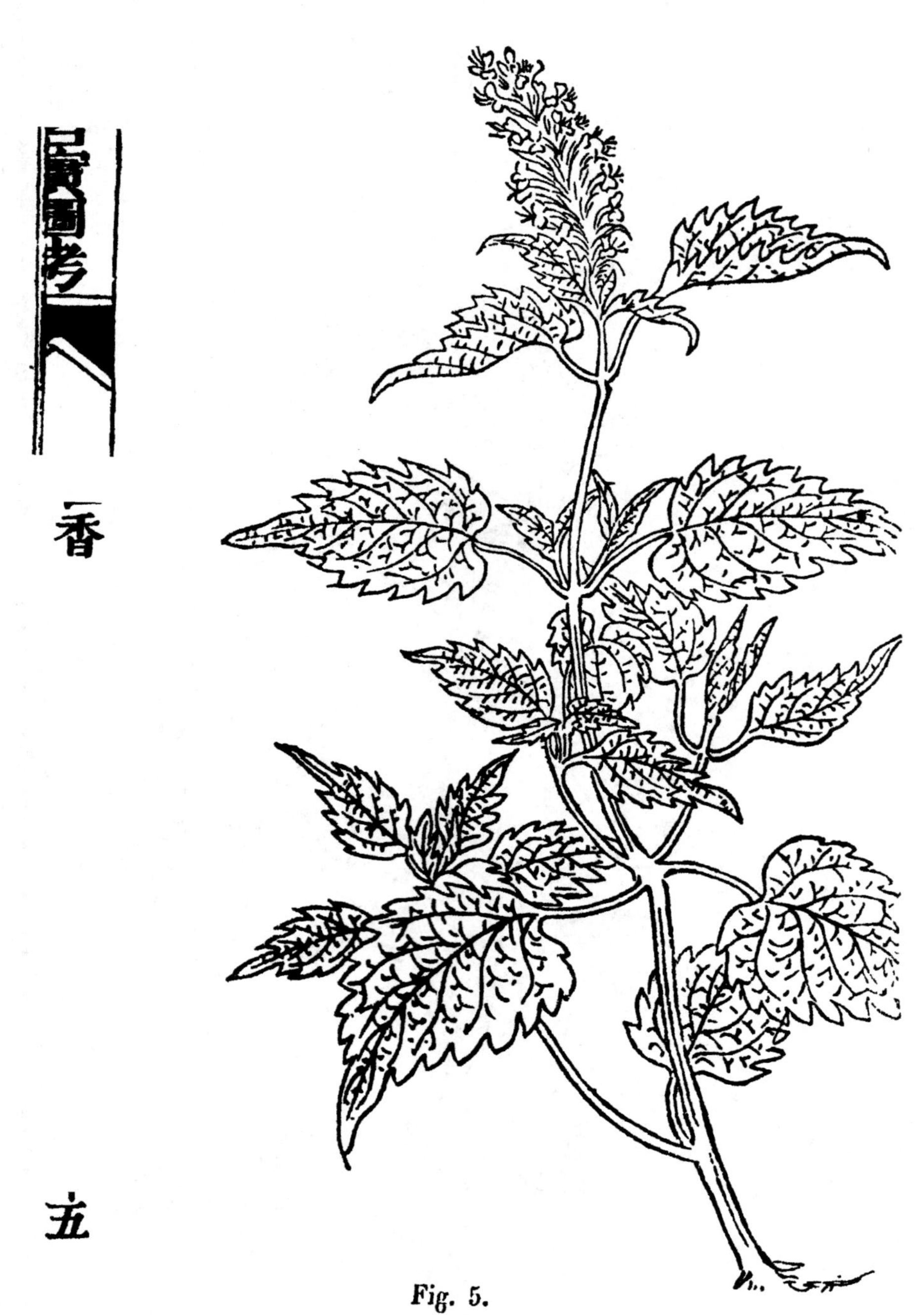

Fig. 5.

Ho hiang. (Tiré du *Tchi wou ming che t'ou k'ao*, publié en 1848.)

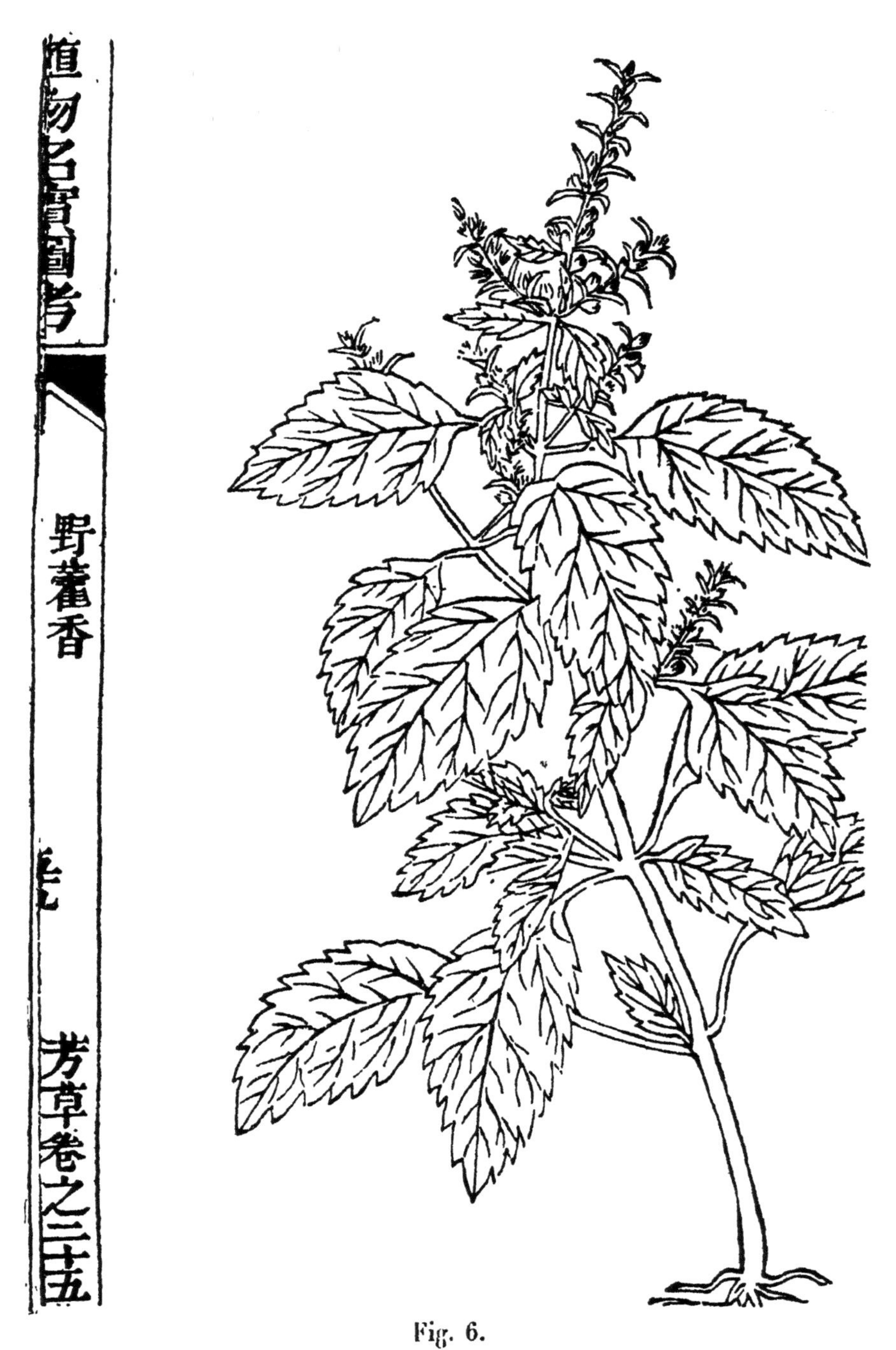

Fig. 6.

Ho hiang sauvage. (Tiré du *Tchi wou ming che t'ou k'ao.*)

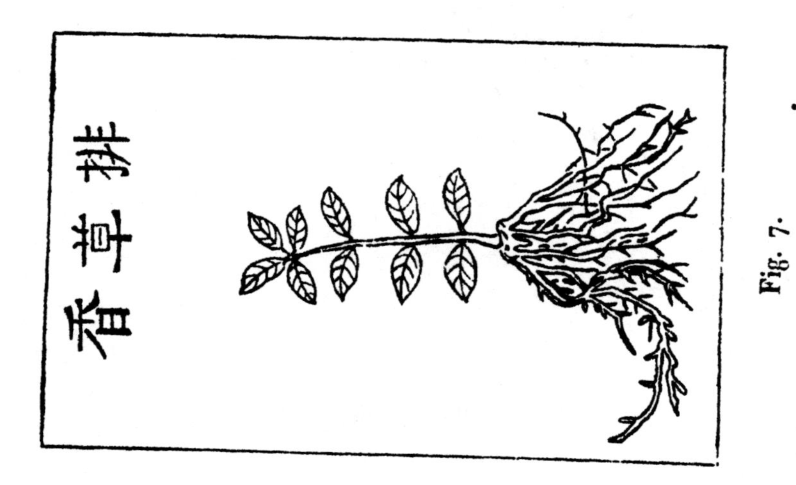

Fig. 7.
Pai ts'ao hiang, japonais *kawamidori*
(*Lysimachia sikokiana*).

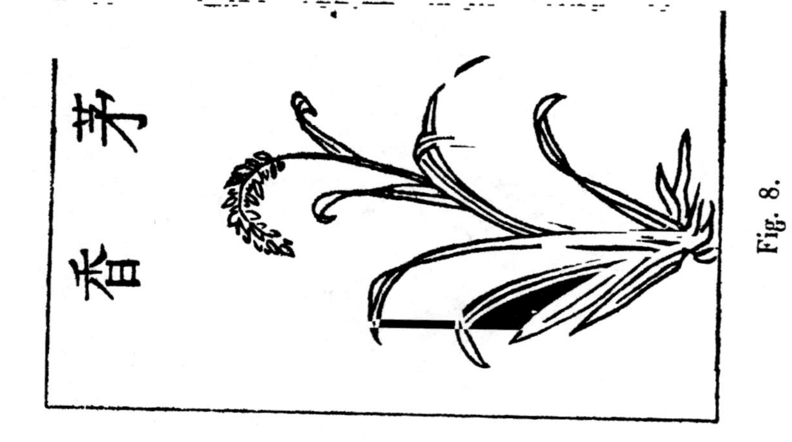

Fig. 8.
Mao hiang (*Andropogon schoenanthus*).

Tiré du *Pen ts'ao kang mou.*

Fig. 9.
Pai mao hiang
(*Andropogon schoenanthus*).

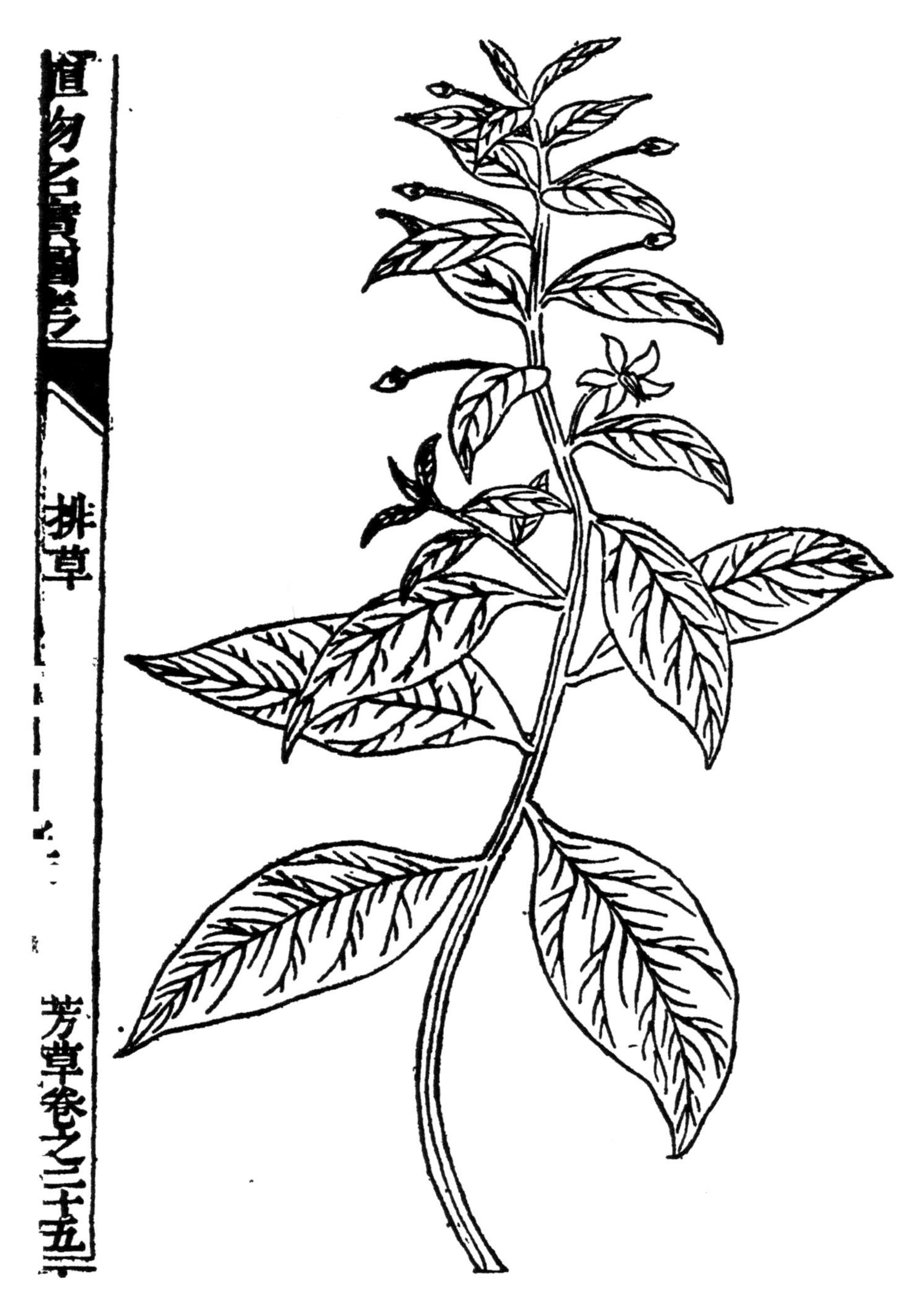

Fig. 10.
Pai ts'ao, *Lysimachia sikokiana.*
(Tiré du *Tchi wou ming che t'ou k'ao.*)

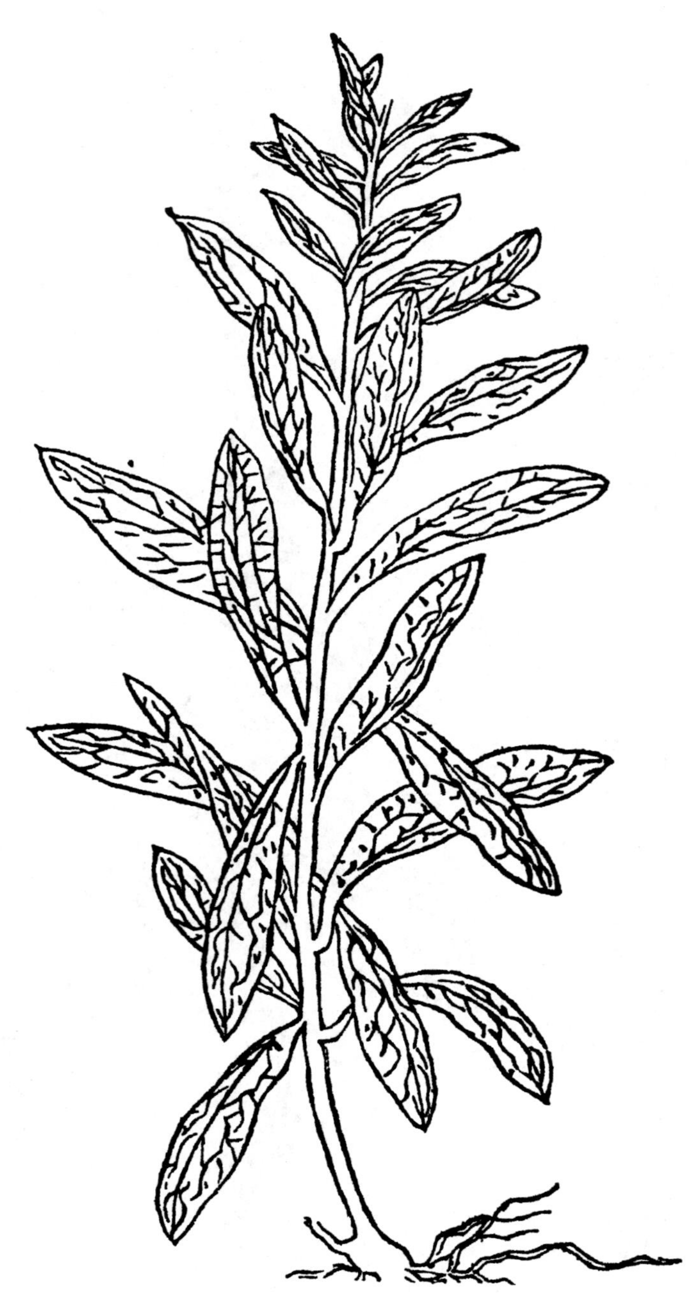

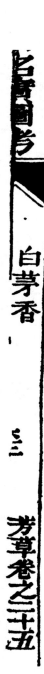

Fig. 11.

Pai mao hiang, *Andropogon schoenanthus.*
(Tiré du *Tchi wou ming che t'ou k'ao.*)

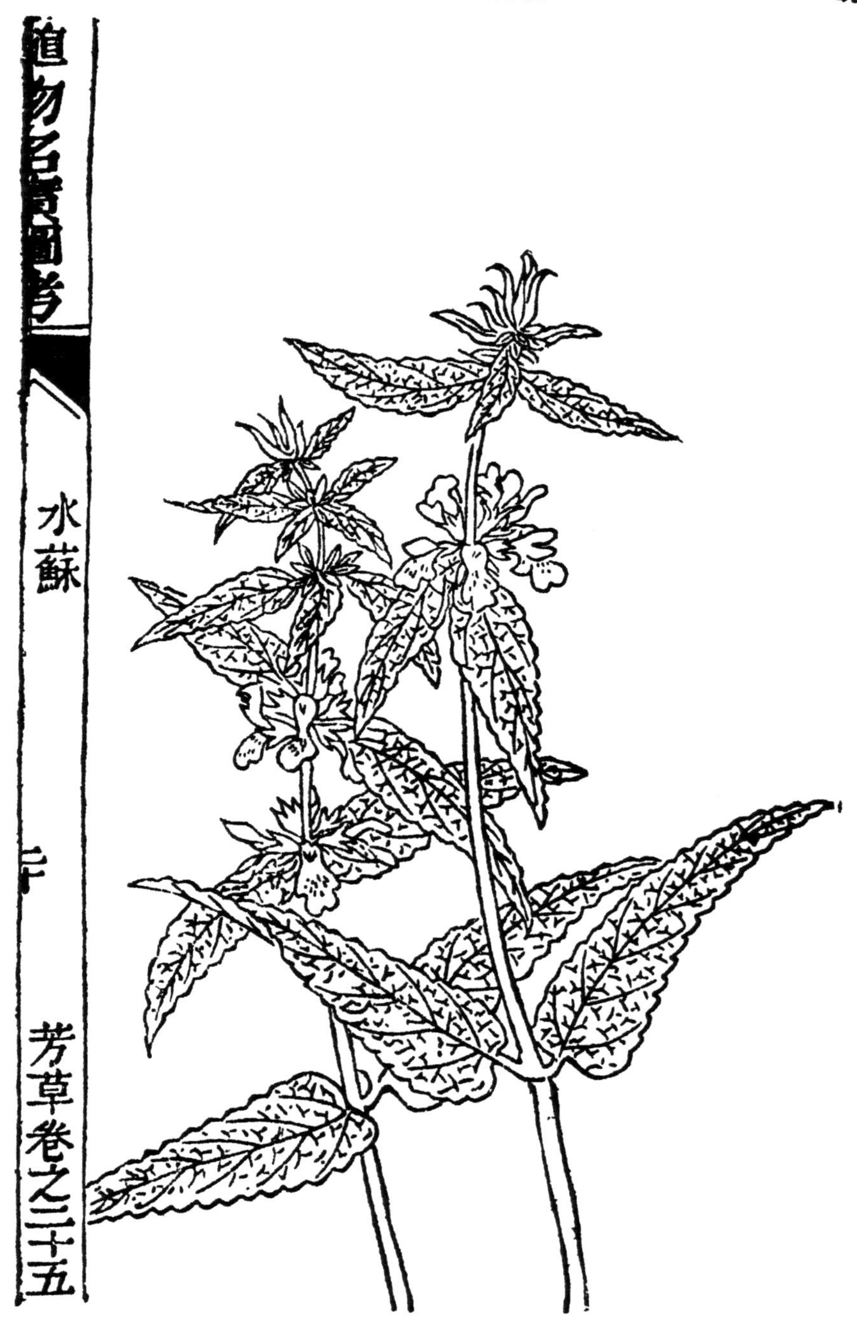

Fig. 12.

Choui sou (japonais *mama-ogara*), *Stachys aspera*.

(Tiré du *Tchi wou ming che t'ou k'ao.*)

149

中国展览品

BULLETIN OF THE ART INSTITUTE OF CHICAGO

DECEMBER-NINETEEN-EIGHTEEN

VOLUME XII NUMBER 9

MAINE WOODS—BY HOWARD GILES
AWARDED HONORABLE MENTION
ANNUAL EXHIBITION OF AMERICAN PAINTINGS

THE CHINESE EXHIBITION

ALL emanations of ancient Chinese art must be interpreted from the religious conceptions and ideals of the nation. Worship of the great elementary forces of nature, deep reverence for the departed, unlimited devotion to the ancestors and their ethical traditions, an insatiable yearning for salvation and immortality, combined with a sound and practical philosophy of life and moral standards, form the keynote of the mentality of Chinese society. Like that of Egypt, the art of ancient China is one of the dead, and the monuments discovered in the graves bear a distinct relation to the beliefs entertained by the people in a future life and simultaneously reflect the actual state which their civilization had reached.

The visitor intent on studying the present Chinese exhibits in the Art Institute should be conscious of the fact that in these collections are represented two great periods which are fundamentally distinct and are separated not only by a vast span of time, but are also characterized by diverse social and religious ideas and accordingly by different means of artistic expression. One is the Han period (206 B. C.—A. D. 220) covering the time around our era and marking the transition from the impersonal art of the archaic epoch to the middle ages; the other is the T'ang period (A. D. 618-906), being China's Augustan age in literature, poetry, painting, and sculpture. The green and brown glazed pottery vases and animals, as well as the rubbings displayed on the walls, are representative of the art of the Han; the clay figures of human beings and animals illustrate artistic achievements of the T'ang.

In the era of the Han, graves were laid out in large sepulchral chambers composed of flat stone slabs. These formed a vault sheltering the coffin and were

QUEENSBORO BRIDGE—BY JOHN F. FOLINSBEE
AWARDED HONORABLE MENTION
ANNUAL EXHIBITION OF AMERICAN PAINTINGS

decorated with scenes in flat relief depicting favorite incidents of ancient history or mythological subjects in a narrative or almost epic style. Somewhat naive and primitive in the representation of human figures and in the expression of emotions, they are nevertheless full of life in their description of battles, hunting scenes, court processions, royal receptions, and domestic affairs. These engravings in stone come down from the second century A. D. and present our most important archaeological source for the study of ancient Chinese civilization.

The so-called hill-jars symbolize the deep-rooted belief of the ancient Chinese in immortality. They were convinced of the existence of three Isles of the Blest, supposed to be far off in the eastern ocean, where grew a drug capable of preventing death and securing immortality. Several emperors eager to obtain this precious drug sent out expeditions over sea in quest of the Fortunate Islands, the mysteries of which held the imagination of the people deeply enthralled. To these religious sentiments of their contemporaries the artists of the time lent visual expression by molding mortuary jars of cylindrical form, with covers shaped like the hilly Islands of the Blest emerging from the sea and surrounded by bands of sea-waves. They are posed on three feet molded into

A LITTLE FAIRY—BY RICHARD W. BOCK
AWARDED HONORABLE MENTION, ANNUAL
EXHIBITION OF AMERICAN SCULPTURE

figures of bears, the bear being an emblem of strength and endurance. Such jars were interred in the graves and implied the mourner's wish that his beloved deceased might reach the land of bliss and attain eternal life on the Fortunate Isles. The meaning of death was to the Chinese a continuation of this life in another sphere. The property dear to the living ones was reproduced in clay and confided to the grave. The likeness of an object suggested a living reality, and the occupant of the tomb was believed to enjoy the durable clay offerings as if they were real things. The spirits of the departed were hence obliged to continue the preparation of their food, and thousands of models of kitchen ranges have been discovered in the graves of the Han period. The green-glazed urns of cylindrical shape with a tiled roof on the top were modeled in imitation of towers which served as granaries, and were filled with cereals to supply the dead with their daily meal. Drink was as necessary to them as food, and jars bearing out the idea of a draw-well were deposited in the grave to furnish a constant supply of fresh water. The roofed well-house contains an opening for the pulley over which the ropes pass for raising or lowering the buckets, and a water-pail is placed on the edge of the well-curb. The large globular vases with two tiger heads on the sides were designed to hold offerings of wine. Some are plain, others are decorated with relief bands displaying hunting scenes with animals in flying gallop and mounted archers aiming at them with cross-bows. These motives are identical with those represented on the rubbings from the contemporaneous tomb-stones. The art of the Han was the great epoch of Chinese idealism expressive of religious ideas and sentiments, faith, hope, and resurrection, in a forceful, straightforward way.

The leading ideas of the great T'ang epoch move along somewhat different lines. In pictorial art realism and naturalism prevail; in sculpture, thanks to the beneficial influence of Buddhism and Hellenistic-Indian traditions, the human figure predominates. In the burial clay figurines of this memorable period we meet a surprisingly personal and human element, which eloquently speaks its own language and testifies to a highly developed individualism as well as to a noble refinement of social customs. The feminine ideal of that epoch is portrayed in numerous graceful statuettes with a large variety of style in costume and hair-dressing, varying according to local usage. The coffin in the grave was flanked at both ends by male or female attendants on horseback. Powerful knights clad in iron armor protected the lord from demons or malignant intruders whose avarice might have disturbed the peace of his burial place. Yama, the Indian god of death, was a favorite conception of the people of the T'ang period. He usually stands over a demon or a reclining bull which is his emblem and appears as a mighty warrior with heavy armor and plumed helmet—an efficient guardian of the grave. Bull-carts were employed to carry the coffin and paraphernalia at the funeral to the burial place, and the bull was modeled with a high degree of realism. The lover of horses had his favorite steeds immortalized in clay. Often they were pictured as if mourning for their deceased masters. The camel loaded with merchandise, emblematic of commerce, adorned the grave of the merchants who carried on a lucrative trade with Central Asia and Persia. Figures of actors and dwarfs took care of the entertainment of the dead, dwarfs being noted in China for their wit and sagacity and being frequently employed as jesters and court fools. BERTHOLD LAUFER.

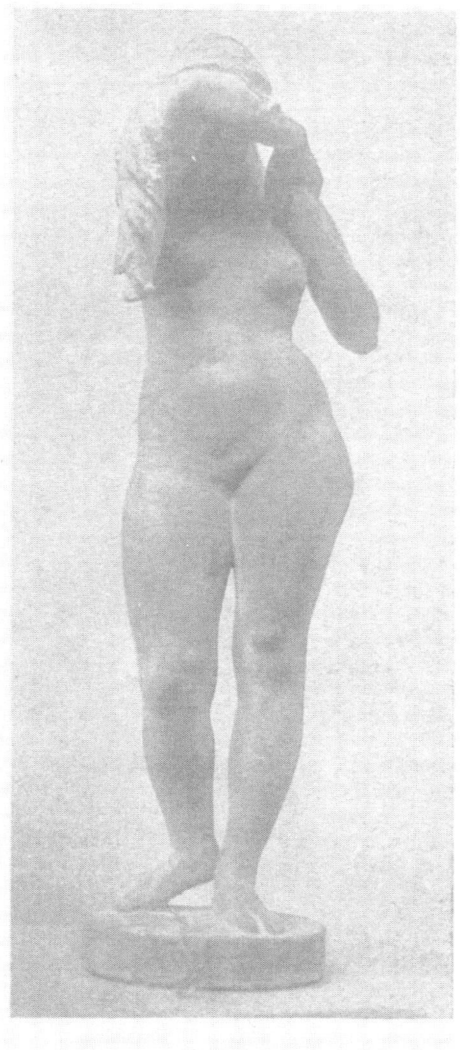

EVE—BY GEORG LOBER
AWARDED HONORABLE MENTION, ANNUAL EXHIBITION OF AMERICAN SCULPTURE